Ultrasonic Imaging and Holography

Medical, Sonar, and Optical Applications

Edited by

George W. Stroke
State University of New York
Stony Brook, New York
and
Harvard University
Cambridge, Massachusetts

Winston E. Kock
Visiting Professor, University of Cincinnati
Cincinnati, Ohio
and
Consultant, The Bendix Corporation

Yoshimitsu Kikuchi
Tohoku University
Sendai, Japan

and

Jumpei Tsujiuchi
Tokyo Institute of Technology
Tokyo, Japan

Ultrasonic Imaging and Holography

Medical, Sonar, and Optical Applications

PLENUM PRESS • NEW YORK–LONDON

Library of Congress Cataloging in Publication Data

U. S.—Japan Science Cooperation Seminar on Pattern Information Processing in Ultrasonic Imaging, 3d, University of Hawaii, 1973.
Ultrasonic imaging and holography.

Includes bibliographical references.
1. Ultrasonics in medicine—Congresses. 2. Electronic data processing—Medicine—Congresses. 3. Holography—Congresses. I. Stroke, George W., ed. II. Title. [DNLM: 1. Holography — Congresses. 2. Ultrasonics — Congresses. W3UN61 1973u / QC244.5 U58 1973u]
R895.A1U54   1 3   610'.28   74-1371
ISBN-13:978-1-4613-4495-7   e-ISBN-13:978-1-4613-4493-3
DOI: 10.1007/978-1-4613-4493-3

Proceedings of the U.S. – Japan Science Cooperation Seminar on Pattern Information Processing in Ultrasonic Imaging, held January 7-13, 1973, at the University of Hawaii, Honolulu, Hawaii

© 1974 Plenum Press, New York
Softcover reprint of the hardcover 1st edition 1974

A Division of Plenum Publishing Corporation
227 West 17th Street, New York, N.Y. 10011

United Kingdom edition published by Plenum Press, London
A Division of Plenum Publishing Company, Ltd.
4a Lower John Street, London W1R 3PD, England

All rights reserved

No part of this book may be reproduced, stored in a retrieval system, or transmitted, in any form or by any means, electronic, mechanical, photocopying, microfilming, recording, or otherwise, without written permission from the Publisher

FOREWORD

This book contains the complete set of papers presented at the Third United States-Japan Science Cooperation Seminar on the Subject of Holographic Imaging and Information Processing, held at Hawaii, January 8 to 13, 1973, under the joint sponsorship of the National Science Foundation and the Japan Society for the Promotion of Science. The papers present the latest advances and state of the art in areas ranging from ultrasonic holography, radar and sonar to ultrasonic pulse-echo imaging, acoustic microscopy and image processing in biomedical engineering. Computer processing of ultrasonic images is extensively discussed. Several papers describe the remarkable applications of ultrasonics in medical diagnostics. One describes new features in R & D management that are particularly relevant to the field as a result of recent changes in national science and technology policies. We also present the most recent advances in the powerful methods of image improvement (sharpening, deblurring) which have been made possible with the aid of holograms used as computing elements; the holographic image deblurring method is capable of an astonishing improvement of images that are still unattainable with even the most powerful digital computer methods. The book follows the 1971 Plenum Publishing Company book, "Applications of Holography", which comprised the papers presented at the Second U. S. -Japan Seminar held in Washington, D. C., October 13 to 18, 1969. In both these seminars, the discoverer of holography, Professor Dennis Gabor who received the 1971 Nobel Prize in Physics for his work, was an active participant. All three seminars (the first having taken place in Japan from October 2 to 6, 1967) were sponsored jointly by the National Science Foundation and the Japan Society for the Promotion of Science. The singular role of Gilbert B. Devey, Program Director, National Science Foundation, Washington, D. C., in stimulating the three meetings is noted with much gratitude.

Since medical ultrasonic diagnostics has recently become a field of major activity in clinical applications as well as in industrial exploitation, a few additional introductory remarks may be useful. The comments may help, also, in drawing attention to the timeliness of the papers which describe the intense current research activity in the field.

Productivity in health care delivery and early screening for possible disease may indeed be considerably improved by use of the new method of ultrasonic diagnostics which is now under intensive development in leading academic, national and industrial laboratories around the world. Advanced prototype ultrasonic imaging already makes it possible to reveal non-invasively (in an X-ray like manner), disease (or the absence of it) in almost all human internal soft tissue organs including the prostate, the bladder, the heart, and the breasts, among others. A widespread use of ultrasonic imaging is also made in obstetrics. Strikingly remarkable images of fetuses are being readily obtained with needle-like sharpness in the best cases, starting even earlier than the 10th week of pregnancy. The method permits one to guide an aspirating needle in order to extract some of the amniotic fluid without injury to the fetus, in view of examining chromosomes for possible abnormalities. These are just a few representative illustrations of the many diverse uses of the method of medical ultrasonic diagnostics which are now being perfected, and of which the principles and details are being presented in this volume by leading authorities in the field. The role of holography in ultrasonic imaging was also pioneered by Professor Dennis Gabor, Inventor of the field, and one of the authors.

The importance of the papers is enhanced by the reports of recent clinical experiments which have demonstrated the usefulness of ultrasonic imaging for diagnostic visualization. The method is complementary to radiography, nuclear medicine and thermography, among others. It may be used safely, notably in cases where other methods are either inapplicable or perhaps hazardous (e.g., as are ionizing radiations in the case of pregnancies).

FOREWORD

In addition to the use of ultrasonics for medical diagnostics, this volume also presents the results of pioneering use of ultrasonics in surgery (e.g., the brain).

The important advances in ultrasonic diagnostics revealed by the U.S.-Japan 1973 Hawaii seminar also resulted in the formation by the National Science Foundation of a special "Blue Ribbon Task Force on Ultrasonic Imaging" in February, 1973. Under the chairmanship of John B. Manniello, a special consultant to NSF (then Vice President of Government Operations at CBS Laboratories, and since, Science Attache to the U.S. Ambassador to Italy); and the direction of C. Branson Smith, Director of the Office of Experimental R & D Incentives at NSF, the Task Force carried out a world survey of the state of the art in ultrasonic medical diagnostics and made recommendations for accelerating more widespread clinical use of ultrasonic instrumentation and increased industrial participation. The Task Force consulted with leading medical and industrial authorities in the world, in countries ranging from the United States, Japan, Germany, Great Britain and Austria, to Denmark, Sweden, Holland and Australia, among others. It reported through C. Branson Smith to Dr H. Guyford Stever, Director of the National Science Foundation (concurrently also, the Science Advisor to the President of the United States). It included Professor Dennis Gabor, Nobel Laureate; Dr Frederick Seitz, President of Rockefeller University; Dr E. D. Pellegrino, Vice President for Health Sciences at Stony Brook (and now Chancellor of the University of Tennessee Medical Center); Dr William E. Glenn, Director of Research at CBS Laboratories in Stamford, Connecticut; Dr Leonard Weiss, Director of Experimental Pathology at Roswell Park Memorial Institute; Dr Lester Goodman, Chief of Biomedical Engineering, National Institutes of Health; Dr Gilbert B. Devey, National Science Foundation; and Dr George W. Stroke, of Stony Brook and Harvard Medical School. An official report has been made by the National Science Foundation.

While the formal editorship of the volume was the responsibility of the co-chairmen, we want to acknowledge with particular gratitude the leading role of Mrs Gladys Hayes, Professor Dennis Gabor's Administrative Assistant at CBS Laboratories, in actively assuming the responsibility for the assembly, organization and preparation of the papers, as well as the editing and retyping of a major part of the manuscripts into their camera-ready form.

The two United States editors also express their sincere appreciation to the two Japanese editors, Professor Y. Kikuchi, and Professor J. Tsujiuchi for their very effective participation in the editorial work; and notably for obtaining manuscripts representative of the latest state of the art in Japan. The historical review, fully documented and beautifully illustrated, as included by Professor Y. Kikuchi in his report, will be found to be as useful to advanced workers as it will be by those wanting to enter the field.

Special thanks are also due Mr Renville H. McMann, President of CBS Laboratories, in supporting this work, notably by providing use of the facilities of CBS Laboratories, and for his particular interest in encouraging the work in the field of ultrasonic imaging, in view of its full implementation on a national and worldwide scale.

The present volume of scientific and technological papers, together with the existing literature in the field, notably by such authorities as Professors P. N. T. Wells and Y. Kikuchi, will be found to be a particularly useful basis for the work on the accelerated development and clinical use of medical ultrasonic instrumentation.

George W. Stroke
Winston E. Kock

CONTENTS

IMAGE PROCESSING IN BIO-MEDICAL
       ENGINEERING · · · · · · · · · · · · · · · 1
  K. Atsumi

ULTRASONIC HOLOGRAPHY: A PRACTICAL
       SYSTEM · · · · · · · · · · · · · · · · · · 87
  B. B. Brenden

NEW DIMENSIONS FOR R & D PROGRAM
       MANAGEMENT · · · · · · · · · · · · · · · 105
  G. B. Devey

ULTRASONIC TISSUE VISUALIZATION AND
       SURGERY IN BRAIN · · · · · · · · · · · · · 125
  F. J. Fry and R. C. Eggleton

A PROJECT OF ULTRASONIC TOMOGRAPHY
       ("SONORADIOGRAPHY") · · · · · · · · · · · 151
  D. Gabor

IMAGE INFORMATION PROCESSING FOR
       PULSE ECHO SCANNING METHODS · · · · 159
  M. Ide

ULTRASONIC IMAGING AT STANFORD
       RESEARCH INSTITUTE · · · · · · · · · · · 191
  E. D. Jones

OPTICAL INFORMATION PROCESSING AND
       ACOUSTO-OPTICS · · · · · · · · · · · · · · 201
  H. Kashiwagi and K. Sakurai

PRESENT ASPECTS OF "ULTRASONOTOMO-
       GRAPHY" FOR MEDICAL DIAGNOSTICS · · 229
  Y. Kikuchi

NEW FORMS OF ULTRASONIC AND RADAR
        IMAGING. 287
  W. E. Kock

ACOUSTIC MICROSCOPY. 345
  A. Korpel

SOME ASPECTS OF OPTICAL HOLOGRAPHY
        THAT MIGHT BE OF INTEREST
        FOR ACOUSTICAL IMAGING 363
  A. W. Lohmann

SIGNAL PROCESSING METHOD IN ULTRA-
        SONIC SCANNING TECHNIQUE 379
  M. Okujima

ULTRASONO-CARDIO-TOMOGRAPHY 425
  Y. Kikuchi, D. Okuyama, and M. Tanaka

COMPUTER PROCESSING OF ULTRASONIC
        IMAGES 455
  M. Onoe

OPTICAL IMAGE IMPROVEMENT IN BIO-
        MEDICAL ELECTRON MICRO-
        SCOPY AND ULTRASONICS. 503
  G. W. Stroke and M. Halioua

ULTRASONIC HOLOGRAPHY BY TRANS-
        DUCER ARRAY AND LIQUID-
        CRYSTAL DEVICE 517
  M. Suzuki, T. Iwasaki, S. Fujiki, and
  A. Hakoyama

HOLOGRAPHIC SYNTHETIC APERTURE
        SONAR SYSTEM. 531
  J. Tsujiuchi, S. Ueha, and K. Ueno

PRESENT STATE OF THE CLINICAL APPLI-
      CATION OF ULTRASONOTOMOGRAPHY· ·553
  T. Wagai

SOME PATHOBIOLOGICAL CONSIDERATIONS OF
      DETECTION OF BREAST CANCER BY
      ULTRASONIC HOLOGRAPHY· · · · · · · ·567
  L. Weiss

BIOGRAPHIES OF CONTRIBUTORS · · · · · · · · · ·587

AUTHOR INDEX · · · · · · · · · · · · · · · · · ·611

SUBJECT INDEX · · · · · · · · · · · · · · · · · 621

IMAGE PROCESSING IN BIO-MEDICAL ENGINEERING

Kazuhiko Atsumi

Institute of Medical Electronics
Faculty of Medicine
University of Tokyo, Japan

I. INTRODUCTION

Computer introduction into all fields of medical science has been rapid lately, and the trend will be more necessarily so in the future, in medical science.

All sorts of medical information are found in all sorts of medical organizations, gathered, transmitted, retrieved, processed and become useful. The kinds of information to which I am referring here are four:

1. descriptive information; case history, etc.
2. digital information; blood count, cholesterol count, etc.
3. analog information; ECG, EEG, etc.
4. pattern information; X-ray film, scinti-(graphic image)   gram, ultrasonic tomogram, etc.

Excepting the pattern information, the other three are relatively easy for computers to process, but computers have a hard time to deal with patterns, while in medical science, patterns are important information.

II. MEDICAL PATTERN INFORMATION

There are ways of processing medical pattern information, as listed below:

1. Patterns are stored on films or optical glasses before they are read for use, such as X-ray films, cell patterns, chromosome patterns, etc.

2. Patterns are directly read on-line by way of sensors. These are: RI image, ultrasonic tomogram, thermogram, etc.

The kinds of patterns are:

1. two-dimensional pattern
2. three-dimensional pattern
3. dynamic pattern

For use of these patterns as information for diagnosis or screening, there is a way of total mechanical processing; another is one for men and the machine to cooperate. Examples such as the analog information, ECG automatic diagnosis are of the former, but in these cases, the recognition of pattern as information is not yet perfect. Therefore, as a provisional method, the latter man-machine cooperation system must be relied upon. This is more effective and less costly at the moment.

There is a preparatory process that is necessary for easy pattern processing such as correction of noise by detectors and/or interfaces which is known as geometric or uniformity correction. The man-machine system follows this process at the present time, and in the future will reduce man's role and become purely a mechanical recognition of patterns or automatic analysis.

In the field of research for the man-machine cooperation system, efforts are being made to:

1. get patterns as discriminative as possible to make physicians' judgment easier. All efforts are made in getting images as close as possible to originals by employment of digital filters for image improvement and enhancement.

2. get measurements of characteristic data from patterns; this is necessary for extraction of measurements such as dimension, density distributions, reading of features, etc., from patterns of cells, chromosomes, etc.

3. get dynamic measurements of patterns; for processing of the pattern changes by time passage, time-differential patterns are taken or curve adjustments at different time indications are done.

Perfect pattern recognition is yet impossible; it is yet in the "screening" stage. Categorization will be for study in the future. By "screening" is meant a process whereby one determines certain conditions as normal or abnormal from the pattern. A typical example is the detection of cancer cells by auto-screening. By "categorization" it is meant that the machine does diagnosis through classification of all cases it handles from the information. This is now an almost virgin field of research.

The present research status in Japan with respect to pattern information processing for medical application is now being recognized more acutely than before as having great importance.

Reports have been made on relative subjects from the below-listed scientific societies:

Japan Society of Medical Electronics (Annex 1)
Japan Society of Image Information Processing
Japan Society of Electronics & Communication

As you will note from the listing of Table (1), the number of papers reported at the annual conferences of Japan Society of Medical Electronics in the last five years has increased greatly, while the coverage of research has enlarged.

The table does not include reports on ultrasonics and the basic problems relative to satisfactory recognition of patterns for medical purposes, but if the field is extended to include all these, the number of papers on the total coverage of the Medical Pattern Information processing will further increase.

Research is done in specific fields as listed below, in Japan:

RI Image   Plaque (Bacteria)

Thermogram   Chromosome

X-ray Image   Vascular Image

Cancer Cell   Ultrasonic Image

I will pick a few from the list and explain how research is done there.

Table (1) The Papers on Pattern Information Processing in Medical Fields Reported in the Annual Conferences (1968 - 1972) of Japan Society of Medical Electronics

Items of pattern \ Year	1968	1969	1970	1971	1972	Total
X-ray film	2	4	2	6	3	17
RI - image	1	1	3	6	7	18
thermogram	0	0	1	0	0	1
cancer cell	3	0	1	2	3	9
blood cell	0	0	0	1	0	1
chromosome	0	0	0	1	3	4
bacteria	0	0	0	1	0	1
vascular image	0	0	0	0	1	1
general problems	0	0	1	1	1	3
Total	6	5	8	18	18	55

III. PROCESSING ON RI IMAGE

The output from the RI Image apparatus is a digital pulse. Moreover, as RI image analysis is itself very poor, it has no value as pattern information. If the pattern is analyzed into 200 x 200 elements or so, it is easy to put the RI image into a computer for processing, and it is useful.

1. RI Imaging Apparatus

There are two types of RI image apparatus[1]; one, a moving type, and the other, a stationary one. The moving type apparatus is used on a human body with its scintillation detector having a focusing collimeter to scan straight across the body while it moves by steps in the right angle to it, repeating scanning to get dwo-dimensional information. It is also called a scintiscanner (Figure 1).

The stationary type apparatus takes a large portion of the body for the two-dimensional pattern in a short period

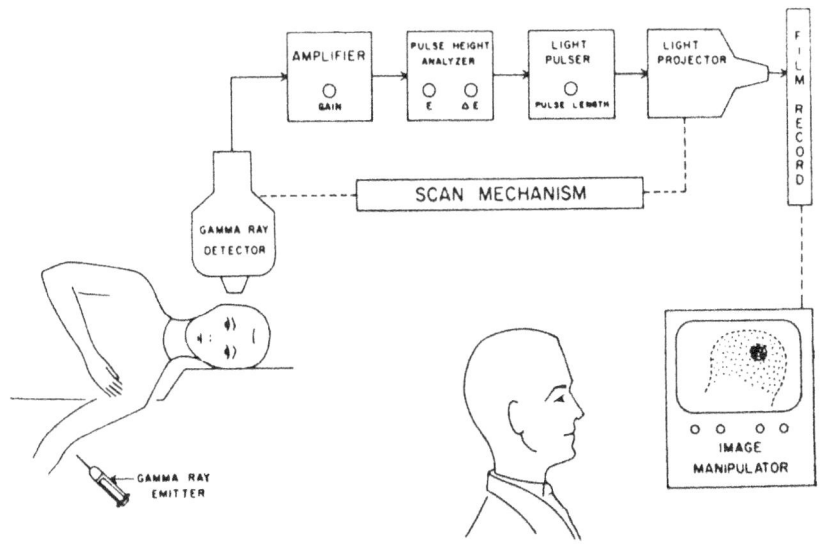

Figure 1. Patient's brain is scanned by a scintiscanner and its image displayed on television.

of time as it remains still. The resolution of the stationary type is usually lower than that of the moving type. The scintillation camera (Auger camera) and image intensifier camera are of this same group.

2. Pattern Information Processing (Figure 2)

There are several kinds of noise to disturb RI patterns, but the most serious ones are those that come from the statistical fluctuation of the measurements and

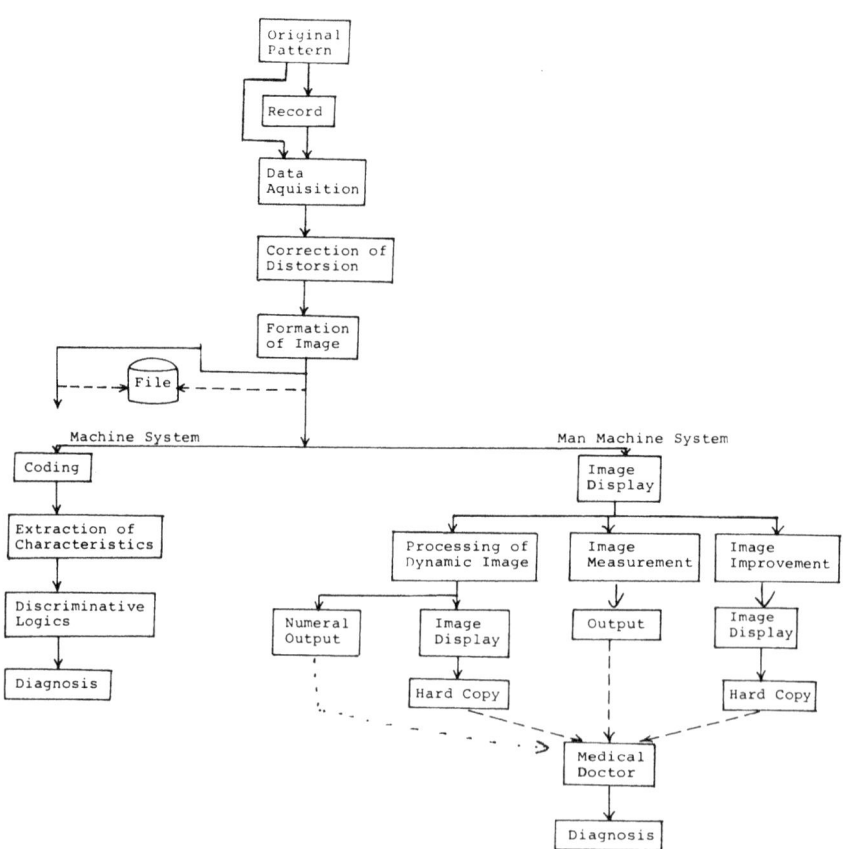

Figure 2. Flow Chart of Pattern Information Processing (Iinuma)

the blurring of the pattern. The former is caused because
the sensitivity of the RI image apparatus is too low and
its result is the mottled effect of the image, making the
pattern indiscernible. The latter is caused because the
apparatus has not enough spatial resolving power, and the
image lacks detail. The former is corrected by smoothing; the latter by a restoration process. In order to make
the image more visible, differentiation and subtraction
processes are also used.

    a.   Smoothing[2,3,4] was carried out by replacing a
central elemental image with an average value
of five or eight neighboring elemental images
(Figure 3). This operation reduces the noise
in the image due to statistical fluctuation in
high spatial frequency, and forms a smoothed
image within a certain range of deviation. If
the smoothing on the image is repeated, all
useful information will be lost from the image
since the elemental image tends to have an
identical value at the limit.

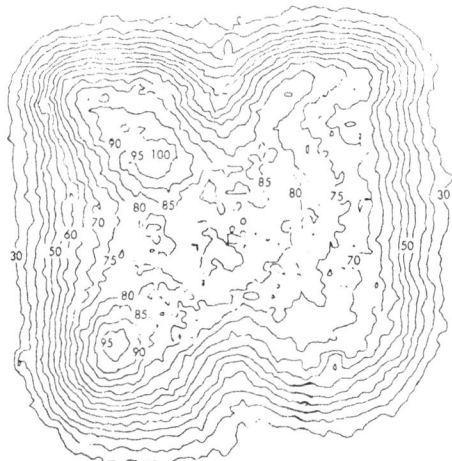

Figure 3.   Iso-contour display of the "smoothed image".
Iso-contour lines are drawn at 5% intervals between 30% and 100%. (Iinuma)

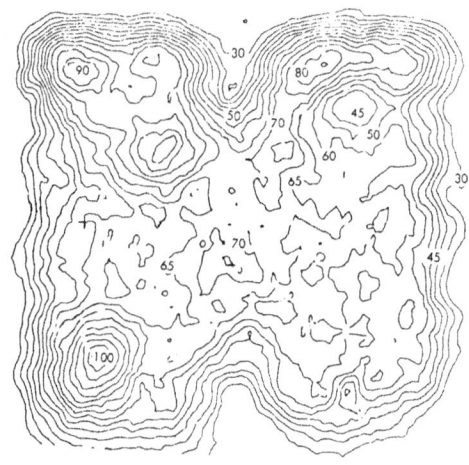

Figure 4. Iso-contour display of the "focused image". Iso-contour lines are drawn at 5% intervals between 30% and 100%. (Iinuma)

b. Restoration (Focusing)[5,6,7,8,9,10] (Figure 4)
Restoration is performed on the observed image in order to compensate for the blur in the image caused by a finite resolving power of the system, for which there are the following methods:

1. iterative approximation[7]
2. differential operator method
3. Fourier transform

c. Differential Imaging[11,12] (Figure 5)
Differential imaging is enhancement of the perimeter of the image, in which the difference of counts in neighboring element images is calculated and plotted in a two-dimensional array.

IMAGE PROCESSING IN BIO-MEDICAL ENGINEERING

Figure 5. "Differential Image" of the "focused image" that is calculated due to "iterative approximation method". The image is plotted by six different symbols of increasing densities according to the values of $D/\bar{G} = 0$-$4, 4$-$6, 8$-$10, 10$-$12,$ and more than 12. (Iinuma)

  d. Subtraction[13]. In order to determine abnormal portions, the pattern obtained is placed on top of the pattern of the normal condition of the patient. An example of this subtraction process is the way the chest X-ray film patterns are handled. The diseased and the normal patterns are processed. Scintigrams of a liver and a pancreas are recorded on top of one another, from which the liver pattern is subtracted to get the image of the pancreas.

3. Image Display.
  a. Hardware - The instruments used for display of images are cathode ray tubes (CRT), X-Y plotters, and line printers. CRT is very speedy. If its main memory capacity is large enough, it can be used as an on-line apparatus for data acquisition and display at the same time.

Iinuma and Fukuhisa, et al. are using a computer system which can do the digital acquisition of RI image and on-line processing. They use CRT, a curve plotter and a line printer for the display. Their cathode ray tube is a unique one which they named the DAC system[14]. DAC is the abbreviation of Direct Access Controller, and it can be operated on-line. This is an apparatus capable of reading and writing core memory directly by the hardware only without program interruption, in which sense it is a kind of process controller (Figure 6).

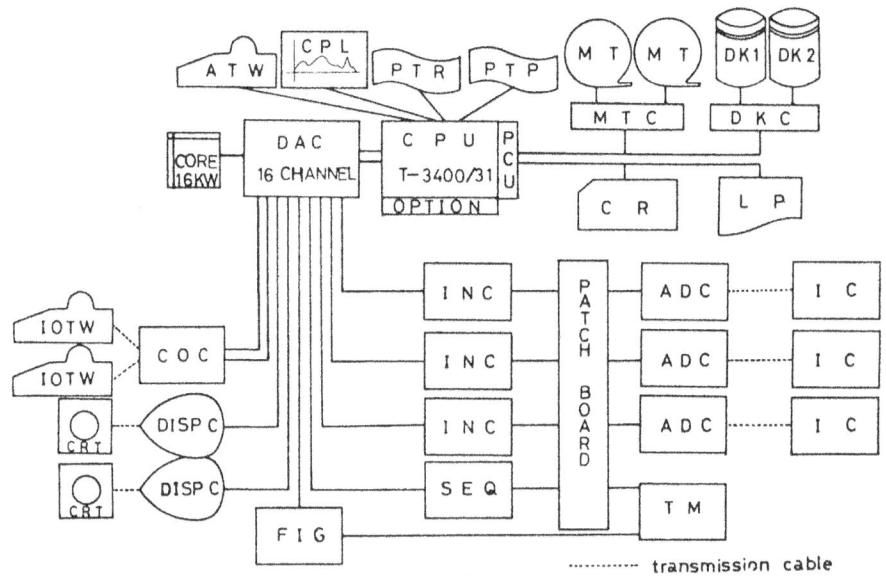

Figure 6.   DAC On-line System   (Iinuma)

b. <u>Software</u>. By dividing the information into several grades by counts, displays of elements of different levels are obtained, the methods for which vary as (1), letters or figures; (2), signs of different densities; (3), isocontour; (4), three-dimensional display; (5), color display.

Concerning each component of the hardware[15]:

CRT
α displays of two-dimensional patterns by brightness modulation

β three-dimensional by count of Z on X-Y dimensions

γ profile display by line to show the diagram of the RI image

Display by a Curve Plotter
α Iso-contour

β Three-dimensional display by piling-up of profile lines on the two-dimensional plane, as a bird's-eye view

Line Printer. By dividing counts of the elements of the digital image at random levels and allotting letters, figures and signs at these levels, thereupon getting two-dimensional printing, the following becomes possible:

α to get a pattern which continuously changes density

β to print equal-count elements on equal levels

The maximum level of display legible to the human eye is 20 density levels, and the maximum intensity made possible by the number of piles is four times. One problem left for future study with respect to display hardware is the quality reject improvement of CRT display apparatus. A brightness level of over 16, size of a CRT face that is 20×20 cm^2, and display elements over 128×128 are hoped to be achieved. In order to reduce noise on the face, an inexpensive sort of supplementary device to the CRT display apparatus and the addition of a light pen as a manmade input are desired.

4. Computer

With the exception of special cases, medium-sized computers are now used most commonly in Japan for processing medical pattern information, but mini-computers are also used. Kimura[16], et al. state requirements for computers for this type of usage are as follows:

    a. multi-purpose usage

    b. on-line and off-line usage

    c. permanent preservation of original data

    d. possible change, addition and development of programs to be processed

    e. acquisition of a speedy and capacious supplementary memory device

    f. various kinds of display apparatus

    g. capable of being connected to a larger-sized computer

    h. Ease of handling

    i. low cost

Merits of using a digital computer are as follows:

    a. The original data detected by an RI apparatus can all be stored with almost no harm.

    b. Various ways of processing enable collection of necessary information in quantities.

    c. Various display methods per image are possible.

5. Man-Machine System

Kojima[17] and Hisata, et al. carry their RI pattern processing research in a dialogue mode by connecting a light pen to a clinical data analyzer, CDS-4096. The light pen operation is done by picking and tracking (Figure 7).

IMAGE PROCESSING IN BIO-MEDICAL ENGINEERING

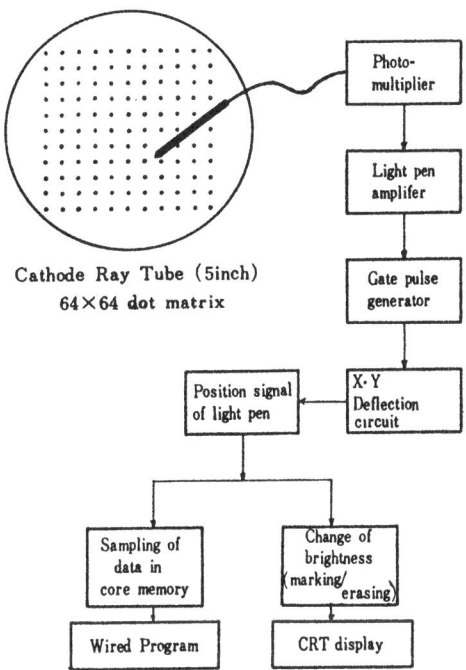

Figure 7. Schematic illustration of operation of light pen accessory. (Kojima)

These applications are possible:
-   a.   selection of four regions of interest
-   b.   calculation of counts accumulated in each region
-   c.   calculation of area
-   d.   display of count profile by arbitrary line
-   e.   assignation of arbitrary threshold levels
-   f.   time histogram of counts accumulated in each part

The merits of this man-machine system are:
-   a.   fast reading of the count distribution curves on the pattern's arbitrary lines, or peak counts

b. detection of abnormal regions by display done by indication of multi-level lines at random

c. four zones at random can be selected at once, and output from organs, abnormal regions, blood vessels and the background can be processed at once, with changes by time passage observed. This is most effective for diagnosis.

6. Utilization of Television

A pattern recognition study of radioisotope images has been done by the combination of a closed circuit television (CCTV) and a small digital computer, by Takizawa[18]. Images of liver photoscan are scanned by a television camera and then the density levels of their images sampled and stored through the input A-D converter in the memory of the computer. The stored images are reconstructed after the necessary correction for the camera characteristics. After the reconstruction, parameters that are considered as useful data for the pattern recognition are extracted, using the computer programs (Figures 8, 9, 10; Photograph 1):

a. maximum width of liver image at high and low levels

b. maximum and minimum height of liver, and its location measured from right edge of liver

c. the ratio of core area (high density) to the whole area

d. size of the shell area (intermediate density area)

e. size of liver at high and low levels and the ratio

f. recognition of the border between left and right lobe

g. space occupying lesion; yes or no; if yes, size and location of lesion(s)

h. recognition of the spleen and its parameters. Results obtained with this device revealed that the method is quite useful for the clinical radioisotope imaging works.

7. Simulation of Pattern Images

Iinuma and Fukuhisa[19], et al. are conducting research to determine the size of a square for human eye

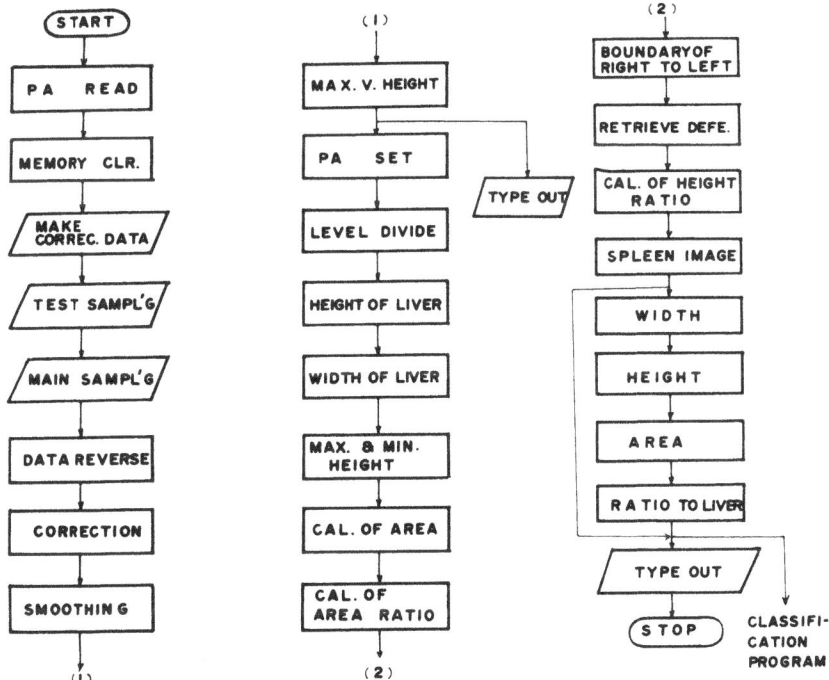

Figure 8. Computer flow chart for automatic measurement of liver RI image. (Takizawa)

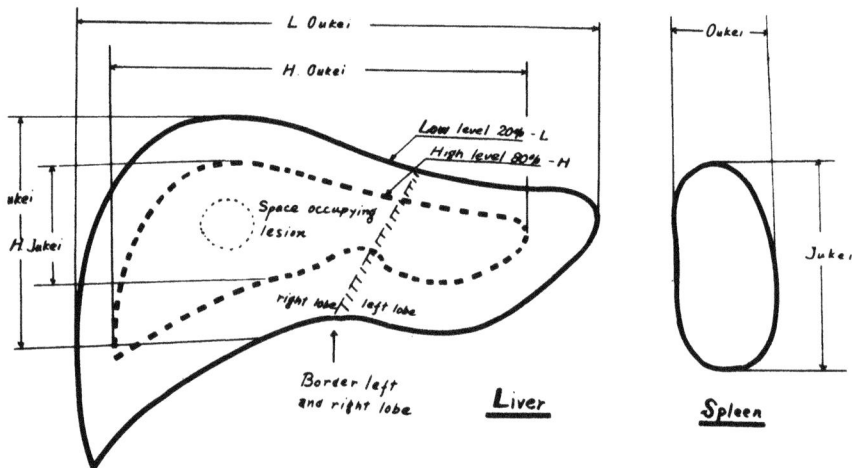

Figure 9. Illustration of parameters for automatic measurement. (Takizawa)

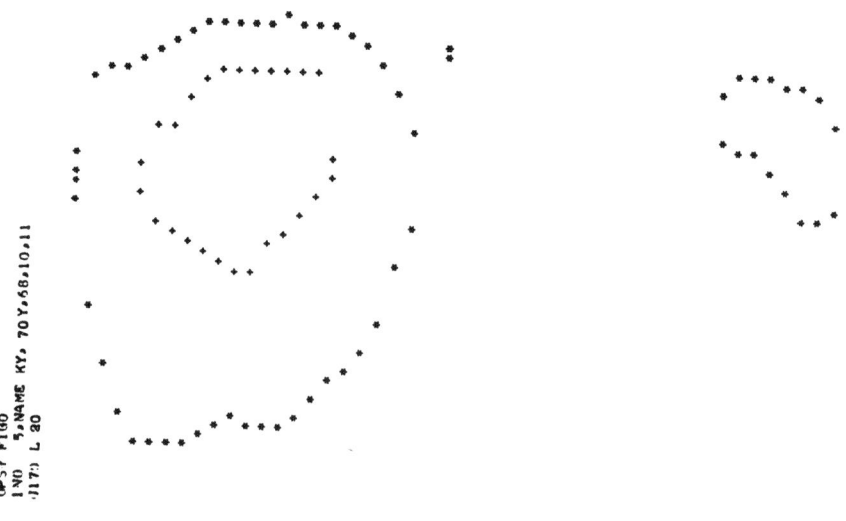

Figure 10. Display of iso-count line for measurement. (Takizawa)
* : 20% level
+ : 80% level

IMAGE PROCESSING IN BIO-MEDICAL ENGINEERING

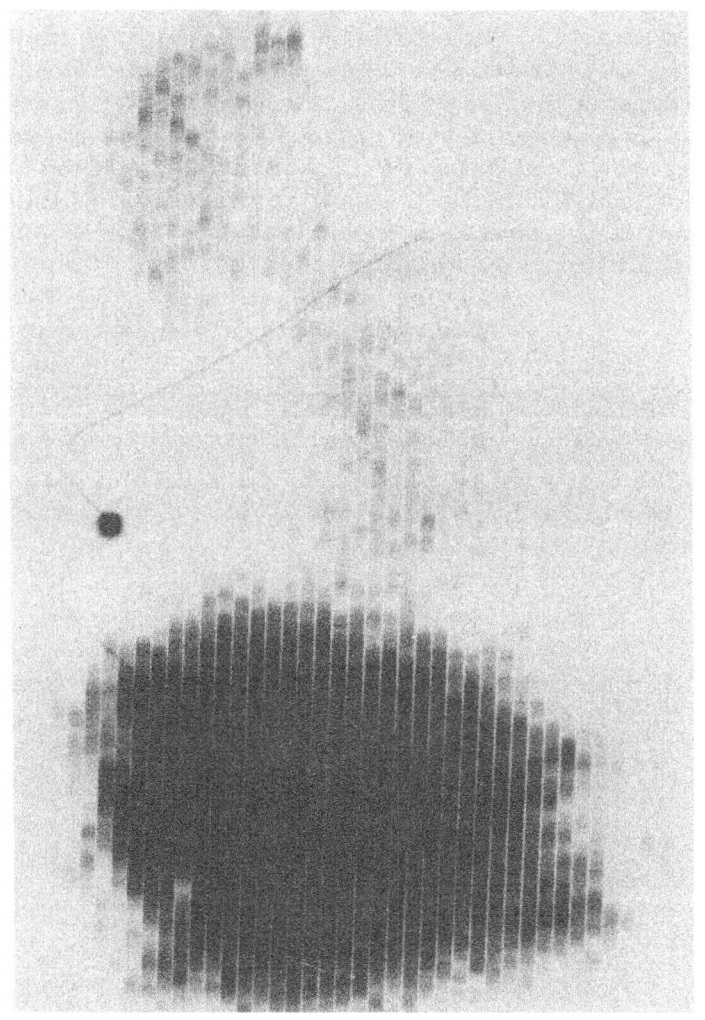

Photograph 1. A case of measurement, K. Y. 70Y, Hepatoma, a Scintiphoto. (Takizawa)

detection (Figures 11 and 12). They are attempting to make by a computer, a simulated RI pattern by using a two-dimensional, homogenous source containing one cold spot and Gaussian point spread function. The pattern is a square of elements 85 x 85 = 7225, and the sizes of defective squares are taken at 0, 3 x 3, 5 x 5, 7 x 7, 9 x 9, and 11 x 11 elements. They revised it to have three kinds of background counts and prepared 30 plates consisting of various defects displayed on a cathode ray tube and line-printer. The categories of judgment are three-stepped as "detected", "not detected", "not easily detected"; the last, "not easily detected" was made the limit of judgment. As a quantitative expression of ease of determination, the following formula was established:

$$S/N = \frac{\text{Maximum depth of defect in a blurred pattern}}{\sqrt{\text{Counts of background}}}$$

The defective RI pattern by CRT display was completely detected at S/N 1.4. At S/N 0.8, it was "not detected" (Tables 2 and 3).

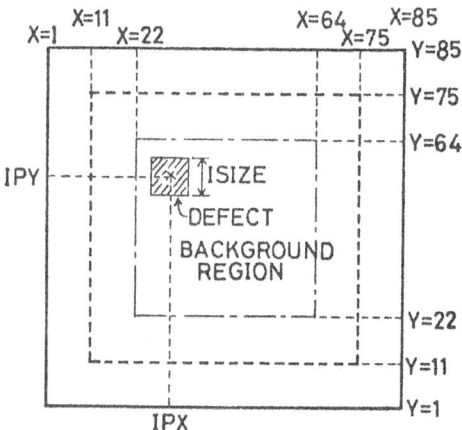

Figure 11. Configuration of a defect in a simulated phantom (Plane Phantom No. 1). (Iinuma)

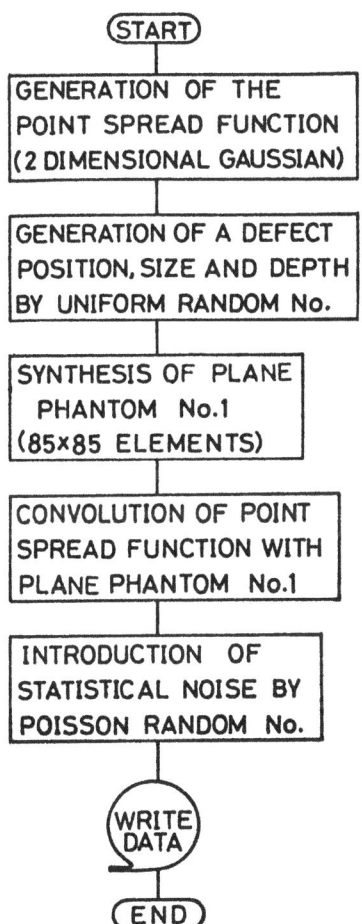

Figure 12. A block flow chart for a simulation program of Plane Phantom No. 1. (Iinuma)

Table (2)
A defect configuration in 30 different planes. Phantom No. 1 (Iinuma)

case No.	center coordinate of a defect		size of a defect (ISIZE)	relative depth of a defect (CSP)
	(IPX)	(IPY)		
1	22	60	7 × 7	30%
2	44	25	7 × 7	50
3	46	36	11 × 11	20
4	41	55	3 × 3	50
5	44	35	9 × 9	50
6	59	47	5 × 5	10
7	57	59	7 × 7	50
8	51	59	3 × 3	20
9	32	52	3 × 3	40
10	47	49	11 × 11	30
11	—	—	0	0
12	31	43	3 × 3	40
13	25	23	9 × 9	20
14	24	51	3 × 3	30
15	—	—	0	0
16	42	25	3 × 3	10
17	61	33	11 × 11	30
18	52	29	5 × 5	10
19	48	46	5 × 5	50
20	58	26	7 × 7	10
21	49	37	3 × 3	50
22	61	48	9 × 9	50
23	61	53	7 × 7	40
24	55	24	5 × 5	20
25	31	35	5 × 5	40
26	27	41	5 × 5	30
27	—	—	0	0
28	48	38	11 × 11	30
29	43	25	3 × 3	50
30	53	28	5 × 5	50

Table (3)
Classification of a defect into three categories according to its detectability by human recognition. "Not Detected" means all persons cannot recognize the presence of a defect. "Not Easily Detected" means some persons can recognize the presence of a defect, but some cannot. "Detected" means all persons can recognize the presence of a defect. Background Count = 100 (Iinuma)

	defect size	defect depth (%)
"Not Detected"	3 × 3	10, 20, 30
	5 × 5	10
"Not Easily Detected"	3 × 3	40, 50
	5 × 5	20
	7 × 7	10
"Detected"	5 × 5	30, 40, 50
	7 × 7	20, 30, 40, 50
	9 × 9	20, 50
	11 × 11	20, 30

Background Count = 25

	defect size	defect depth (%)
"Not Detected"	3 × 3	10, 20, 30, 40, 50
	5 × 5	10, 20, 30
	7 × 7	10
"Not Easily Detected"	5 × 5	40
	9 × 9	20
	11 × 11	20
"Detected"	5 × 5	50
	7 × 7	30, 40, 50
	9 × 9	50
	11 × 11	30

8. Clinical Application

Torizuka[20], Kimura[21], Yasukochi[22], Hisada[17], Takizawa[18], et al. are now engaged in the processing of scintigram information of the thyroid, liver, pancreas, kidneys, and cerebral blood flow.

The pattern information processes now employed are: smoothing, image enhancement, and imaging as an analog pattern. A nine-point average smoothing and an iterative approximation after one or two practices seem effective.

The subtraction scintigram[13] was obtained by subtracting ^{198}Au-colloid activity from ^{67}Ga activity in each matrix which was fed into the magnetic tape (Photograph 2). Similar subtraction techniques have been employed in pancreatic scans using ^{198}Au for liver and ^{75}Se-selenomethionine for liver and pancreas. In this procedure, detection of cancer in the liver was useful; however, in the pancreas it was not useful.

IV. PROCESSING ON A THERMOGRAM PATTERN

Medical thermography is a method of detecting a temperature distribution on the skin or surface of a living body and displaying it as a thermal pattern.

Thermograms shown as black-and-white, or color patterns are obtained by scanning with an infrared camera.

In the viewpoint of human engineering, man can discriminate only five level changes in black and white patterns and ten in color patterns, so it is difficult to detect fine temperature abnormality in usually taken thermograms. Fujimasa[23] et al. have been trying to solve this problem by utilization of digital data processing. Thermal signals that are the output of an infrared camera are transmitted to magnetic tape and converted with an A-D converter. These data are typed out in two-dimensional thermo-matrix (Figure 13).

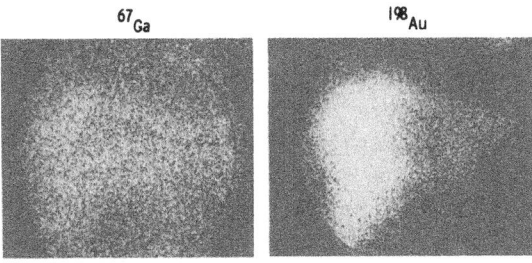

Photograph 2. Subtraction Scintigram of metastatic liver cancer. (Torizuka)

THERMO- MATRIX OF THORACIC SURFACE
(A.K., ♂ 20y. healthy)

$31.0 + a \times 1/10$ (°C)

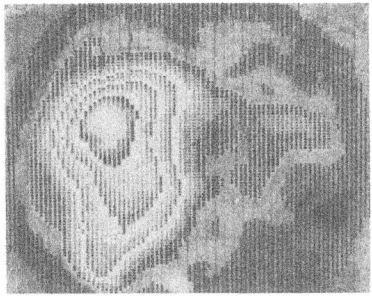

Figure 13

A thermogram in the human body has characteristic symmetry through the body's midline in front and back of its surface. Utilizing this phenomenon, processing of a thermographic image has been carried out by subtracting a normal half-pattern from the other (abnormal) half-pattern of the opposite side with a digital computer (Figures 14, 15, 16). However, in recent studies, this symmetry in the human body has been revealed to be contradicted except in extremities; so if a method of subtraction would be applied in the processing of thermograms, the same patient's thermogram pattern in a healthy condition would be required as a standard pattern to subtract.

V. PROCESSING OF X-RAY IMAGES

Diagnosis by X-ray images is basic in medical practice, either on static ones such as films or television displays, or on dynamic ones showing changes by time. At any rate, an X-ray pattern is a complicated thing by itself, while morphological differences of individuals further complicate the work of processing them as information. The work requires the highest sort of technicality. The following techniques are known today in Japan:

1. To improve image clarity techniques such as enhancement and correction, three-dimensional displays are employed.
2. To have images of normal and abnormal conditions mounted on each other to detect abnormal changes; the subtraction technique.
3. To find abnormal regions by spatial frequency distribution.
4. To detect abnormal changes from image irregularity, or an irregular correlation function.
5. Image extraction - to extract one organ's image out of a multiple pattern, either for enhancement or subtraction.

IMAGE PROCESSING IN BIO-MEDICAL ENGINEERING

SUBTRACTED THERMO - MATRIX

$(T_L - T_R) \times 10$

```
-12 00-17-11-17-13 00 00 00 00        3-1
+06-08-06-06-09-20 00 00 00 00        3-2
                                      3-3
+05+03+13+08+02-10+06+07+05 00        3-4
+02+09+12+09+07+05+06+13+17 00        3-5
+05+01+05+06+07+08-04+02 00 00        3-6
-05 00 00+04+05+02-05-08 06 00        3-7
 00-01-03+01+01+02-06-07 00 00        3-8
 00-01-03+02-01-02-06-07 00 00        3-9
+02+03-01+02 00-01+05-08 00 00        3-10
+03+03+02+02+04+04+03 00 00 00        3-11
 00+06+01+01+01+05+07 00 00 00        3-12
+07+04+06+01-06+03+03 00 00 00        3-13
-02-03 00+13 00+02 00 00 00 00        3-14
-01-06+03+03-01-03 00 00 00 00        3-15
-02 00+01+01-07-02 00 00 00 00        3-16
-01+04+04+02-05-07 00 00 00 00        3-17
+03-01-02+01-09-12 00 00 00 00        3-18
+06 00-02-04-07-13 00 00 00 00        3-19
 00 00-03-02-04-13 00 00 00 00        3-20
-02-09-10-06-11-13 00 00 00 00        3-21
 00-03-07-11-18-12 00 00 00 00        3-22
                                      3-23
+05+11+05+19-01-02+19-13-27 00        3-24
-01+11+15+12+05-19-05+19 00           3-25
```

Figure 14

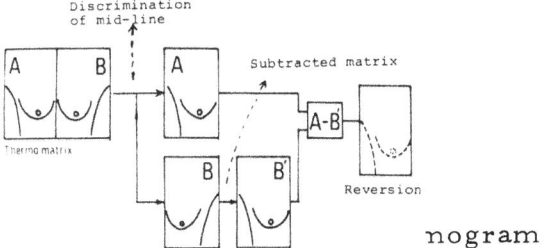

Figure 15.   nogram

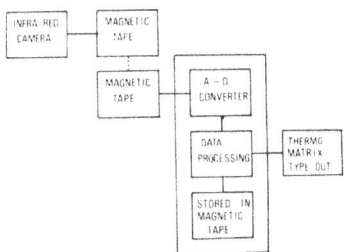

Figure 16.   Data Processing of Thermogram

6. To analyze relativity of space and time information.

Umegaki[24] has forecast that the above (6) will progress over all others in the future as it enables diagnosis at dynamic functions of the body.

In Japan today, X-ray image processing is employed for diagnosis of esophagus, stomach and chest diseases.

1. <u>The Esophagus</u> - Umegaki[25] et al. used a television sampling apparatus and extracted and recorded esophagus patterns as information for processing. The method was to analyze the curves made by passing barium through an esophagus. First he set the barium quantity to pass through the esophagus and made an assumption that the peak of the curve would coincide with the time of the flow of barium at that spot. He then obtained the characteristic response pattern by comparison of the barium-passing curves at the oral side and at the anal side. Data obtained were useful for the quantitative calculation of the barium passing through the esophagus. The barium-passing curves across the body over the region of the esophagus were sampled every two millimeters, and the following calculations obtained:

a. Correlation function between neighboring curves (Figures 17 and 18)

b. Duration of the peak

c. Barium quantity at each of the spots

d. Power spectrum curve

He reports that in a normal esophagus, the pattern is regular, with barium running toward a certain direction, but this regularity gets lost over an abnormal portion of an esophagus. Moreover, the correlative function becomes lower over a diseased portion than over healthy parts.

Kobayashi and Takizawa[26] et al. have developed what they call "multichannel videodensitogram" (video-

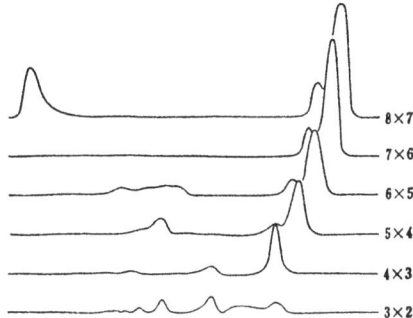

Figure 17. Differentiation product of neighboring curves in maximum correlation function between (Umegaki)

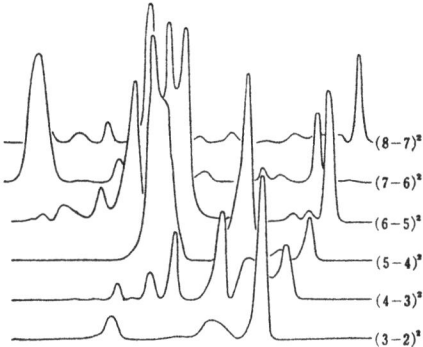

Figure 18. Difference of differentiation of neighboring curves in maximum correleation function between (showing irregularity in cancer function (Umegaki)

kymogram) which is a dynamic analysis of patterns using an X-ray television. They synchronized the numerous spots of the television monitor with barium running by, using a minicomputer. Data were recorded at 50 spots on the monitor at 60 prints per second, maximum. They could control measuring time at a desired duration of 0.5 second. The display is three-dimensional by placing the esophagus X-ray image on the X-Y dimensions and plotting the depth (Z) on it (Photograph 3 and Figure 19).

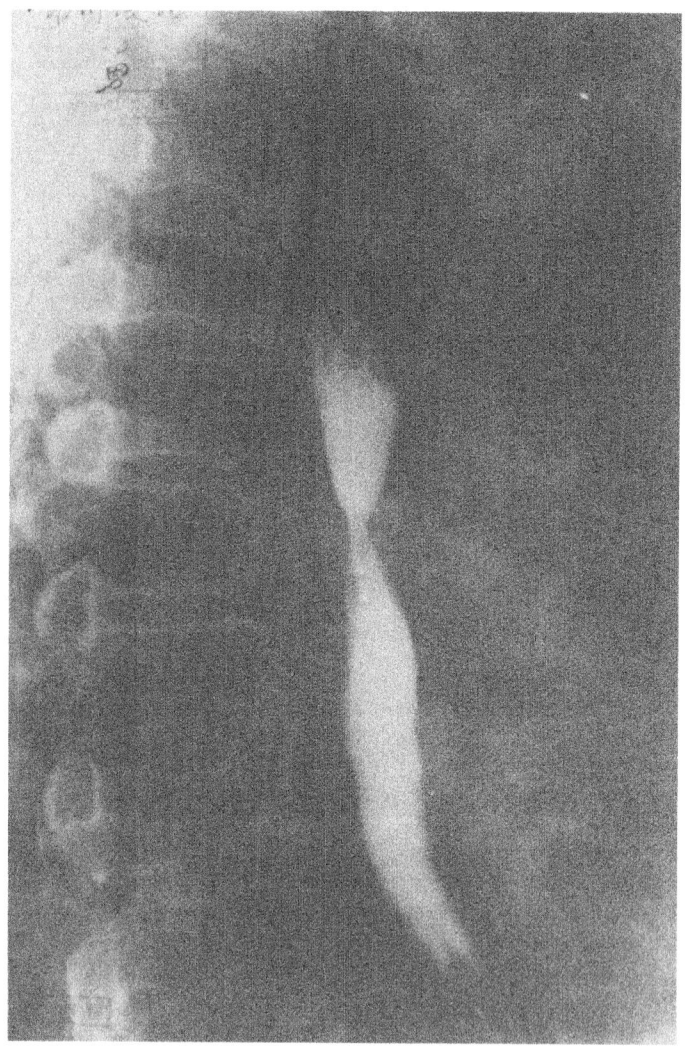

Photograph 3. An X-ray photograph of esophagal cancer. (Kobayashi and Takizawa)

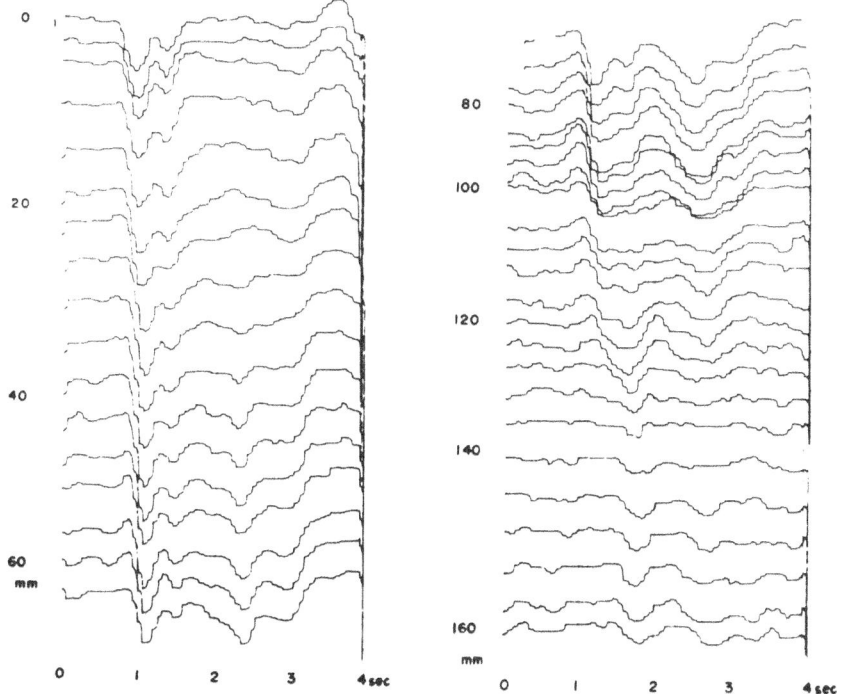

Figure 19. Multichannel videodensigram at the supine position.(Kobayashi and Takizawa)

2. The Stomach - Akatsuka, Takatani and Isobe[27,28], knowing that input of all the information in the computer after complete scanning of the pattern would make all sorts of processing possible, thought there would be a more effective method of saving memory and time required for the foregoing. They paid attention to the contour only for the foregoing listed, and they paid attention to the contour only for processing the shadow images.

They developed a device to extract all the features of the contour images, and are now working on stomach X-ray images with it.

Their hardware consists of a flying spot scanner (i.e., CRT, a lens system, and a photomultiplier), a control system and a data processor. With the assumption that the continuous boundary line between the brightness and the darkness would be the contour line, they irradiated the CRT spot on the film, saw in what direction the brightness changed by the circular search motion of the spot, and controlled this circle to follow the contour. The small circle was generated by the sine wave and cosine wave in the X-Y coordinate plane. The direction of the contour line is detected by the starting pulse as soon as the spot passes over the contour line. These sinusoidal waves are sampled as the increment of the X-Y coordinate; thus, the center of the circle is advanced toward this direction by piling up these increments. For noise rejection at the first cycle, it seeks only the direction of $\pm 45°$ in the previous direction. In the case of failing to find the contour in such a direction, it seeks the whole direction at the second cycle (Figure 20). The traced directions were digitized in all four directions: north, south, east and west. The 3,000 sampled points were stored in 500 words (12 bits per word). As shown in Figure 21, these four directional sequences converted to three modulated values; i.e., M(direct); L (turn left); and R (turn right). Each point of the contour line can be expressed as

$M^r R M^l L$, meaning M r-times, R once, M l-times and L once, RM R or LM L.

Processing these expressions, the contour expressed by these letters having two parameters; i.e., length and curvature. Index I was employed to stand for the straight line; A for the right turn; and B for the left turn. For example, Figure 22 shows IB IA-----. In this manner, the the total shape of the image and details of the contour live and are identified. Figure 23 shows further classification of the detail into six types of lines as illustrated. These are applied to identify the stomach's cardia, pylorus and angle, and also, by a finer procession, to find the contour line defect corresponging to such as tumors; and so on.

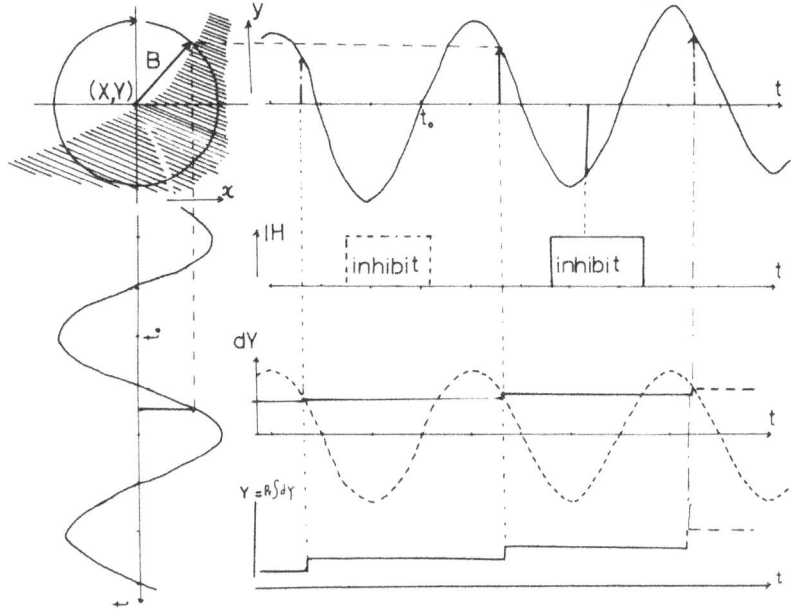

Figure 20. Detection of direction of contour line. (Akatsuka)

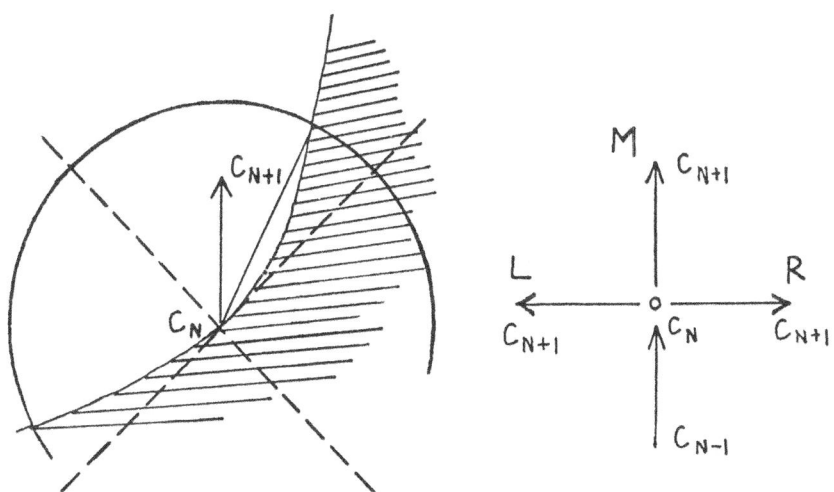

Figure 21. Three directions of M, R and L. (Akatsuka)

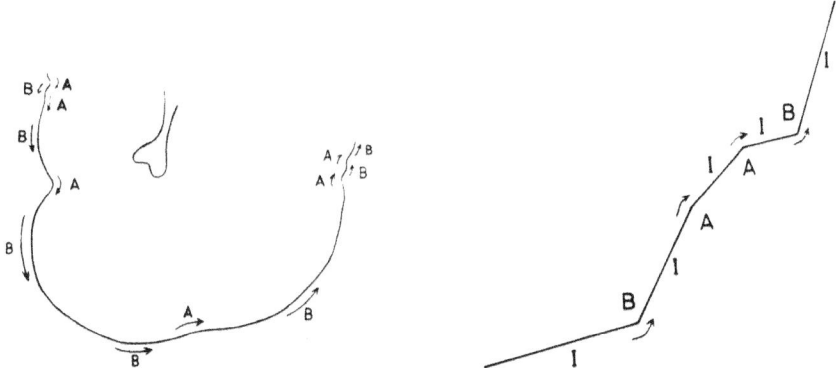

Figure 22. Expression of contour line with I, A and B. (Akatsuka)

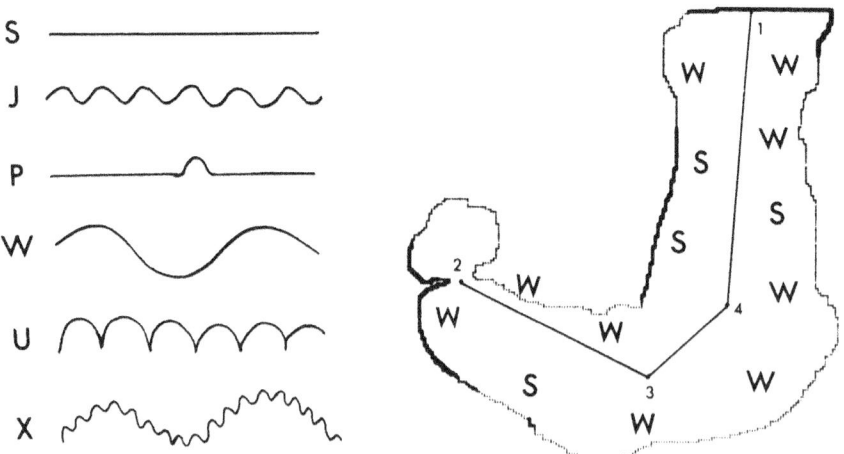

Figure 23. Expression of contour line with six feature line. (Akatsuka)

3. The Chest

   a. Coding - Ishiyama and Yamamura[29], et al. intend to establish an automatic diagnostic data processing system of respiratory diseases centering on lung cancer. In this report they deal with a roentgen diagnosis of lung cancer and pulmonary tuberculosis using a digital electronic computer.

   In processing posteroanterior chest roentgen findings by a digital electronics computer, the findings are divided into minimum shadow units concerning nine points such as the site and nature for coding (Table 4). This method prevents discrepant readings by different viewers and faulty interpretation by inexperienced examiners. Having a favorable reproducibility, it is found satisfactory.

   Ten coded basic roentgen types of lung cancer and pulmonary tuberculosis (Table 5) were fed into the memory of a digital electronic computer and then coded roentgen findings of 66 confirmed cases of lung cancer and 184 confirmed cases of pulmonary tuberculosis were put so that only those roentgen types corresponding to the confirmed diagnostic data might be typed out through operation. The time required for operation was 0.4 second on an average.

   An accurate diagnostic rate for confirmed cases of primary lung cancer was 37%, while that of pulmonary tuberculosis was 36%. Those cases corresponding to roentgen types of both cancer and tuberculosis accounted for 53% in either primary lung cancer or pulmonary tuberculosis. It was considered that at least 90% of the primary lung cancer cases tested had a possibility of lung cancer, as did the same percentage of pulmonary tuberculosis cases.

   The present method in which posteroanterior chest roentgen findings were employed in diagnosing lung cancer was found favorable, since it revealed a comparatively few overlooked cases; however, about half the cases tested were found to correspond to roentgen types of both

cancer and tuberculosis. This handicap, however, might
be improved by less usage of so-far accepted morbid pattern classifications and by inductively deriving roentgen
criteria based on actual data of cases observed.

      b. <u>Subtraction</u> - Ishiyama and Yamamura[30], et
al. are trying to remove the noise coming from bone and
vascular images in chest X-ray film automatic screening
processes. For this purpose they try video subtraction
between old and new films of one person. Their aim in
research is to get a computerized system of diagnosis by
A-D conversion of the optical density of the subtracted
images. In this research, they report that 122 pairs of
apparently no significant differences were 25% misread,
owing to density irregularity caused by poor mounting of
the pair on each other, and/or poor correction; while on
the other hand, 54 pairs having decisive differences were
video-subtracted and 75% showed definitely the unsubtracted
spots on the subtracted images.

      c. <u>Spatial Frequency Analysis</u> - Takenaka[31] et al.
determine the spatial frequency of X-ray films by developing an apparatus consisting of a flying spot scanner, a
direct current corrector, a blanking level suppressor, a
frequency analyzer and a recorder. They have observed
that spectra show differently as scanning is done from the
hilum toward the peripheral regions in X-ray films, but
spectra are found to be the same from the same spot in
the same area from person to person. In patterns suffering from lobar pneumonia, the spectrum amplitude lowers
in the low frequency area. However, X-ray films receive
influences from many factors. Unless, therefore, all
these factors are controlled, clinical diagnosis by spatial
frequency will not be possible, they find. It is possible to
classify patterns into diffusive, sclerotic, and dotty kinds,
and it is useful for followup of the stages of variation of
one diseased case. Takizawa[32] studied to calculate spatial frequency spectra of the pulmonary roentgenograms
by the digital Fourier transform with a small computer.

Table 4. The coding of posteroanterior chest roentgen findings. (Ishiyama)

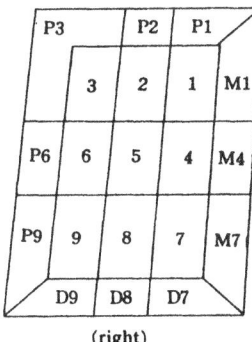
(right)

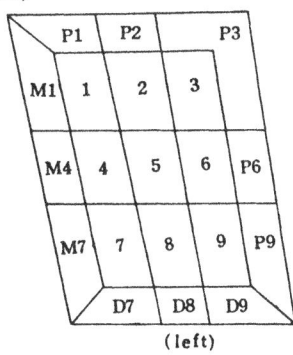
(left)

O: Absence of lesion
I, II, III,....: Lesions numbered
General lesions
  A: Site
    r : Right l: Left;
    1, 7: Medial area
    4 : Hilar area
    2, 5, 8: Median area
    3, 6, 9: Lateral area
    M: Mediastinum
    P : Pleura
    D : Diaphragm
  B: Shape
    0 : Most of the margin corresponding to the thoracic wall or mediastinum
    1 : Round
    2 : Quasi-round
    3 : Multiangular
    4 : Triangular
    5 : Streaky
  C: Size
    1 : Miliary (less than 1 mm in diameter)
    2 : Acinous (1-5mm in diameter)
    3 : Lobular (6-15mm in diameter)
    4 : Segmental (more than 15 mm in diameter)
    5 : Lobar
    6 : A-lung-sized
  D: Density
    1 : Calcified
    2 : Intermediate (same with or more than lung-marking)
    3 : Thin (thinner than lung-marking)

  E : Margin
    0 : In cases of B_0
    1 : Regular
    2 : Irregular
  F : Homogeneity
    1 : Homogenous
    2 : Nonhomogenous
    3 : Nonhomogenous with rarefaction
  G: Number
    1 : Single
    2 : A few (2-5)
    3 : Considerable (6-10)
    4 : Abundant (more than 10)
  H: Distribution
    0 : In cases of G_1
    1 : Distributed homogenously
    2 : Distributed nonhomogenously
  K: Concentricity
    0 : In cases of G_1
    1 : Concentric
    2 : Eccentric
Particular lesions
  a : Diaphragm
    1 : Elevated
    2 : Irregular
  b : Mediastinum
    1 : Displaced rightward
    2 : Displaced leftward
  c : Pleura
    1 : Thickened
    2 : Calcified
  d : Deformity of thorax
  e : Displacement of trachea
  f : Defect of ribs
  g : Abnormal cardiac silhouette

Table 5. The coding manifestations of the roentgen diagnostic criteria and their expression by binary numbers. (Ishiyama)

		B C D E F G H K
		01234512345612301212312340 12012
1. Lung cancer (LC)		
a. Tumorous type	$B_{123}C_{3456}D_2E_{12}F_{12}G_1H_0K_0$	01110000111010011110100010 0100
b. Infiltrative type	$B_{1235}C_{12345}D_2E_2F_2G_{234}H_{12}K_{12}$	01110111100100001001101 1011
c. Atelectatic type	$B_{04}C_{56}D_{23}E_{12}F_1G_1H_0K_0$	10001000111011011100100010 0100
d. Pleuritic type	$B_{04}C_{456}D_2E_{01}F_1G_1H_0K_0$	10001000111010110100010 0100
e. Cavitary type	$B_{123}C_{3456}D_{23}E_{12}F_3G_1H_0K_0$	01110000111011011001 10000 0100
f. Specific type*		
g. Metastatic type*		
2. Pulmonary tuberculosis (TB)		
a. Exudative type	$B_{124}C_{3456}D_{23}E_2F_2G_1H_0K_0$	01111000111011001010100010 0100
b. Caseoinfiltrative type	$B_{123}C_{123}D_2E_2F_2G_{234}H_{12}K_{12}$	01110011100001000100110 1011
c. Fibrocaseous type	$B_{1235}C_{12345}D_2E_2F_{23}G_{234}H_{12}K_{12}$	01110111110001011001 10 1011
d. Cirrhotic type	$B_{1235}C_{12345}D_{12}E_2F_{12}G_{234}H_{12}K_{12}$	01110111110111000111001 1011
e. Disseminated type	$B_{123}C_{12}D_2E_2F_{12}G_1H_1K_2$	01110011000001000111000001 0001
f. Far-advanced mixed type*		

* Excluded from the present study

The film density is detected by a scanning microphotometer, and scanning distance per each sampling location on the film is 10 mm, excluding the rib and clavicular shadows.

Calculations have been made by Simpson's rule, and the calculated spectrum is displayed from 0.1 to 10.0 line pairs/mm. In order to find out the true image spectra, the noise spectra contained in the original spectra are suppressed by means of digital filtering. Clinical applications are tried by this method; normal cases, spontaneous pneumothorax, pulmonary emphysema, bronchial asthma, and lung cancer. Some features of patterns are observed, and the possibility of the pattern recognition by spatial frequency spectrum analysis of the chest X-ray film is proved by this method.

d. Dynamic Functioning Analysis - Kobayashi and Takehisa[33], et al. are obtaining a lung's functional information by a mini-computer data processing on the findings of mutually synchronized scanning of two X-ray films taken with a time interval, and have put the results into A-D conversion.

They developed the phase selection controller which can take chest X-rays in inhaling-exhaling stages, pulmonary artery's systolic-diastolic periods, etc. The density signal for the film is transferrable to an eight-bit digital signal by the two-channel A-D converter. After the density correction, smoothing and bustraction processes, the result is displayed in the five-level color pattern by the digital plotter.

e. Pattern Recognition - A little experiment has been reported concerning automatic recognition or classification of radiographic images, with the exception of image enhancement. Fukumura and Toriwaki, et al.[34,35,36,37,38] have been developing an automatic interpretation system applicable to chest radiograms. Their goal is the classification of given posteroanterior chest photofluorograms into normal or abnormal categories.

The usual formulation of the pattern recognition problem; that is, schematic division of the process into three steps; preprocessing (normalization), feature extraction, and classification; is also relevant to some extent with respect to analysis of chest radiograms. In the processing of chest radiograms, it is most important in the foregoing procedure to recognize various images contained in the film such as heart, ribs, vessels and abnormal shadows in the lung.

The automatic interpretation system of chest roentgenograms; Version 2 (AISCR-V2) recognizes heart ribs, great vessels, and some kinds of abnormal shadows in the lung field from chest photofluorograms.[34,37] This system consists of four subsystems including the sampling and quantization system and the recognition systems of heart shadow,[37] rib images,[35] and abnormal shadows in the lung field.[36] They are shown in Figures 24, 25, 26 and 27.

In AISCR-V2, many linear filters are utilized that differ in density and have average pattern matching and bridge filter qualities. In the conversion of gray pictures into line figures, the wave propagation method[38] is applied.

In the experiment,[36] chest photofluorograms (70 mm x 70 mm) taken in mass screening in which the area to be processed was about 50 mm x 50 mm, were used as the input to the system. The sampling interval was 0.2 mm on the film, so that one sample contained 250 x 250 points. The value of gray level (film density) for each sample point is represented by a decimal number of three significant digits. The digitized picture obtained is stored as the data file in the general purpose digital computer. Examples of the results of chest photofluorograms of lung cancer are shown in Photograph 4 and in Figures 28 and 29.

IMAGE PROCESSING IN BIO-MEDICAL ENGINEERING 39

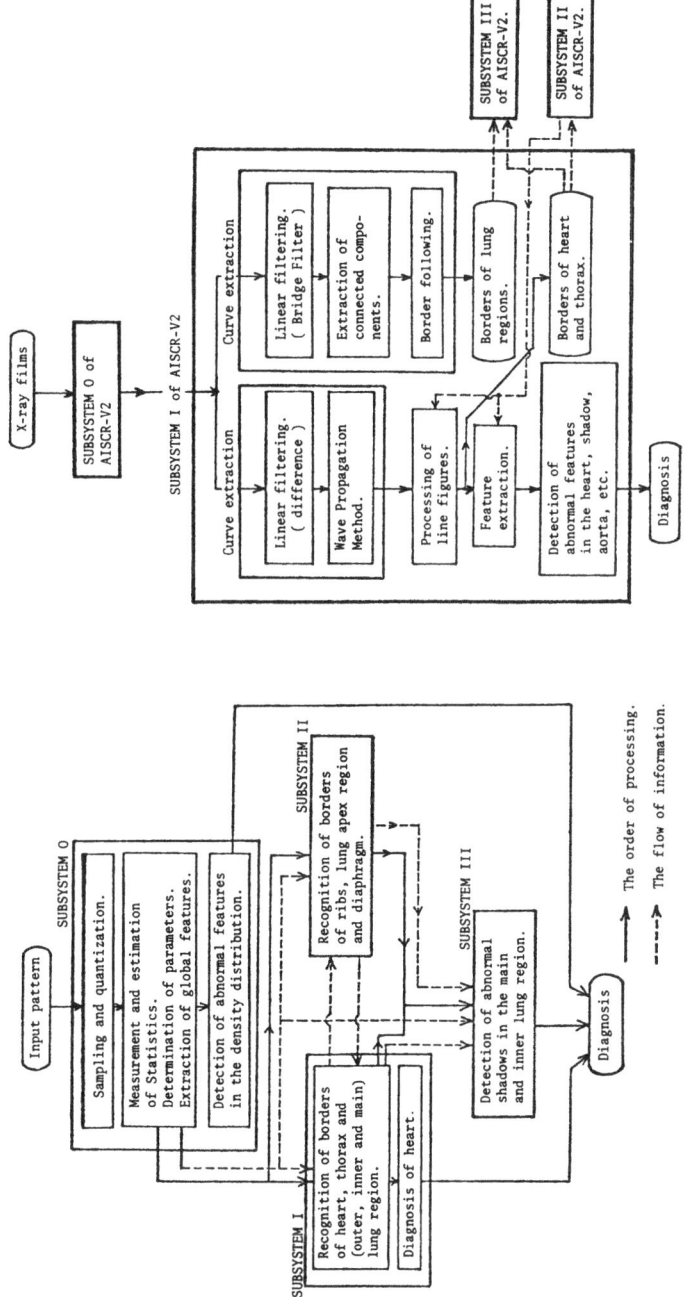

Figure 24. Block diagram of AISCR-V2. (Fukumura, Toriwaki)

Figure 25. Block diagram of Subsystem I of AISCR-V-2. (Fukumura, Toriwaki)

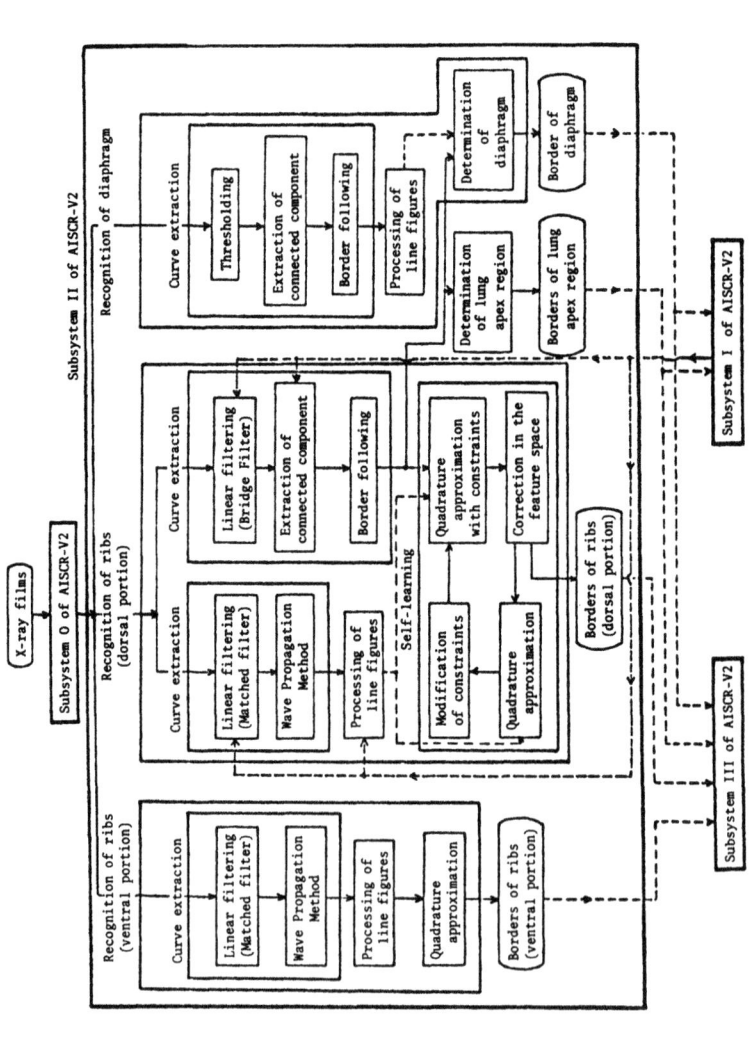

Figure 26. Block diagram of Subsystem II of AISCR-V2. (Fukumura, Toriwaki)

IMAGE PROCESSING IN BIO-MEDICAL ENGINEERING

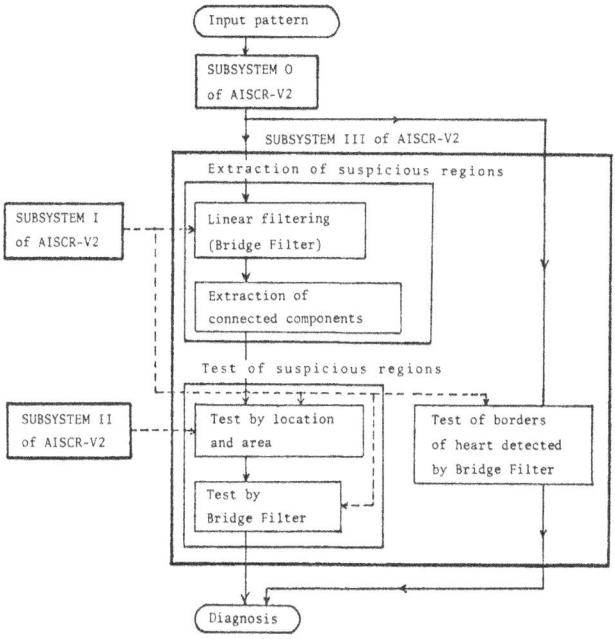

Figure 27. Block diagram of Subsystem III of AISCR-V2. (Fukumura, Toriwaki)

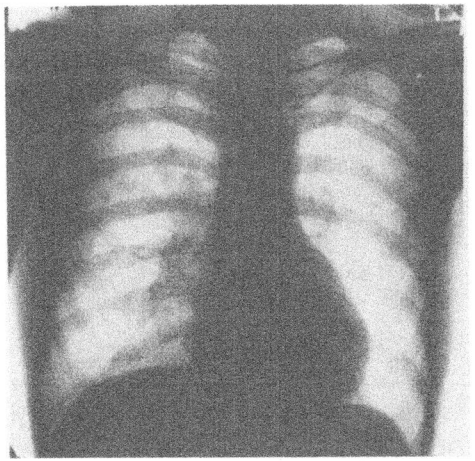

Photograph 4. Chest photofluorogram (lung cancer) (Fukumura, Toriwaki)

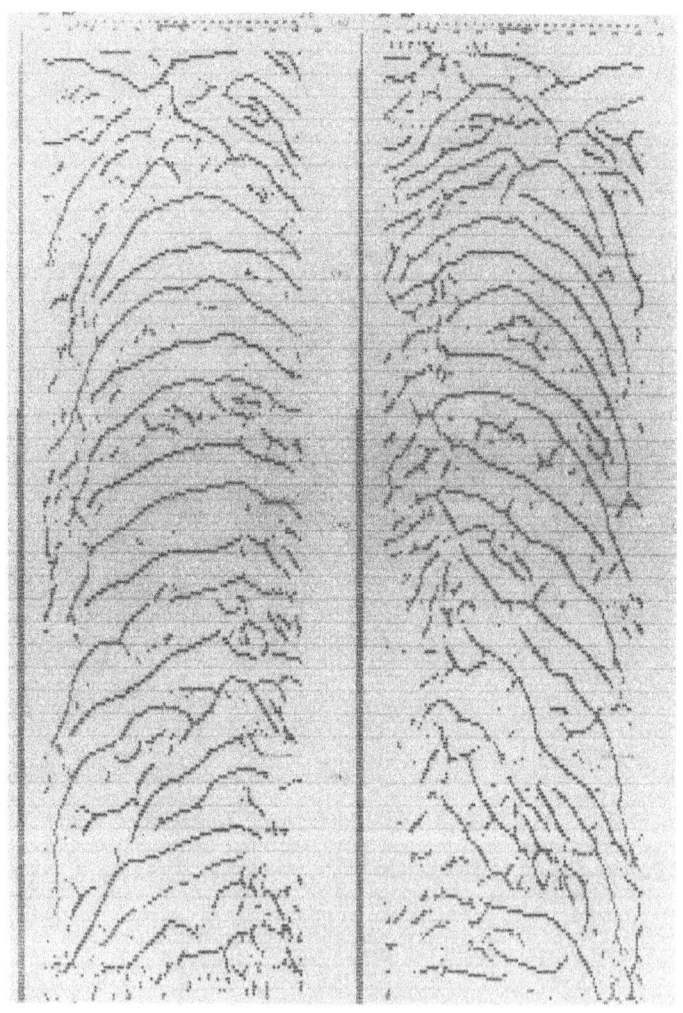

Figure 28. Line figures recognized by the Subsystems I and II. Contains borders of heart, thorax, dorsal portions of ribs, vessels, etc. (Fukumura, Toriwaki)

IMAGE PROCESSING IN BIO-MEDICAL ENGINEERING

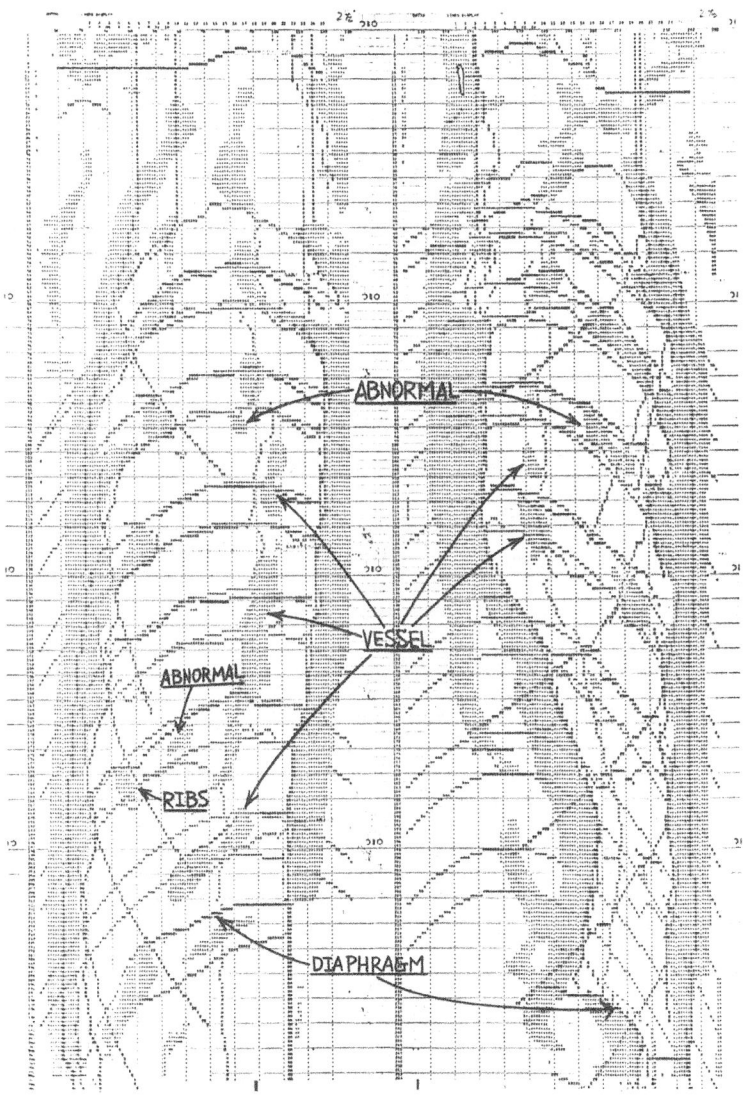

Figure 29. Output of the total system. Curves are borders of heart, thorax, boundaries among outer, main and inner lung region, and rib borders (dorsal and ventral) approximated by quadratures. Shaded areas in lung field are shadows suspected of being abnormal. Final decisions concerning them also given in the figure. (Fukumura, Toriwaki)

VI. PROCESSING ON CANCER CELL PATTERNS

1. In 1967, Ishiyama and Yamamura,[39] et al. developed the Automatic Cytoscreener and measured sizes of cell nuclei automatically (Figure 30). Luminous flux radiated from a flying spot tube is focused with a lens assembly. This spot of light (two to five microns in diameter) is thrown upon a glass slide of the type used for microscopy. Scanning is performed in 200 unit regions, each of which occupies a square 1.14 by 1.14 mm.

The spot of light, modulated by the cell nucleus, is detected by a photomultiplier. The detected signals are amplified, reformed, digitized and fed into a computer which recognizes each of the cell nuclei scanned as a single nucleus, calculates the sectional area of the nucleus, classifies it into one of five area levels previously defined, and totals the number of nuclei in each level class (Figure 31).

In the next research step, they studied automatic measurement of density in addition to the automatic measurement of area using the Photomicroscanner. Eight parameters are calculated in the measuring of nuclear and cytoplasmic density and area as follows (Figure 32):

Cell area (A_C)

Nuclear area (A_N)

Nucleo-cytoplasmic ratio (A_N/A_C)

Total cytoplasmic (staining) density (D_C)

Total nuclear density (D_N)

Nucleo-cytoplasmic density ratio (D_N/D_C)

Average cytoplasmic density (aD_C)

Average nuclear density (aD_N)

IMAGE PROCESSING IN BIO-MEDICAL ENGINEERING

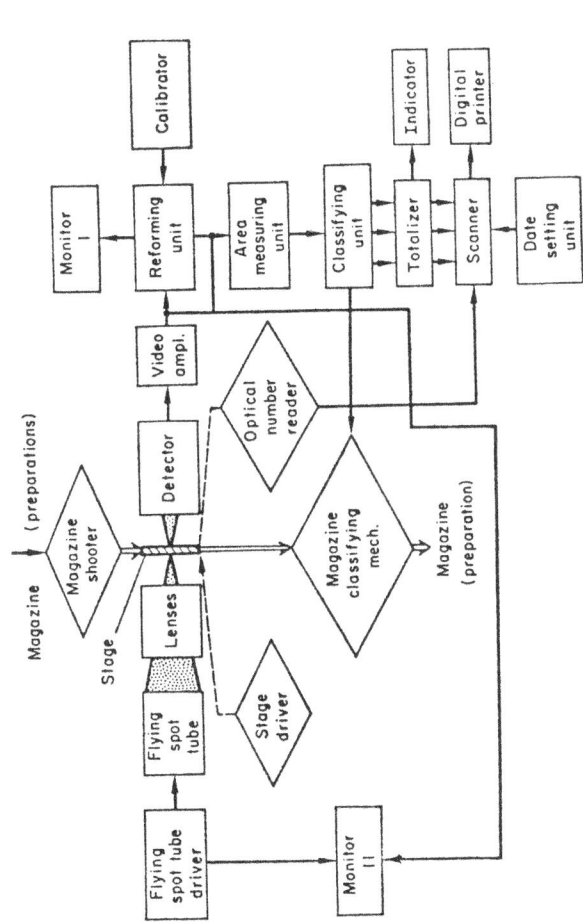

Figure 30. Block diagram of the Automatic Cytoscreener. (Ishiyama)

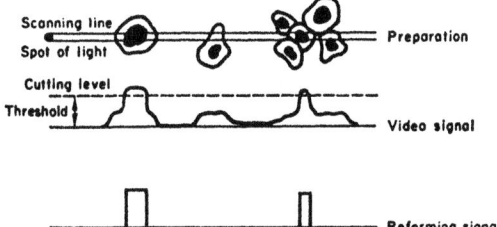

Figure 31. Relations between the preparation and the detected signals. (Ishiyama)

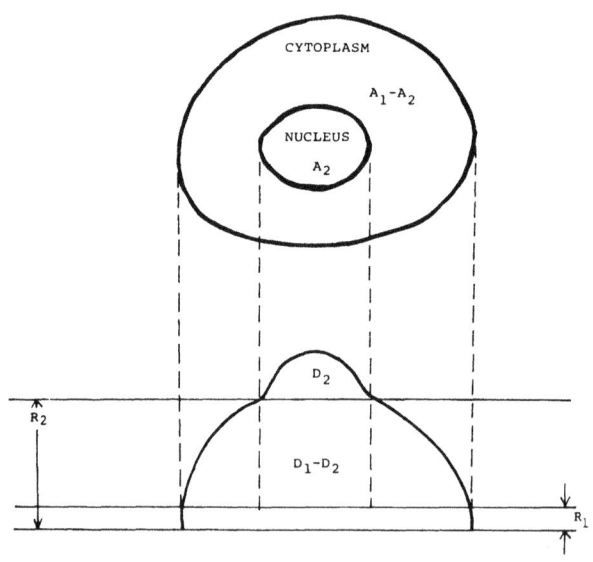

CELLULAR AREA : $A_C = A_1$ NUCLEAR AREA : $A_N = A_2$
NUCLEAR CYTOPLASMIC RATIO = A_2 / A_1
CYTOPLASMIC TOTAL OPTICAL DENSITY : $D_C = (D_1-D_2)(A_1-A_2)/A_1$
NUCLEAR TOTAL OPTICAL DENSITY : $D_N = D_2 + (D_1-D_2)A_2/A_1$
AVERAGE CYTOPLASMIC OPTICAL DENSITY : $aD_C = D_C/A_1-A_2$
AVERAGE NUCLEAR OPTICAL DENSITY : $aD_N = D_N/A_2$

Figure 32. Method of calculation of individual parameters from outputs of photomicroscanner. (Ishiyama)

The correlation of all pairs between two parameters out of eight was analyzed and the sensitivity and specificity were calculated. Results were adopted in logic to discriminate cancer cells from normal cells.

According to the results of experiments to detect smears (Table 6), correct diagnoses for cancer cells were 97.2% in the internal sample and 89.2% for normal cells. As for the external sample, diagnostic accuracy was rather satisfactory.

Table (6) Results of Automatic Diagnosis of Cancer Cell (Ishiyama)

(Internal Sample)			
	Correct Diagnosis	Wrong Diagnosis	Correct Diagnosis Ratio
Cancer Cell	35	1	97.2 %
Normal Cell	25	3	89.2 %
Total	60	4	93.8 %
(External Sample)			
	Correct Diagnosis	Wrong Diagnosis	Correct Diagnosis Ratio
Cancer Cell	40	5	88.8 %
Normal Cell	123	17	87.8 %
Total	163	22	88.1 %

2. Hashimoto[40,41] also developed the auto cytoscreener which is a light microscope having an automatic stage driven by a controller; an industrial television camera which served as an image sensor; a processor for the computation of the size and density of the nucleus; an automatic changer of specimens; a monitor television; display and recording device (Photograph 5; Table 7).

The image of a nucleus stained with Feulgen stain is transformed into a video signal by the television camera. The distribution of concentration of DNA of any cross-section of the nucleus is indicated by the pulse on the video

Photograph 5. The Auto-cytoscreener developed by Hashimoto, et al.

Table (7) Various methods for automatic cyto-screening (Y. Hashimoto)

Method to extract biological characteristics	1) Feulgen reaction * 2) spectrophotometry 3) fluorospectrophotometry
Treatment of materials	1) slide glass * 2) floating fluid
Detecter	1) television camera * 2) flying spot 3) Nipkow disk 4) Coulter counter 5) spectrophotometer
Deciding parameter	1) combination of nuclear size and its concentration* 2) nuclear concentration 3) nuclear size
Objective of processing	1) processing of measuring results of single cell 2) processing of simultaneous measuring results of multiple cell groups *
Method of processing	1) on - line processing * 2) off - line processing

* The method used by Hashimoto.

signal. The height of the pulse increases in proportion to the concentration of DNA in the nucleus, while the width of the pulse increases in proportion to the diameter of the nucleus. The pulse width in 10% concentration is set at parameter X; in 40% it is set at Y; and in 60%, at U, respectively (Figure 33). The number of the pulse whose width is within certain limits is designated as Z_X or Z_Y, respectively. X, Y, U and Z are parameters which express the size of the nucleus in certain concentrations. An object having a density higher than 60% is almost always a non-cellular component such as dust, while an object having a density lower than 10% usually represents the nucleus of a non-malignant squamous epithel cell. Consequently, measurements are automatically omitted on objects of which U is not equal to zero, or X equals zero; so measurements are made on objects of between 10% and 60% density. Concerning the object with between 10% and 40% density, those of which X is less than eight microns and greater than twenty microns are excluded from measurement because they are either too small or too large to be the nucleus of a cancer cell (Figure 34; Table 8).

According to the experimental results on 575 cases of mass screening for smears utilizing this system, ninety percent of the non-malignant cases were screened out.

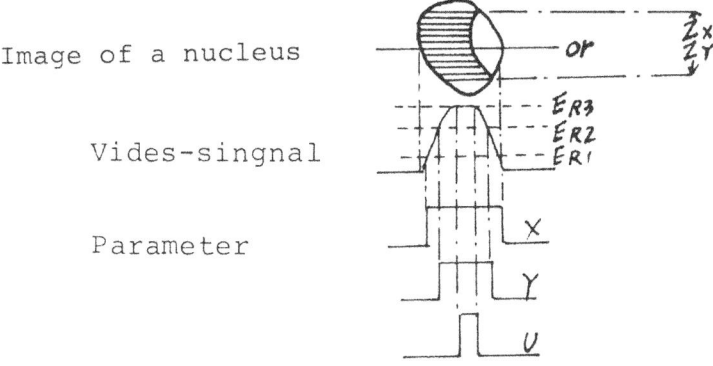

Figure 33. Parameters used for diagnosis. (Hashimoto)

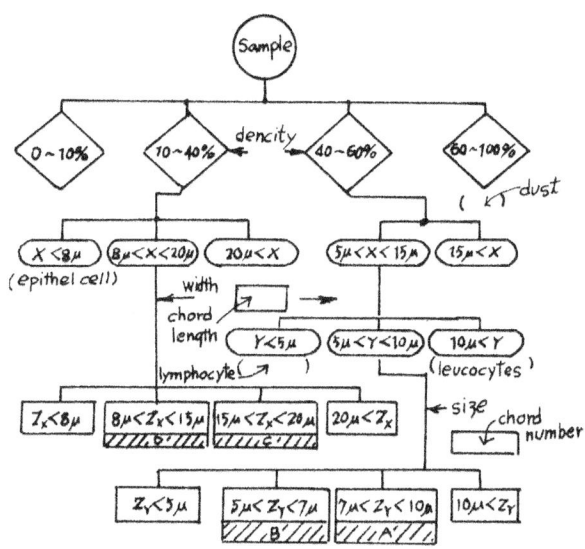

Figure 34. Diagnostic branch. (Hashimoto)

Table (8) Results of Diagnosis per Cells (Hashimoto)

Density	40 - 60 %		10 - 40 %		< 10 % > 60 %	Total
Diagnosis	A	B	C	D		
Dust	1	5	2	4	28	40
Agglutinated leukocyte			25	24	32	81
Flat-epithel Cell			2	5	966	973
Benign-atypical Cell			5	12	223	240
Cancer Cell	20	6	93	71	13	203

VII. PROCESSING OF BLOOD CELL PATTERNS

Takagi and Kaihara,[42] et al. studied a method of automatic classification of white blood cells. First they separated the white cell group from the blood samples containing both red and white cells, then classified these separated white cell groups of each type according to the index of their nuclear features.

Patterns of blood cells photographed with a microscope are digitized by pattern reader and normalized into 32 levels. The frequency distribution is shown in Figures 35 and 36. Utilizing these results, levels of background, red cells, cytoplasma and nuclei are determined.

The running lengths in each level per one line are calculated and statistics of a run length are taken in total lines. The distribution of run length of lymphocyte (L), monocyte (M), neutrophile leucocyte (N) and eosinophile leucocyte (E), are carried out. Area of nucleus, maximum run length and total run length are adopted as parameters of an index to classify white cells. The total run length is divided into three equivalent parts. The shortest division is designated as (1), the medium division as (2), and the longest division as (3). The distribution of each division and the ratio of (2)/(3) are shown in Table 9.

According to these results, lymphocytes are distributed in the field of long run length and monocytes distributed in the field where the total run length is large. Length of the run length in neutrophile leucocytes is short and has a tendency to decrease to the right. Many run lengths are short in eosinophile leucocytes due to granules of cytoplasma.

Automatic classification of white blood cells will be possible with the program based upon the above-mentioned parameters.

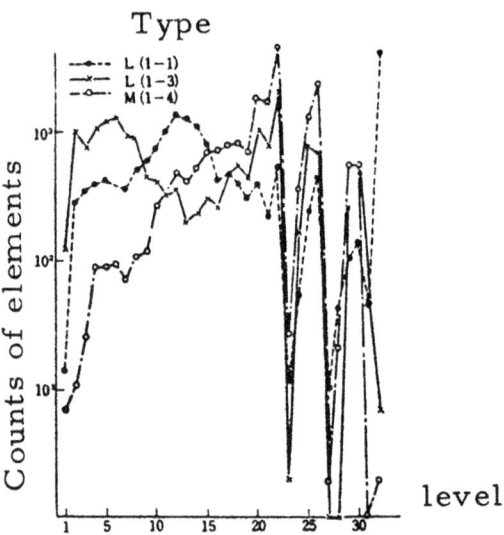

Figure 35. Distribution of density of cells (I). (Takagi and Kaihara)

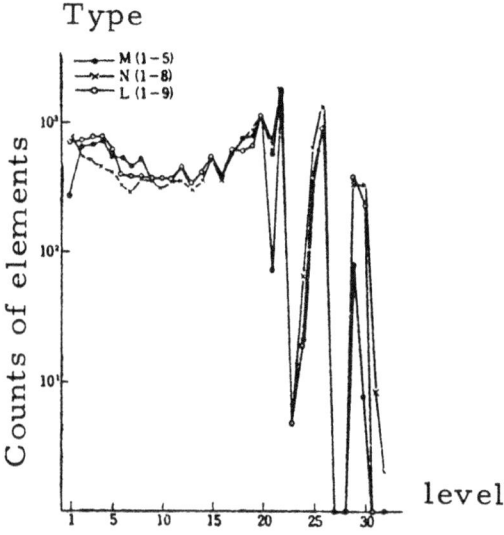

Figure 36. Distribution of density of cells (II). (Takagi and Kaihara)

Table 9. Distribution of run length of white blood cells and ration of (2)/(3). (Takagi and Kaihara)

Name	R.L(max)		R.L(total)		Distribution (%)						Ratio (%) (2)/(3)		Nucleus (area)
	(X)	(Y)	(X)	(Y)	(1)	(2)(X)	(3)	(1)	(2)(Y)	(3)	(X)	(Y)	
E1	48	34	156	166	0.692	0.211	0.096	0.590	0.204	0.204	2.200	1.000	1851
L1	36	40	44	40	0.204	0.227	0.568	0.150	0.275	0.575	0.400	0.478	1055
L2	29	31	37	38	0.243	0.216	0.540	0.342	0.131	0.526	0.400	0.250	663
L3	38	54	57	60	0.192	0.210	0.596	0.450	0.200	0.350	0.352	0.571	1497
L4	31	38	43	39	0.186	0.232	0.581	0.307	0.205	0.487	0.400	0.421	892
M1	56	53	127	124	0.590	0.244	0.165	0.564	0.177	0.258	1.476	0.687	2462
M2	34	40	184	176	0.744	0.179	0.076	0.801	0.130	0.068	2.357	1.916	1383
M3	49	47	49	59	0.122	0.285	0.482	0.288	0.220	0.491	0.482	0.448	1629
M4	42	56	172	190	0.674	0.162	0.162	0.815	0.084	0.100	1.000	0.848	2048
M5	36	38	281	281	0.836	0.120	0.042	0.850	0.177	0.032	2.833	3.666	1748
N2	22	16	90	108	0.722	0.200	0.077	0.731	0.111	0.157	2.570	0.705	487
N3	26	46	101	81	0.584	0.336	0.079	0.654	0.320	0.024	4.250	13.000	920
N4	32	51	117	109	0.581	0.213	0.205	0.669	0.256	0.073	1.040	3.500	1320
N7	42	40	74	89	0.500	0.229	0.270	0.528	0.258	0.213	0.850	1.210	1219
N8	30	21	84	126	0.535	0.357	0.107	0.539	0.277	0.182	3.330	1.520	898
N9	42	42	147	156	0.693	0.204	0.102	0.743	0.160	0.096	2.000	1.600	1464
N10	24	40	81	62	0.395	0.345	0.259	0.520	0.387	0.122	1.330	3.428	894

VIII. PROCESSING OF PLAQUE PATTERN IN BACTERIAL CULTURE

Watanabe[43], et al. studied automatic measurement of numbers and size of plaques in a dish of bacterial culture. Patterns of plaques are digitized to 64 black-and-white levels by pattern reader and display of these elements of 21 K (144 x 144).

In order to eliminate the noise due to framing of the dish for bacterial culture, and that due to being digitized, processing is necessary. The former noise is eliminated by differential filtering and the latter one by low pass filtering.

Plaque overlaps are separated by statistical treatment of frequency distribution in the area; however, exact separation of plaque overlap will be carried out successfully by discrimination according to characteristic figures of each pattern.

IX. PROCESSING OF CHROMOSOME PATTERN

1. Iisaka, et al.[44,45] have been trying automatic analysis of chromosomes with a medium-size computer. PHOSDAC-1000 is utilized for the input system. Patterns of photographed chromosomes are read out by a television camera, A-D converted with points of 420 x 420, and these data read as an eight bit word. These results are displayed as color images according to density levels on a television monitor through an IC buffer (Figure 37, and Photographs 6a and 6b. Software for analysis is composed of image correction, filtering, segmentation, measurement, analysis, and module of output. The read data are stored in a magnetic disk and analysis is carried out in dialogue mode by a light pen.

The function of processing includes filing of the digital image, preprocessing for binary pattern and extraction of the chromosome's contour. Array, measurement, classification, karyotyping, and statistics are also included in

IMAGE PROCESSING IN BIO-MEDICAL ENGINEERING

the function of processing. Total processing consists of six steps, selected by a light pen, or commanded by the computer (Photographs 7 and 8):

a. Cords of data are stored with input.

b. Preprocessings are performed by level normalization, cutting off data within the threshold, and high or low filtering.

c. By monitoring and tracking, arbitrary regions are selected.

d. Counts are measured and the area is calculated.

e. Karyotyping is performed.

f. Analyzed results and output of image on hardcopy are displayed.

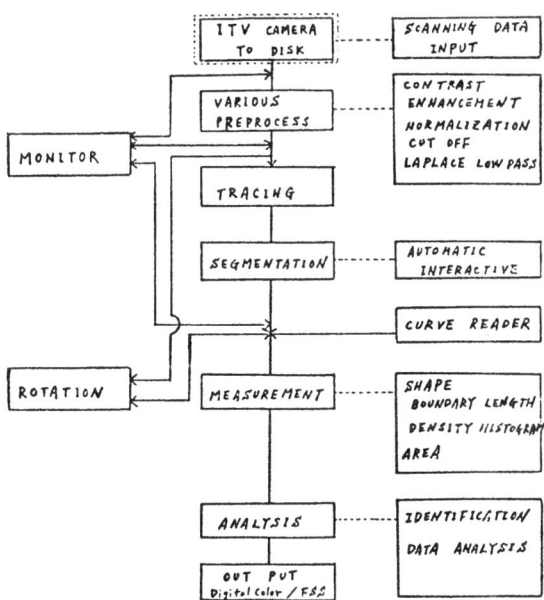

Figure 37. Block Diagram of interactive chromosome image analysis. (Iisaka)

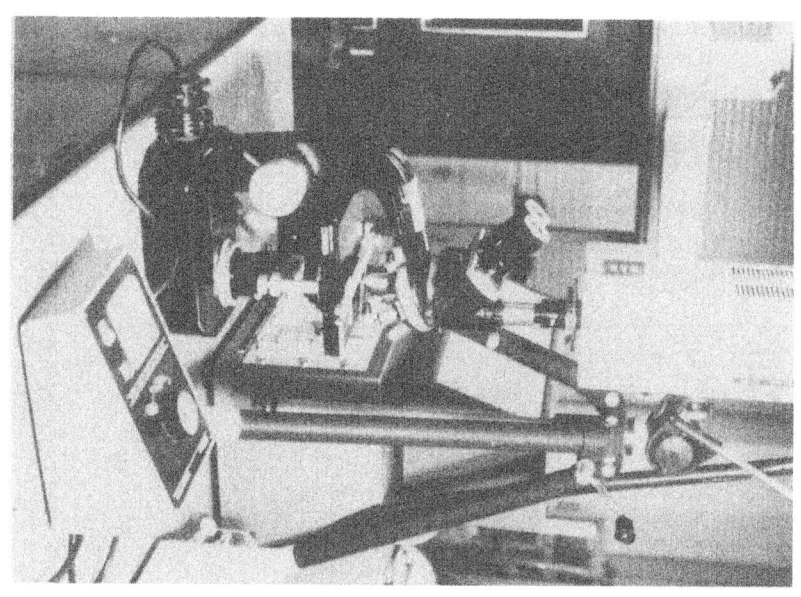

Photograph 6a.   Scanning System.   (Iisaka)

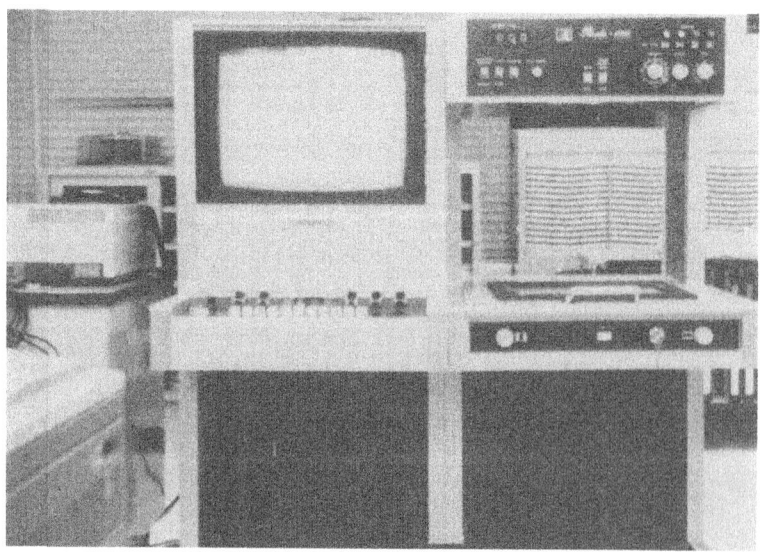

Photograph 6b.   Display System.   (Iisaka)

IMAGE PROCESSING IN BIO-MEDICAL ENGINEERING

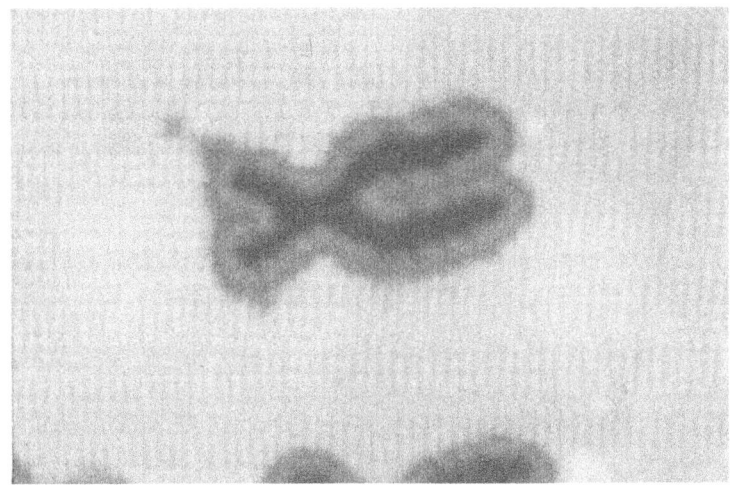

Photograph 7. A black and white reproduction of a chromosome originally displayed with color.

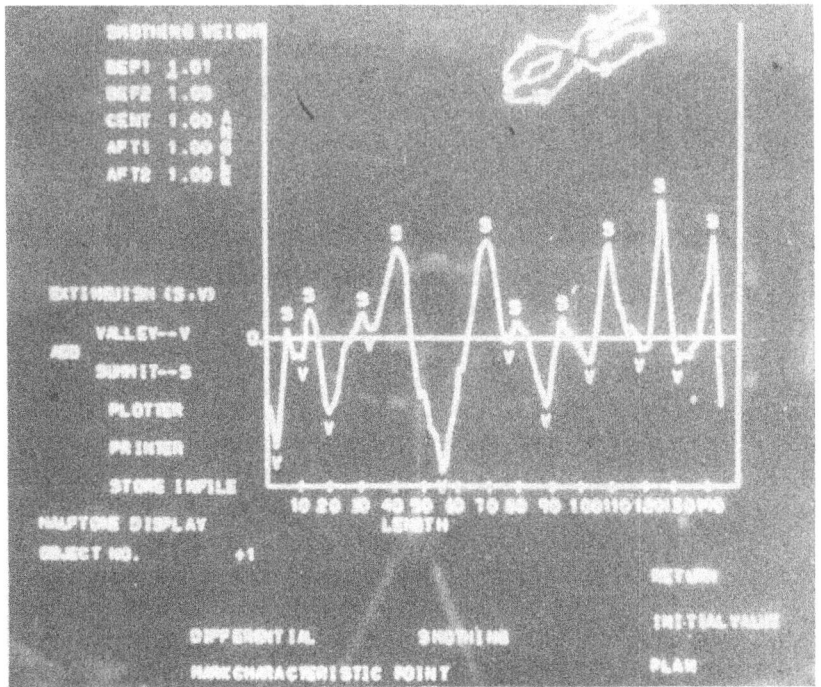

Photograph 8. Display of differential smoothing of a chromosome pattern. (Iisaka)

2. The feasibility of a minicomputer for automatic classification of chromosomes has been shown by Onoe, et al.[46] Figure 38 shows their hardware system, of which details are described in Chapter III of this book. The attachment of a magnetic disk is under construction, and upon completion, will replace the time-consuming handling of paper tapes.

The entire process is divided into the following five steps shown in the flowchart in Figure 39:

    a. <u>Digitization of original picture</u> - An original picture is digitized into 350 x 350 picture elements by a mechanical scanner. A string of picture elements on a scan line with gray levels above an appropriate threshold is named "segment". Only the coordinates of end points of each segment are stored in the computer.

    b. <u>Separation of chromosomes</u> - Individual chromosomes are separated, using information on the overlap of segments. Such parameters as location, inclination of principal axis, area, etc. are calculated. The contour and principal axis of each chromosome are displayed on a storage cathode ray tube as shown in Figure 40a, and may be corrected, if necessary, by a joy stick.

    c. <u>Revolution of the chromosome and calculation of the histogram</u> - Each chromosome is revolved to normalize the direction of its principal axis and a histogram of dispersion of the sampled points perpendicular to the principal axis is calculated.

IMAGE PROCESSING IN BIO-MEDICAL ENGINEERING

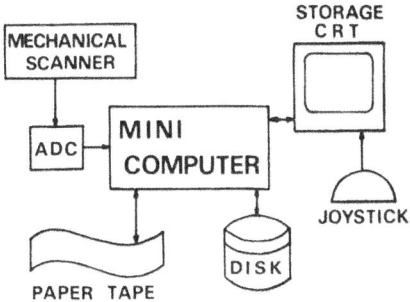

Figure 38. Hardware System (Onoe)

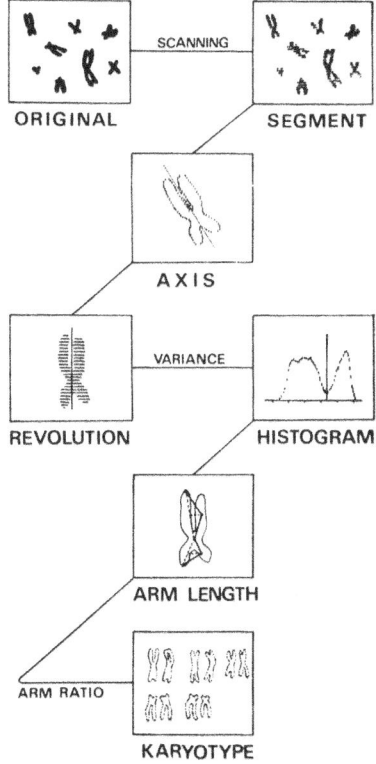

Figure 39. Flow Chart of Chromosome Analysis (Onoe)

d. Determination of centromere and determination of arm length and arm ratio - The minimum point of the histogram yields the location of centromere;arm length and ratio are calculated.

e. Karyotyping - Chromosomes are arranged first in order of total length and then partially rearranged, taking arm ratio into account. A result of karyotyping is shown in Figure 40b.

3. Momoi[47] studied automatic classification of chromosomes according to the index of arm length and arm ratio as shown in Figure 41.

"A_1", having the longest arm length, overlaps with the arm length of "A_2", but it is possible to separate A_1 from A_2 according to the index of arm ration. Then A_1 becomes one of the two chromosomes according to the index of arm ration that are extracted from four chromosomes having a long arm length. The arm length of A_2 does not overlap with that of A_3. Two A_2's are derived from four chromosomes having the longest arm length except A_1.

In the same sequence, four or six chromosomes having a long arm length are extracted from the remaining chromosomes and they become A_3, B, C(X), D, E, G(X), successively, according to the index of arm ratio.

The discrimination method between G and Y was investigated with the finding that the parallelism of the long arms of G was higher than that of Y.

IMAGE PROCESSING IN BIO-MEDICAL ENGINEERING 61

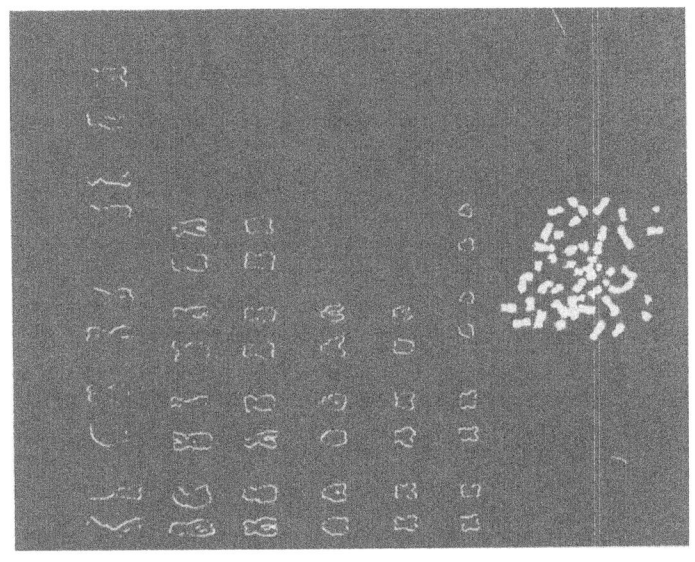

Figure 40b. Karyotyping of Chromosomes (Onoe)

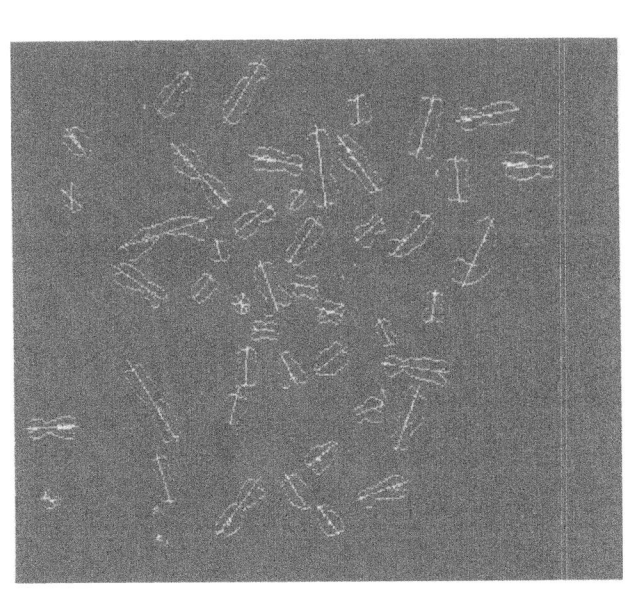

Figure 40a. The Contour and Principal Axis of Chromosomes (Onoe)

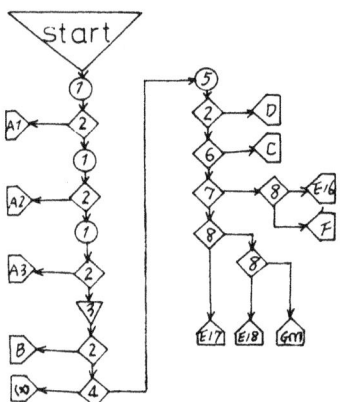

1. Extraction of high-ranked four chromosomes
2. Length of arm ratio
3. Extraction of six high-ranked chromosomes
4. Maximum arm length
5. Extraction of 24 high-ranked chromosomes
6. Length of arm
7. Extraction of six chromosomes with small arm ratio
8. Extraction of high ranked two chromosomes

Figure 41. Flow Chart of Automatic Classification of Chromosomes (Momoi)

X. PROCESSING OF VASCULAR PATTERN

"Crossing Phenomena," meaning the crossing of an artery and a vein in the retina, is a useful finding for diagnosis in ophthalmology. Yamamoto, et al.[4e] studied automatic identification of crossing phenomena in a color photograph of ophthalmo-fundoscope. For this procedure, development of color information processing, extraction of contour line from a black-and-white pattern, classification of crossing phenomena by information regarding color, and quantification are required.

Three approaches are considered: Variation of venous diameter; disappearance of the vein; and meandering of the vein to quantify crossing phenomena.

Yamamoto tried to measure the diameter of the Y line which would be cut by the vein of X line automatically. The relationship between minimal diameter of the vessel and maximal gradient was carried out and the result coincided with the medical doctor's judgment; namely, that a minimal vascular diameter was useful as an index to classify crossing phenomena. He reported that it was necessary to enhance vascular portions and suppress the background of the retina in order to get good conversion from a color image of ophthalmofundoscope to a black-and white pattern.

XI. PROCESSING OF ULTRASONIC PATTERN

1. <u>Digital Simultaneous Tomogram Method</u> - The fundamental reason for not being able to complete diagnosis of a patient is because the input voltage to brightness detail characteristics of the cathode ray tube soon saturates and the dynamic range is narrow. Therefore, a very ingenious method of sensitivity graded tomography pairs has been devised. It goes without saying that research is being actively conducted to improve this method to reproduce colored photographs, and to develop a method by which a logarithmic amplifier is to be used.

Yokoi and Ito[49] thought that eliminating the fundamental defect of ultrasonic diagnostic equipment could never be realized as long as the cathode ray tube is used, and therefore, they developed a new method: The "digital simultaneous tomogram method"(Figure 42). It became possible to complete diagnosis of a patient with only scanning by this method which adopts memory technique and enables the display in color, for a long time, of the image monitored on the CRT. In this method, the output of a receiver obtained in the form of an analog signal is directly transmitted to an A-D converter, converted into digital data, and memorized. Therefore, they would be able to use the memorized data repeatedly and it could be watched as a still image, even when the persistent CRT is not used. Figure 43 shows the schematic diagram of the A-D converter of a three-bit, seven-color system and the attached color coder.

As shown in Figure 42, input signals are converted into digital signals by means of the A-D converter. For example, if the voltage of Figure 43 is 1.0 V, input signals are converted into three bits of b_1, b_2, and b_3, as shown in Figure 43. If b_1, b_2, and b_3 thus obtained are applied to three electrodes, of R(red), G (green), and b (blue) of the color cathode ray tube, the colored images shown in Figure 43 would be obtained. It goes without saying that a more accurate image would be obtained if the number of bits would be increased to four, five, and so on.

2. How to Memorize the Output of an A-D Converter

Now let us suppose that the range of horizontal A_0, and vertical B_0, as shown in Figure 44a is scanned and this is displayed in a rectangle of horizontal A and vertical B on the cathode ray tube, as shown in Figure 44b. We would like to describe here the dividing of A direction into M, and B direction into N squares, and the corresponding ultrasonic diagnostic image under B mode to each square. We should now think about a way of storing the data of the two directions, A and B. Suppose it takes T seconds to scan the probe the width of A_0, from Point A to Point Q,

IMAGE PROCESSING IN BIO-MEDICAL ENGINEERING

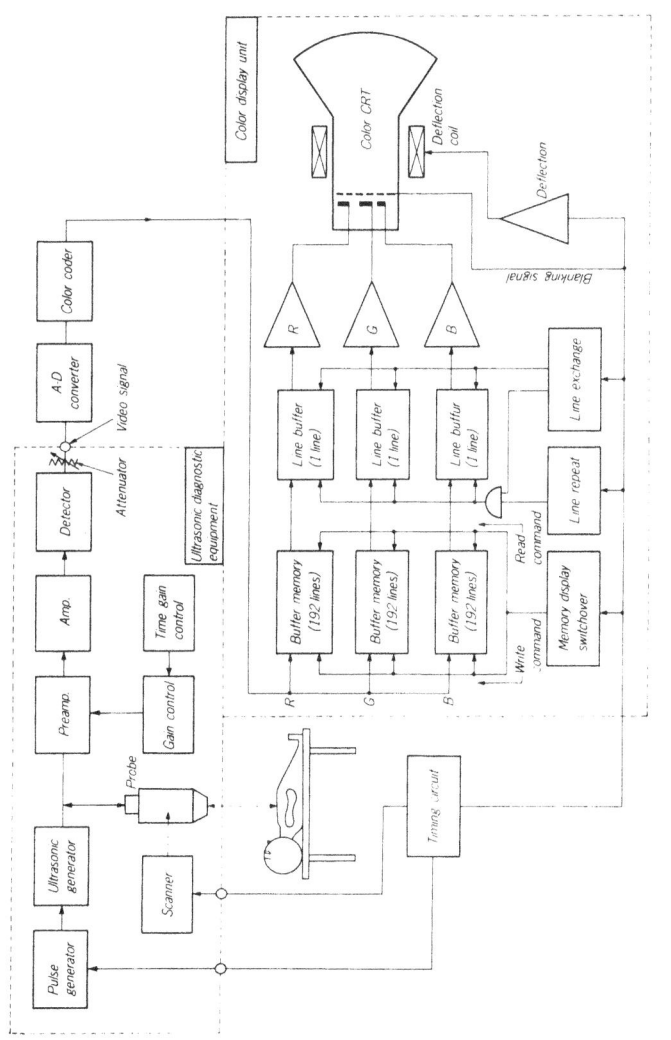

Figure 42. Schematic Diagram of Simultaneous Tomogram Method (Yokoi and Ito)

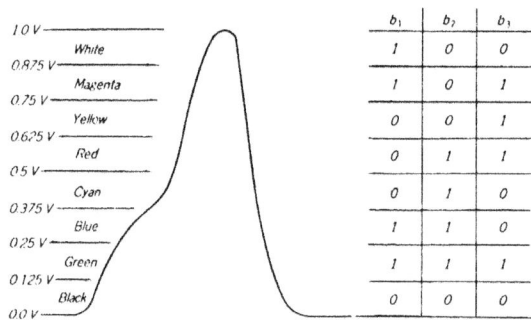

Figure 43. Functions of A-D Converter (Yokoi and Ito)

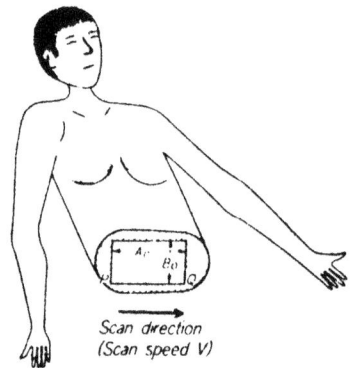

Figure 44a. Scan Area (Yokoi and Ito)

B. Corresponds to 133 microseconds when 10 cm-t tissue(figure 7)

A-D converter output should be memorized.
Every 208 μs when M=N=64
Every 104 μs when M=N=128.

Figure 44b. Display Format of Cathode Ray Tube (Yokoi and Ito)

shown in Figure 44a. It is necessary, therefore, to emit ultrasonic waves M times in T seconds, for Width A, shown in Figure 44b, should be divided into M parts and data should be memorized M times. Width A_0, shown in Figure 44b, should correspond to A width.

Let us explain this relation by referring to Figure 45. Ultrasonic waves emitted at T_1 time shown in Figure 45 return to the probe according to their positions of reflection; i.e., the distance from the probe received with the receiver and transmitted to the A-D converter. After all data of ultrasonic waves emitted at Time T_1 are memorized at Time T_2, about 40 ms after T_1, the second ultrasonic pulses are emitted by means of the next trigger signal. Reflected waves of T_2 are accordingly memorized by the A-D converter, as in the case of T_1.

The same thing is repeated at T_3, T_4, ... About every 40 ms, the data of Figure 45 are converted from analog into digital data and are memorized in the memories shown in Figure 42 in the form of a digital signal, not as an analog signal.

In this way, Division M in Direction A, shown in Figure 44b can store in good order the oscillating interval and the M times of oscillation of the pulse generator, while Division N in Direction B samples N times the output of the A-D converter each time ultrasonic waves are oscillated, and the signals are stored in good order in memory.

3. <u>Display</u> - Figure 46 shows the display format of 128 x 128. When the entire equipment is switched over to display mode, the data of one line of horizontal scanning lines of television are read into line buffers by read command and stored there temporarily. Only one line of scanning lines of television is applied to R, G and B, of the color CRT upon request of the CRT. In consequence, the image colored in seven colors as shown in Figure 43 is displayed on the color cathode ray tube. One line is taken out at a time and displayed on the color CRT just

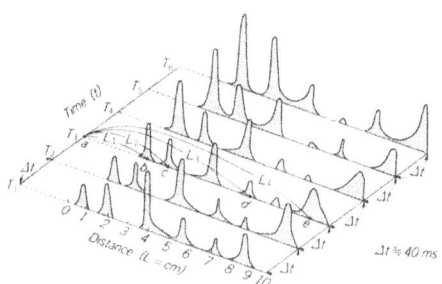

Figure 45. Principle of Simultaneous Tomogram (Yokoi and Ito)

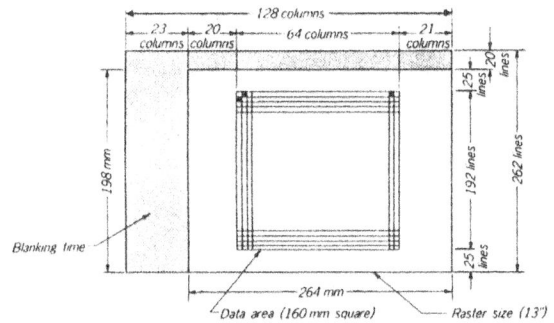

1. Figures to be displayed are divided into 128 x 128 = 16.384 picture elements.

2. The picture element size is 160 mm - 128=1.25 mm square.

3. No interlacing. The number of lines is 525 - 2=262.

4. Each picture element consists of two lines.

5. 128 picture elements correspond to 256 lines.

6. Horizontal scan time is 63.5 microseconds.

7. The number of pictures is 57 frames.

8. 8 and 14 colored pictures (B&W) can be displayed.

Figure 46. Display Format. (Yokoi and Ito)

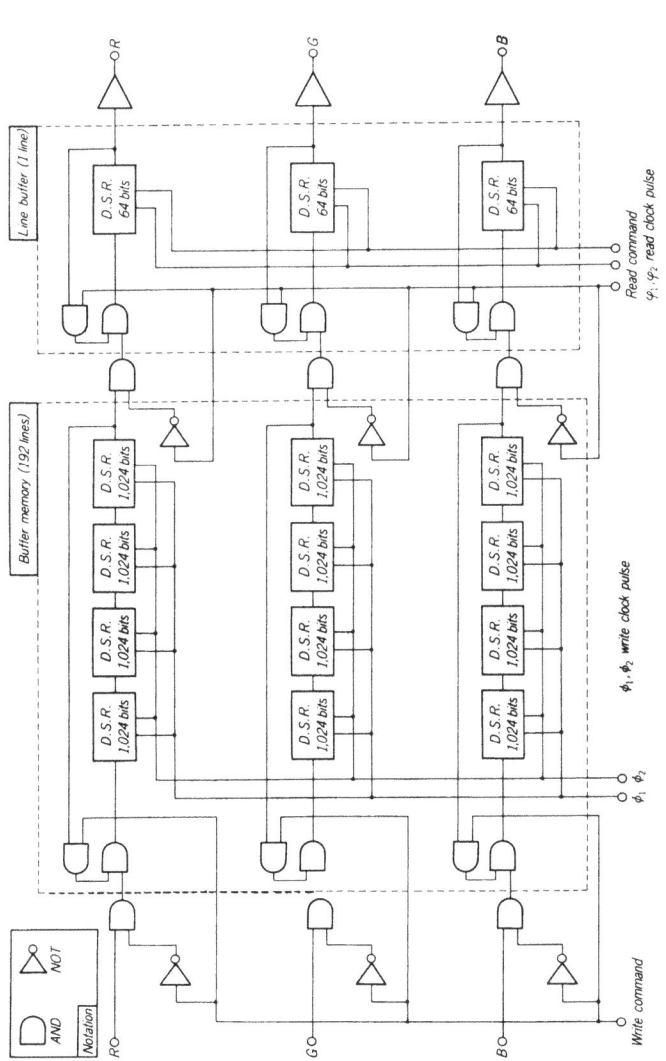

Figure 47. Memory Device of Display by using Dynamic Shift Register(DSR) (Yokoi and Ito)

like bailing out water, one cup after another, from a jug of memory with a cup of line buffer. It would be easy to imagine that there are three jugs of R, G and B in which colored water of R, G and B is filled, respectively. Each signal of R, G and B is thuss well timed, with horizontal or vertical synchronous signals of the cathode ray tube, by means of movement of the timing circuit, taken out, led to the cathode ray tube, and displays a colored image. A color television unit thus displays in color the entire ultrasonic reflected image.

4. <u>Dynamic Shift Register</u> - It goes without saying that the core could be used as memory, but it would be impossible to realize it from the point of finance. They have, therefore, decided to use a temporary memory circuit provided with dynamic shift registers as shown in Figure 47. The way of using it is as follows:

    a. Data from the A-D converter are directly memorized in dynamic shift registers.

    b. The way of using dynamic shift registers is exactly the same as a desktop calculator where feedback loop is formed and data circulated.

    c. The circulating speed of dynamic shift registers is changed between "memory mode" and "display mode."

5. <u>Results Obtained with Simultaneous Tomogram Method</u> - Figures 48a and 48b show the simultaneous tomograms of carcinoma and fibroma of the breast, obtained with the developed equipment.

Figures 48c and 48d show the simultaneous tomograms of normal midline echo and subdural hematoma.

Figures 48e and 48f show the UCG and the fetus, respectively.

IMAGE PROCESSING IN BIO-MEDICAL ENGINEERING 71

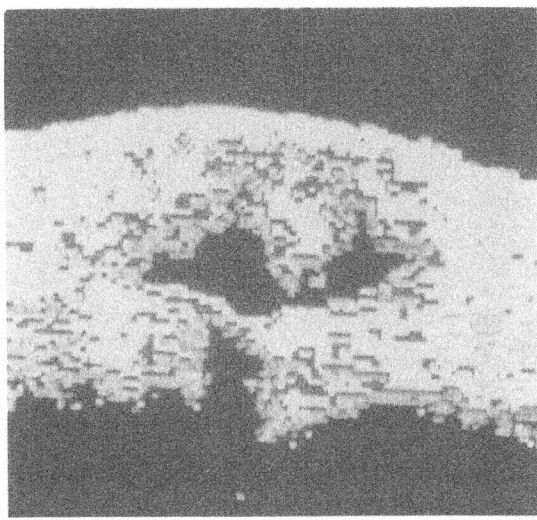

Figure 48a. Simultaneous Tomogram showing Carcinoma of the Breast (Yokoi and Ito)

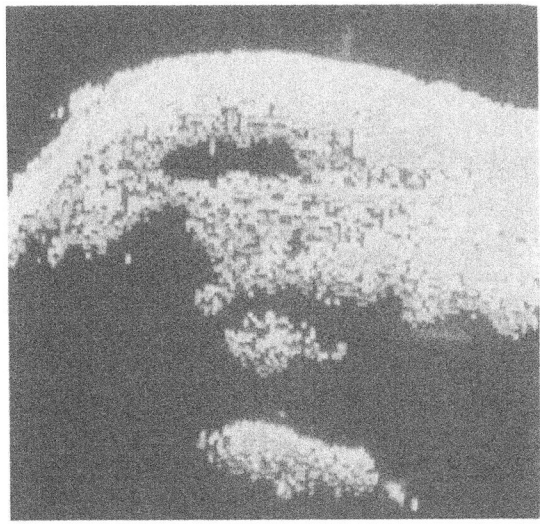

Figure 48b. Simultaneous Tomogram showing Fibroma of the Breast (Yokoi and Ito)

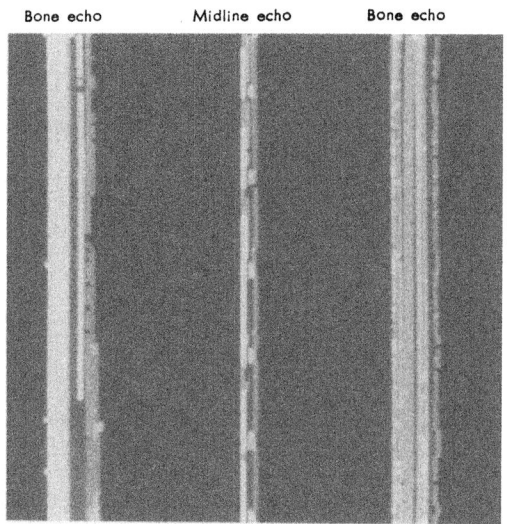

Figure 48c. Simultaneous Tomogram showing Normal Midline Echo (Yokoi and Ito)

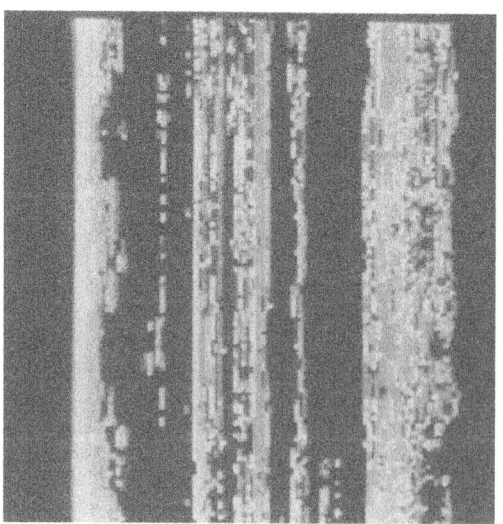

Figure 48d. Simultaneous Tomogram showing Subdural Hematoma (Yokoi and Ito)

IMAGE PROCESSING IN BIO-MEDICAL ENGINEERING 73

Figure 48e. Simultaneous Tomogram showing UCG (Yokoi and Ito)

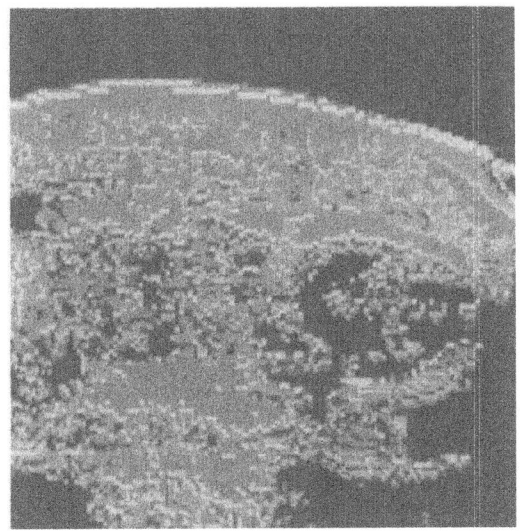

Figure 48f. Simultaneous Tomogram Showing the Fetus (Yokoi and Ito)

The characteristics of the simultaneous tomogram method are to be able to detect even the slightest change in tissue as shown in the six figures as described. They also came to know that it was possible to display the desired range in a certain color by changing the operating point of the A-D converter with a variable attenuator, as shown in Figure 42.

ACKNOWLEDGEMENTS

I would like to express my gratitude to:

    Takeshi Iinuma, Ph.D., National Institute of Radiological Science, Chiba

    Kazubumi Kimura, M.D., Osaka University Medical School, Osaka

    Masaomi Takizawa, Shinshu University, Faculty of Medicine, Matsumoto

    Kanji Torizuka, M.D., Kyoto University School of Medicine, Kyoto

    Takao Akatsuka, Ph.D., Tokyo University School of Technology, Tokyo

    Junichiro Toriwaki, Ph.D., Nagoya University School of Technology, Nagoya

    Taro Ishiyama, M.D., Osaka University, School of Medicine, Osaka

    Yoshio Hashimoto, College of Medical Technology, Amagasaki

    Shigekoto Kaihara, M.D., Tokyo University School of Medicine, Tokyo

    Joji Iisaka, Ph.D., International Business Machines Corporation, Tokyo

    Hiromu Yokoi, M.D., Nissei Hospital, Osaka

    Kenichi Ito, Ph.D., Tokyo Shibaura Electric Company, Ltd, Tokyo

    Morio Onoe, Professor, Institute of Industrial Science, University of Tokyo

REFERENCES

1. T. Iinuma, Data Processing of RI Images with Computer, Ouyo-Butsuri (Applied Physics) 40(6):94, 1971 (in Japanese).

2. T.A. Iinuma, T. Nagai and N. Fukuda, Digital Data Processing and Display in Radioisotope Imaging, Medical Radioisotope Scintigraphy, Vol. I, PP 715-729, 1969, IAEA, Vienna.

3. K. Torizuka, K. Hamamoto, T. Mukai, M. Morita, T. Kosaka and T. Suzuki, Studies on Digital Computer Processing of the Radioisotope Image through the Scinticamera, Nippon Acta Radiologica, 31(2), 119, 1971 (in Japanese).

4. K. Torizuka, K. Hamamoto, T. Mukai, T. Mori, Y. Ishii, T. Kosaka, Y. Takeda and K. Ikekubo, Studies on the Digital Computer Processing in the Thyroid Scintigraphy, Nippon Acta Radiologica, 31(7):63, 1971 (in Japanese).

5. T. A. Iinuma and T. Nagai, Image Restoration in Radioisotope Imaging Systems, Physics in Medicine and Biology, 12, PP 501-509, 1967.

6. T. Nagai, T.A. Iinuma and S. Koda, Computer-Focusing for Area Scans, Journal of Nuclear Medicine, 9, PP 507-516, 1968.

7. T.A. Iinuma, Image Enhancement by the Iterative Approximation Method, Quantitative Organ Visualization in Nuclear Medicine, PP 549-580, University of Miami Press, 1971.

8. E. Tanaka and T. A. Iinuma, Approaches to Optimal Data Processing in Radioisotope Imaging, Physics in Medicine and Biology, 15, PP 683-694, 1971.

9. Y. Okumura, The Application of the Modulation Transfer Function to Reduce Unsharpness in Scintigrams, International Journal of Applied Radiation and Isotopes, 22, PP 49-52, 1971.

10. T. Inouye, Spectral Resolution Improvement by Non-damping Filter, Nuclear Instruments and Methods, 104, PP 541-544, 1972.

11. T. Nagai and T. A. Iinuma, Comparison of Differential and Integral Scans, Journal of Nuclear Medicine, 9, Page 202, 1968.

12. T. Nagai, N. Fukuda and T. A. Iinuma, Computer-Focusing Using an Appropriate Gaussian Function, Journal of Nuclear Medicine, 10, PP 209-212, 1968.

13. K. Hamamoto, K. Torizuka, T. Mukai, T. Kosaka, T. Suzuki and I. Honjo, Usefulness of Computer Scintigraphy for Detecting Liver Tumor with 67_{Ga}-Citrate and the Scintilation Camera, Journal of Nuclear Medicine, 13(9):667, 1972.

14. T. A. Iinuma, E. Tanaka, T. Hiramoto, T. Ishihara, T. Matsumoto, K. Fukuhisa and S. Yashiro, An On-line Computer System for Data Acquisition, Processing and Display of the Radioisotope Image, Quantitative Organ Visualization in Nuclear Medicine, PP 449-464, University of Miami Press, 1971.

15. K. Fukutake and T. Iinuma, RI Image Display Utilizing Computer, Kaku-Igaku (Nuclear Medicine) 8, 431, 1971 (in Japanese).

16. K. Kimura, Processing Apparatus for RI Data and Image Processing, Sogo-Rinsho, 20 (8), 1784, 1971.

17. K. Kojima and K. Hisada, Application of Light Pen Accessory for Analyzing the Data of Radioisotope Image, Radioisotopes 21 (7):426, 1972 (in Japanese).

18. M. Takizawa, Sampling and Processing for Liver Scintigram using a CCTV and a Small Computer, Nippon Acta Radiologica, 31, PP 655-659, 1971 (in Japanese).

19. T. Iinuma and K. Fukuhisa, Digital Simulation of Radioisotope Imaging, (1) on the Recognition of a Defect in Plane Source by Human Observer, Nippon Acta Radiologica, 31 (12), 1270, 1972. (In Japanese)

20. K. Torizuka, K. Hamamoto, R. Morita, T. Mukai, T. Kosaka, J. Handa and H. Nishitani, Regional Cerebral Blood Flow Measurement with ^{133}Xenon and the Scinticamera, American Journal of Roentgenology, Radium Therapy and Nuclear Medicine, 112 (4), 691, 1971.

21. K. Kimura, Y. Sugitani, Y. Yamanouchi, H. Inada, T. Furukawa, H. Abe, H. Matsuo, Y. Hisazumi, M. Inokuma, F. Kajitani, H. Nishimura and Y. Yamano, Digest of the 10th Conference Japan Medical Electron, Page 253, 1971 (in Japanese).

22. H. Kono, K. Suzuki, H. Yasukochi and T. Iino, Automatic Analysis of Scintigram of Thyroid, Digest of the 10th Conference of Japan Society of Medical Electron, Page 245, 1971 (in Japanese).

23. I. Fujimasa, Y. Sakurai and K. Atsumi, Digital Processing on Patterns of Medical Thermograms, Digest of 9th Conference Japan M. E., Page 290, 1970 (in Japanese).

24. Y. Umegaki, Computer and Cancer Clinic, Gan-no-Rinsho (Cancer Clinic) 16(4), 301, 1970 (in Japanese).

25. Y. Umegaki, T. Machida and O. Takatani, Data Processing of X-ray TV and Its Application, Digest of 8th Conference Japan M. E., Page 210, 1969 (in Japanese).

26. T. Kobayashi, M. Takizawa, Application of the Small Computer for Esophageal Examination, Japan Journal of Clinical Radiology, 16, PP 507-510, 1971 (in Japanese).

27. T. Isobe, T. Akatsuka, Y. Ono and O. Takatani, Feature Extraction from Medical Images by Tracking Method (I), Digest of 10th Conference Japan M. E., Page 263, 1971 (in Japanese).

28. T. Isobe, T. Akatsuka, Y. Yasuoka and O. Takatani, Feature Extraction from Medical Images by Tracking Method (II), Digest of 11th Conference Japan M. E., Page 55, 1972 (in Japanese).

29. Y. Yamamura, E. Tsubura, T. Ishiyama, F. Hirao and K. Tokui, A Diagnostic Data Processing of Respiratory Diseases Centering on Lung Cancer, Medical Journal Osaka University, 15 (4), 403, 1965.

30. Y. Yamamura and T. Ishiyama, Computer Diagnosis on Respiratory Diseases, Saishin-Igaku 22(4), 730, 1967 (in Japanese).

31. E. Takenaka, Spatial Frequency Spectrum of X-ray Films, Digest of 9th Japan M. E., Page 286, 1970. (in Japanese)

32. M. Takizawa, Spatial Frequency Analysis of Radiographic Images of the Lung, Nippon Acta Radiologica, 32, PP 540-550, 1972 (in Japanese).

33. M. Takizawa, Sampling and Processing of X-ray TV Image Using a Small Computer, Japan Journal of Medical Electronics, 10, PP 394-400, 1972 (in Japanese).

34. T. Fukumura, J. Toriwaki and Y. Suenaga, Pattern Recognition of Chest Photofluorograms, Proceedings of First USA-Japan Computer Conference, PP264-271, October, 1972.

35. T. Negoro, Y. Suenaga, J. Toriwaki and T. Fukumura, Ribs Identification in Chest Photofluorographic Films, Papers of Technical Group on Image Technology, IECE, Japan, IT72-24, October, 1972 (in Japanese).

36. Y. Suenaga, J. Toriwaki and T. Fukumura, Recognition of Abnormal Shadows in the Lung Field of Chest Photofluorographic Films, ibid., IT72-25, October 1972 (in Japanese).

37. J. Toriwaki, T. Fukumura and H. Ono, Automatic Recognition and Feature Extraction of Heart Shadow in the Chest Photofluorogram, ibid., February, 1973 (in Japanese).

38. K. Shikano, J. Toriwaki and T. Fukumura, Wave Propagation Method-A Method for Conversion of Grey Pictures into Line Figures, Trans. Institute Electronics Comm. Engineers, Japan, Pt D, 50-D, 10, PP 668-675, October 1972 (in Japanese).

39. T. Ishiyama, E. Tsubura, F. Hirao and Y. Yamamura, A Study of the Automation of Cytodiagnosis, Medical and Biological Engineering, 7, 297, 1969.

40. Y. Hashimoto, T. Yokouchi, A. Sugiyama, K. Miyawaki, Y. Kishigami, T. Takahashi, Y. Nomura and S. Hattori, Automation in Cytology, Japan Journal Medical Electronics and Biological Engineering, 7 (1), 35, 1969 (in Japanese).

41. Y. Hashimoto, Y. Kishigami, S. Noda and I. Makino, Automatic Cyto-Screening Device for Uterine Cancer, Digest of the 9th International Conference of Medical and Biological Engineering, Page 17, 1971.

42. M. Takagi, T. Matsumoto, S. Kaihara, Studies on Classification of Blood White Cells Utilizing Computer, Digest of 10th Japan Medical Electronics, Page 233, 1971. (In Japanese)

43. F. Watanabe and J. Iisaka, Basic Studies on Automatic Measurement on Plaque, Digest of 10th Japan Conference Medical Electronics, Page 241, 1971 (in Japanese).

44. M. Takagi, S. Kaihara and J. Iisaka, Basic Studies on Data Processing of Medical Imaging Information, Digest of 9th Japan Conference M. E., Page 288, 1970 (in Japanese).

45. J. Iisaka, S. Kaihara and M. Takagi, Automatic Analysis on Chromosomes Utilizing Computer in Dialogue Mode, Digest of 11th Japan Conference M. E., Page 53, 1972 (in Japanese).

46. M. Onoe, M. Takagi and K. Yukimatsu, Automatic Classification on Chromosomes Utilizing Mini-Computer, Digest of 11th Japan Conference M. E. 49, 1972 (in Japanese).

47. H. Momoi, K. Watanabe, R. Kato, A. Taniguchi, K. Sawai, K. Suzuki, S. Yamamoto and Y. Yamanouchi, Automatic Classification of Human Chromosomes, Digest of 10th Japan Conference, M. E., Page 235, 1971 (in Japanese).

48. S. Yamamoto, K. Suzuki, M. Matsui and K. Kato, Quantitative Classification of Vascular Crossing Phenomena in Colour Photograph of Ophthalmo-fundoscope, Digest of 10th Japan Conference M. E., Page 437, 1971 (in Japanese).

49. H. Yokoi and K. Ito, Ultrasonic Diagnostic Equipment with Color Display Unit-for Simultaneous Tomogram Method, Toshiba Review, 76, 14, 1972.

ANNEX

PAPERS ON PATTERN INFORMATION PROCESSING IN MEDICAL FIELDS, 1968-1972

Reported in Annual Conferences of Japan Society of Medical Electronics

1968:

1. K. Miyawaki, Y. Hashimoto, et al., Pattern Recognition in Automatic Cytoscreening

2. K. Miyawaki, Y. Hashimoto, et al., Automatic Cytoscreening in Gynecology Fields

3. Y. Yamamura, T. Ishiyama, et al., Studies of Automatic Cytoscreening

4. J. Toriwaki, T. Fukumura, et al., Distribution of Density Value in X-ray Films

5. Y. Yamamura, T. Ishiyama, et al., Studies of Automatic Processing of Pattern Information of Chest X-ray Films

6. T. Fukuda, K. Torizuka, et al., Observation on Hemodynamics with Scintillation Camera

1969:

1. K. Miyawaki, Y. Hashimoto, et al., Analyzing Apparatus for X-ray Films

2. Y. Umegaki, et al., Information Processing on X-ray TV Pattern

3. T. Kobayashi, M. Takizawa, Functional Analysis on X-ray Films Utilizing the Computer

4. N. Fukuda, Pattern Information Processing in Nuclear Medicine-Sharp Focusing of Scintigram by Correction

5. J. Toriwaki, T. Fukumura, et al., Simulation on Automatic Screening Process on Chest X-ray Films With Computer.

1970:

1. Y. Hashimoto, K. Miyawaki, et al., Automatic Cytoscreening in Gynecology Fields.

2. K. Fukuhisa and T. Iinuma, DAC On-Line System of National Institute of Radiological Sciences.

3. K. Kimura, et al., Pattern Information Processing in Nuclear Medicine.

4. M. Kuwahara, K. Torizuka, et al., Data Processing on Radioisotope Scintiscanning.

5. T. Ishiyama, Y. Yamamura, et al., Studies on Automatic Pattern Information Processing of Chest X-ray Films.

6. E. Takenaka, Spatial Frequency Analysis of X-ray Images.

7. M. Takagi, S. Kaihara and J. Iisaka, Basic Studies of Medical Pattern Information Processing.

8. I. Fujimasa, Y. Sakurai and K. Atsumi, Pattern Information Processing on Thermogram.

1971:

1. M. Takagi, T. Masutomo and S. Kaihara, Classification of White Blood Cells with Computer.

2. H. Momoi, Y. Yamanouchi, et al., Automatic Classification of Human Chromosomes.

3. Y. Hashimoto, et al., Automatic Cytoscreener.

4. T. Ishiyama, Y. Yamamura, et al., Studies on Automatic Cytoscreening.

5. F. Watanabe and J. Iisaka, Basic Studies on Plaques Measurement.

6. S. Kaihara, M. Iio and H. Kameta, Automatic Measurement on Pattern Information of RI-Scintigrams.

7. H. Kono, H. Yasukochi, et al., Automatic Analysis of Thyroid Scintigrams.

8. K. Fukuhisa, DAC On-line System of National Institute of Radiological Sciences.

9. T. Mukai, K. Hamamoto and K. Torizuka, Digital Processing of RI Pattern Information.

10. K. Torizuka, et al., Functional Imaging with Scinticamera.

11. K. Kimura, H. Abe, et al., Processing System of RI Pattern Information.

12. H. Ono, J. Toriwaki, T. Fukumura, et al., Identification of Heart Pattern in Chest X-ray Film.

13. J. Toriwaki, T. Fukumura, et al., Identification of Rib Pattern in Chest X-ray Films.

14. Y. Suenaga, J. Toriwaki, T. Fukumura, et al., Basic Studies on Distribution of Density in Chest X-ray Films.

15. Y. Umegaki and T. Matsumoto, Frequency Analysis and Image Enhancement of Patterns in X-ray Films.

16. T. Isobe, T. Akatsuka, O. Takatani, et al., Method and Apparatus for Tracing of Contour Lines of Medical Patterns.

17. M. Takizawa, Subtraction of X-ray TV Patterns with Computer.

18. Iisaka, S. Kaihara, M. Takagi, et al., Processing on Medical Pattern Information with Dialogue Mode.

1972:

1. T. Yoneyama, et al., Medical Pattern Information Processing System.

2. K. Ikeda, K. Minato, T. Ishiyama, et al., Studies on Automatic Cytoscreening.

3. Y. Imazato, R. Kashida, et al., Data Acquisition and Analysis for Autocytoscreening.

4. Y. Hashimoto, et al., Autocytoscreening-Autocytoscreener and Autostainer.

5. M. Onoe, M. Takagi and K. Yukimatsu, Automatic Classification on Chromosomes with Mini-computer.

6. K. Suzuki, R. Kato and H. Momoi, Automatic Classification Apparatus for Human Chromosomes.

7. J. Iisaka, S. Kaihara and M. Takagi, Analysis of Chromosomes Utilizing Computer with Dialogue Mode.

8. T. Isobe, T. Akatsuka, O. Takatani, et al., Method and Apparatus for Tracing of Medical Patterns.

9. J. Toriwaki, T. Fukumura, et al., Characteristics of Frequency Spectrum Pattern of Chest X-ray Films.

10. Y. Suenaga, J. Toriwaki, T. Fukumura, et al., Identification of Abnormal Image in Chest X-ray Films.

11. A. Kobayashi, S. Nomura, et al., Processing of γ-Camera Image Information with Computer T-300M.

12. M. Yamamoto, K. Suzuki, et al., Detection of Vascular Image in Color Photography of Ophthalmo-fundoscope.

13. H. Abe, K. Kimura, et al., Digital Simulation on RI Pattern Information Processing.

14. H. Kono, H. Yasukochi, et al., Studies on Auto-Analyzing of Thyroid Scintigrams.

15. A. Hirakawa, H. Ueyama, and M. Kuwahara, Auto-Analysis of Pattern Information of RI Renogram with BASIC Language Utilizing Mini-computer.

16. T. Mukai, et al., Drawing up of Image Pattern Concerning Lung Ventilation and Renal Blood Flows with 133-Xe Clearance Curves Utilizing Scinticamera.

17. Y. Ishii, et al., Renal Model with Characteristics of Renal Urine Flow.

18. K. Saito, et al., Color Display on Pattern in Nuclear Medicine.

ULTRASONIC HOLOGRAPHY: A PRACTICAL SYSTEM

Byron B. Brenden

Holosonics, Inc.
2950 George Washington Way
Richland, Washington 99352

INTRODUCTION

The use of a liquid-air interface as a device for imaging sound (*) was first described by Sokolov (1) in 1935. Thus the liquid-air interface is probably the first device to be considered for imaging sound. Other methods of imaging sound have received much more attention. For instance, an ultrasound camera involving the use of a piezoelectric plate as a detector, which was also first suggested by Sokolov (2), was brought into practical use by several investigators (3-6) and is now commercially available (7,8). A recent discussion of this camera is provided by J. E. Jacobs (9). Other ultrasonic imaging methods are discussed in a review article by Berger (10) and in a book by Hildebrand and Brenden (11) as well as in the proceedings of the three international symposia on acoustical holography (12-14).

None of the aforementioned techniques of ultrasonic imaging have received the attention given those techniques in which pulse echo methods have been used to generate A, B or C-Mode displays.

---

*The word 'sound' is used in this article to connote vibrations of all frequencies whether audible or ultrasonic. The word acoustic is also used in the same broad sense.

Recent discussions of these methods are given by Uematsu and Walker (15), White (16), and Garrett and Robinson (17). Commercial equipment is available from a number of companies including Picker, Siemens, Nuclear Enterprises and Unirad. Most of these systems are discussed briefly by Garrett and Robinson (17). It may be assumed that many people, especially those attending this seminar, are familiar with one or more of these systems. Ultrasonic holography, and in particular the practical acoustical imaging systems based upon the principles of acoustical holography which are discussed in this paper are not, as yet, so familiar to their potential users. Thus it is the purpose of this paper to bring these systems to the attention of the participants in this seminar and to the readers of the published procedings. Acoustical imaging systems of the type described in this paper are currently commercially available through Holosonics, Inc.

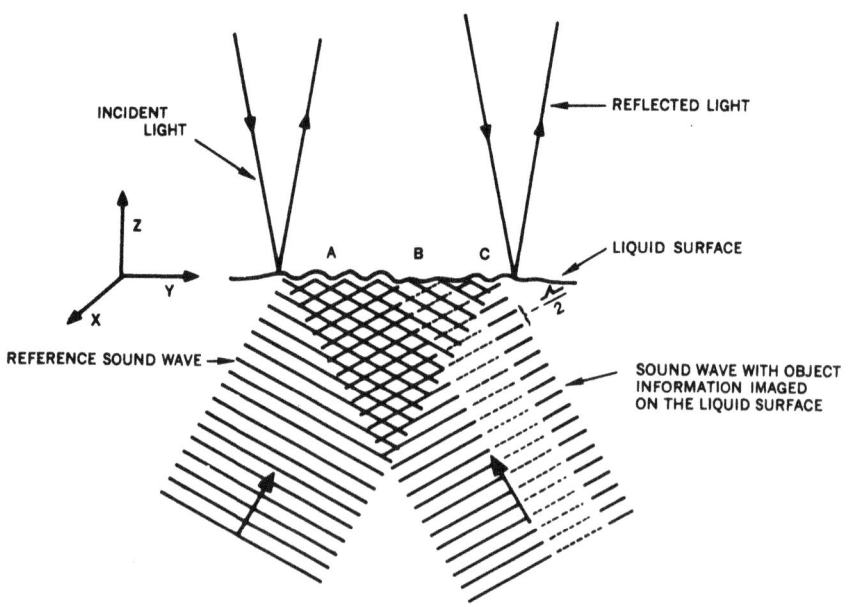

Figure 1 Effect of interfering sound waves upon a liquid surface.

LIQUID SURFACE IMAGING

Certain comparisons and contrasts should be made to characterize these systems. First of all, they are based upon the use of a liquid surface (liquid-gas interface) as a sound detecting and imaging device. However, the method of imaging is quite different from that suggested by Sokolov (1). The Sokolov system did not transform different intensities of sound into corresponding intensities of light as does the holographic liquid surface system. The manner in which this is accomplished is illustrated in Figure 1. Two beams of sound are incident upon the liquid surface. One beam has interacted with the object to be examined. This interaction is represented in Figure 1 by a reduced wave amplitude over region B as compared to regions A and C. The sound field of interest in the object is imaged upon the liquid surface where it is mixed with a reference wave of uniform intensity. The resulting interference pattern is impressed upon the liquid surface as variations in elevation. Thus the liquid surface becomes a phase grating with respect to reflection of light. The amplitude $A(x,y)$ of the standing surface ripples thus formed is linearly proportional (18) to the pressure amplitude $Po(x,y)$ in the object beam, i.e.,

$$A(x,y) = \text{constant} \bullet Po(x,y).$$

Furthermore, the amplitude of the light $V_1(x,y)$ diffracted into the first order upon reflection from the liquid surface is linearly proportional (18) under certain conditions to the amplitude $A(x,y)$ of the standing surface ripples, i.e.,

$$\begin{aligned}V_1(x,y) &= \text{constant} \bullet A(x,y) \\ &= \text{constant} \bullet Po(x,y)\end{aligned}$$

Thus, if the liquid surface is viewed in first order diffracted light it produces an image of the object in which the sound intensities are faithfully reproduced in light.

In all the examples of imaging given in this paper, the object was back illuminated and imaged by through transmission. The resulting images are therefore subject to different interpretation from those resulting from a B-Mode scan in which the primary interaction mechanism is reflection. Images produced by through transmission are primarily the result of the absorption characteristics of

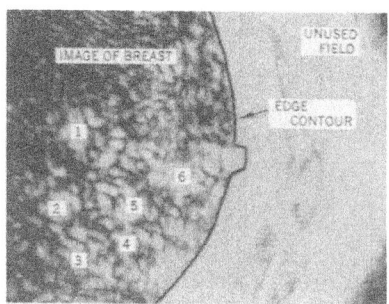

Figure 2 Acoustical image at 3MHz of a fluid filled cyst of a 40 year old woman. Picture by courtesy of M. D. Anderson Hospital and Tumor Institute.

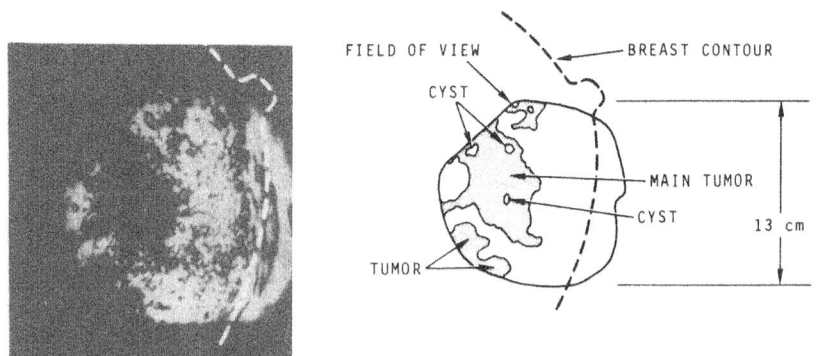

Figure 3 Acoustical image at 3MHz of tumors in an excised breast. Picture by courtesy of M. D. Anderson Hospital and Tumor Institute.

the object although scattering and reflection play important roles also. This difference in interaction mechanism is quite apparent in comparing images of hard tumors with liquid filled cysts in the breast. Figure 2 is a typical image of a fluid filled cyst. The volunteer whose breast was imaged was approximately 40 years old and had a previous history of fibrocystic disease. She could feel a cyst in position 1. This same cyst was much more highly visible in direct viewing than were the other areas which are probably cysts also. The photograph was incapable of displaying the difference in brightness actually displayed in the image.

In contrast with the image of the cyst in Figure 2, the tumor of Figure 3 is shown to be highly absorbing. Both of these pictures were taken using an ultrasonic frequency of 3MHz. Figure 3 is the image of the excised breast of a woman in her mid-fifties. The presence and the size of the tumor were confirmed by biopsy and histo-radiography.

In comparing through transmission acoustic images with X-rays the first major difference is that the acoustic image is normally presented as a positive whereas the X-ray image is presented as a negative. Bones, being both highly reflective and highly absorptive to ultrasound appear dark on the acoustic image but light on the X-ray. Edge

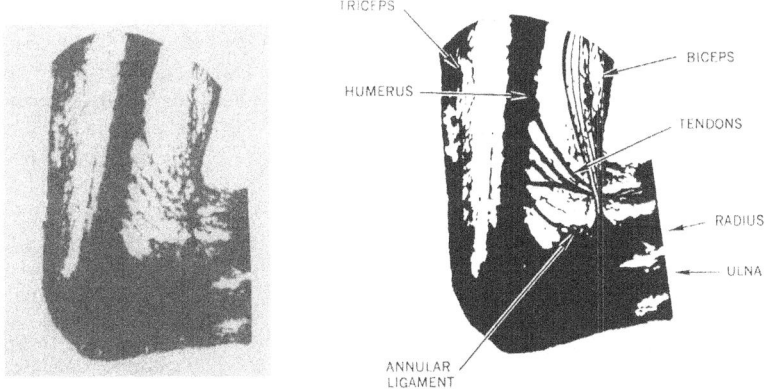

Figure 4 An acoustical image of the upper arm at 3MHz.

contours are much less distinct in acoustic images first of all because the resolution capabilities are less but also because sound interacts with soft tissue much more strongly so that bone edge contours are obscured in places by images of connective tissue. This point is illustrated in the acoustic image of the upper arm shown in Figure 4 which clearly shows the attachments of the extensor carpi and brachioradialis to the humerus. The humerus appears as a broad dark band with rather ragged edges especially in the region where the muscles are attached. There the attachment pads glow with sound in some areas and deflect sound out of the acceptance cone of the system in other areas making the latter areas appear dark. The tendon of the biceps brachii and its attachment to the radius appear as dark features. The annular ligament appears to be imaged and it further appears that the lowest tendon attached to the humerus passes under or is strapped down by the annular ligament.

PRACTICAL CONFIGURATIONS

The previous discussion has served to introduce the concept of using liquid surfaces for sonic imaging and to illustrate the type of images that may be expected. Figure 5 is used to explain the imaging features more clearly and completely. An object, such as an arm, is placed in the beam of ultrasound generated by the object beam transducer. Sound scattered by features in the object is collected by the acoustic lenses which are adjusted to image features within the object upon the liquid surface where the object beam mixes with the reference beam and forms an interference pattern. Deformation of the liquid surface occurs as a result of the radiation pressure of the sound field upon the surface. The interference pattern is registered as variations in the elevation of the liquid surface. Of course, the entire acoustical subsystem thus far described is mounted in a tank of water so that the sound at a typical frequency of 3MHz is readily propagated from the transducer to the liquid surface. Typically the field of view of the system is 12 cm in diameter.

Light from a coherent source such as a laser is used to read out the image that is registered in the hologram on the liquid surface. Light from the laser is first focused to a point located approximately one focal length

away from the optical lens. The cone of light emerging
from the point fills the aperture of the optical lens and
is collimated by the lens. About 3% of this light is reflected from the liquid surface and diffracted by the phase
hologram existing on the liquid surface. This reflected
and diffracted light is focused to several spatially separated images of the point source. Since the light in
the various diffracted orders is spatially separated in
this focus plane, a pinhole may be used here to transmit
the light in one first order while blocking all other
light. The viewing system, whether it be a closed circuit
television system as shown in Figure 5, a camera or a lens
and ground glass screen, is focused upon the plane of the
liquid surface as seen through the collimating lens. Both
the transducers and the laser are pulsed so that holograms
are formed at any rate up to about 100Hz. By this means
a real time acoustical image is obtained. Other operating
characteristics of the Holosonics, Inc. imaging system are
discussed elsewhere in this Symposium by Weiss.

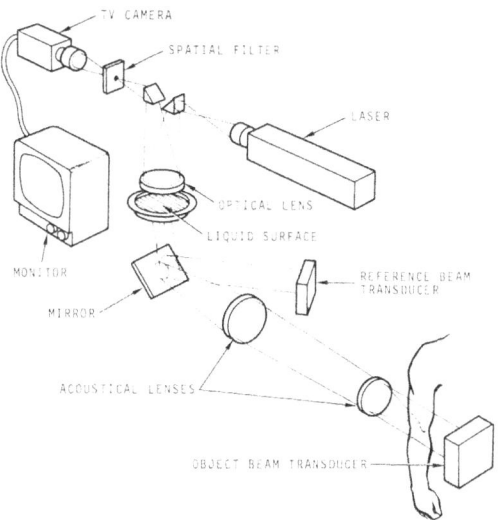

Figure 5 Schematic drawing of the acoustical and optical
components of the Holosonics, Inc. Model 100
acoustical imager.

The general purpose Model 100 imaging system shown in Figure 6 was used to produce most of the images shown in this paper. The television monitor normally used to view the images is shown to the left of the console. The image may also be viewed on the ground glass screen located below the control panel. Polaroid camera backs or cut film holders may also be used at this position. Still shots recorded on film are a valuable record but motion and the real time aspects of the imaging are useful in interpretation, so a video tape record is often made.

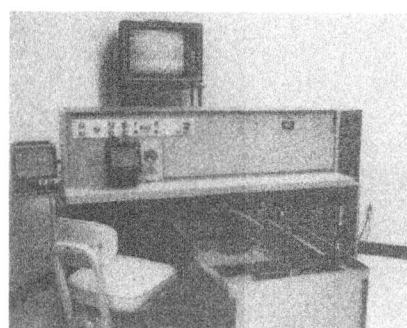

Figure 6   Holosonics, Inc. Model 100 acoustic imager.

When a hand is placed in the tank the typical image which results is shown in Figure 7. The best focus in this image is upon the metacarpal bone of the thumb. Certain features of the adductor pollicis, the flexor pollicis and the extensor muscles of the index finger are also seen in this type of image.

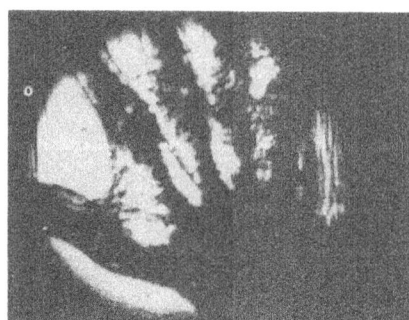

Figure 7   Image of the hand at 3MHz.

ULTRASONIC HOLOGRAPHY: A PRACTICAL SYSTEM

Some interesting imaging of excised organs has been done by various investigators. Figure 8 shows several tumors in the excised mammary tissue of a dog. In this case, however, some of the fatty tissue has been dissolved away. This is the only example shown in this paper in which this has been done. The picture does clearly illustrate the absorbing nature of tumor tissue. There are, however, some portions of the tumor in the lower right hand side that appear to be quite transparent. This is again illustrated in Figure 9 where the tumor image has dark, well defined edges but the center is quite transparent. Variations in absorption such as this need further study to be fully understood.

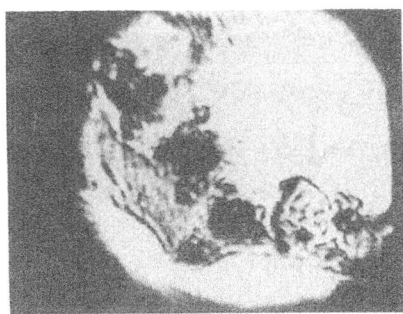

Figure 8 Tumors in excised mammary tissue of a dog. Acoustical image at 5MHz by courtesy of Dr. David Holbrooke, Children's Hospital, San Francisco.

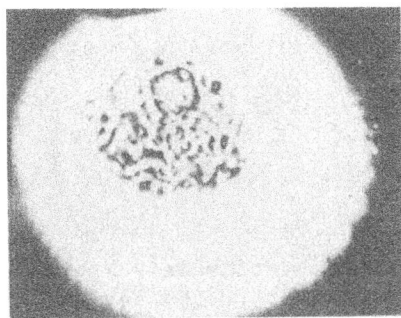

Figure 9 Tumor in excised human mammary tissue. Acoustical image at 5MHz, by courtesy of Dr. David Holbrooke, Children's Hospital, San Francisco.

Figure 10 is an image of an excised human uterus done at 3MHz using the equipment shown in Figure 6. It is interesting to note that the endometrial cavity appears quite transparent relative to the surrounding walls. This was somewhat unexpected since in earlier work with a human uterus (19) the endometrial cavity had to be filled with air to clearly define its edges. In Figure 10, the larger myoma or fibroid is well focused but the smaller myoma is slightly out of focus. Regions A are areas of the cornu where part of the dissection occured. A Copper T intrauterine device, inserted to determine the system's capability for showing such a device, is well imaged. The cervix is just out of the bottom of the field of view.

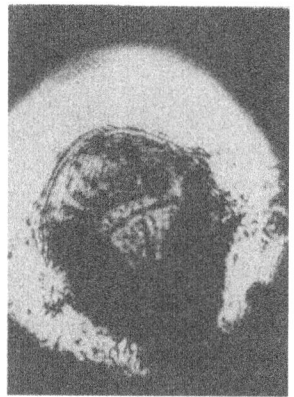 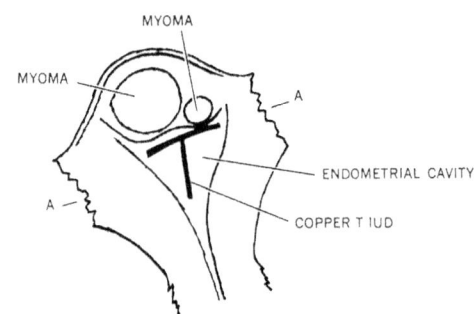

Figure 10   Human uterus. Acoustical image by W. W. Taylor, M.D., Holosonics, Inc.

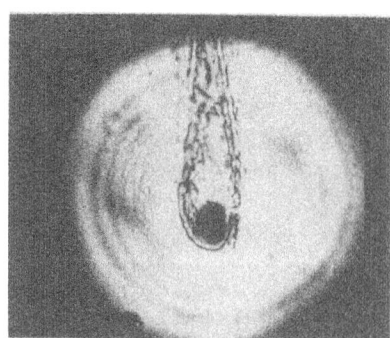 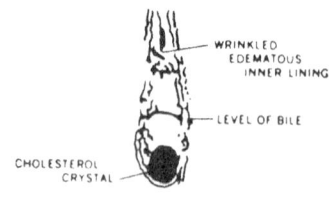

Figure 11   Gall bladder with single stone of pure cholesterol crystal. Picture by W. W. Taylor, M.D., Holosonics, Inc.

The acoustical image at 5MHz of an excised gall bladder which appears in Figure 11 shows a smooth thin wall near a single stone of pure cholesterol crystal and a ragged wall where the bladder was dissected away from the liver. Wrinkled edematous inner lining is correctly imaged in the upper portion of the picture.

Portions of the circulatory system are also well imaged as illustrated in Figures 12 and 13. In the case of Figure 12 the structure is thought to be the radial artery. The radius and ulna lie one behind the other above the artery. The dark material below the artery is muscle tissue. The artery appears to branch with the lower branch receding from the plane of focus.

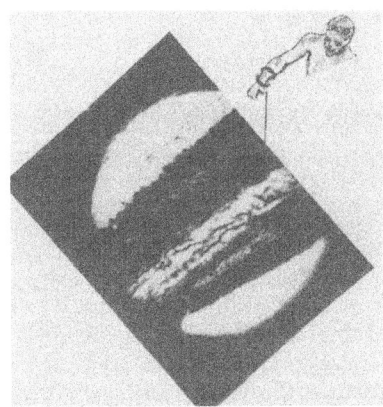

Figure 12 Radial artery showing a bifurcation imaged at 3MHz.

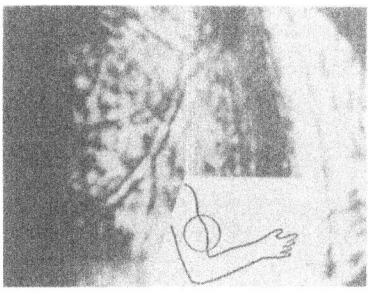

Figure 13 Branching vascular structures in the region of the biceps. Imaged at 3MHz. Sketch shows field of view.

In through transmission imaging a blood vessel is imaged as though it were being viewed in cross section. The image consists of two generally parallel dark walls with a clear region between them. An explanation of the interaction of the sound wave with a blood vessel can be given with the help of Figure 14. Distances a, b, c, d and e are assumed equal. Sound ray 1 passes through only two thicknesses, namely thicknesses a and e whereas sound ray 2 passes through three thicknesses b, c and d. If the

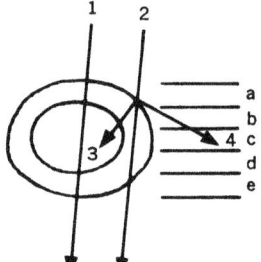

Figure 14 Interactions of sound rays with a blood vessel.

wall of the blood vessel is more absorbing than the surrounding tissue or the blood within the vessel, ray 2 will be attenuated to a greater extent than ray 1. Ray 2 also suffers scattering losses due to refraction, as illustrated by ray 3, and reflection, as illustrated by ray 4.

MAMMOGRAPHY

The problem of early diagnosis of carcinoma of the breast is treated elsewhere in this symposium by Weiss. Very early in the development of liquid surface holography it was felt that this acoustical imaging technique would be useful in diagnostic examination of the breast. As reported at a recent meeting of the American College of Surgeons (20) there is now strong evidence that early detection can sharply reduce the death rate. Our first tests aimed at this problem were conducted in 1967 at Roswell Park Memorial Institute in Buffalo, N. Y. in cooperation with Weiss and Holyoke (21). These tests involved the use of rats to determine how well tumors could be imaged. After the completion of the rat tests suitable equipment was designed to examine the human breast. The first piece of equipment built for this purpose had an effective numerical aperture of 0.08 and lacked sufficient resolution to yield conclusive results. Several schemes for compressing the breast were tested and a very satisfactory method was evolved. More recently new equipment was designed having an effective numerical aperture of 0.16. Clinical tests of this equipment were conducted at M. D. Anderson Hospital and Tumor Institute of Houston, Texas, with the cooperation of Dodd and Hevezi (22). The resolving capabilities were much improved but a new method of breast compression was tried which was not satisfactory. Currently our acoustical mammography equipment is set up in our own laboratory at Richland, Washington under the direction of W. W. Taylor. Figure 15 is a schematic drawing of the equipment used at M. D. Anderson Hospital showing the manner in which the breast was brought into the acoustic beam. Only the breast of the patient was immersed in water since the acoustic beam propogated close to and parallel with the surface of the water in the tank. The water was warmed and constantly recirculated and filtered.

We have concluded from our experiments to date that a numerical aperture approaching 0.2 and suitable compression of the breast are both necessary for adequate imaging. The optimum operating frequency appears to be 3MHz.

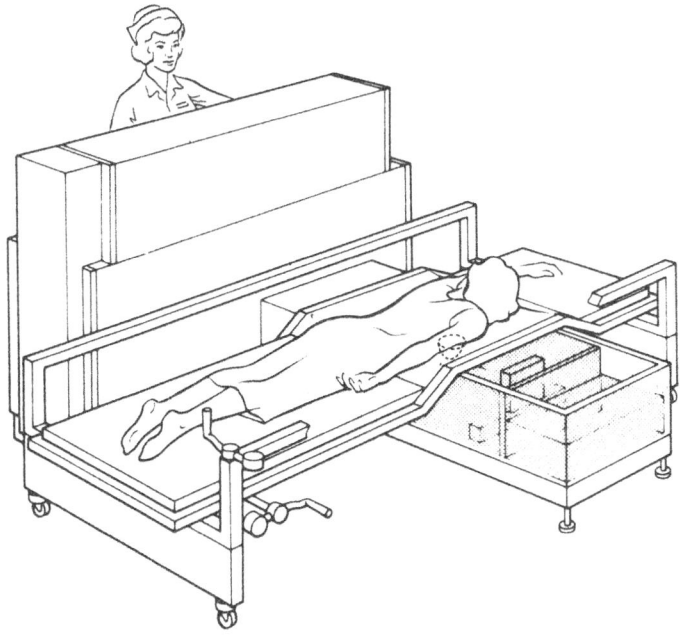

Figure 15   Liquid surface acoustical holography equipment adapted for imaging the breast.

APPLICATIONS

Holosonics and its predecessor Holotron Corporation have invested a considerable amount of money in bringing liquid surface acoustical holography to its present state of development. The quality of imaging that has been achieved provides a certain amount of satisfaction. Even more interesting, however, are the potentials for further improvement and for adaptation to specific applications. Much time and effort has been invested in adapting the technique to early diagnosis of breast disease and yet the job is not complete. What we have seen, however, strengthens our conviction that the technique will have a major impact on this problem.

The opacity of the cholesteric crystal shown in Figure 11 together with the visibility of blood vessels illustrated in Figures 12 and 13 suggests that plaques of cholesteric material building up within the circulatory system could be imaged. Other circulatory system malfunctions can also be observed. Holbrooke, at Children's Hospital in San Francisco has, for instance, imaged pools of blood formed in areas where the vessel wall has weakened and ballooned.

Although bones are quite opaque to ultrasound, it seems quite likely that acoustical imaging of skeletal structure can be useful especially, where real time imaging without exposure to ionizing radiation is important.

In order to realize the full potential of acoustic imaging in these and other applications, suitable manipulative devices will have to be developed. These are needed in order to controllably bring the areas of interest into the focal plane of the system.

As Vilkomerson (23) has pointed out, the sensitivity limit of this type of system may make deep lying abdominal structures in adults difficult to image. The situation is not so severe with children, however, so in their case abdominal imaging should be possible.

ULTRASONIC HOLOGRAPHY: A PRACTICAL SYSTEM

Figure 16 is the final example of through-transmission liquid-surface acoustical imaging to be given in this paper. It is the image of a 26 week, stillborn, human fetus done at 5MHz. Figure 16 is a composite of several pictures taken with equipment similar to that shown in Figure 6. One of the most striking features of this type of image is that the nonossified portions of the tibia, the femur and other bones are so transparent. Another striking feature, not so apparent in this composite picture but quite evident in direct viewing and in more suitably exposed pictures, is the soft tissue structure.

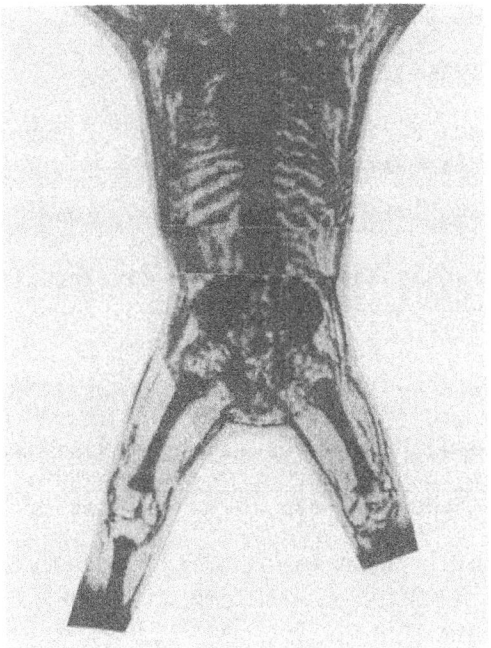

Figure 16   Acoustical image at 5MHz of a 26 week, stillborn, human fetus. Picture by courtesy of Dr. David Holbrooke, Children's Hospital, San Francisco.

REFERENCES

1. S. Y. Sokolov, "Ultrasonic Oscillations and their Applications", Techn. Physics, USSR, Vol 2, pg. 522 (1935).

2. S. Y. Sokolov, "Means of Indicating Flaws in Material", U. S. Patent 2,164,125 (1939).

3. P. K. Oschepkov, L. D. Rozenberg and Ya. B. Semennikov, "Electron-acoustic converter for the visualization of acoustic images", Soviet Phys-Acoustics $\underline{1}$:362 (1955).

4. W. R. Turner, "An Evaluation of Image Techniques in the Ultrasonic Inspection of Material", Naval Ordnance Laboratory, Silver Springs, Md. Report NAVORD-4090 (July 2, 1956).

5. C. N. Smyth, F. Y. Poynton and J. E. Sayers; Proc. IEE $\underline{110}$:16-28 (1963).

6. J. E. Jacobs, Science Journal, pg. 60-65 (April 1965).

7. Manufactured by James Electronics, Inc., 4050 N. Rockwell St., Chicago, Illinois

8. Manufactured by Tokyo Shibaura Electric Co.

9. J. E. Jacobs, "Ultrasound Imaging Systems", Research Techniques in Nondestructive Testing, R. S. Sharpe (editor) Academic Press 1970, Chapter 3.

10. H. E. Berger, "A Survey of Ultrasonic Image Detection Methods", Acoustical Holography, Vol 1, A. F. Metherell et.at., Eds. Plenum Press, New York, 1969.

11. B. P. Hildebrand and B. B. Brenden, An Introduction to Acoustical Holography, Plenum Press, New York, 1972, Chapter 7.

12. A. F. Metherell, H. M. A. El Sum and Lewis Larmore, Eds. Acoustical Holography, Vol 1, Plenum Press, New York, 1969.

13. A. F. Metherell and L. Larmore, Eds. Acoustical Holography, Vol 2, Plenum Press, New York, 1970.

14. A. F. Metherell, Ed., Acoustical Holography, Vol 3, Plenum Press, New York, 1971.

15. S. Uematsu and A. E. Walker, A Manual of Echoencephalography, The Williams and Wilkins Company, Baltimore, 1971.

16. D. N. White, Ultrasonic Encephalography, Medical Ultrasonic Laboratory, (Publisher) Queen's University, Kingston, Ontario, Canada, 1970.

17. W. J. Garrett and D. E. Robinson, Ultrasound in Clinical Obstetrics, Charles C. Thomas (Publisher) Springfield, Illinois, 1970.

18. Reference 11 Eq. 6.21 and 6.48.

19. Reference 11, Figure 8.17 c, d, e.

20. Reported by David Perlman in the San Francisco Chronicle, Wednesday, Oct. 11

21. L. Weiss and E. D. Holyoke, "Detection of Tumors in soft tissues by ultrasonic holography", Surg. Gyn. and Obstet. 128 (s):953, 1969.

22. A report of this work to be published soon, probably in the Journal of the Association for the Advancement of Medical Instrumentation.

23. D. Vilkomerson, "Analysis of Ultrasonic Holographic Imaging Methods", Acoustical Holography, Vol 4, 401-429, Glen Wade, Ed., Plenum Press, New York, 1972.

NEW DIMENSIONS FOR R&D PROGRAM MANAGEMENT

Gilbert B. Devey

National Science Foundation

Washington, D. C.

INTRODUCTION

Changes in science and technology policies have confronted R&D program managers with new challenges. Until recently much research supported by government could be characterized as "science for the sake of science." Today a study of national science policies clearly shows that research is supported by government on the basis of "science for the sake of society." In the past researchers conducted scientific investigations largely because they had fun doing so; now the emphasis is on doing things demanded by society. The extent of the recent dramatic shifts of emphasis in national science policies to goal-oriented research and development is apparent in a number of documents and speeches. Consider, for example, the following excerpts from the reports of the Central Policy Review Staff and the Council for Scientific Policy in the United Kingdom[1] the so-called Rothschild and Dainton reports.

From the Dainton report:

> ...[there] is the increasing pervasiveness of science and technology into all aspects of national activity...(p.9)

> R&D programmes must be scrutinized and assessed by scientists of wide knowledge...[and] those making these judgements should be continually aware of national needs and objectives. (p.13)

And, from Rothschild's report:

> ...applied R&D...is not an activity in its own right; it is part of a wider process...(p.2)

> ...scientists cannot be so well qualified to decide what the needs of the national are...as those responsible for ensuring that those needs are met. (p.4)

The shift in emphasis is also noted in the UNESCO report "Science Policy and the European States"[2]

> Science policy...calls for the active participation of everyone, from the responsible political authorities to the research workers. (p. 9)

> ...modern States find it increasingly necessary to resolve...research... well defined and, scientifically speaking, promising. (p.61)

> ...greater R&D productivity would be achieved by concentrating efforts... where research is...already flourishing. (p.62)

A milestone in the development of U.S. science policy was reached on March 16, 1972 when President Richard M. Nixon delivered his Message on Science and Technology[3]. In his concluding comments the President noted that "The years ahead will require a new sense of purpose and a new sense of partnership in science and technology. We must define our goals clearly, so that we know where we are going. And then we must develop careful strategies for pursuing those goals, strategies which bring together the Federal Government, the private sector, the universities, and the States and local communities in a cooperative pursuit of progress. Only then can we be confident that our public and private resources for science and technology will be spent as effectively as possible."

Four major goals of the National Science Foundation for FY 1973 are in keeping with the policy found in the Presidential Message. These goals are:[4]

-- To increase the Nation's base of scientific knowledge and maintain the health and vigor of the economy and welfare of the U.S. through science.

-- To encourage research in selected areas that can underpin efforts to stimulate the Nation's economic growth and productivity; that can lead to improvements in environmental quality; and that can enhance the ability of government and private institutions to use science and technology in problem solving;

-- To revitalize science education programs to make them serve more effectively the needs of students, instructors, institutions and society in general; and

-- To increase the exchange of scientific knowledge on a national and global basis for the benefit of the U.S.

Japanese policy is explicit about the role science and technology is to play on behalf of society; it also provides clear guidelines for R&D project management. Let us consider the following statements found in the report of the Council for Science and Technology, <u>Japanese Science Policy in the 1970's</u>.[5]

...orient research and development to the expected changes in the social and economic conditions. (p.2)

...establish objectives of research and development and manage research and development prescriptively in order to meet social and economic needs... (p.2)

...there is growing demand for scientific and technical exchanges with advanced nations to meet the need for internationalization of economy. (p.3)

This last statement, the need for internationalization of economy, is emphasized and reinforced in the Summary of White Paper on Science and Technology, where we find that "...international competition using science and technology as a weapon is expected to be further intensified..." [6]

The purpose of this paper is to show how the scientific community - both scientific institutions and scientists - is responding to these dramatic changes in emphasis. Indeed, this meeting, the third in a series that began in Tokyo in 1967, serves as a good example. Your first United States-Japan Science Seminar on "Holography" examined the theoretical underpinnings of a fast growing area of scientific inquiry. The second meeting in 1969 was somewhat more specialized and focused on "Information Processing by Holography." Now, in 1973, you have adopted an objective of the Seminar: "to formulate as explicitly as possible areas of research and development of national importance for which funding is required in academic laboratories and in university-industry cooperative projects, for the purpose of now bringing about, as quickly as possible, the practical implementation of the urgently needed, medically more effective ultrasonic instrumentation." This is an excellent illustration of the new sense of purpose in science and technology.

NEW CHALLENGES TO R&D PROJECT MANAGERS

Over a period of many years, the U.S.Government has played a major role in the support and direction of large scale, highly sophisticated research and development programs for defense, atomic energy and space exploration. Project management schemes are well developed in these areas. The total systems approach as applied by the National Aeronautics and Space Administration (NASA) to such major undertakings as Project Apollo is an excellent example of the maturity reached in project management of large scale technological programs.

Project managers for government supported projects related to economic health and social well-being, however, generally have little control over how or whether the end product will be accepted by the customer. This is in sharp contrast to project management for R&D in defense and space exploration, and to a considerable extent in atomic energy, which has the good fortune to be concerned with a total system, a closed-loop network, in which the characteristics of the component parts are largely determined by a single organizational entity. That is, producer and user are one and the same institution. System specification, research, development, production, service, product performance and marketing are all controlled by a single institution. The major advantage that such project managers have is that they are responsible for all aspects of the project from conception to final application. They are dealing with a complete technological package.

In the parlance of industrial managers, this total systems approach is a "business plan." The numerous factors that must be considered in the total systems approach - the development of a business plan - are illustrated by figure 1. Basic and applied research, project conception and systems definition are but a small portion of the business plan. As shown by figure 2, much of the business plan (shaded area) is beyond the control of a nonmission agency that provides support only for the research, or basic science, phase. So new dimensions appear in the management of government supported projects related to the economic and social well-being of a nation. Project management must devise programs which will develop a base of knowledge from which others can extract the information required to furnish a wide variety of goods and services, products and techniques.

As we have seen, R&D is a small part of the total technological innovation system, albeit a vital part. A successful R&D program often comes to naught because development of the all-important business plan has been ignored or neglected. Successful R&D without such a plan is insufficient incentive to encourage the business community to invest scarce financial and manpower resources to exploit the results of that R&D. A fully developed business plan, on the other hand, provides a measure of risk against which management can compare available options for investment and subsequent profit potential from new products or for

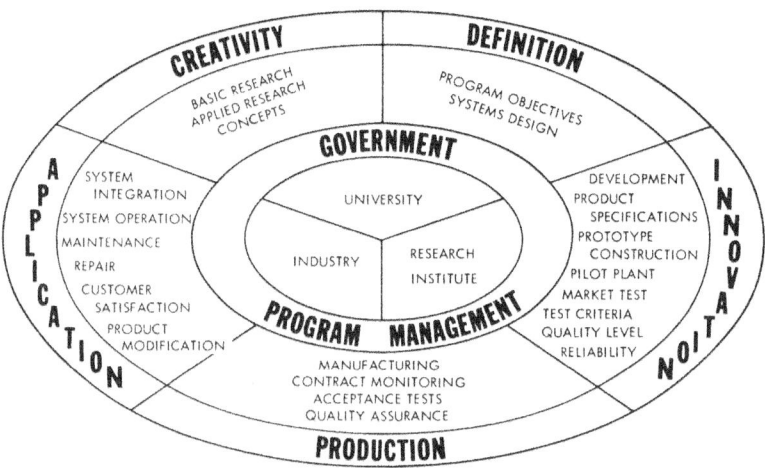

Figure 1. Producer-User Project Management

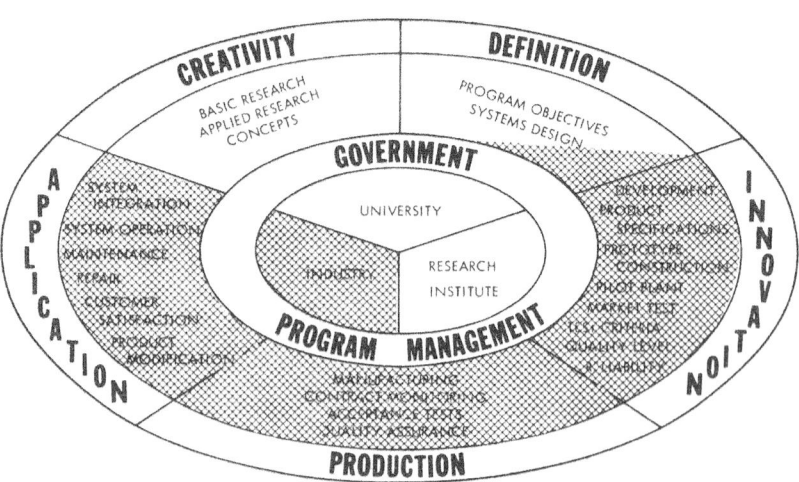

Figure 2. Nonmission Project Management

improved and more efficient services. New mechanisms for project management are required in this demanding situation. New techniques in project management must be developed to couple more effectively the programs of the university, industry and government. What is needed is to devise ways to erase - or at least to influence effectively - the shaded portions of figure 2. This, then, is the challenge. Before noting several of the ways in which this challenge is being met, it is important that we show how scientific activities relate to the continuum of creativity, definition, innovation, production and application in a technological system (figure 1). The Dainton report mentioned above provides us with three categories of scientific work which are useful for this purpose. These categories are BASIC SCIENCE, STRATEGIC SCIENCE and TACTICAL SCIENCE[7], as described below.

> basic science: research and training which have no specific application in view but which are necessary to ensure the advance of scientific knowledge and the maintenance of a corps of able scientists, upon which depends the future ability of the country to use science.

> strategic science: the broad spread of more general scientific effort which is needed as a foundation for tactical science. For this "strategic" work to be successful, it is necessary to maintain the vigour of the underlying scientific disciplines and to deploy these disciplines with due regard to national goals.

> tactical science: the science and its application and development needed by departments of state and industry to further their immediate executive or commercial functions. The extent and nature of this activity may vary widely according to the functions served and to the degree that they involve science. At one extreme it may contain a significant element of sophisticated research over a long period; whilst at the other extreme it amounts to little more than a modest intelligence and advisory activity.

Since its inception, NSF has adjusted its organizational structure and program emphasis in keeping with shifts in national science policy. The support of basic science was stressed initially, and in this budget year (FY 1973) the largest single program increase is for basic scientific research. The support of strategic science received an impetus when the National Science Board adopted a resolution in 1962 on the nature of engineering research programs within NSF.[8] Provisions contained in the amendments to the NSF Act (Public Law 90-407), followed by the establishment of Interdisciplinary Research Relevant to Problems of our Society (IRRPOS) and the formation in 1971 of Research Applied to National Needs (RANN) (into which IRRPOS was merged) introduced programs which put the Foundation solidly behind the support of strategic science and extended its activity far in the direction of tactical science. The category of tactical science was fully incorporated in NSF activities by three recent actions.

First, on April 13, 1972, in accordance with provisions of the NSF Act, the President authorized NSF to support applied research relevant to national problems involving the public interest in profit-seeking institutions when the use of such institutions is advantageous in accomplishing the program objectives.

Second, in response to the President's Message on Science and Technology[3], the Foundation established an Experimental Research and Development Incentives Program. As noted by NSF Director Stever, "An important aim of this program is to identify the blockages -- informational, technical and marketing -- which hamper the innovative process. At the same time, we are seeking various incentives for innovation that can effectively overcome the blockages."[9]

The third action was the establishment of the National R&D Assessment Program.[10] This program will provide an analytical capability for objective study and assessment of how science and technology contribute to the achievement of national goals. Through this program the National Science Foundation will analyze patterns of R&D and technological innovation; the incentives and decisions that underlie the existing patterns; and the implications that our choice of options will have in shaping future patterns of R&D and technological innovation.

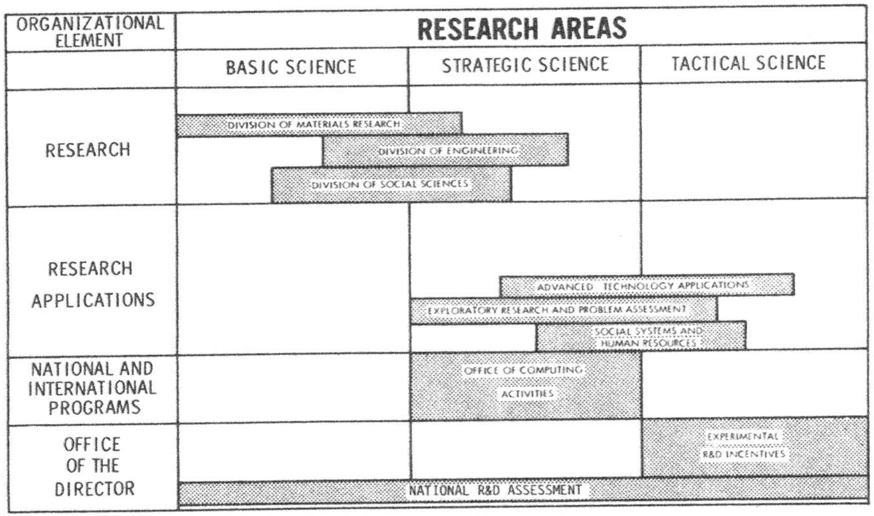

NEW DIMENSIONS FOR R&D PROGRAM MANAGEMENT

Figure 3. Typical Program Emphasis

It is interesting to note how present NSF activities relate to basic, strategic and tactical science. A typical distribution is shown in figure 3 -- in this case the area of Pattern Information Processing as it relates to Advanced Automation. Similar arrays could be constructed for other areas of science and technology. Programs having the potential to influence the continuum that ranges from the creative through to the application phase now exist and a variety of mechanisms are being evaluated to maximize the benefits to society from the achievements of scientific inquiry. As expressed by Dr. Raymond L. Bisplinghoff, Deputy Director, National Science Foundation, "NSF activity extends from basic research through proof-of-concept ... The formulation and carrying out of projects moves within a fairly close clear-cut procedural path from concept to proof-of-concept. At the conclusion of that phase, the results will be ready for transfer to industry or to a mission agency for prototype development and subsequent commercial availability. If earlier transfer is practicable, it will be undertaken. The idea is to bring a concept along to the point where an observer from industry or a mission agency can look at it and carry enough data and drawings back to his office in his mind or his briefcase to decide whether to proceed toward development and use."[11] It is useful to examine a few of the mechanisms that are being used.

CLOSER UNIVERSITY-USER INTERACTIONS

Much attention currently is being given to the development of closer and more effective university-user interactions. Many people believe that the trickle-down theory for the utilization of research results (via the published literature) is no longer adequate. We have been coupling science to technology (university to industry) for many, many years; the U.S. has perhaps done this far better than any culture in the history of mankind. But today new dimensions have been added; the tremendous amount of new information available, and the factor of TIME. We must learn to do faster the things we have been doing for these many years just to maintain technological leadership, and to increase productivity.

The RANN (Research Applied to National Needs) program at NSF requires grantees to identify potential users of research results and to plan for effective information transfer to them. It is essential that the user be involved in the planning and/or implementation of the research in all appropriate and practical ways. The key factor governing eligibility for support under RANN (in addition to scientific merit) is the <u>potential national impact</u> of the anticipated results and their relevance to a <u>particular national need</u>. Guidelines for proposal preparation for consideration under RANN programs are found in publication NSF 71-21.

Some of the RANN coupling procedures to encourage utilization are:

- Regular reporting of results to users

- Cooperative research arrangements between university, industry, nonprofit and government researchers

- Collaborative efforts with trade associations and professional societies

- Workshops conducted by the grantee for user representatives

- Establishment by the grantee of advisory groups to coordinate interactions

These procedures and other aspects of NSF programs have been presented at some length by Bisplinghoff.[12]

The new Experimental R&D Incentives Program seeks to determine, by experiment, the actual effects of various incentive mechanisms for increasing the rate of transfer of new technology to goods and services manufactured and consumed in the private and public sectors. In most cases, it is expected that experiments conducted under this program will involve close cooperation between two or more of the following institutions: universities and colleges, profit and nonprofit research institutes, industry and industrial trade associations, professional societies, Federal laboratories, State and local governments, and public service industries and organizations. Such cooperative projects will be encouraged. More information on this program is to be found in publication NSF 72-26.

Rustum Roy recently analyzed past models of university-industry patterns of interaction, described some recent experiments and recommended future courses of action in an article in SCIENCE[13]. Roy has selected several patterns that appear to be most promising. They are:

1) The multi-performer, coherent area program. Such a model, centered around the workshop concept, is used by NSF with the cutting tool and grinding materials industry under the aegis of the Division of Materials Research.

2) The single university coupled to an industry association in two variations, viz. (a) a large project funded jointly by the government and the association for performance on the campus, and (b) a physically separate laboratory near the campus.

3) The single university coupled to a single industry.

Roy contends that university-industry coupling has never been tried seriously and that more widespread discussion is needed among those who have experienced such activity.

The Division of Engineering has supported a number of specialized university-industry-government workshops of 2-5 days duration for the purpose of determining the most profitable direction of fundamental research in broad fields of engineering and to acquaint faculty members with important problems of U.S. industry and to identify areas of opportunity where the conduct of research may help to solve these problems. Workshop organizers are asked to adopt the following objectives:

1) Identify the important and critical unsolved problem areas.

2) Compile a priority list of problems and distinguish and emphasize the priorities in fundamental studies vis-à-vis the shorter term priorities pertinent to problem oriented projects.

NEW DIMENSIONS FOR R&D PROGRAM MANAGEMENT

3) Where appropriate, identify individual researchers and institutions capable of conducting high quality research on the problems identified in 1 and 2. Alternatively, specify the necessary qualifications in training and experience for individual researchers and the institutional resources necessary to conduct research in this area.

4) Develop criteria for the evaluation of research proposals in this area.

The Engineering Foundation Conference on "Pattern Information Processing" (February 23-27, 1972), supported by NSF through the Division of Engineering, is of special interest to participants in this Seminar. Biomedical pattern processing was examined in considerable depth at one of the workshops held at that conference. Deliberations at this Seminar are a logical extension of that discussion. A "Summary Report" of the Conference was prepared and distributed by the Engineering Foundation. One of the recommendations made by the "Automation Systems" workshop was that the area of automation technology needs an organization to serve as an assembly point for government, university, professional society, and user groups in order to integrate effort and enhance information transfer. This recommendation was implemented through a grant to the American Automatic Control Council (AACC) to support the activities of its Automation Research Council (ARC). The membership of the ARC includes experts in various aspects of industrial, biomedical and services automation as well as the related social and economic impacts. The Council has set as its primary goal the formulation of a national plan for research and development in automation. By holding frequent meetings and through the conduct of a number of industry and technology workshops the ARC will develop the essential infrastructure and rapport between the academic and industrial researchers and managers that is needed for rapid progress.

The several coupling patterns described above are typical of the mechanisms currently being utilized in developing closer researcher-user interactions in projects that move from the concept phase to the proof-of-concept phase. These mechanisms in themselves do not entail support for the development of prototype equipment nor the introduction of new products in the market place. These mechanisms are, however, expected to provide far better opportunities for innovation and an environment conducive to the generation of the new knowledge to increase the probability that the results of research will be exploited usefully.

HEALTH-RELATED RESEARCH

National R&D programs in health-related fields have not been closely coupled to the vast resources of our high technology industries. In 1969 Charles D. Flagle[14] noted that there is no formal counterpart in health to the mechanisms "in which defense planners and representatives of advanced technology industries engage in a process of speculative systems development." More recently, others have examined this area in considerable detail [15][16]. There is, however, an important development at the Social and Rehabilitation Service (SRS) of the Department of Health, Education and Welfare (DHEW) which may be a harbinger of change in this pattern.

The newly formed and developing national network of Rehabilitation Engineering Centers (REC), supported by the SRS, appears to have all of the characteristics essential to the effective application of technological developments for the improvement of health service. Let us briefly examine these characteristics and review a little of the background that led to the implementation of this program.

The Federal government for 25 years has supported research and development in limb prosthetics, orthotics and sensory aids. From its beginning the Committee on Prosthetics Research and Development (CPRD) of the National Academy of Sciences (National Research Council) has attempted to coordinate and correlate all the work in these areas and to assist in the introduction of the results to clinical application. This long term support pattern developed "the subculture, technostructure, or infrastructure which must exist before the full-scale application of technologies is possible."[17] Thus, when a comprehensive and detailed plan of action was made (1971) for the next 5-10 year period for the support of special centers devoted to integrated rehabilitation engineering programs, the SRS was a receptive listener. This plan is included in the report "Rehabilitation Engineering - a Plan for Continued Progress."[18]

A network of five Rehabilitation Engineering Centers (REC) now exists with principal funding provided by the Social and Rehabilitation Service. Each center must have strong teaching affiliations with medical and engineering schools and have a substantial patient load. Each of the Centers has been assigned specific areas of study as central objectives. Eleven different institutions are involved in the five centers. The centers and their central scientific/technological objectives are shown in Table I. Also of importance is the operation of Patient Engineering Service Departments in the program.

The third objective of the REC program is vital to the implementation of the "business plan" depicted in figure 1. That is, each center will collaborate with laboratories and industry to carry new devices and techniques through all phases of research, development, and clinical evaluation to active production and patient use.

First year support for these centers ranges from $175,000 to $350,000; significant matching funds are required (20% - 25%). A total of ten centers could be in operation at the end of fiscal year 1973 (June 30, 1973) and it is hoped that the program can grow to support a network of fifteen centers with each receiving from $750,000 to $1,000,000 annually from the Federal government.

Table 2 shows the objectives of the rehabilitation engineering centers.

The institutional arrangements and organization mechanisms that have evolved over 25 years and that are now being employed in the direction of the Rehabilitation Engineering Centers program may have overcome a major deficiency that for many years has limited the effectiveness of Federal support of health related research programs.

TABLE 1. Rehabilitation Engineering Centers
Social and Rehabilitation Service Oct. 1972

CENTER	MAJOR MISSION
1. RANCHO LOS AMIGOS HOSPITAL UNIVERSITY OF SOUTHERN CALIFORNIA	FUNCTIONAL ELECTRICAL STIMULATION
2. MOSS REHABILITATION HOSPITAL TEMPLE UNIVERSITY DREXEL UNIVERSITY	LOCOMOTION AND POSTURE
3. TEXAS INSTITUTE FOR REHABILITATION AND RESEARCH BAYLOR UNIVERSITY TEXAS A&M RESEARCH FOUNDATION	EFFECT OF PRESSURE ON HUMAN TISSUE
4. HARVARD UNIVERSITY MASSACHUSETTS INSTITUTE OF TECHNOLOGY	SENSORY AIDS FOR THE DEAF AND BLIND
5. NORTHWESTERN UNIVERSITY	INTERNAL STRUCTURE PROTHESIS

TABLE 2. Objectives of Rehabilitation Engineering Centers

1. TO IMPROVE THE QUALITY OF LIFE OF THE PHYSICALLY HANDICAPPED THROUGH A TOTAL APPROACH TO REHABILITATION, COMBINING MEDICINE, ENGINEERING, AND RELATED SCIENCE.
2. TO PERFORM RESEARCH AND DEVELOPMENT IN PIONEERING AREAS WHEREIN A CENTER HAS DEVELOPED UNIQUE CAPABILITIES.
3. TO COLLABORATE WITH LABORATORIES AND INDUSTRY TO CARRY NEW DEVICES AND TECHNIQUES THROUGH ALL PHASES OF RESEARCH, DEVELOPMENT, AND CLINICAL EVALUATION TO ACTIVE PRODUCTION AND PATIENT USE.
4. TO MAKE AVAILABLE NEW DEVICES AND TECHNIQUES TO ALL PATIENTS REFERRED TO THE CENTER.
5. TO EDUCATE OTHERS TO PROVIDE THESE DEVICES AND TECHNIQUES TO PATIENTS THROUGHOUT THE NATION.
6. TO COOPERATE WITH OTHER CENTERS IN FITTING AND EVALUATING THEIR DEVELOPMENTS WHENEVER THE NEED IS INDICATED.
7. TO PROVIDE AN ENVIRONMENT FOR EDUCATION OF PHYSICIANS, ENGINEERS, AND OTHER TECHNICAL PERSONS IN RELATED LIFE AND PHYSICAL SCIENCES.
8. TO COMMUNICATE EFFECTIVELY WITH OTHER CENTERS THROUGH RECOGNIZED MEANS AND COOPERATIVE EFFORT.

THE DEDICATED UNIVERSITY

A number of observers believe that the present university structure is not suited for the conduct of multidisciplinary, problem-oriented, research. Roy contends that "The 'relaxation times' of universites are measured in decades." [19] Alvin Weinberg is more forceful, averring that "society will have to invent new institutions that can apply science to the broad socio-technological problems of the future." [20] Peter L. Berger believes that universities need to introduce "structured curricula instead of the 'cafeteria' style of education ... confused with intellectual freedom ... [and] an understanding of the values of specialization instead of ... 'interdisciplinary' chitchat." [21] NSF Director H. Guyford Stever is not so pessimistic concerning universities, noting that "Our experience in the Foundation shows that enthusiasm for multidisciplinary science exists in universities and that traditional barriers to its pursuit can be overcome." [22] Some of the approaches to removing these barriers have been mentioned in this paper. Let me now suggest a dramatic departure from tradition and introduce the concept of the Dedicated University.

A dedicated university would continue to be an institution of higher learning and to conduct its affairs in a manner conducive to the pursuit of scholarly endeavors. But there would be an important shade of difference between the educational aspects of a dedicated university and academia as we know it today. Studies in the fundamentals essential to the solid foundation of knowledge needed for future accomplishments would still be stressed. In the dedicated university, however, the creative aspects of the learning experience would relate closely to the area of dedication. Term papers, essays, theses, research projects, laboratory experiments and field activities would be coupled to the area of dedication. Thus, students would be aware that they were personally contributing to the attainment of a socially worthy goal through a program of synchronized action.

All of the various disciplinary departments in a dedicated university would collaborate in a major attack on a scientific/technological problem of importance to society. In addition to basic research, applied research and concept development, the program of the dedicated university would include all of the non-research factors of the total technological system (figure 1), and would extend coverage to a demonstration to test the social, economic and technical worth of the results in a real-world environment.

The concept of the dedicated university likely would be rejected by conformists as being radical, impractical or perhaps simply "too new." But the character of the university does change, albeit slowly, and the examples cited in this paper suggest that change is continuing. A proposal for the establishment of a dedicated university was debated in the British Parliament[23] and steps have been taken to develop the proposal.[24] Interestingly, the British plan centers on rehabilitation engineering and home health care delivery and encompasses much of the program area of the Rehabilitation Engineering Centers described above.

Both the concept of the "dedicated university" and that of the "business plan" approach to R&D program management share an element that is vital to success -- that of a synchronized action plan. Synchronized action, then, is one way to describe the new dimensions for R&D program management.

AUTHOR'S NOTE

This paper is an elaboration of the "business plan" approach to R&D program management developed in my lecture "Project Management for Automation Technology" and my participation in the Japan Industrial Technology Association (JITA) Symposium on "Pattern Information Processing Systems (PIPS)," Tokyo, March 1972. My short report about the PIPS project appears in the November 24, 1972 issue of NATURE.

The remarks in this paper are the responsibility of the author and do not represent official positions of the Federal Government or the National Science Foundation.

REFERENCES

1. United Kingdom, The Lord Privy Seal, A Framework for Government Research and Development, (London: H.M. Stationery Office, November 1971).

2. United Nations Educational, Scientific and Cultural Organization, Science Policy and the European States, No. 25, (Paris, UNESCO, 1971).

3. U.S., President, Weekly Compilation of Presidential Documents, March 20, 1972.

4. U.S., Congress, Senate. Special Subcommittee on National Science Foundation, Committee on Labor and Public Welfare. Statement of H. Guyford Stever, Director of the National Science Foundation. May 4, 1972.

5. Japan, Prime Minister's Office, Japanese Science Policy in the 1970's, A Report for the Council for Science and Technology, (Tokyo: Science and Technology Agency, April 1971).

6. Japan, Summary of White Paper on Science and Technology, (Tokyo: Science and Technology Agency, April 1971), p.4.

7. A Framework for Government Research and Development, (Dainton) op.cit.,pp.3-4.

8. U.S.,National Science Foundation, resolution of the National Science Board, May 1962.

9. Stever, H. Guyford, "Are the Institutions of Science Adequate?", The Robert A. Welch Foundation Conferences on Chemical Research: XVI Theoretical Chemistry, Houston, Texas, November 21, 1972.

10. U.S.,National Science Foundation, National R&D Assessment Program, Summary of Program Plans, Washington,D.C., November 1, 1972.

11. Bisplinghoff, R.L., "The Role of Government in Electric Power Research; The View from the National Science Foundation," Conference on Research for the Electric Power Industry, Washington, D.C., December 11, 1972.

12. Ibid.

13. Roy, R., "University-Industry Interaction Patterns," SCIENCE, Vol. 178, December 1, 1972 (pp.955-960).

14. Flagle, C.D., "Technological Developments in the Health Services," Proceedings IEEE, vol. 57, pp. 1849-1854, November 1969.

15. Devey, G.B., "Government-Industry Interrelationships in Automated Multiphasic Health Testing and Services," Volume 3 Proceedings of the Invitational Conference on AMHTS, DHEW Publication No. (HSM)72-3011, pp. 181-190, January 1970.

16. "An Assessment of Industrial Activity in the Field of Biomedical Engineering," National Academy of Engineering, Washington, D.C., 1971.

17. Devey, G.B., Toward Automated Health Services," Proceedings IEEE, vol.57, p. 1831, November 1969.

18. "Rehabilitation Engineering - A Plan for Continued Progress," National Academy of Sciences, Washington,D.C., April 1971.

19. "University-Industry Interaction Patterns," op.cit.

20. Weinberg,A.M., "The Scientific University and the Socio-Technological Institute in the 21st Century," The Graduate Journal, vol.8, no.2, pp. 311-316, 1971.

21. Berger, P.L., "A Lot of Beautiful People," U.S.News & World Report, Dec. 4, 1972, p. 58.

22. "Are the Institutions of Science Adequate?", op.cit.

23. Great Britain, Parliament, Sessional Papers (House of Lords), Official Report: 9 April 1970, vol. 309, no. 60, "Chronically Sick and Disabled Persons Bill."

24. Wolff, H.S. Personal communication.

ULTRASONIC TISSUE VISUALIZATION
AND SURGERY IN BRAIN

F. J. Fry and R. C. Eggleton

Indianapolis Center for Advanced Research and
the Indiana University Medical School
Indianapolis, Indiana 46202

Combining the tissue visualization capability of ultrasound with the method of ultrasonic focal lesion production in tissue permits a non-invasive approach to diagnosis and surgery. Any body tissue which can be visualized ultrasonically is a potential candidate for application of an intense ultrasonic lesioning beam when an ablative surgical procedure in the tissue is the method of choice for therapy. This combination method has been first applied to brain. Ultrasonic visualization and lesioning in brain is conducted transcutaneously with an underlying section of skull bone removed by conventional surgical techniques as a prelude to the ultrasonic procedures. Only those cases have been treated thus far in the human in which the physician has provided a craniectomy as part of a more conventional brain surgery procedure for therapy. Although initially applied to brain, the methods are not limited to this structure and exploration of the more general application of the technique is envisioned.

INTRODUCTION

Development of a practical system for non-invasive visualization of brain structures and surgical intervention (in the sense that no physical device is passed into the brain to implement the visualization or surgery) has followed from combining the soft tissue visualization capability of ultrasound and the precise lesion generating capability arising from use of focused ultrasonic beams. Ultrasound exhibits tissue selective threshold effects which are of considerable value in brain tissue modifications. Focused ultrasound used in brain tissue can produce individual small lesions (.001 cubic mm to several cubic mm are typical volumes) which in combination can generate a wide variety of lesion complexes permitting matching of a composite lesion to a selected normal or abnormal (tumor) brain structure. These lesions are produced at sites in brain without damage to tissue not in the target zone. A number of tissue selective effects in brain will be discussed under the Lesion Generation section. Acoustic parameters of brain tissue are such that predictable placement of the focal zone is possible (beam displacement after transmission through the full human brain is less than .5 mm). Additionally, the acoustic intensity absorption coefficient for brain tissue at a frequency of 1 MHz is approximately 20% per centimeter so that with a transducer aperture angle of 56° the transducer intensity gain is such that a focal lesion can be generated after passage through the full human brain laterally without damage to tissue along the beam entry pathway.

In order to capitalize on the unique lesion generating capability of ultrasound in terms of building a practical surgical unit, it is necessary to visualize the surgical field insofar as possible, with the complement of structures normally presented to the surgeon under direct optical line of site viewing. Fortunately, ultrasound can be used to visualize a broad spectrum of normal and abnormal structures in brain, and these will be more fully discussed under the section on Tissue Visualization. With a 56°total included angle beam transceiver operated in a pulse echo mode, the azimuthal resolution approximates one wavelength for the ultrasonic frequency used (typically 0.6 mm for 2.25 MHz). Range resolution approaching one wavelength has been achieved by appropriate transceiver design and electronic pulse drive system. These items will be discussed under the Methods Section.

FUNCTIONAL REQUIREMENTS AND SYSTEM DESIGN CRITERIA

Formulation of a practical set of functional requirements and system design criteria in this case is based on integrating an idealized set of requirements and state of

TABLE I

Functional Requirements and System Design Criteria for Ultrasonic Visualization and Surgery in Brain.

Idealized Requirements	Possibilities with Ultrasonic Methods	Practical Realization
1. Non-invasive (no penetrating probes in brain tissue) visualization of all normal and abnormal brain structures and tissue differentiation capability (the imposed arbitrary resolution limitation being set at structures 5.5 mm in size with a spacing distance of at least .5 mm).	1. Non-invasive visualization and characterization of tissues with different acoustic impedances. An ultrasonic frequency of 2 MHz has a wave length in tissue of .75 mm. Resolution capability is set for a given frequency.	1. Non-invasive visualization, conducted transcutaneously without interfering skull bone, of interfaces having an acoustic impedance mismatch of 0.2%. Resolution capability essentially at one wavelength.
2. Structural features should be viewed in true spatial relationship.	2. Soft tissue velocity differences are in general very slight (a few %) so that geometric distortion is not a serious problem.	2. Structures viewed ultrasonically have been identified in their true spatial orientation with accuracies of the order of the resolution capability.

3. Visualization should present minimal hazard to the patient in either short or long term temporal sequences.

4. Non-invasive (no penetrating probes in brain tissue) lesioning of appropriate brain structures (traumatizing brain pathways or invasive tissue growths).

5. No change produced in tissue outside the lesioning zone.

3. Sonic intensity levels needed to visualize many tissues present no hazard to the patient. No cumulative dosage effects have been shown to be operative with appropriate control.

4. Ultrasound under controlled dosage conditions provides focal lesions in tissue.

5. Due to focusing the ultrasonic beam and tissue selective effects, only tissue in the focal zone is affected.

3. No side effects have been observed which can be related to the visualization procedures.

4. Lesions of the required size or shape have been produced in a large number of experimental animal brains. (Approximately 2,500 cats and monkeys). Approximately 100 human patients have had ultrasonically produced focal lesions in brain. Lesions produced transcutaneously without interfering skull bone.

5. No change in tissue outside the focal zone has been seen by histological processing of

6. Individual lesions should be as small as .1 cubic mm with the capability of generating lesion size and shape to correspond with desired structure.

7. Non-hemmorhagic lesion.

6. The coalescence of many individual small lesions provides a composite lesion of desired size or shape.

7. Tissue selective thresholds exist for ultrasound under controlled dosage conditions.

brain tissue on approximately 2,500 experimental animals.

6. Using a 56° total included angle ultrasonic beam, small individual brain lesions of 1 cubic mm volume can be produced with 1 MHz sound frequency. At 4 MHz, brain lesions of .001 cubic mm volume can be produced. Using appropriate focused ultrasonic beams and sonic frequencies, focal lesions can be produced to suit the circumstance.

7. In normal brain tissue, blood vessels are the most resistive to ultrasound when compared with any of the other brain tissues.

8. Accurate control of lesion size, placement and possibility of tissue selective effects through precise dosage control.	8. Sound velocities in soft tissues are only slightly different and most of the energy is in the transmitted wave. Ultrasonic absorption in different tissues varies considerably (white matter in brain is 50% greater than gray matter).	8. Studies involving approximately 2,500 experimental animals and approximately 100 humans have shown that ultrasonically generated lesions can be placed where desired; lesions of any complex size or shape can be generated and tissue selective effects do occur.
9. Possibility of directly visualizing the induced lesion.	9. Changes in acoustic impedance of the order of 0.2% can be observed.	9. Ultrasonically generated focal lesions have been visualized by ultrasound in both experimental animals and man.
10. Temporal sequential visualization and lesion production over long time period	10. The idealized set of requirements appears to be possible.	10. Experimental animals have had lesions produced in sequence up to

(days to years) without compromise to the patient and with undiminished capability.

11. Atraumatic procedures for the patient.

12. Overall economy in time and money.

one year apart. Some human patients have had brain visualization procedures extending to the order of one year.

11. The possibilities have been practically realized.

12. Some few patients have been treated on an outpatient basis.

11. For visualization and lesioning, patients need not be anesthetized, be rigidly held for periods of more than a few minutes, nor experience pain.

12. Patients can be treated on an outpatient or very minimal care basis with potential savings to the overall aspects of health care delivery.

the art in a number of areas of technology (machine tools, ultrasonic transceivers, computers, ultrasonic brain structural visualization and brain lesioning and neurological surgery). Table I presents a summary of these formulations.

Before proceeding, it is important to discuss an item which is peculiar to the rationale of brain lesioning as a therapeutic modality. Surgical therapeutics for brain does not necessarily involve lesioning only abnormally identified structures, but can involve lesioning in grossly normal structures in order to bring about a more appropriate balance condition to alleviate an apparently intolerable condition (hyperkinesia, pain, etc.). At the present time structural localization with ultrasound and lesioning are done most precisely in those cases where a section of skull bone has been removed and replaced by a rigid stainless steel open mesh system. This prosthetic device provides acoustic transparency and a rigid replacement for the skull. Only those cases have been so prepared in which it is judged there exists no other way to achieve maximum benefit to the patient. The possibility of achieving a comparable transkull capability, particularly with respect to structure localization, appears feasible. Recent studies involving transmission of focus ultrasonic beams through skull sections with appropriate approach angles (approximate normality of the central ray with respect to the skull) shows that the beam focus is retained and that targets viewed through the skull can be precisely defined and located.

MATERIALS AND METHODS

The system block diagram for ultrasonic visualization and surgery is shown in Figure 1 and a schematic representation of the system is shown in Figure 2. Practical implementation of the functional requirements listed in Table I is achieved through utilization of the system composed of the devices shown.[1] Focused ultrasound operating in a pulse echo mode for visualization possesses the necessary spatial resolution, sensitivity to display an appropriate array of structures in a thin slice (1 mm thick) two dimensional format which can be overlaid on appropriate two dimensional atlas material and time economy (of the order of 1 second) for generation of each section presentation. Localization of the lesioning beam with respect to the presented structures is readily implemented and, with a combined transducer-

ULTRASONIC TISSUE VISUALIZATION AND BRAIN SURGERY

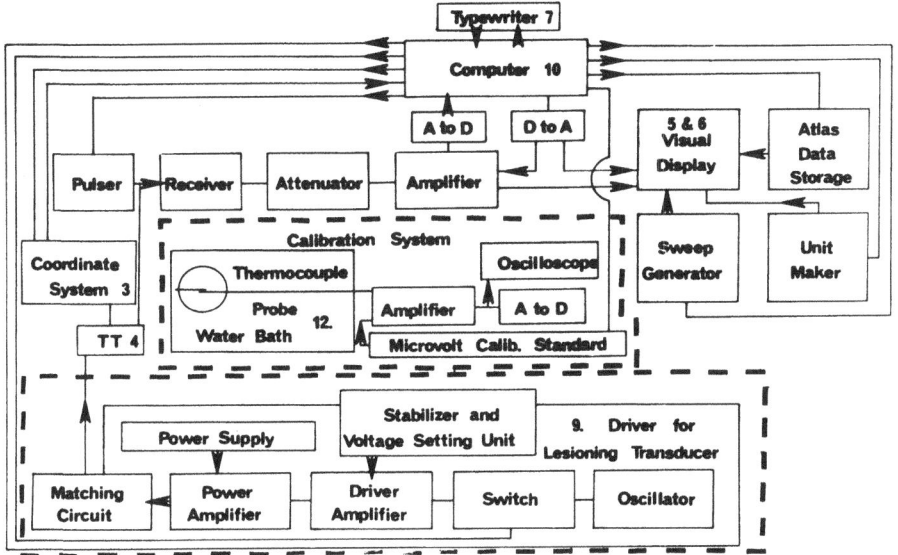

Figure 1. Block diagram of system for ultrasonic visualization and surgery. All units are under computer control. The computer is commanded to carry out the operator's wishes through the typewriter interface. Visual display information of the patient's tissue structural status, his functional status and other information are continually available to the physician operator who determines the subsequent sequence of events. An acoustic beam field plotting and sound intensity calibration system is incorporated so that precise control of the requisite physical parameters can be maintained.

transceiver unit for both visualization and lesioning, is uniquely free of transducer transposition errors.

The patient (1) lying on table (2) in a relatively unconstrained awake state is shown positioned for a left lateral approach to the head. Either coronal or horizontal plane brain scans at any angle with respect to the standard zero brain reference planes (Frankfort plane, etc.) can be taken, but since internal brain structures are directly visualized these planes are of significance primarily for use of brain atlas referencing. Since brain structural landmarks

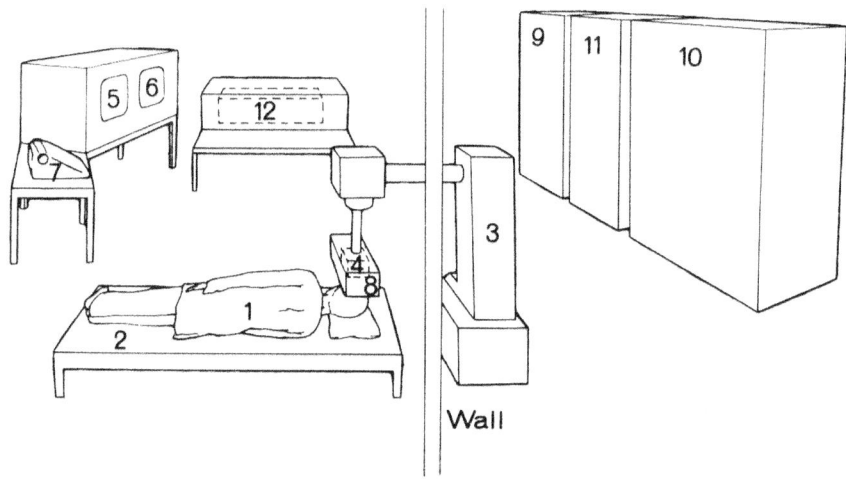

Figure 2. Schematic diagram of system for ultrasonic visualization and surgery. A wall separates the patient from most of the equipment. An arm supporting the ultrasonic transducer extends through the wall so as to reach the patient area. The physician operator in close proximity to the patient sits at the visual display and typewriter console. The calibration system is not used with the patient in the irradiation area.

can be generated repeatedly within a matter of seconds and brain lesioning performed in tenths of a second, there is minimal requirement for long term (hours) rigid constraint to the head (several minutes for any one period is adequate.)

Mechanical movement of the transducer-transceiver (T-T(4)) system for scanning (with any desired path contour) is provided through 5 degrees of freedom. Three orthogonal motions are generated by the computer controlled machine tool (3) which can be programmed to follow a desired scan pathway in three dimensions. Two rotational degrees of freedom with appropriate drives and angular encoders permit specific target regions to be viewed from T-T as motion in X, Y, Z is generated. Ultrasonically generated two dimensional images of brain structures are viewed in full size on a CRT monitor (5) which has a slave (6) from which photographs can be taken. With a 56° total included angle transceiver acoustic beam operating at 2.25 MHz, the system can differentiate two nylon monofilaments (.003" diameter) suspended in water and spaced 0.5 mm apart.

Figure 3 shows a transverse plot of the visualization beam at the center of the focal zone. This beam is axisymmetric in a plane transverse to the direction of sound propagation. Resolution capability is visually presented by scanning a two element monofilament nylon target arranged in a "V" shape. The filament divergence is 1 mm per centimeter. Two distinct echos can be obtained when the .003" filaments are 0.5 mm apart, but they coalesce into a single echo below this spacing distance. Figure 4 shows a graphic representation of the resolution capability.

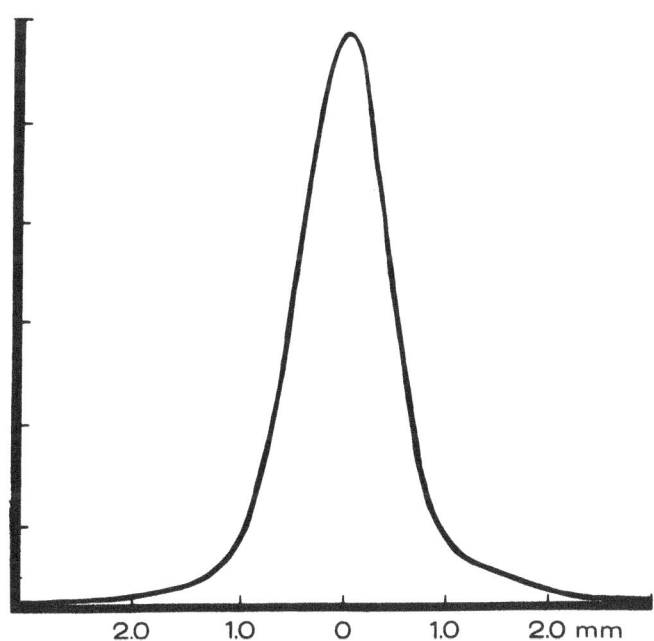

Figure 3. Sound field configuration of visualization beam at the center of the focal zone. Vertical scale is relative sound intensity. Horizontal scale is in mm.

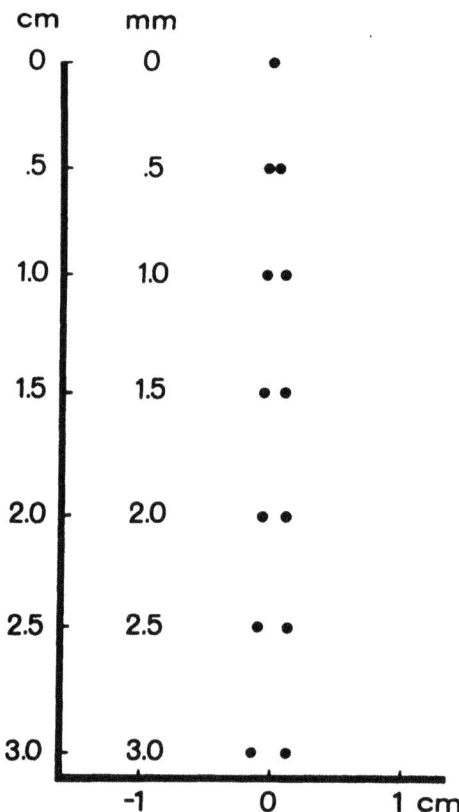

Figure 4. Copy from echogram used to define resolution capability of the visualization system. The black dots represent signals received from scanning of two monofilament nylon strings arranged in a V shape configuration. The bottom of the V is at the top of the figure and the filaments are divergent at the rate of 1 mm per centimeter length. Vertical column labeled mm shows the spacing distance of the filaments for each echo set. Two distinct echos are obtained for 0.5 mm spacing.

Overall system sensitivity is such that 110 db of attenuation is required to extinguish the received echo at normal incidence from a glass target in water (essentially 100% reflection). The linear amplifier provides 85 db gain. A 185 volt spike is delivered to the transceiver and this produces 50 milliwatt average acoustic intensity at the center of the beam focus when operating the system at 1 KHz repetition rate. A typical electrical driving pulse wave form and the transceiver electrical output waveform generated from the acoustic pulse reflected from a glass plate is shown in Figure 5. Range and azimuth resolution achieved with this system are approximately equal and are essentially at one wavelength limitation.

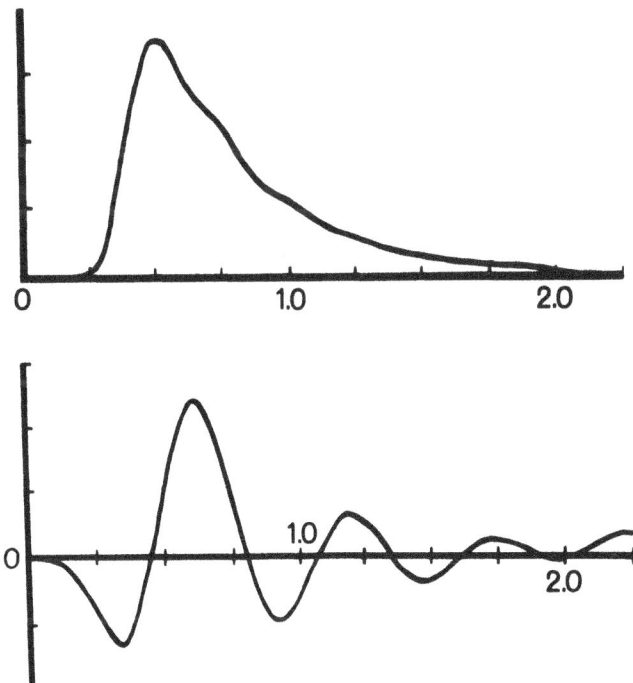

Figure 5. Top. Voltage pulse waveform delivered to transceiver. Horizontal scale is in μ sec.
Bottom, voltage waveform output from transceiver as a result of acoustic input received from acoustic pulse reflected from a glass plate. Horizontal scale is in μ sec.

With this type of system it is possible to put the received echos of interest in the limited dynamic range of the display units so as to have optimum presentation over the full system dynamic range. Accurate attenuation calibration provides information on which tissue absorption profiles can be generated. Tissue impedance profiling, after the method suggested by Jones,[2] can also be readily incorporated into the system as a further aid in tissue typing.

Commands to the computer (10) through the typewriter (7) permit operation in an interactive mode so that a dynamic interchange between the operator's desires and the patient's tissue profile (spatial relationships and acoustic properties) is possible.

Coupling between the patient and T-T is made through a water bath (8) in which T-T is submerged. The water is supported and restrained by a thin plastic (.010" thick) polyethylene drape which is brought into contact with the lubricated scalp surface (commercially available Groom and Clean has been used for this purpose).

Lesion generation is provided by T-T operating at the same frequency as for transception, but at much higher power. The focal zones for the high power CW and low power pulse echo regimes are coincident so that no transposition errors occur. Electrical power is provided by a 4 KW source (9), (frequency range .1 to 100 MHz) which, when operated under pulsed conditions, achieves rise times of less than 10 nsec. Feedback control is provided on the r.f. voltage level to T-T so that variations in acoustic intensity are maintained within a few percent. The computer clock controls the preselected time on period for any delivered ultrasonic dose to 10^{-5} sec. accuracy. Control of acoustic intensity in the lesion site is provided by setting of the r.f. voltage delivered to the transducer with appropriate compensation for absorption due to varying path tissue lengths from the port of entry to the lesion site. Predetermined absorption values for the appropriate tissues are inserted into the computer program. Interface equipment is lumped under Item 11 in Figure 2.

In order to maintain a precise knowledge of the ultrasonic intensity and beam configuration which is essential to the present practice of ultrasonic surgery, an automated beam plotting and calibration system (12) is provided. Central to the beam plotting and calibration procedure is the thermocouple probe (Figure 6)[3]. The probe is operated in the transient mode. A burst of sound (.1 sec. for beam plotting and 1 sec. for calibration) impinges on the thermocouple junction (linear dimensions of the order of .001"). Transducers capable of producing intensities of at least one watt per square centimeter in a CW mode can be calibrated and their field plotted. The thermocouple probe is calibrated against a radiation pressure method (small rigid sphere suspended in water). Focal zones having half power widths of the order of 1 to 0.1 mm can be precisely defined by this probe when the ultrasonic frequency is below 10 MHz. Automatic beam plotting is under computer control and after the transducer beam is automatically centered on the probe, the sound intensity calibration for the maximum intensity region is determined. A typical field plot of the sound beam configuration for a lesioning transducer operating at 1 MHz is shown in Figure 7.

In order to make the overall system convenient for physician usage in the interactive mode, the display monitor will ultimately be capable of displaying both brain atlas information and the ultrasonically derived structural details from the patient. These can then be intercompared and relative shifts in structures as well as abnormal features identified. It will also be possible to display the focal beam contour on the monitor so that the focal zone to be lesioned and the entire tissue pathway traversed by the beam will be seen in its correct anatomical location. A most fortuitous circumstance occurs in ultrasonic lesions generated in brain since they can usually be ultrasonically visualized as they are produced. This visualization in brain is additionally possible for electrolytically induced lesions and small needle tracts (.020" diameter needle tracts).

TISSUE VISUALIZATION

In brain, with a transcutaneous approach without interfering skull bone, using the focused ultrasound transceiver and system previously mentioned, it is possible to distinguish all fluid filled soft tissue interfaces within the resolving power of the system (at 2.25 MHz this is approximately .6 mm).[4-7] Such interfaces have acoustic impedance mismatches of at least 1% which are readily detected. A variety of abnormal features are also detectable because they have

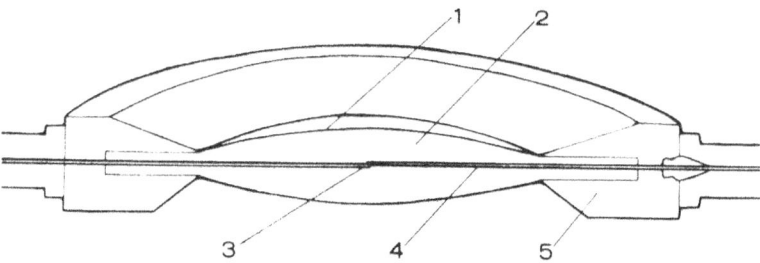

Figure 6. Schematic cross section diagram of calibrated thermocouple probe. Sound is transmitted through the thin polyethylene window (1) and through absorbing oil medium (2). Transient heating of medium (2) by the sound pulse raises the thermocouple junction (3) temperature in a predictable manner. Thermocouple wire (4) is .003" in diameter. Wires at the junction are etched down to .0005" in diameter. A rigid stainless steel housing (5) provides support for the system. The window dimension is very large compared to the beam dimensions (typically 3.0" in diameter).

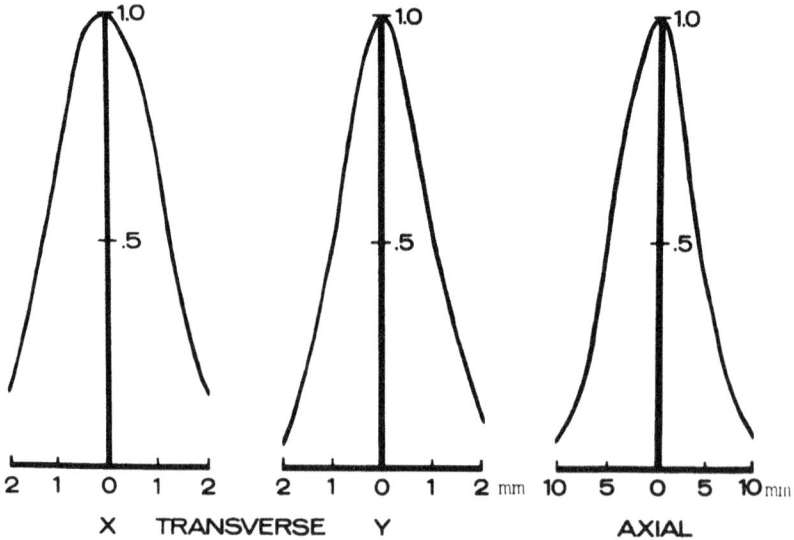

Figure 7. Transverse and Axial Plot of ultrasonic field configuration for lesioning transducer operating at 1 MHz.

comparable acoustic impedance mismatches. Some gray-white matter interfaces have also been detected, but here the acoustic impedance mismatch apparently is of the order of 0.2 to 0.4%[8]. Table 2 gives a listing of structures which have been visualized. The ability to directly visualize brain structural detail is of value in directing ablative surgery of an abnormal tissue as well as providing structural localization for a variety of procedures classified as stereotaxic neurosurgery.

As an indication of the structural detail and localization potential of the method, a series of coronal section echograms spaced at various intervals is shown in Figure 8. Structural details are indicated in the legend and the specific region of interest, because of its importance as a reference landmark, is the region of the splenium of the corpus callosum. This white matter structure crosses the brain midplane immediately anterior to the termination of the junction between the falx and tentorium. Falx and tentorium are clearly defined here because they are represented as interfaces between soft tissue and cerebral spinal fluid (CSF) filled spaces. The midplane crossing of the corpus callosum can be clearly seen as an echo free midplane passage region. Brain atlas reference diagrams of the same planar brain views shown by the echograms are included for comparison. The spectrum of structural detail available as far as we are aware has never been seen by any other method or combination of existing methods including the recently developed EMI scanner.[9] Tumor growth rate and/or regression in human brain has also been followed.[10] An example of a particularly favorable circumstance (for the patient and the applied therapy) is seen in the series of echograms (Figure 9) taken 10 months apart during which time linear accelerator and ultrasonic irradiations were conducted. There is no implication that the ultrasonic irradiation was a contributor to this tumor (glioblastoma multiforme) regression since there is no way of separating these two therapeutic modalities for this patient. The patient's neurological status improved concurrently with the indication of tumor regression as visualized ultrasonically.

Ultrasonic visualization of ultrasonically induced lesions has been demonstrated in both animal and human brains. A single focal lesion placed in the region peripheral to a tumor (glioblastoma multiforme) in a human brain to test the potential for blocking tumor proliferation in a

TABLE 2

Brain Structures Already Visualized with Ultrasonic Transcutaneous Approach Without Interfering Skull Bone

Normal Structure	Type of Complex	Interface
Ventricles	Fluid (CSF) filled spaces	Fluid-soft tissue
Falx (Interhemispheric Fissure)	Fluid (CSF) filled spaces	" "
Tentorial Region	Fluid (CSF) filled spaces	" "
Sylvian Fissure Complex	Fluid (CSF) filled spaces	" "
Various Sulci	Fluid (CSF) filled spaces	" "
Internal Carotid Arteries	Blood vessel	" "
Great Cerebral Vein	Blood vessel	" "
Middle Cerebral Artery	Blood vessel	" "
Pineal Body	May be calcified	Tissue-tissue
Colliculi	Mid-brain nuclei	Fluid-soft tissue
Splenium of Corpus Callosum	Cerebral Hemisphere connecting link	Outlined by fluid in adjacent ventricle
Columns of Fornix	Connection from posterior brain regions to hypothalmus, etc.	" "
Posterior Commisure	Hemisphere cross connection	" "
Anterior Commisure	Hemisphere cross connection	" "
Lateral border of thalamus	Gray matter	Gray-white matter
Internal capsule	White matter	" "
Medial border of Globus Pallidus	Gray matter	" "

Table 2 (Continued)

Abnormal Structure	Type of Complex	Interface
Glioblastoma Multiforme	Tumor	Tissue-tissue
Astrocytoma	Tumor	" "
Hemmorhage	Clotted or fresh blood	Fluid-tissue
Ultrasonically generated lesion	Modified tissue	Tissue-tissue
Electrolytic lesion	Modified tissue	" "
Needle tracts (.020" dia. needle used to produce the tract)	Modified tissue	" "

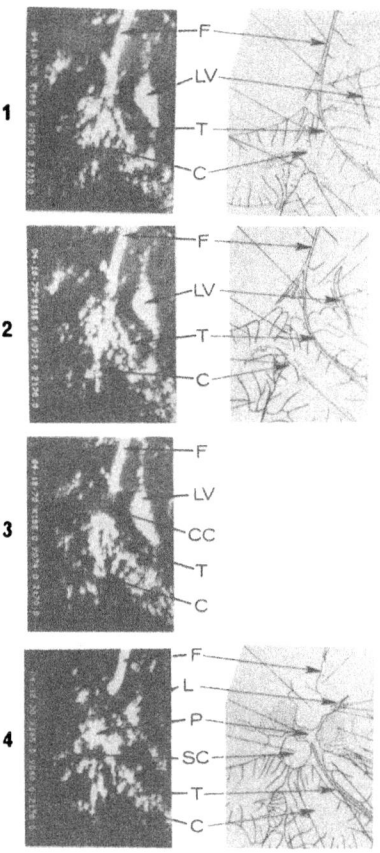

Figure 8. Left Side: echograms of coronal brain sections in live human patient. Right Side: human brain atlas diagrams.

F - Falx
LV- Lateral ventricle
T - Tentorium
C - Cerebellum
CC- Corpus Callosum
SC- Section of Superior Colliculus
P - Pineal

1. Most posterior section
2. 1 mm anterior to (1)
3. 2 mm anterior to (2)
4. 6 mm anterior to (3)

This selection of echograms taken from a set with 1 mm spacing per section shows the considerable detail which is obtainable in comparison with standard brain atlas information. The structural details and spacial accuracies are such as to be essentially ideal for any localization requirements including ultrasonic lesioning in brain.

ULTRASONIC TISSUE VISUALIZATION AND BRAIN SURGERY 145

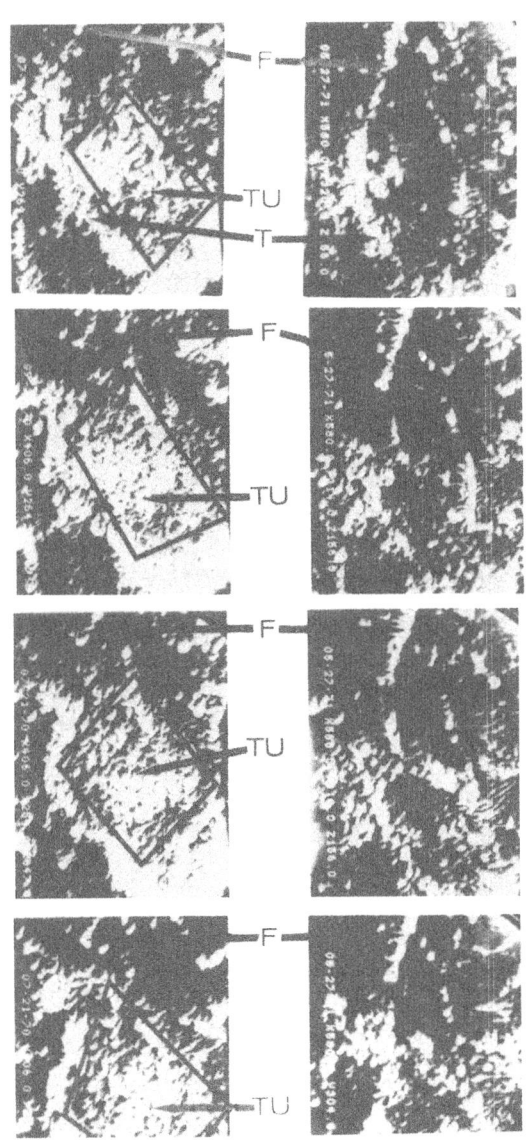

Figure 9. Coronal section echograms of human brain. Left Side: echograms taken 7-21-70 showing evidence of massive tumor infiltration. Right Side: echograms taken 5-27-71 showing almost complete elimination of abnormal echos.
F - Falx, T - Tentorium, TU - Tumor.

given direction is shown ultrasonically visualized in Figure 10. The target site chosen for lesioning is indicated in the figure by the two right angle intersecting electronically generated lines. Within a few seconds after lesioning, the visualization scan was made showing the appearance of a new echo in the appropriate location. Changing the attenuator setting in the system to bring the echos from the tumor region to extinction revealed that the ultrasonic lesion was more intensely reflecting than the tumor tissue, or, for that matter, any other structures in this region.

LESION GENERATION

The ultrasonic dosage (time and intensity relationship) needed to produce a histologically observable lesion in gray matter in brain is shown in Figure 11. These data are for sound frequencies in the 1 to 9 MHz range.[11-13] There is a fine structure related to frequency for this dosage graph, but here this structure has not been shown. For any selected frequency the threshold region is well defined. A change in ultrasonic dose of 10% in the threshold region moves from the condition of no lesion to 100% probability of lesion formation. White matter in brain under normal physiological conditions is more sensitive than gray matter and this

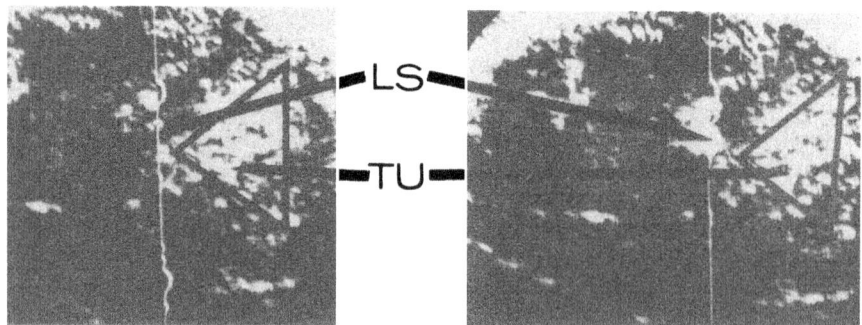

Figure 10. Ultrasonic visualization of ultrasonically produced focal lesion in human brain. Left: echogram before lesion production. Right: echogram immediately after lesion production. LS - Lesion Site. TU - Tumor.

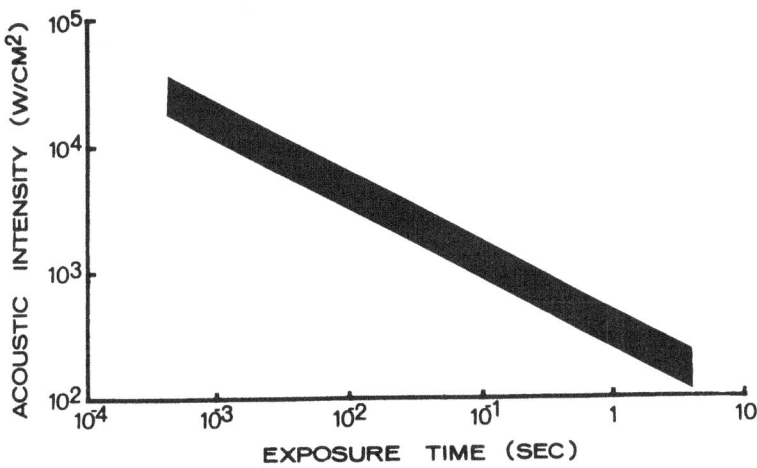

Figure 11. Threshold region for production of focal lesions in gray matter of adult animal (rat, cat, monkey) and human brain. Abscissa is in seconds and ordinate is in watts per square centimeter peak sound intensity in the beam at the lesion site. Frequency range is from 1 to 9 MHz.

feature can be used to selectively disrupt fiber tracts with respect to surrounding gray matter. [14]

For the first time there is available a non-invasive technique for producing lesions of any desired size or shape in brain at any number of selected sites in any selected temporal sequence. That this is a highly desirable mode for modification of brain to bring about a therapeutic benefit in a variety of abnormal conditions is of course not totally and readily predictable. There are presently some circumstances in which this type of approach appears to be beneficial (stereotaxic surgery for intractible pain, various hyperkinetic disorders, and severe psychological disorders involving violent aggressiveness). Potentially the possibilities appear to be quite profound if some presently evolving models of neuronal circuit relationships based on maintenance of neuron population and connection rates can be given adequate testing.[15] Readjustment of neuron population ratios by lesioning in brain subsequent to a trauma (stroke, accident, etc.) which may have disturbed the normal balance may provide the basis for an appropriate therapy.

SUMMARY

Ultrasound used for both tissue visualization and lesioning offers the potential for non-invasive, atraumatic, minimally hazardous and economically feasible delivery of diagnosis and surgery for certain specific disease entities. The system described herein is being initially explored with respect to brain, but a wide variety of body tissues and organs would appear to be within its scope of diagnostic and therapeutic possibilities.

REFERENCES

1. F. J. Fry and R. C. Eggleton, Ultrasonic Device for Human Disease Diagnosis and Surgery, Ultrasonics Symposium Proceedings, IEEE Catalog No. 72CHO708-8SU, (33-37) 1972.

2. J. P. Jones, Impediography: A New Approach to Non-Invasive Ultrasonic Diagnosis, Proceedings of the 25th Annual Conference on Engineering in Medicine and Biology, The Alliance for Engineering in Medicine and Biology, Vol. 14, LC61-24788, 1972.

3. W. J. Fry and R. B. Fry, Determination of Absolute Sound Levels and Acoustic Absorption Coefficients by Thermocouple Probes - Theory, J. Acoust. Soc. Am., 26, 294-310, 1954.

4. F..J. Fry, R. F. Heimburger, L. V. Gibbons, and R. C. Eggleton, Ultrasound for Visualization and Modification of Brain Tissue, IEEE Journal of Sonics and Ultrasonics, Vol. SU-17, No. 3, July, 1970.

5. F. J. Fry, Ultrasonic Visualization of Human Brain Structure, Inv. Radiol., Vol. 5, No. 2, 117-121, March-April 1970.

6. F. J. Fry, Ultrasonic Visualization of Ultrasonically Produced Lesions in Brain, Confia. Neurologica, 32, 38-52, 1970.

7. F. J. Fry, Intracranial Anatomy and Ultrasonic Lesions Visualized by Ultrasound, Ultrasonographia Medica, Vol. 1, Ed. J. Bock and K. Ossolnig, Verlag der Wiener Medizinischen Akademie, Vienna, 1971.

8. R. F. Heimburger, F. J. Fry, and R. C. Eggleton, Ultrasound Visualization in Human Brain: The Internal Capsule, a preliminary report, Accepted for publication in Surgical Neurology, 1972.

9. New EMI Machine for Diagnosing Brain Disease and Computerised Soft Tissue Tomography - EMI, Central Research Laboratories, Shoenberg House, Trevor Road, Hayes, Middlesex, U.K.

10. R. F. Heimburger, R. C. Eggleton, and F. J. Fry, Ultrasonic Visualization in Determination of Tumor Growth Rate, Accepted for publication in the Journal of the American Medical Association, 1972.

11. F. Dunn and F. J. Fry, Ultrasonic Threshold Dosages for the Mammalian Central Nervous System, IEEE Trans. Biomed. Eng., BME-18, 253-256, 1971.

12. F. J. Fry, G. Kossoff, R. C. Eggleton, and F. Dunn, Threshold Ultrasonic Dosages for Structural Changes in the Mammalian Brain, J. Acoust. Soc. A., 48, No. 6, (part 2), 1413-1417, December, 1970.

13. F. J. Fry, J. E. Lohnes and F. Dunn, Fine Structure of the Ultrasonic Threshold Region in Mammalian Brain, Accepted for publication in Journ. Acoust. Soc. Am., 1973.

14. J. W. Barnard, W. J. Fry, F. J. Fry, and R. F. Krumins, Effects of High Intensity Ultrasound on the Central Nervous System of the Cat, J. Comp. Neurol. 103, 459-484, 1955.

15. F. J. Fry, Quantitative Organization in the Mammillary Bodies of Adult Cat, Brain Research, 37, 115-122, 1972.

A PROJECT OF ULTRASONIC TOMOGRAPHY ("SONORADIOGRAPHY")

Professor Dennis Gabor, Imperial College

Staff Scientist, CBS Laboratories

London, England and Stamford, Connecticut USA

In X-ray radiography there exists a technique, known as tomography, in which a sharp image of <u>one</u> section of the body is obtained by moving the X-ray tube and the photographic plate in opposite directions. Can we not produce a similar, or even better effect with ultrasound?

Something almost as good has been, in fact, already achieved by large-aperture focusing of ultrasound images, with or without holography. Focused holograms have the advantage of the amplifying effect of the reference beam, but they have the disadvantage, which always arises in the case of coherent sound, of very strong speckle noise. The layers of the body outside the one that is focused also scatter sound wavelets into the picture, and as their amplitudes add or subtract at random they produce a strong noise which in fact reduces the resolving power by one or two orders of magnitude. Ultrasonic imaging, if carried out with incoherent sound, is free from this effect; so in principle it could attain the quality of X-ray tomography.

But could we not do <u>better</u>? Could we not cut out the effect of the sections, other than the one observed, by a method that is not available in the case of X-rays: by <u>gating</u>? That is to say, by admitting to the image formation, only sound wavelets which have issued from the section to be singled out? In fact, a type of sound-tomography already exists:

the B-scan method. Unlike X-ray tomography, this produces images not of sections at a certain depth in the body, but at right angles to the surface, by sorting the pulses according to their return time. But a section is recorded line after line by a scanning method. This method has been developed to admirable perfection; but one must ask, would it not be possible to obtain gated images simultaneously, in one shot?

The project which we have called "sonoradiography" is an attempt toward this. At present, it is only partly developed, but we believe that it is timely to bring it to the attention of experts.

Figure 1 shows the basic principle. Its characteristic feature is that it operates with holograms, but these are not produced in the usual way, with coherent waves, but with very short single pulses. Indeed, if one wants perfect depth discrimination, one must not use short wavetrains, but pulses, as short as possible. But how to produce holograms with a single pulse? Figure 1 explains how this is done. Let us take a very thin membrane as the detector surface, immersed in the water, metallized to make it visible. Send a short, sharp pressure pulse through the membrane into the body at the other side. Let there be a scattering point somewhere in the section, that is selected for seeing. This point will send back a more or less spherical wavelet (somewhat distorted by the refraction in the body) and this will produce on the membrane a rapidly spreading ring-shaped fine bulge. We now make this into a hologram by illuminating it with a high-frequency, stroboscopic light source, preferably a laser, but only during a short interval. The trace of the spreading ring will now appear exactly like the hologram of a point-object: like a system of Fresnel zones. It is easy to see that the gating can be made very selective. For instance, illuminate only during the time in which the fringes spread from zero to 15 degrees. As the cosine of $15°$ is only 1/16 less than unity, this means that only 1/32 of the depth can contribute to the diffraction figure, because the layer is traversed twice, once by the ingoing, once by the returning wave, and this is good depth definition.

AN ULTRASONIC TOMOGRAPHY PROJECT ("SONORADIOGRAPHY") 153

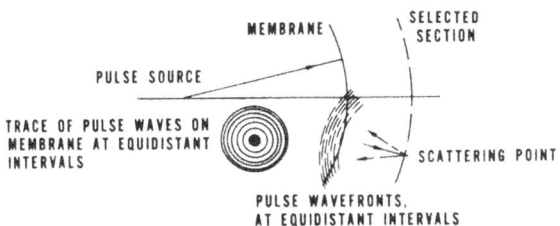

Figure 1. The principle of sonoradiography producing the equivalent of a hologram by periodic illumination.

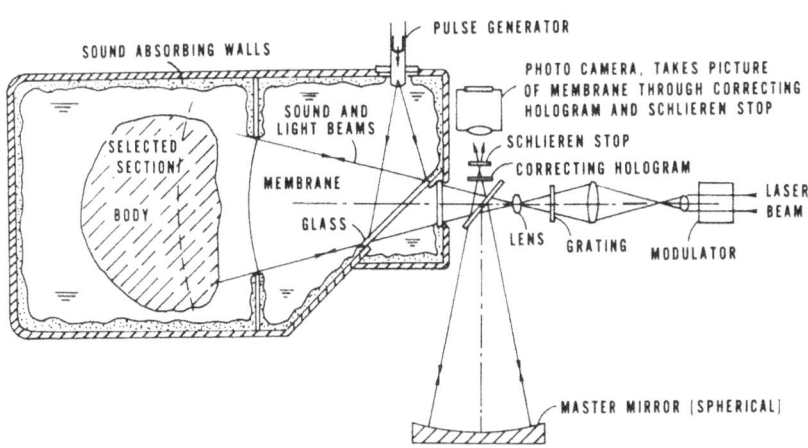

Figure 2. Scheme of sonoradiography.

On the other hand, a spread of 15°, which corresponds to f:2, gives a diffraction limit of 2.4 wavelengths, which is, for instance, 0.7 mm for five kilohertz; also good enough.

The substitution of a pulse for a sinusoidal wavetrain has two other important advantages. It is well known that the membrane is a rather insensitive detector, because the dynamic amplitude of sound waves in water is so small. At one watt per cm square, the maximum amplitude is only 28 Angstroms, which is only about 1/200 of a light wavelength; too small to detect by ordinary interference methods. But if we pack the whole energy of a wavetrain into a single pulse, we can go much higher with the power, and we make multiple use of this power, just as if it were a wave train. Moreover, a single pulse can be a pure pressure pulse, and this will not damage living tissues. In sinusoidal waves, it is the expansion phase that is dangerous; this is what tears tissues apart. Of course, a compression wave is reflected as an expansion wave from a medium of lower refractive index such as an air cavity, so there is a limit in the application to chest and stomach, which we do not yet know. But these organs are anyway not very suitable subjects for ultrasound.

Our project is based on the expectation that with a pressure pulse lasting only a fraction of a microsecond, we can go up with the power per square centimeter to 10 kilowatts or more. For areas of the order of a few humdred centimeters square, this means an energy of the order of a joule. How can we pack such an energy into a pulse of, say 100 nanoseconds duration? The pulse generator which we have developed for this purpose is shown in Figure 2, which also shows the whole scheme.

A tube of two cm^2 area is closed at one end by a thin, spherical aluminum cap, in contact with the water. This tube is evacuated, and an aluminum thimble of the same spherical shape and the same thickness is held at 10 cm from the end. A membrane is burst at the other end of the tube, and the atmospheric pressure propels the thimble against the cap. As the masses are matched and the impedance of water is very large, practically the whole kinetic energy of the thimble, about two joule, is converted into a short, sharp pressure pulse. Theory shows that this wave has an extremely steep front and an exponential tail. The mass of water in the whole wave is just equal to that of the aluminum cap. Thus, the depth of the wave can be made about 0.3 mm, and its duration about 200 nanoseconds. The wave will be nearly spherical, and will be spread in an angle which is determined by the curvature radius of the cap and its diameter, approximately as shown in the figure.

The wave is reflected at a glass mirror, and propagates toward the spherical membrane as if it came from its curvature center. If the membrane is thin (say, 25 microns), the reflection from the membrane will be negligibly small. The wave penetrates it almost without a loss and goes into the body, from where it is back-scattered. For the purpose of simple explanation, we can neglect the distortion of the spherical wave in the body.

We want to have a tomogram of a section, at a certain depth. This is achieved by illuminating the membrane at the instant at which the wave returns from that section stroboscopically for a short time, of the order of a few microseconds, sufficient for, say 20 fringes, with a laser beam. The laser focus is at or near the curvature center of the membrane, so the reflected beam comes to a sharp focus. The zero order light, reflected by the undistorted membrane, is cut out by a schlieren stop, so that only the distortion is photographed.

At this point a question arises. We cannot expect the membrane to keep a perfectly spherical shape so as to produce a perfect focus; hence, we cannot expect the zero order light to be cut out by a very small, fixed schlieren stop. We have overcome this difficulty by a <u>correcting hologram</u> in front of the stop. We make use of the 45° semireflecting mirror which we require anyway for separating the ingoing light beam from the reflected light beam. We place below this a good spherical "master mirror" which has the shape the membrane ought to have. By the interference of the two beams, from the membrane and from the master mirror, we produce a hologram. Illuminated by the imperfectly spherical membrane, this will then produce the perfect focus of the master mirror. This has been verified in the laboratory of Professor G. W. Stroke, as shown in Figure 3. One must tilt the master mirror slightly, lest the three foci overlap. It is seen that this has converted the very imperfect membrane into a perfect spherical mirror. Professor Stroke has shown that the membrane, undisturbed, keeps this shape for many days. But when there is a live body in the tank, one can expect the membrane to be moving; so our suggestion is to take the correcting hologram on photochromic material an instant before the sound wave exposure.

This photograph is a hologram, preferably on a very small scale, which is then reconstructed as usual, in laser light.

The apparatus as shown in Figure 2 contains another trick which may greatly contribute to its efficiency and sensitivity. It is proposed to illuminate the membrane, not with a simple spherical wave but with one which is corrugated, with a high spatial frequency, with a phase differential of half a wavelength of light between crests and troughs. Such a wave can be produced by projecting on the membrane the image of a phase grating with a phase difference of π between tops and bottoms of equal width. Such phase gratings have been produced with great perfection by Dr W. E. Glenn and his collaborators, Dr Charles Ih, and Donald Lamberty, of CBS Laboratories. They have the property that they diffract nothing into the zero order. In order to image them, one must therefore use a lens with sufficient power to take in at least the two first orders. If now in the viewing of the membrane one would use a lens system of lesser power which does not take in the side lobes and does not resolve the image of the grating, the undisturbed membrane will appear dark, and one may be able to operate without a schlieren stop. But even more important is the gain in

Figure 3. Experimental verification; G. W. Stroke's laboratory at Stony Brook, New York.

sensitivity which may be obtainable by this method. The phases between crests and troughs are now accurately balanced. Any local displacement of the membrane will upset this balance, and produce a phase shift proportional to the difference in height, as in a phase contrast microscope. In this arrangement, the maximum sensitivity is at zero amplitude, while in ordinary schlieren methods, it is at an amplitude which produces a phase shift of $\pi/4$.

This increase in sensitivity is highly desirable because in reflection imaging we can count only on a small fraction of the power returning in a direction opposite to the ingoing wave. It must be mentioned though, that this effect is by no means as strong in the present method as in the B-scan method where the attenuation is reported to be of the order of 10 decibels per cm penetration. The reason is that in the B-scan method, the receiver is a small probe which collects the back-scattered energy only in a very small, solid angle. In the present method, the solid angle can be much larger, of the order of 0.2.

This is a provisional report on the development which is now going on in CBS Laboratories, Stamford, Connecticut, USA, in collaboration with the New York State University at Stony Brook.

ACKNOWLEDGEMENTS

The author wishes to thank the National Science Foundation under whose grant, No. G. K. 28302, the project of Sonoradiography was developed.

He also wishes to thank CBS Laboratories where the actual work was performed, and his collaborators:

> Professor George W. Stroke, State University of New York at Stony Brook; Dr William E. Glenn, Director of Research; Dr Charles Ih, and Mr Donald Lamberty, all of CBS Laboratories.

IMAGE INFORMATION PROCESSING FOR PULSE ECHO SCANNING METHODS

Masao Ide

Musashi Institute of Technology

Tamazutsumi, Setagaya-ku, Tokyo, Japan

I. INTRODUCTION

At present, ultrasonic imaging systems are widely used in many fields of medicine, mainly for medical diagnoses, though they are partly used as flow detectors for industrial use. A medical ultrasonic imaging system depicts the images due to acoustic differences of a cross-section of an examined anatomy, and the display patterns change greatly according to the characteristics of each part of the system.

In this paper I shall describe the factors affecting display patterns of an ultrasonic imaging system for medical applications, improvement of display characteristics, color display, direct recording of display patterns with Video Tape Recording (VTR), and matters to be developed in the future.

II. OUTLINE OF PULSE ECHO IMAGING SYSTEM

An ultrasonic imaging system for medical applications is called an ultrasono-tomograph, or B mode ultrasonic diagnostic equipment. Figure 1 shows a block diagram of general pulse echo ultrasonic diagnostic equipment.

1. <u>Frequency Range</u> - Approximately one to fifteen MHz has been used as a carrier frequency of ultrasound; usually two MHz (approximately) for internal cranial diagnosis; two to five MHz for soft tissues such as the mammary gland, liver and fetus; and 10 to 15 MHz for the eyes. The pulse width which differs according to carrier frequency, is approximately 0.1 to 5 microseconds.

2. <u>Scanning</u> - The ultrasono-tomogram system by pulse reflection scans the ultrasonic beam irradiated from the transducer, and at the same time, the position and direction of the electron beam of the cathode ray tube sweep according to the scan of the ultrasonic beam. Scannings are used such as:

    a. automatic mechanical scanning
       Linear, circular, arc, sector, radial, compound

    b. Manual scanning
       Contact compound

    c. Electronic scanning (high speed)
       Linear, sector

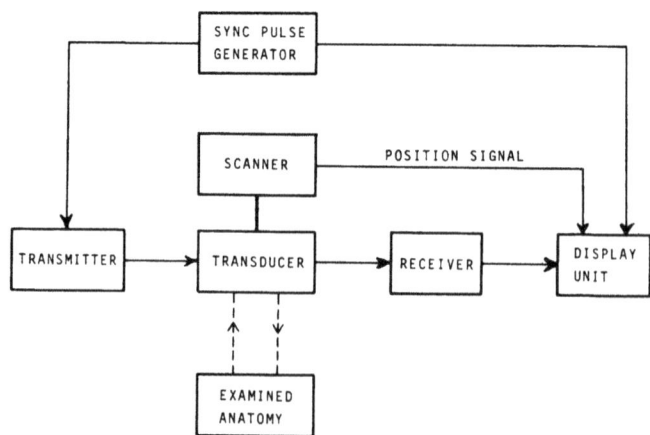

Figure 1. Ultrasonic Diagnostic Equipment

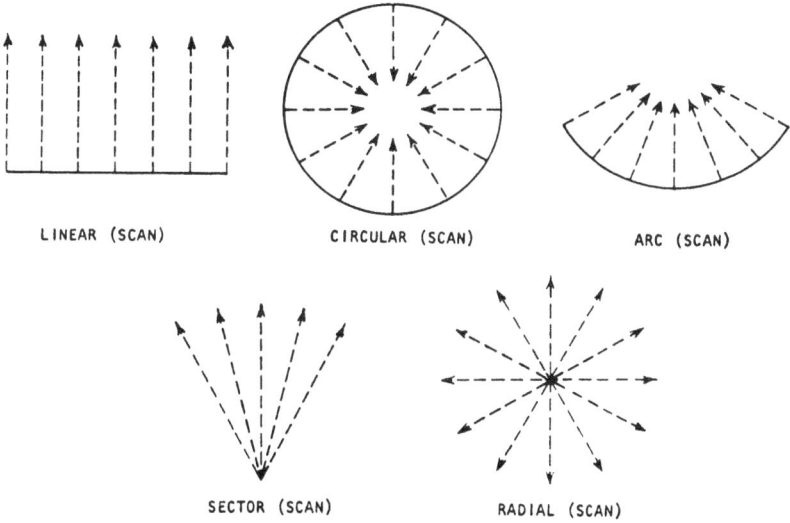

Figure 2. Scannings

Representative scannings are shown in Figure 2. Any combination is called compound scanning. A combination of linear and sector, circular and sector, is used to great extent.

3. <u>Ways of Display and Record</u> - Cathode ray tubes such as persistent, storage, color, etc., are used mainly as a way of display. Photographic methods using film are employed as the means of recording displayed images by cathode ray tubes. On the other hand, electrical recording of the output of ultrasonic diagnostic equipment with video tape has been studied recently.

Examples of ultrasonic diagnostic equipment in Japan are shown in Figures 3, 4 and 5. Figure 3 shows an arc scan especially designed for the breast and the thyroid gland, with five MHz PZT $10\phi 75$ R concave transducer. It is manufactured by the Toshiba Company. Figure 4 shows a manual contact compound scan made by Aloka. Figure 5 also shows manual contact compound scan equipment made by NEC.

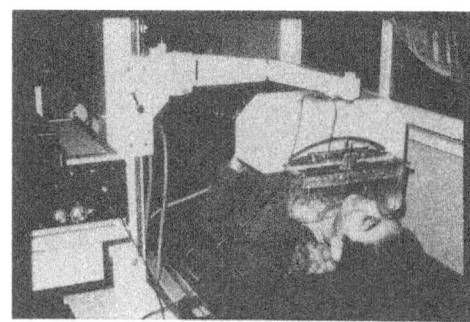

Figure 3. Arc Scan Diagnostic Equipment

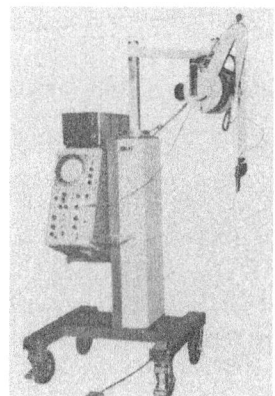

Figure 4. Manual Contact Compound Equipment

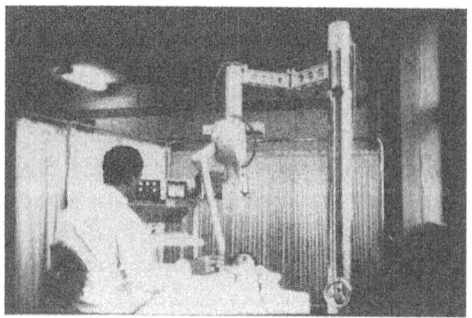

Figure 5. Manual Contact Compound Equipment

III. IMPROVEMENT OF DISPLAY CHARACTERISTICS

1. <u>Factors Affecting the Display</u> - Factors affecting the display of ultrasonic diagnostic equipment could be:[1]

   a. Transducer
      Sensitivity, damping characteristics of acoustic field

   b. Amplifier
      Bandwidth, linearity (linear, logarithmic), rejection, clipping, STC, FTC

   c. CRT (Cathode Ray Tube)
      Spot size, gradation (gray scale), distortion of deflection, suppression of background, display time, persistence characteristics, storage characteristics

   d. Photographic materials
      Resolution, gradation (gray scale, gamma characteristics)

   e. Scanning
      Suppression of artifacts by compound scanning

2. <u>Improvement of Amplitude Display Characteristics</u> - It is usual that the gradation of the equipment in general use is of a lesser grade, especially the equipment using a storage type CRT tube, which display black and white images, having no half-tone. By using equipment of such display characteristics, display images greatly change according to the sensitivity of the equipment, as in the case of display of cross-sectional images of the inner part of the body. In actual diagnosis, echo pulses reflected from the inner part of a body have various ampludes, from a very small one to a very large one. However, in narrow display equipment, only a definite part can be displayed in a definite sensitivity.

Figure 6 shows a circumstance in which display patterns change greatly by a change of sensitivity of the equipment. In this figure, one scanning line is represented in the A-mode. Accordingly, sensitivity settings of the equipment is a great problem for quantitative diagnoses. In order to avoid this disadvantage, sensitivity-graded tomography, so to speak, was introduced. According to this method, diagnosis is carried out by a series of patterns obtained by changing the sensitivity in several levels.[2] On the other hand, the studies not only on narrow dynamic range equipment but also on wide dynamic range equipment have been made.[3,4] Such equipment having a high power of expression in amplitude has a high frequency logarithmic amplifier in the receiving part, and compresses and displays reflected pulses logarithmically. Moreover, a cathode ray tube having a high power of expression of brightness is employed. Actually, by using a logarithmic amplifier, ultrasono-tomograms have been obtained. In the case of linear amplification, amplitude of reflected pulses decreases abruptly, but in the case of logarithmic amplification, the rate of decrease is easy, and small signals are received. Accordingly, images full of gradation were obtained. Examples of actual results according to a logarithmic amplifier are shown in the report of Professor T. Wagai in these proceedings.

3. <u>Signal Processing</u> - Before displaying reflected pulses obtained by ultrasonic diagnostic equipment by a cathode ray tube, it is necessary to improve the quality of the images by adding signal processing according to a diagnostic purpose, thereby making diagnoses easier. Here I shall introduce an example of analog signal processing.[5,6,7] Figure 7 shows a block diagram of the equipment of this system, which displays amplitude information of reflected pulses in a wide amplitude range as a gradation in one tomogram. Figure 8 shows the principle of behavior of the equipment. After being compressed logarithmically in amplitude by the high frequency logarithmic amplifier, input signals are displayed by a cathode ray tube of good

IMAGING INFORMATION PROCESSING FOR PULSE ECHO SCANNING

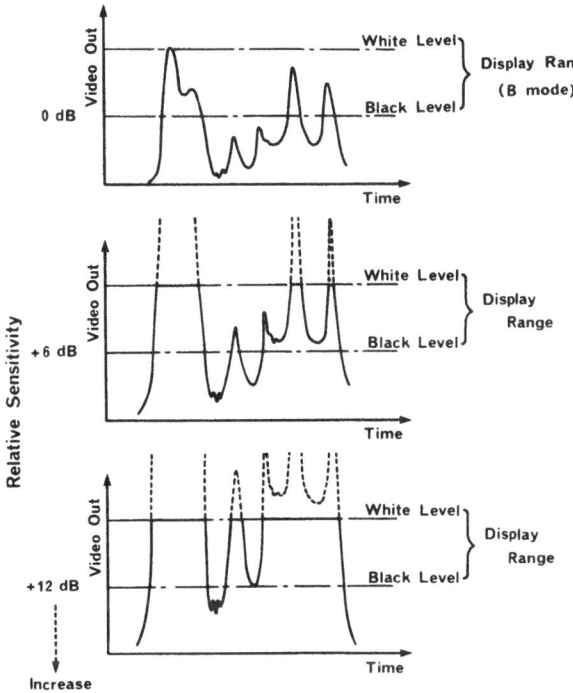

Figure 6. Principle of Sensitivity Graded Tomography

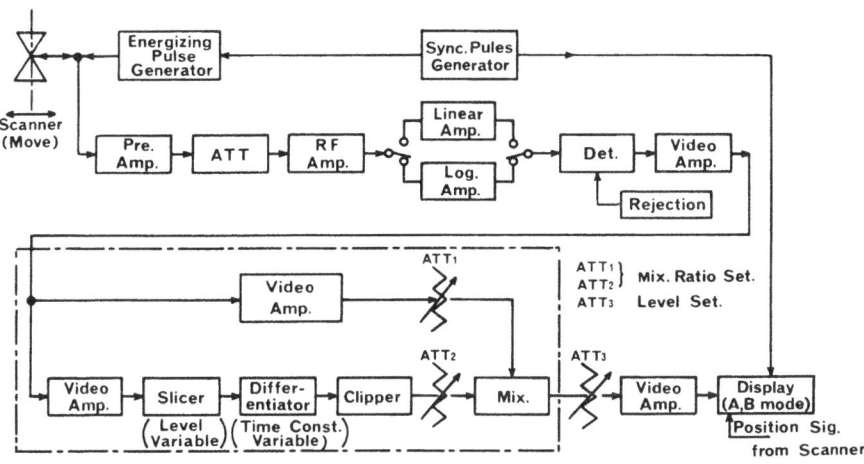

Figure 7. Contour Emphatic Display

gradation characteristics. Moreover, this system is able to add contour if necessary to the image which is full of information by compression.

After detecting the compressed signal received and picking up a strong signal over a level, we lead it up to a differentiating circuit so as to take a sharply built-up signal (contour signal). Then the contour signal and the original video signal are mixed again with an appropriate ratio. This complex signal is displayed on a cathode ray tube.

An example of experimental results obtained by the equipment is shown in Figure 9, where the frequency is 2.25 MHz, with a 30 mm ϕ, 100R focused transducer. The displayed part is that of the calf of the leg of an adult male. It is obvious that the bottom tomogram has both information of amplitude and position.

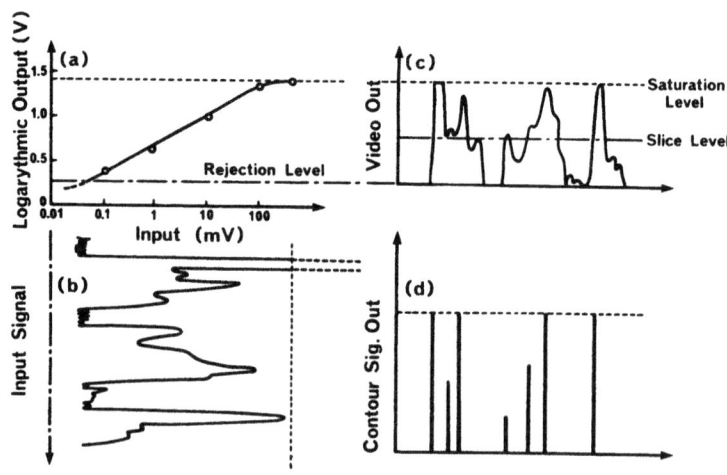

Figure 8. Behavior of Contour Circuit

IMAGING INFORMATION PROCESSING FOR PULSE ECHO SCANNING

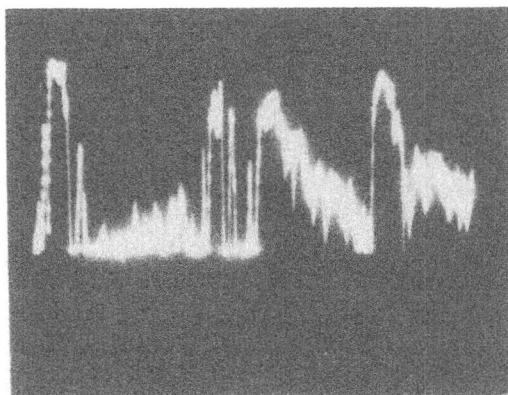

A-MODE

Log Compressed Image

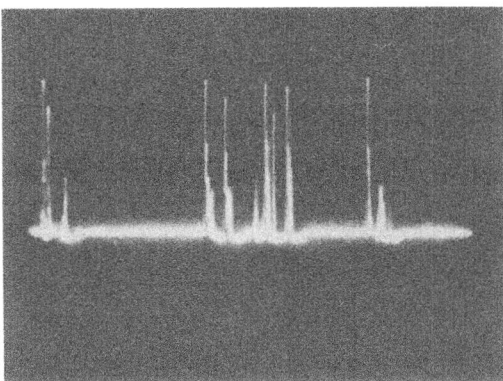

Contour Image

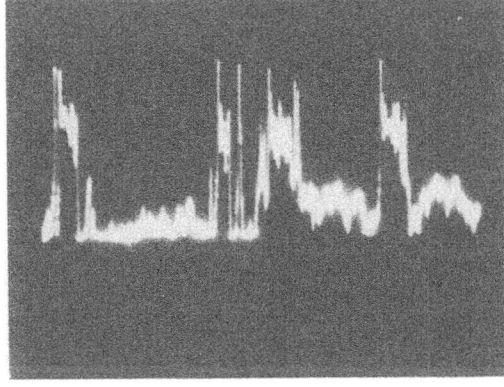

Contour Emphatic Image

Figure 9. An Example of Contour Emphatic Display

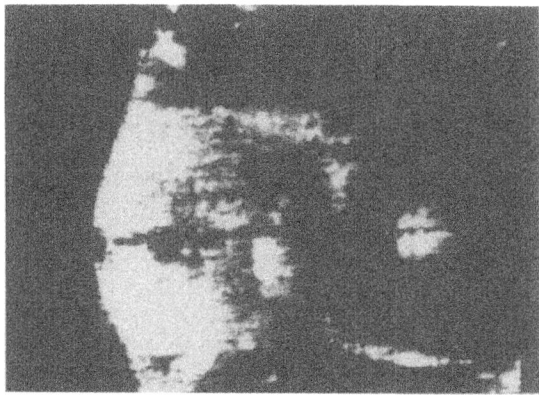

B-MODE

Log Compressed Image

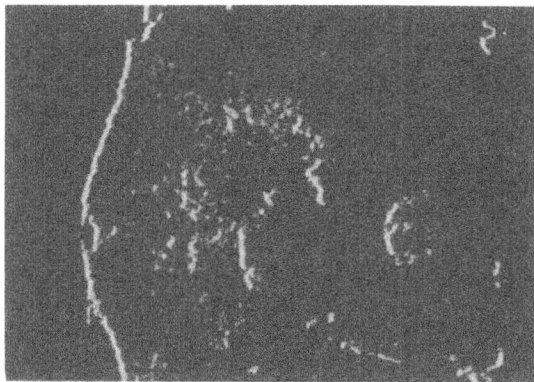

Contour Image

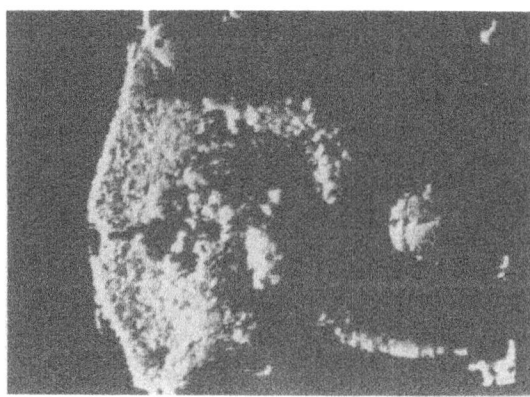

Contour Emphatic Image

Figure 9. An Example of Contour Emphatic Display

4. Color Display System for Ultrasono-Tomography

At present, output of ultrasonic diagnostic equipment; that is, diagnostic images, are displayed mainly by persistent or storage type cathode ray tubes. As the power of discernment of the hue is superior by far to that of brightness in the human eye, readings of ultrasono-tomograms become easier if several tomograms can be displayed as one colored image. Figure 10 shows a block diagram of the color display system.[8,9]

At first, output signals of diagnostic equipment or reproductive signals once stored with video tape recordings are fed to the electronic scan conversion circuit and stored. After this, they are read out as standard television signals. Then they are fed to an amplitude-color conversion circuit where amplitude-color conversion is made. After this they are displayed by color monitor as color tomograms.

Scan conversion can be carried out by a digital technique, but in this system we employed a scan conversion system which uses a silicon target storage tube. The storage tube has approximately 10 logarithmic gray levels. Figure 11 shows a block diagram of a scan converter that can convert any scanning into a standard television scan. Scan converted diagnostic signals are fed to n-step (n=7 in this experiment) amplitude discrimination circuit, and converted into digital signals. These digital signals are converted into R, G, B color signals and are displayed by a color monitor, or they are converted further into NTSC standard color signals and displayed by a standard television monitor. Figure 12 shows an example of the relation between hue and the amplitude of the input signal. The relation between hue and amplitude levels can be chosen freely by a change in the combination of digital signals being fed to the logic circuit. This relation should be further investigated in order to make actual diagnoses easier. By the NTSC conversion, record and transmission become easier. The advantage of this system is that many kinds of scan conversion containing a compound system can be accomplished by a relatively simple system.

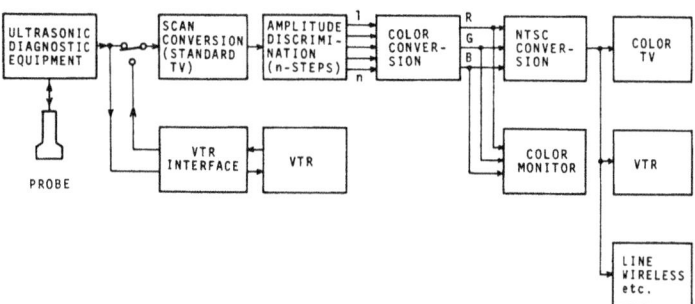

Figure 10. Color Display System for Ultrasono-Tomography

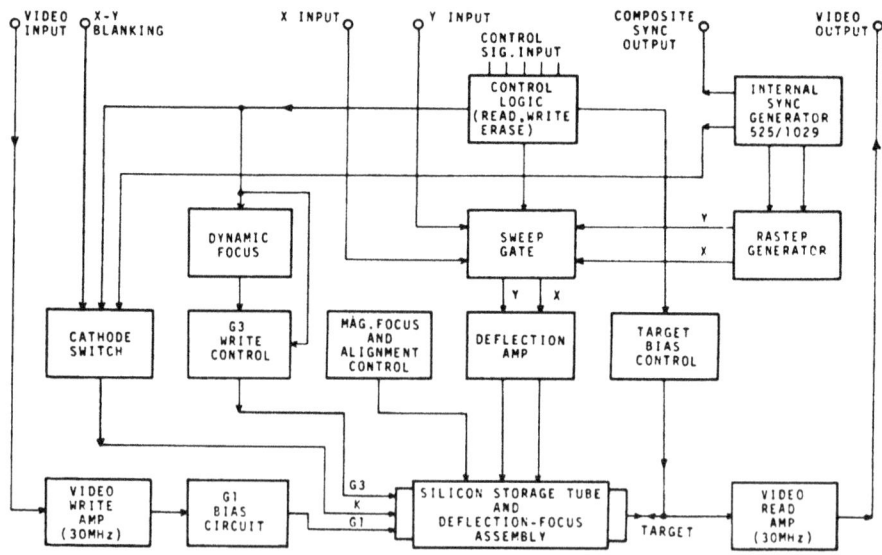

Figure 11. Block Diagram of Scan Converter

IMAGING INFORMATION PROCESSING FOR PULSE ECHO SCANNING 171

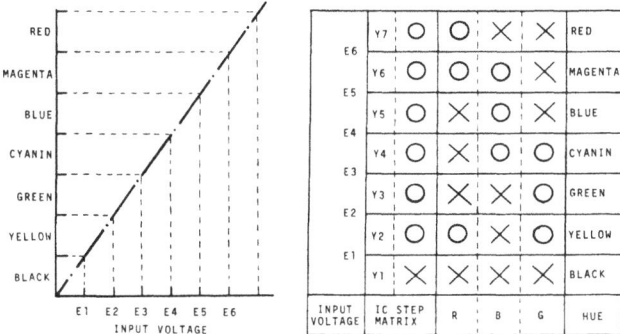

Figure 12. Example of the Relation between Hue and Amplitude

Figure 13 shows a whole view of the system. Figures 14 and 15 show examples of color display in the case of three hues. According to our experiment, the number of hues should not be too great from the point of view of simplification and quantization. Figure 14 shows a color display in hyperthyroidism. The order of amplitude level, from a high, is red, blue, green and black. Relative gain between each hue is 6 dB. Figure 15 shows a cross-section of a throat of another case.

Figure 13. Whole View of the Color Display System

IMAGING INFORMATION PROCESSING FOR PULSE ECHO SCANNING 173

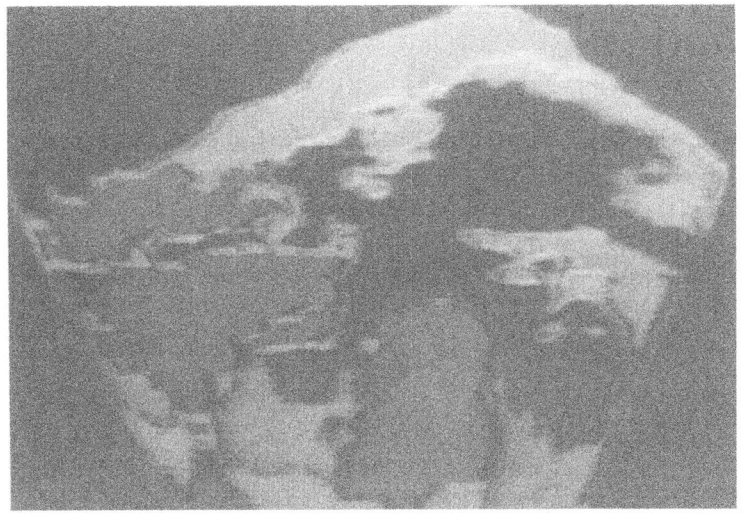

Figure 14. A Black and White Reproduction of a Color Display (Hyperthyroidism)

Figure 15. A Black and White Reproduction of a Color Display (Throat)

IV. IMPROVEMENT OF WAYS OF RECORDING AN
ULTRASONO-TOMOGRAM (Electrical Recording and
Reproduction with VTR)

By recording ultrasono-tomograms obtained from diagnostic equipment directly with a VTR (Video Tape Recorder), not only repeated reproduction but also analytic diagnosis becomes possible by doing some signal processing, or by changing the display unit according to diagnostic purposes from the reproductive signals.[10-14] Advantages of VTR recording in clinical applications are as follows:

1. Reduction of the test time
    Lightens the duty of patients
    Shortens the time of ultrasonic irradiation

2. Recording of a great deal of information at a time by:
    Logarithmic amplifier
    Multiple frequency tomography
    Multiple section tomography

3. Plural display by:
    Sensitivity-graded tomography
    Multiple frequency tomography
    Multiple section tomography
    Color display

4. Availability of signal processing by a reproductive signal
    Diagnoses will become easier, etc.

Consequently, accurate diagnoses can be expected.

As the signal form from diagnostic equipment differs from that of a television signal, diagnostic signals cannot be recorded as they are; accordingly, some interface circuit is necessary for the connection of VTR and the diagnostic equipment.[15]

IMAGING INFORMATION PROCESSING FOR PULSE ECHO SCANNING 175

Figure 16 shows the standard television scanning in Japan, which is the same as in the United States. One frame of television consists of 525 scanning lines. As interlaced scanning is employed, one frame consists of two fields. There are 30 frames in every second. Horizontal sweep frequency is 15,750 Hz and vertical sweep frequency is 60 Hz. Figure 17 shows movement of video heads in a two-headed helical scanning type VTR for home use. Figure 18 shows the switching of video heads. Switching noise appears at the switching position. This noise should be rejected in some way. A brief specification of video tape recording for home use is as follows:

Type:	Helical scanning type for standard television signal in Japan and the United States
Tape:	12.7 mm (1/2 inch)
Tape Speed:	19.05 cm/sec. (7 1/2 inch)
Video Track:	Bandwidth more than three MHz; S/N 40 dB
Audio Track:	Bandwidth, 60-10,000 Hz; S/N, 40 dB

These are enough for the recording of ultrasono-tomograms. However, a video signal of ultrasonic diagnostic equipment is different from that of a television signal; for example, in repetitive frequency. Therefore, it cannot be recorded as it is. Some interface circuit is necessary for the connection. Figure 19 shows a schematic block diagram of the interface circuit[18] which we made. At first, we made a timing pulse. With this pulse, an energizing pulse of ultrasonic diagnostic equipment is locked. Then this pulse is frequency divided down to 60 Hz in order to synchronize the VTR heads and VTR circuit. With this, the switching noise of heads can be rejected from the reproductive image.

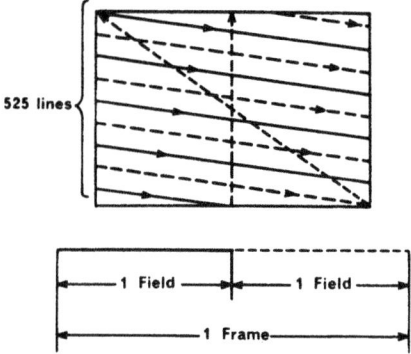

Figure 16. Interlaced Scanning of Television

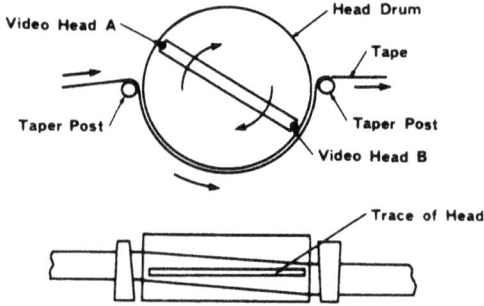

Figure 17. Video Heads (Two-Head Type)

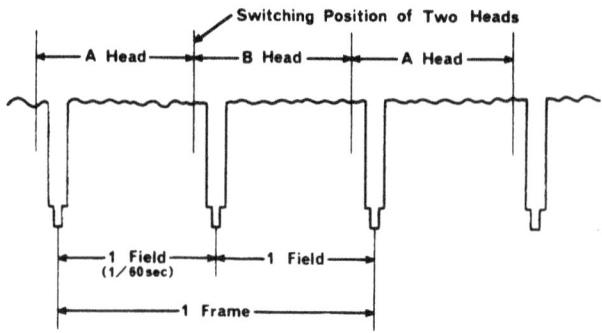

Figure 18. Switching of Video Heads

IMAGING INFORMATION PROCESSING FOR PULSE ECHO SCANNING

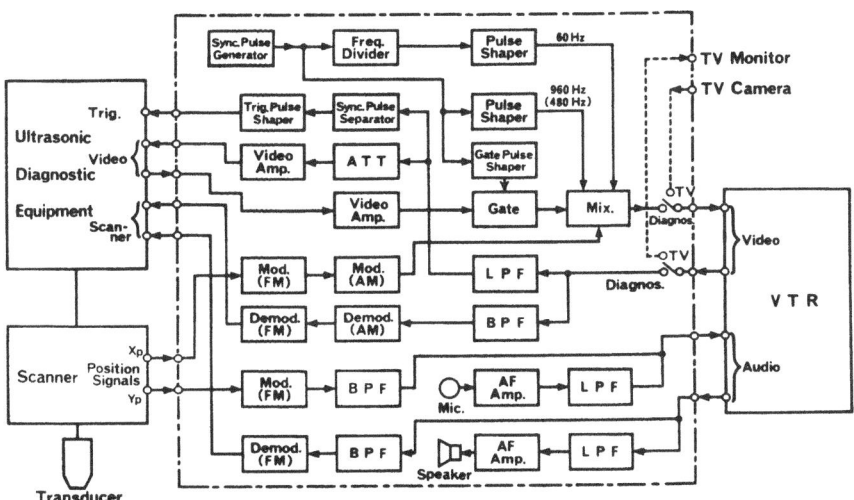

Figure 19. Interface Circuit for VTR Recording (Arc Scan)

Figure 20 shows an experimental result of a VTR recording,[13,18] and these photographs are the sensitivity graded tomograms of hyperthyroidism. The left column is made up of conventionally obtained, five dB step tomograms, and the right column is five dB tomograms obtained by the reproductive signal, recorded in 0 dB condition in the left column.

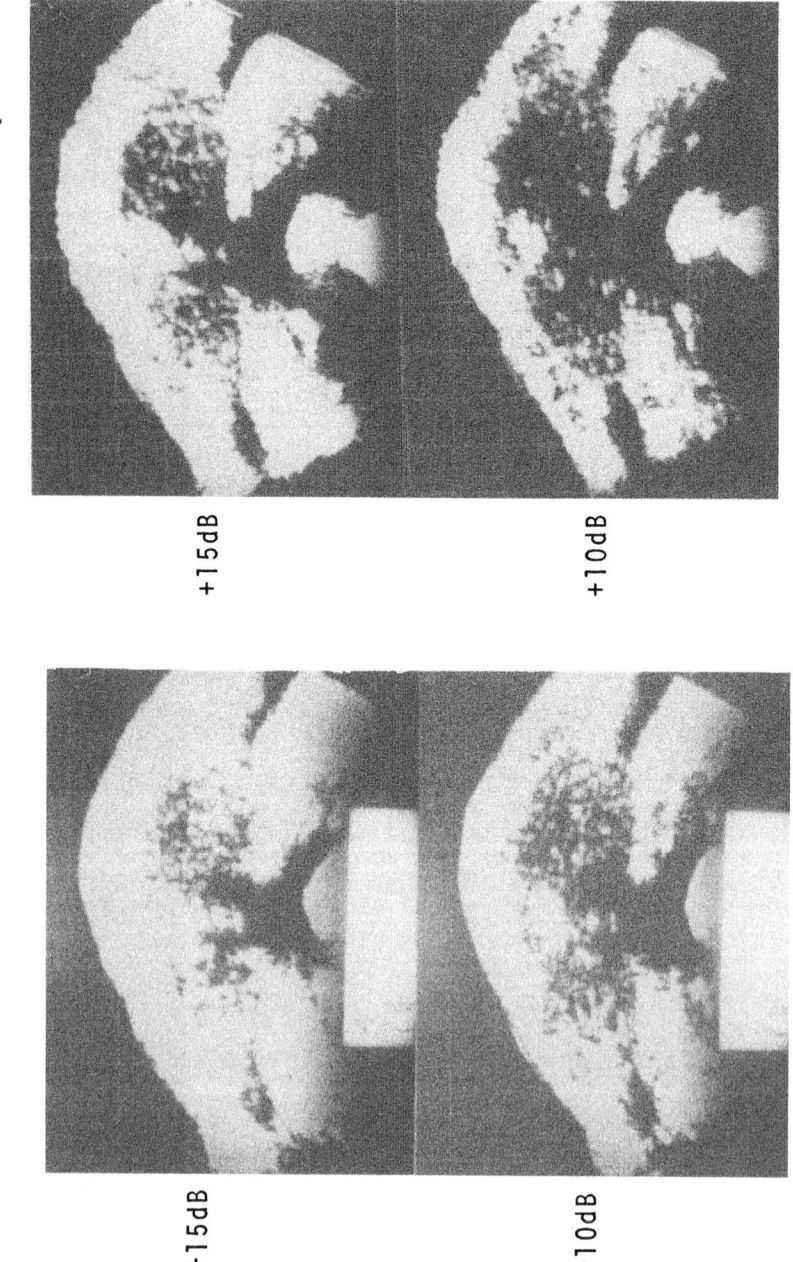

Figure 20, Sheet 1 of 2. An Example of Experimental Results

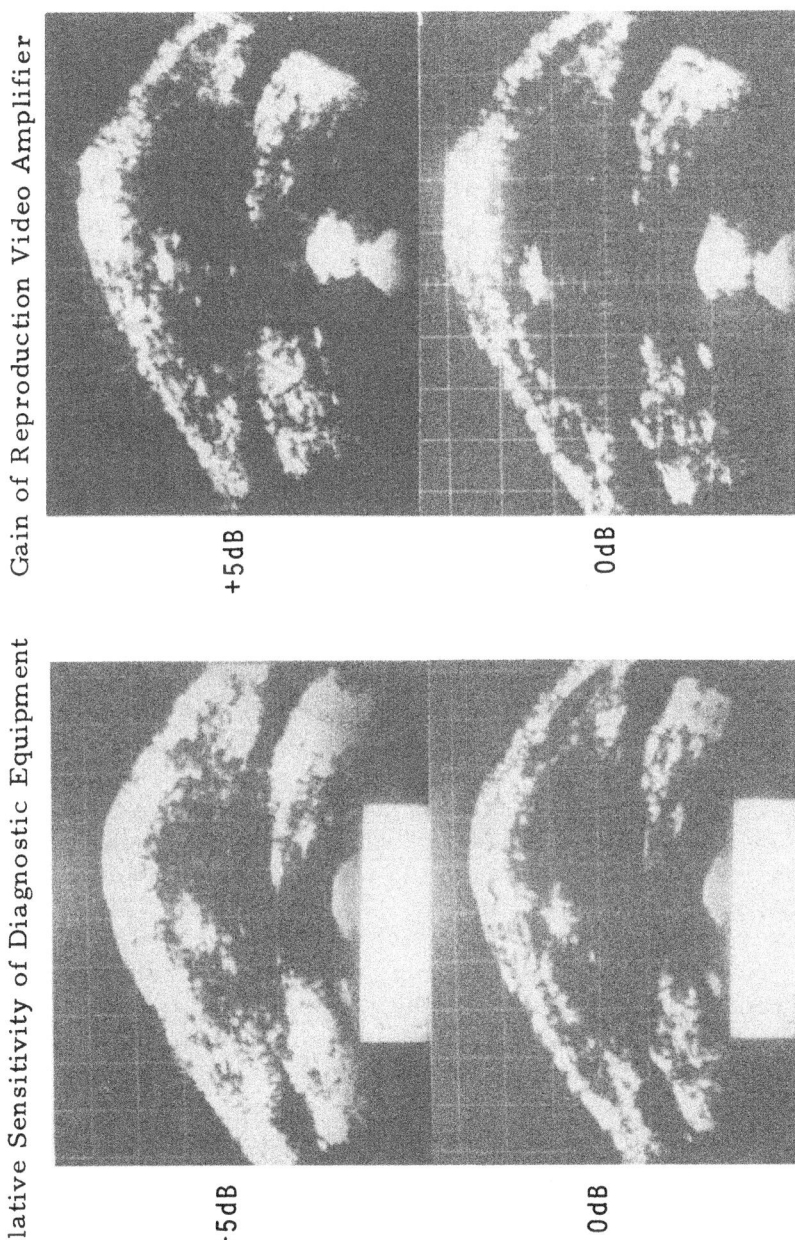

Figure 20, Sheet 2 of 2. An Example of Experimental Results

V. SYSTEMS EXPECTED FOR FUTURE DEVELOPMENT

Some practical systems used at present were described. Here I shall describe a few systems that will be developed and used in the future; also, matters to be investigated.

1. <u>Multiple Frequency Tomography</u> - As characteristics of the tissues differ according to frequency, it is desirable that diagnoses should be carried out by several frequencies of ultrasound in relation to the same object. Figure 21 shows a schematic block diagram of the ultrasonic diagnostic equipment for the multiple frequency tomography[17] for this purpose. Plural transducers of different frequencies are closely fitted up to the scanner and one multiple frequency diagnosis at a time can be provided. A harmonic drive transducer will be available for this purpose.

As scanners used at present are mechanically scanned, their speeds are very slow. Consequently, high density (more than one scanning line density of standard television) display images can be obtained even by time sharing. At the same time, it is possible to record and reproduce these time-sharing complex signals with video tape recording.

2. <u>Multiple Section Tomography</u> - Figure 22 shows a schematic block diagram of multiple section diagnostic equipment.[17]

Since mechanical scanning is slow, plural section tomograms can be obtained simultaneously by means of plural transducers fitted up to the scanner, and by switching them in time sharing. In this system, as a matter of course, the complex signals can be recorded and reproduced with VTR. Further, as this system can be used jointly with multiple frequency tomography if necessary, a great deal of diagnostic information can be obtained at a time. These systems will be very convenient in case of mass medical examinations, or in case of a test by a medical technician.

IMAGING INFORMATION PROCESSING FOR PULSE ECHO SCANNING

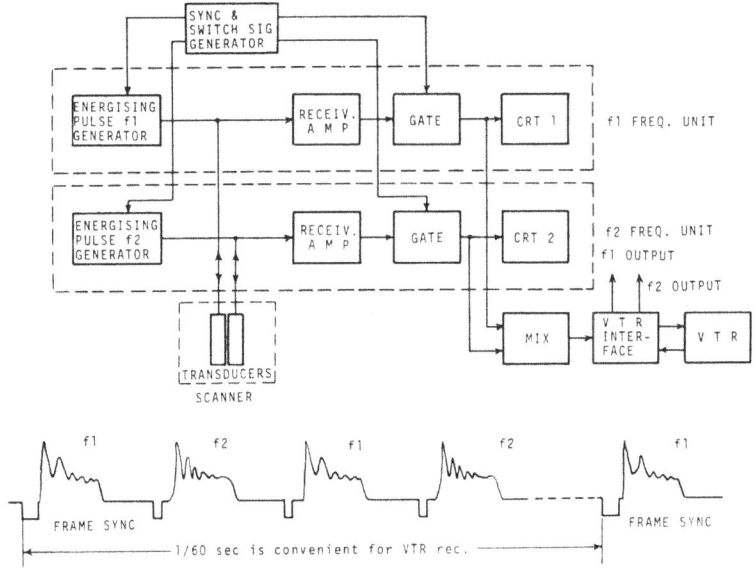

Figure 21. Multiple Frequency Diagnostic Equipment

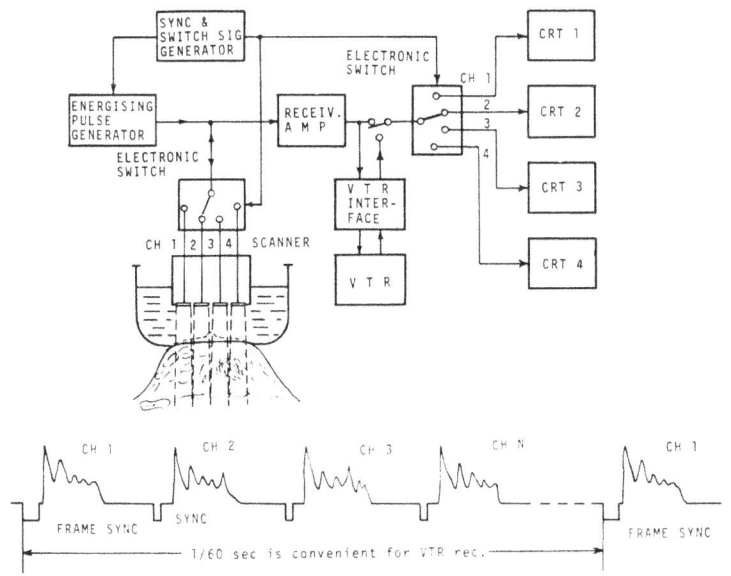

Figure 22. Multiple Section Diagnostic Equipment

3. **A Method for Measurement of Overall Sensitivity of Ultrasonic Diagnostic Equipment** - Fortunately, hazards due to the ultrasound of the diagnostic equipment have not yet been reported. However, we must continue investigation of the equipment in this respect. Studies of the dosages have been made in every place in the world. For determination of dosage, it is necessary to establish the measurement method of acoustic intensity. Also, standardization of the equipment should be hastened. As a means of standardization, I shall here describe a method for measurement of overall sensitivity of ultrasonic diagnostic equipment.

It is necessary to know the overall sensitivity of the equipment or to be able to set it at any desirable sensitivity according to diagnostic purposes, which is necessary for the universal diagnosis or for the preservation of the equipment. A standard attenuation system or a standard test piece, so to speak, is convenient for the measurement of sensitivity. However, since the adjustable range of sensitivity of diagnostic equipment is not so wide, con-

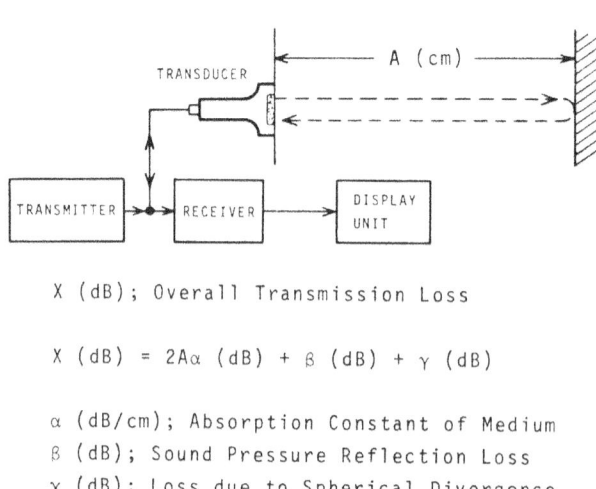

X (dB); Overall Transmission Loss

X (dB) = $2A\alpha$ (dB) + β (dB) + γ (dB)

α (dB/cm); Absorption Constant of Medium
β (dB); Sound Pressure Reflection Loss
γ (dB); Loss due to Spherical Divergence

Figure 23. Measurement of Overall Sensitivity

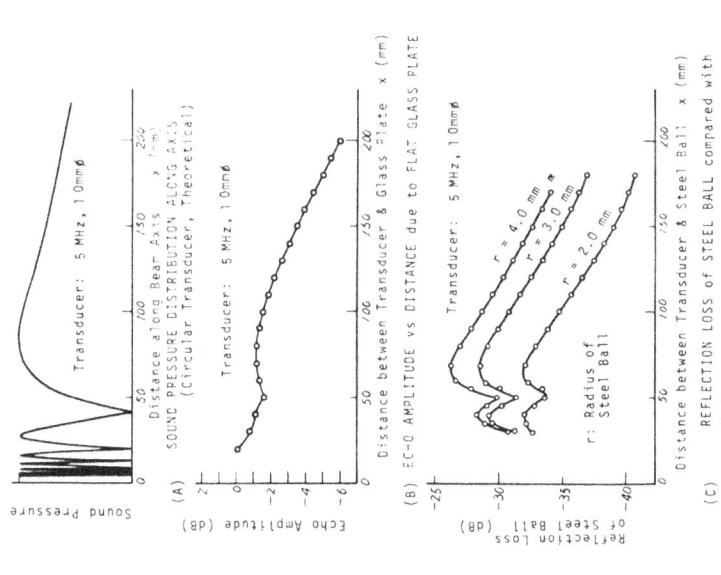

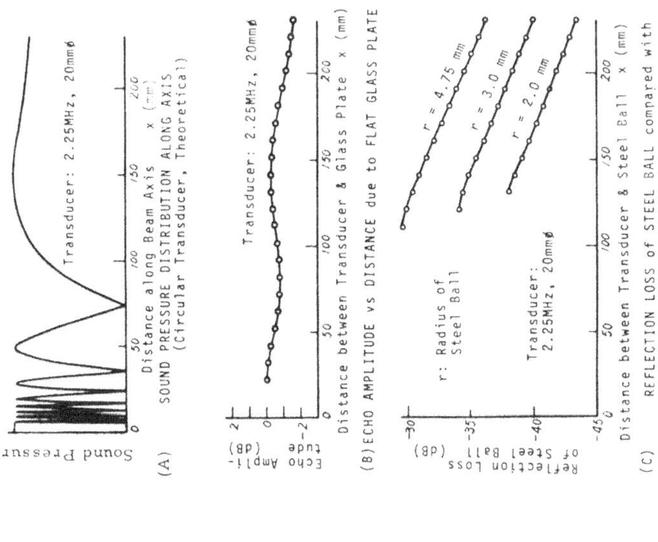

Figure 24. Theoretical Acoustic Field and Measured Results in 2.25 Megahertz

Figure 25. Theoretical Acoustic Field and Measured Results in Five Megahertz

siderably large attenuation is required for the test piece in order to carry out the test under nearly actual conditions.

We investigated a method of measurement of overall sensitivity of ultrasonic diagnostic equipment by using a reflection loss of a steel ball as a test piece. As shown in Figure 23, we define that if the fixed echo level can be obtained at the output of a piece of equipment by a reflection system which has X dB transmission loss, then the overall sensitivity of the equipment is X dB.[18]

The theoretical acoustic field (continuous wave a) and measured results (Pulses b and c in 2.25 MHz are shown in Figure 24. Results in 5 MHz are also shown in Figure 25, and these are almost the same as those in 2.25 MHz.[19,20]

Reflection loss of a steel ball was measured in comparison to the echo level reflected from a flat glass plate. The steel ball should be placed away from the last maximum distance of the acoustic field from the transducer, and it is practical to put it near the last maximum.

Measurements were performed by using transducers of 10 to 30 mm diameter in the frequency 1, 2.25, 5, 10 MHz, and steel balls of 2 to 10 mm. Figure 26 shows experimental results of the reflection loss of steel balls for this method. After calibrating diagnostic equipment by the method, detection of reflected echoes from the inner part of a body was actually carried out as the study for clinical application.

Figure 27 shows an actual view of the reflection system.

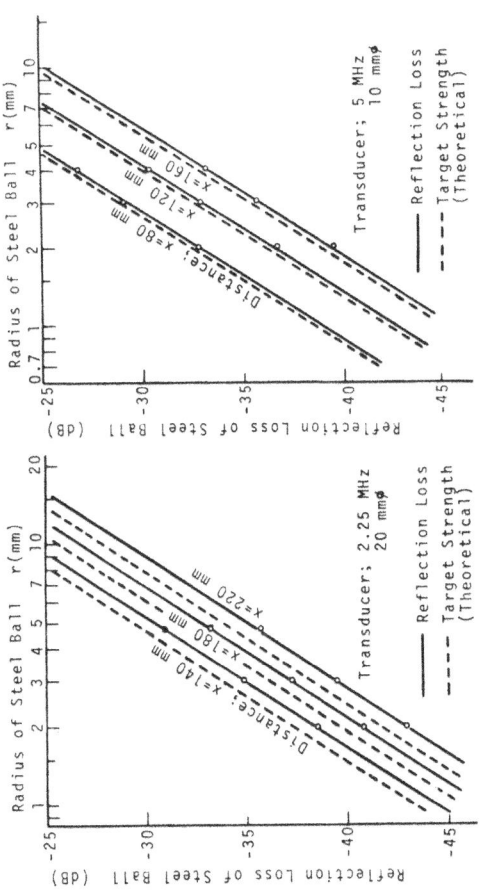

Figure 26. Measured Results of Reflection Loss of Steel Balls

Figure 27. Reflection System

VI. CONCLUSION

As mentioned herein, I have described the problems of an imaging system for medical application, mainly on the studies performed in my laboratory.

ACKNOWLEDGEMENTS

I wish to thank Mr N. Masuzawa who is a lecturer of our college, for his work on these studies.

REFERENCES

1. M. Ide, "Specification of Ultrasonic Diagnostic Equipment and the Display Pattern," Proceedings of Japan Society of Ultrasonics in Medicine (JSUM), Page 126, May, 1966.

2. Y. Kikuchi, "Physical Background of Ultrasono-Tomography," JSUM, Page 131, May, 1968.

3. M. Ide, "Investigation of Power of Expression in Amplitude in Brightness Modulation," JSUM, Page 49, May, 1967.

4. M. Ide and N. Masuzawa, "Improvement of Gradation Characteristics of Ultrasonic Diagnostic Equipment," JSUM, Page 49, November, 1967.

5. M. Ide and N. Masuzawa, "Improvement of Display Characteristics of Ultrasono-Tomogram," JSUM, Page 21, May, 1969.

6. M. Ide and N. Masuzawa, "Improvement of Display Characteristics of Ultrasono-Tomogram (2)," JSUM, Page 21, November, 1969.

7. M. Ide and N. Masuzawa, "Improvement of Visualization Characteristics in Ultrasono-Tomography," Institute of Electronics and Communication of Japan (IECJ), Report on Ultrasonics, US 69-26, Nov. 1969.

8. M. Ide and N. Masuzawa, "Electronic Color Display System for Ultrasono-Tomography," JSUM, Page 123, May, 1972.

9. M. Ide and N. Masuzawa, "Color Display System for Ultrasono-Tomography," IECJ, Report on Ultrasonics, US 72-25, November, 1972.

10. M. Ide and N. Masuzawa, "Electrical Recording and Reproduction of Ultrasono-Tomogram with VTR," JSUM, Page 33, October, 1968.

11. M. Ide and N. Masuzawa, "Electrical Recording and Reproduction of Ultrasono-Tomogram with VTR," IECJ, Report on Ultrasonics, US 68-23, Feb., 1969.

12. M. Ide and N. Masuzawa, "Electrical Recording and Reproduction of M Mode Signal of Heart," JSUM, Page 21, May, 1970.

13. M. Ide and N. Masuzawa, "Electrical Recording and Reproduction of Ultrasono-Tomogram with VTR," IEEE, to be published.

14. M. Ide and N. Masuzawa, "Electrical Recording and Reproduction of Ultrasono-Tomogram with VTR (2)," IECJ, Report on Ultrasonics, US 71-28, Dec., 1971.

15. M. Ide and N. Masuzawa, "Recording and Reproduction of Position Signals in Compound System with VTR," JSUM, Page 73, November, 1971.

16. M. Ide, N. Masuzawa, T. Wagai, et al., "Recording and Reproduction of Arc Scan Ultrasono-Tomogram with VTR," JSUM, Page 75, November, 1971.

17. M. Ide, "Present Problems and the Future of Ultrasonic Diagnosis," JSUM, 13, November, 1971.

18. M. Ide, "A Method for Measurement of Overall Sensitivity of Ultrasonic Diagnostic Equipment," JSUM, Page 53, November, 1967.

19. M. Ide, "A Method for Measurement of Overall Sensitivity of Ultrasonic Diagnostic Equipment," IECJ, Report on Ultrasonics, US 71-12, July, 1971.

20. M. Ide, "A Method for Measurement of Overall Sensitivity of Ultrasonic Diagnostic Equipment (2)" JSUM, Page 7, November, 1972.

ULTRASONIC IMAGING AT STANFORD RESEARCH INSTITUTE

Earle D. Jones

Electronics and Bioengineering Laboratory
Stanford Research Institute
(SRI)

I. BACKGROUND

For the past seven years, Stanford Research Institute has been active in the area of ultrasonic imaging. This paper is intended to give a brief description of some of the past work, a look at our current projects, and a projection for future research.

During the summer of 1967, George Eilers of SRI's Electronics and Optics Group designed and built a very accurate two-dimensional mechanical scanner. This consisted of two orthogonal lead screws that could be programmed to propel a scanning head over a rectangular raster automatically. The scanner was mounted in a fluid-filled tank and worked over a 20 cm square area with a selectable scanning resolution up to approximately 10 lines per mm. This very flexible scanner, together with a simple modulated light and film camera, could be and was, in fact, used to make acoustic holograms, transmission and reflection focused images, and phase images. It was also used as a test bed for ultrasonic components; many acoustic lenses, source and detector transducers, acoustic deflectors and gratings were tested, using this simple but very handy scanner. After many improvements and modifications, it is still being used in 1973.

Early in 1968, Philip Green joined SRI, and working with George Eilers, used the mechanical scanner to produce a few acoustic holograms and transmission images, typically of the sort that would be of significance to non-destructive testing (NDT). The scanner was designed for accuracy, not for speed. The scan rate was about one line per second and the time required for a complete frame was of the order of 10 or 15 minutes. With such long scanning time, the device was obviously not a clinical instrument, but a research tool. What was needed was a real time scanner capable of achieving the image quality that was available from the mechanical scanner.

Later in 1968, Dr Albert Macovski suggested an acoustic imaging scheme whereby a Twyman-Green interferometer could be used to convert the surface deformations caused by an incident acoustic field on a solid into an optical image. Figure 1 shows schematically the arrangement. The orthogonal reference beam in the interferometer was frequency offset with a position-modulated mirror. The perturbed surface was optically imaged onto an image dissector camera tube. The processed output of the image dissector was displayed on a conventional CRT monitor. The system produced ultrasonic images but was limited in usefulness by its very low sensitivity and the ever-present speckle in a coherent optical system. The approach was abandoned in favor of a liquid-surface system that promised improved performance. Liquid-surface imaging systems were reported as early as 1955, and offered the great advantages of simplicity, good resolution, and real-time visualization. Their chief drawback appears to be the phase-contrast optical techniques used to convert the phase variations at the surface to intensity variations. Phase contrast optical systems commonly in use do not adequately reproduce low spatial frequency variations; the resulting images typically show a "differentiated" appearance. Brendon and others have used a coherent acoustic reference wave, angularly offset from the incident object wave, causing the image information to be modulated onto a spatial frequency carrier and therefore, separable from undesired low frequency terms.

ULTRASONIC IMAGING AT STANFORD RESEARCH INSTITUTE 193

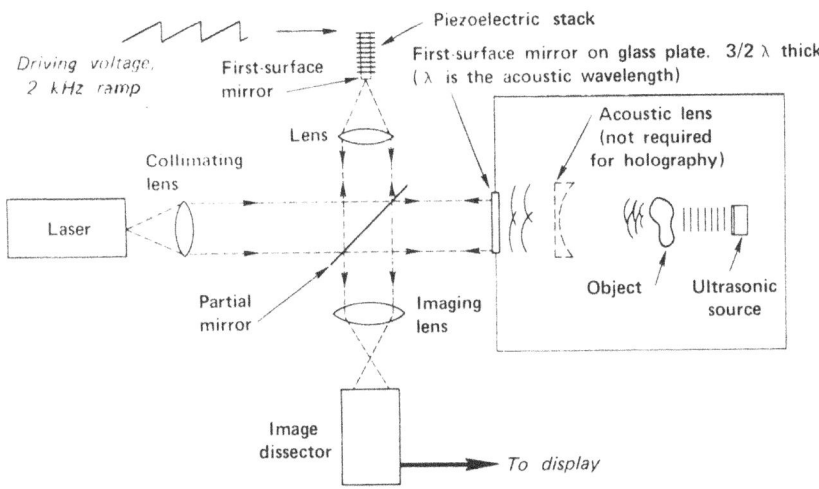

Figure 1. Arrangement for coherent optical detection of ultrasonic images or holographic fields by means of electronic scanning.

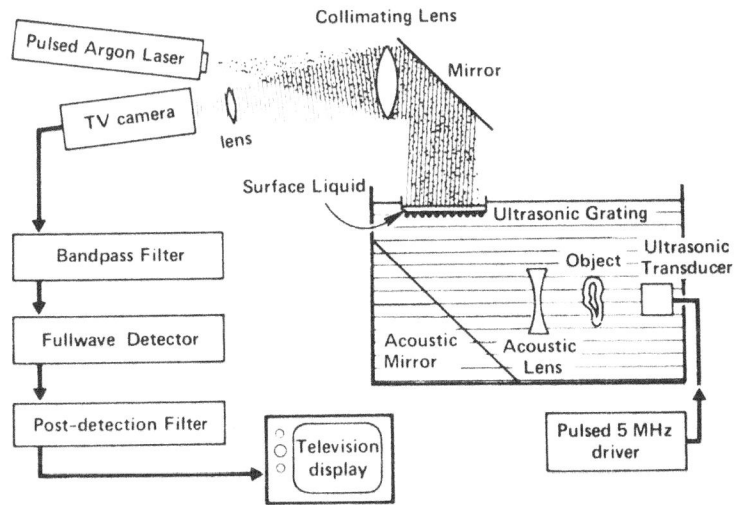

Figure 2. SRI Liquid Surface Acoustic Imaging System

The SRI liquid-surface system employed an ultrasonic grating, alternate strips of acoustically transmissive and attenuative material, at the liquid surface. Such a grating directly modulates the image onto a high frequency carrier and thereby avoids the shortcomings of the phase-contrast optics. In addition, it permits the use of incoherent acoustic energy and allows one to use large image conversion areas. Its sensitivity is somewhat reduced, however, from a coherent system. Figure 2 shows the liquid-surface system in schematic form, and Figure 3 shows the acoustic image of a plastic transistor radio case.

Beginning in 1970, ultrasonic images were produced of biological tissues using the different scanners. Many transmission images were made of excised animal organs; kidneys, brains, blood vessels, etc. The quality of these images was very high, and more than ever, the need for a clinically useful acoustic imaging camera was evidenced. Figures 4 through 6 show typical ultrasonic images of biological materials; Figure 4 is a kidney from a lamb, Figure 5 is the brain of a monkey in sagittal section, and Figure 6 is a human fetus of about 17 weeks' fetal age. In this remarkable picture, the skeletal structure is clearly visible, even though the calcification of the bone is far from complete. X-ray radiographs of this fetus show much less contrast because of the low calcium content of the skeleton at this state of fetal development.

II. CURRENT PROGRAMS

In 1970, Stanford Research Institute proposed to the National Institutes of Health the development of a general-purpose ultrasonic camera that would produce medically significant images in real time. A three-year grant was awarded and the program started in September, 1971. The proposed program called for a one-year design phase, a fabrication, packaging and test phase of about one year. This camera will be a self-contained unit employing a fluid-filled chamber and a flexible membrane for contact to the patient. Broadband insonification is used at a freq-

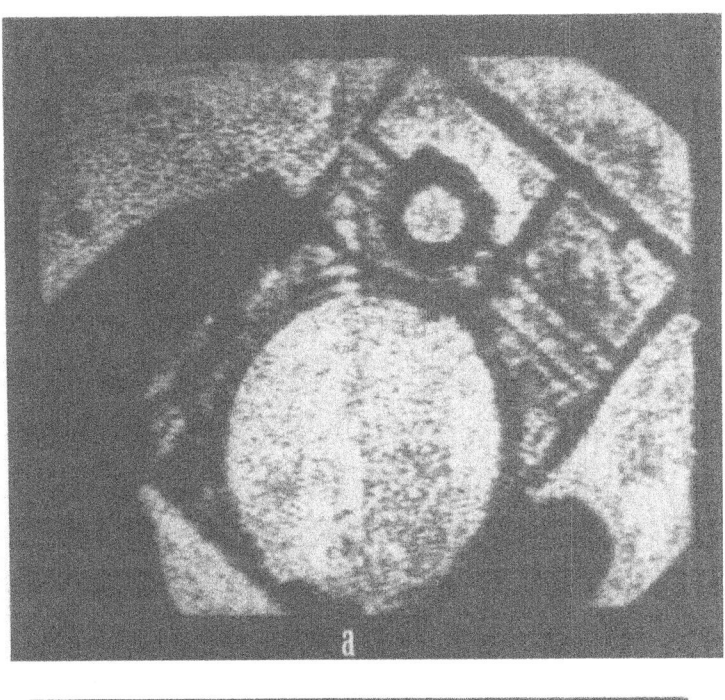

Figure 3. Acoustic Image of a Transistor Radio Case

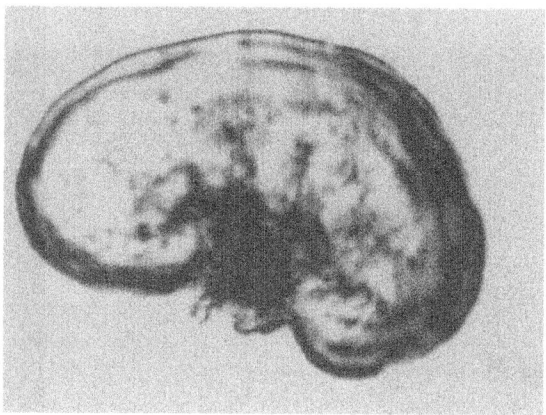

Figure 4. Acoustic Image of a Lamb Kidney

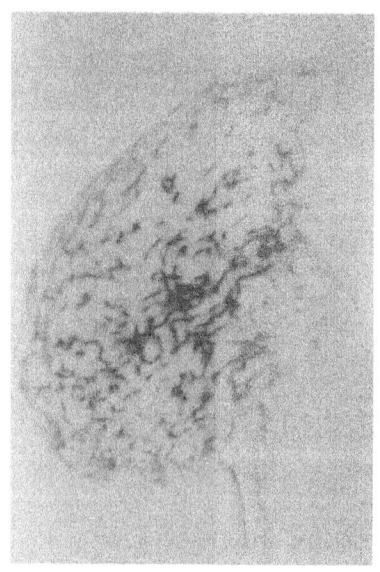

Figure 5. Acoustic Image of a Monkey Brain

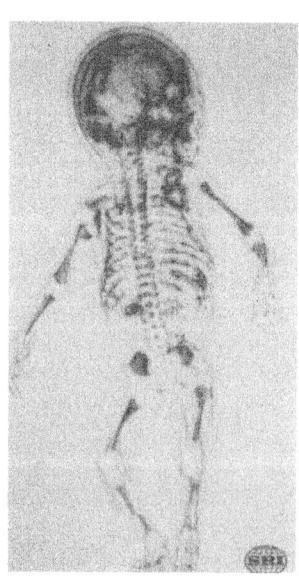

Figure 6. Acoustic Image of a Fetus

uency of approximately 3.5 MHz. A three-element acoustic lens operating at F:1 images a 15 cm square onto a linear array of piezoelectric transducers. A microcircuit preamplifier (logarithmic or linear) and integrator are employed at each transducer. The transducer array is scanned electronically in the fast direction. A novel acoustic deflector is employed to provide the slow scan component. The system is currently under test and images of high quality are being made at 15 frames per second. A Vidicon/CRT scan converter is used to produce 30 frames per second output at NTSC television standards. Because integration is provided at each transducer, the camera shows remarkable sensitivity. Very clean, low-noise images of biological tissue have been produced with acoustic power levels in the tissue of the order of one megawatt per centimeter squared. The philosophy behind the design of this ultrasonic camera is described in a paper by Green, et al.

Studies are underway whose objectives are to better understand the physical properties of biological specimens as they relate to propagating ultrasound. Other research programs are aimed at designing better acoustic components. For example, an IBM optical lens design program has been converted for use on SRI's CDC-3600 computer to design multi-element acoustic lenses. The F:1 triplet lens used in the real-time imaging camera was designed, using this program.

III. FUTURE RESEARCH

Over the next few years, SRI will continue to improve the real-time acoustic camera and develop new related systems for both clinical and NDT applications. The camera currently under development is not designed specifically for any particular portion of the human anatomy or any particular ailment. It is intended to be a general purpose clinical tool that will guide the design of future systems. New camera systems will be designed for diagnosis of specific disease states. For example, Stanford Research Institute proposed a program of research to the National

Institute of Health in 1972, with the objective of using ultrasonic imaging techniques to detect and assess the state of atherosclerotic disease. This proposal was submitted jointly with a medical clinic whose staff would provide clinical guidance during the design phase and evaluation of prototype equipment, the hope being that this very ambitious program would start early in 1973.

It will be necessary for engineers and physicists who design ultrasonic equipment to maintain the closest liaison with clinicians who must necessarily participate actively in all phases of the design. Equally important, it is necessary that those of us who work in academic and nonprofit research institutions to establish and maintain close working relationships with manufacturing organizations. The results of sophisticated research in acoustic imaging cannot influence the wellbeing of any patient until the equipment is manufactured and made available to hospitals and clinics.

IV. CONCLUSIONS

Let us recall that the objective of this seminar is.... "to formulate as explicitly as possible areas of research and development of national importance for which funding is required in academic laboratories and in university-industry cooperative projects, for the purpose of bringing about, as quickly as possible, the practical implementation of the urgently needed, medically more effective ultrasonic instrumentation." Stanford Research Institute intends to pursue vigorously, three areas of research:

1. Ultrasonic components and basic technology

2. Basic measurements on biological tissue

3. Ultrasonic imaging systems

We have already actively begun to concentrate on development of improved ultrasonic components: lenses, gratings, coatings, sources, etc. The acoustic lens designer has really only begun to use some of the sophistication at his disposal and should be able to design very high performance acoustic components. When compared to equivalent optical design problems, the acoustic designer has some clear advantages such as lack of chromatic aberrations, even with broadband ultrasound, and a very wide variety of sonically transparent materials (our lenses have index ratios of four to one at interfaces; optical lenses have about three to two).

SRI will continue to improve and miniaturize the electronic scanning circuitry to achieve lower cost and higher reliability in its systems. More funding will be sought in the area of basic tissue studies. It would be extremely helpful to the system designer to know quantitatively the physical characteristics of all types of biological tissue; absorption characteristics, acoustic impedance, the nature of the interfaces, et cetera.

Only as our knowledge grows in both ultrasonic components and tissue characteristics will we then be able to produce the clinical instrumentation to assist the physician in the attack on the diseases that beset man.

REFERENCES

1. P. S. Green, A. Macovski, and S. D. Ramsey, Jr, Coherent Optical Detection of Ultrasonic Images Using Electronic Scanning, Applied Physics Letters v 16-7, April 1, 1970.

2. B. Brenden, et al., Refinements and Variations in Liquid Surface and Scanned Ultrasonic Holography, IEEE Symposium on Sonics and Ultrasonics, New York City, September, 1968.

3. P. S. Green, A New Liquid-Surface-Relief Method of Acoustic Image Conversion, Acoustic Holography, Vol. 3, Plenum Press, 1971.

4. P. S. Green L. F. Schaefer, and A. Macovski, <u>Considerations for Diagnostic Ultrasonic Imaging</u>, Acoustic Holography, Vol. 4, Plenum Press, 1972.

OPTICAL INFORMATION PROCESSING AND ACOUSTO-OPTICS

H. Kashiwagi and K. Sakurai

Electrotechnical Laboratory, Agency of Industrial Science and Technology, Ministry of International Trade and Industry, Tanashi, Tokyo, Japan

During the present period of a shift away from a post-industrial society and the rise of an information rich society, information science, physical science and energy science are emerging as basic fundamentals which may be called the main pillars of the science of the new age; thus, the standpoint of elucidating a given object scientifically is not necessarily a single approach. Rather, these three pillars can offer three entirely different points of view.

Science is said to be the most advanced "art of knowledge" ever reached by man. Information science is a branch of learning that reflects the actual state of knowledge in its true form. The structure of information science is composed of the following three aspects: elucidation of information phenomena, which is an attempt to understand various phenomena occurring in the external world; information theory, which attempts to clarify the internal structure of the universal arrangement of information; and information technology, which is used to realize in the external world a proposed information theory with any other accumulated information. This information technology consists of four subroutines called extraction, transmission, processing and presentation of information.

The present age requires key punch tasks for man-to-machine communication information transmission; and CPU is the important field of study, although by no means a new study, having been conducted in Japan as well as in a number of other countries.

This study is aimed at finding a means of directly converting information of the external world into machine languages and artificial intelligence.

The Agency of Industrial Science and Technology at the Ministry of International Trade and Industry selected advanced information technology as one of its major projects. The name of this project is "Pattern Information Processing System." This project, which began in July, 1971, is aimed at the study and development of new information processing techniques by which pattern information such as characters, figures, object shapes and vocal sounds are to be recognized and processed. The project is to last eight years and requires a total investment of some 35 billion yen. It should be noted that the research and development will be conducted by the Electrotechnical Laboratory, with assistance from experts in the industry, academic circles and the government.

On the other hand, it is for a long-term gain rather than short-term benefit that an advanced information technology, perfected through the aggressive development of optical information processing using laser and holographic techniques; i.e., acousto-optics; is expected to play an important role.

Research and development of these fields has been conducted in the Electrotechnical Laboratory for several years. The present article is intended to review the present situation of the generation and detection of acoustical surface waves and the application of ultrasound diffraction in ETL (Electrotechnical Laboratory). These research works will contribute to the "Pattern Information Processing System."

I. GENERATION AND DETECTION OF ACOUSTIC SURFACE WAVE

The interaction of light with acoustic waves has been adopted to superimpose an electrical signal on light. Recently, attempts have been made to make acoustic surface waves in image processing by light. The experimental results of the generation and detection of acoustic surface waves are presented here.

1. Generation of an Acoustical Wave by Gun Oscillator

The most popular method of the generation of an acoustical surface wave is to feed an rf signal to an interdigital transducer deposited on the piezoelectric crystal. In this case, it is necessary that the period of the electrode is about a quarter wavelength of the acoustic wave. At high frequency, it is difficult to form an interdigital transducer, and a very advanced technique is needed.

It is well-known that the electromagnetic wave is generated from n-type GaAs when the applied electric field exceeds a threshold field. This phenomenon is attributable to the periodic motions of high field domains from cathode to anode in the diode. Since GaAs is in a piezoelectric semiconductor, the high field domain produces a strong mechanical strain, provided that the direction of domain propagation is properly chosen. Then it is expected that the mechanical strain might be extracted from the diode as ultrasonic waves with the frequency of Gunn oscillation and their higher harmonics.

In Figure 1, the structure of the sample for the generation of a bulk ultrasonic wave is shown. In Type (A),[1] the detected acoustical power was 0.4 watt/cm^2 at fundamental mode, and the second and third harmonics were lower by 14 dB and 23 dB, respectively, at more than the fundamental mode when a voltage of 270 DC was applied.

In Type B, a domain runs through along the surface of a piezoelectric medium (LiNbO$_3$). The leakage of high field domain makes stress, and it propagates in the medium. In this case, the carrier concentration, the mobility, the length and the cross-section of the diode was $1-3 \times 10^{14}$/cm^3, 6,800-7, 100 cm^2/V. sec., 300 μm and 0.7×0.7 mm^2, respectively, and the frequency of Gunn oscillation was 355 MHz and propagated longitudinally to <111> direction.

In the case of the acoustical surface wave also, there are two types of oscillator displacement, as shown in Figure 2 (A) and (B). In this case, Type (A) is more effective than Type (B); so here we present the experimental data concerning Type (A).[2] The Gunn oscillator used was n-type GaAs, whose carrier concentration and mobility were 2×10^{14}/cm^3 and 6000 cm^2/V seconds, respectively. When the supplied electric field exceeded a critical field of about 3,000V/cm, oscillation of the frequency of 105 MHz was obtained. The diode was put on a Y-Z LiNbO$_3$ crystal. In order to put the diode as close as possible to the surface of the substrate, both the diode and the substrate were mirror-lapped. The supplied pulse was 0.3-1 second width, and 30-100 Hz repetition frequency.

The experiments were performed in the following four cases: The Gunn oscillator was placed, (A), on the surface of the substrate directly; (B), across, on the edge of an Au-film evaporated on the substrate surface; (C), on periodic Au-film strips with the width and spacing of 25 µm (in this case, the periodicity is 50 µm); and (D) on periodic Au-film strips with the width and spacing of 50 µm (in this case, the periodicity is 100 µm). These configurations are illustrated in Figure 3.

In Figure 4, the acoustic amplitude is plotted as a function of the bias voltage for the case of (D). The acoustic amplitude increased with the increase of the bias voltage above the threshold and tends to saturate with a further increase of the bias voltage. When the polarity of the field was reversed, no significant change in the acoustic power was observed. For the other case, similar results were obtained. For Case (A), by moving the Gunn oscillator against the detector and measuring the delay time of the signals, the velocity of the waves was estimated at 3400 ± 50 m/sec. For Case (B), the Gunn diode was moved across the boundary of Au-film as shown in Figure 5. When the Gunn diode was put inside the boundary of Au-film ($D \leq 0$; D is the distance between an electrode of diode and the edge of the film) the acoustic signal could not be detected. When the diode was placed across the boundary ($0 < D < 1.0$ mm), strong signals were obtained and the amplitude did not change with D. When the diode was placed completely outside the film, the detected signal became considerably smaller than in the former case.

The relative acoustic powers detected for four cases at an applied electric field of 350 volts are tabulated in Table 1, by taking the output of the case of (A) as 0 dB. From this table it is seen that when the diode is placed on a metallic array with a width of one and a periodicity of L chosen to satisfy the following condition, the output of the surface waves is effectively enhanced.

$$\left. \begin{array}{l} \dfrac{1}{\lambda} = \text{half-integer} \\[1em] \dfrac{L}{\lambda} = \text{integer} \end{array} \right\} \quad (1)$$

OPTICAL INFORMATION PROCESSING AND ACOUSTO-OPTICS

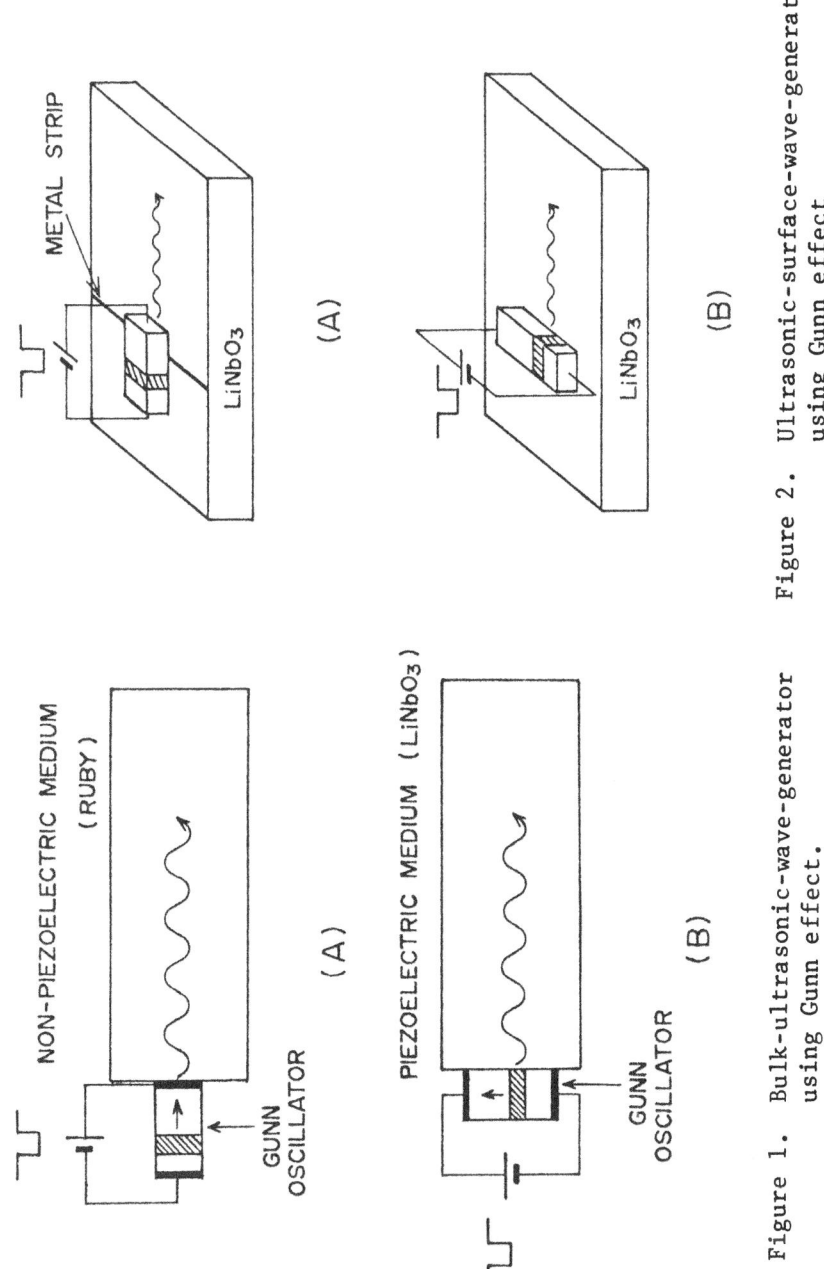

Figure 1. Bulk-ultrasonic-wave-generator using Gunn effect.

Figure 2. Ultrasonic-surface-wave-generator using Gunn effect

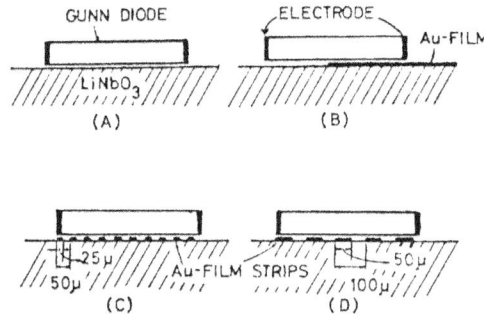

Figure 3. Four configurations used in the experiments

© "Generation of Surface Waves by Gunn Oscillator"
Supplement to the Journal of the Japan Society
of Applied Physics Vol.40, 1971, p.144
Hisao Hayakawa, Susumu Takada, Takehiko Ishiguro
and Noboru Mikoshiba
Electrotechnical Laboratory, Tanashi, Tokyo

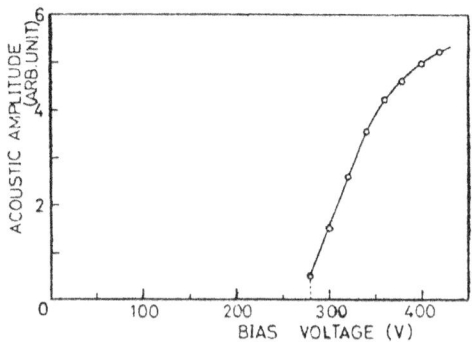

Figure 4. Acoustic amplitude (arbitrary unit) as a function of the bias voltage of the Gunn oscillator

© "Generation of Surface Waves by Gunn Oscillator"
Supplement to the Journal of the Japan Society
of Applied Physics Vol.40, 1971, p.144
Hisao Hayakawa, Susumu Takada, Takehiko Ishiguro
and Noboru Mikoshiba
Electrotechnical Laboratory, Tanashi, Tokyo

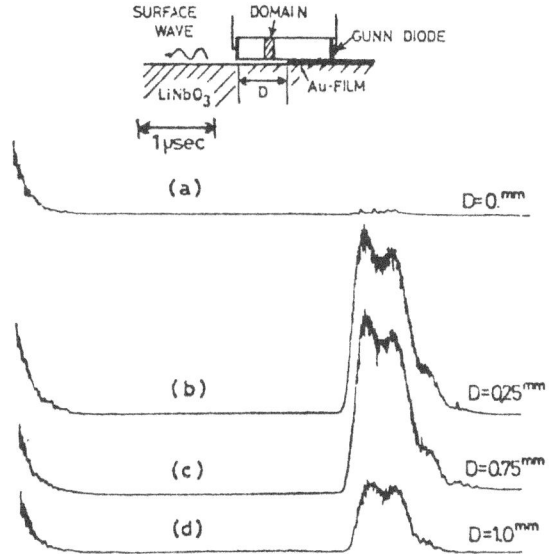

Figure 5. Acoustic signals obtained by moving the Gunn oscillator near the edge of the metallic film coated on the $LiNbO_3$ surface.

© "Generation of Surface Waves by Gunn Oscillator"
Supplement to the Journal of the Japan Society
of Applied Physics Vol.40, 1971, p.144
Hisao Hayakawa, Susumu Takada, Takehiko Ishiguro
and Noboru Mikoshiba
Electrotechnical Laboratory, Tanashi, Tokyo

Table 1. Comparison of relative acoustic power radiated from four configurations

Configuration	Number of Strips	Relative power	Frequency	Bias Voltage
(A)	∕	0 dB		
(B)	∕	15 dB	105 MHz	350 V
(C)	15	23 dB		
(D)	9	35 dB		

© "Generation of Surface Waves by Gunn Oscillator"
Supplement to the Journal of the Japan Society
of Applied Physics Vol.40, 1971, p.145
Hisao Hayakawa, Susumu Takada, Takehiko Ishiguro
and Nobuo Mikoshiba
Electrotechnical Laboratory, Tanashi, Tokyo

Here, λ is the wave length of acoustic wave. For the case (B), (C), (D), the propagation direction of the acoustic waves is confined to the direction normal to the strips. One of the characteristic features of this acoustic wave generation is that the pulsed rf acoustic waves are obtained merely by feeding a DC pulse; also, the size is small. Another feature is that the generator is movable on the surface along the propagation direction of the waves and is capable of variable delays of the signals.

2. <u>Active GaAs Diode Ultrasonic Detector</u> - A subcritically doped GaAs diode (carrier concentration (n) times length (L) of the diode; i.e., n.L products are less than $10^{11} cm^{-2}$) has differential negative conductivity.

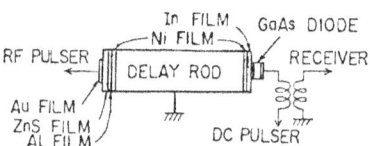

Figure 6. Experimental arrangement.

© "Active GaAs Ultrasonic Detector"
Supplement to the Journal of the Japan Society of Applied Physics Vol.39, 1970, p.56
Hisao Hayakawa, Takehiko Ishiguro, Susumu Takada, Nobuo Mikoshiba and Makoto Kikuchi
Electrotechnical Laboratory, Tanashi, Tokyo

We examined this type of diode to use as the ultrasonic detector. For the bulk ultrasonic detector,[3] the experimental arrangement is shown in Fig. 6. The diode used was an oxygen doped GaAs whose resistivity, mobility, length and cross section were 1,070 Ω-cm, 6,000 cm^2/V. sec, 470 μm and 0.5x0.5 mm^2, respectively. Feeded acoustical wave was longitudinally propagated and their frequency was varied from 200 MHz to 700 MHz. D.C. bias was applied from 40 volt to 400 volt between the two ends of the diode.

In Fig. 7, the bias-voltage dependence of the output signal levels are shown. Above a critical frequency f_c ($f_c = \frac{V_p}{L}$; V_p is almost the same as the velocity of the domain.

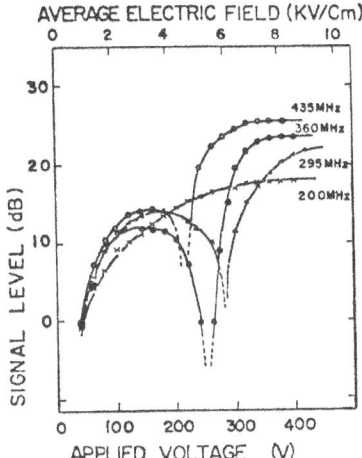

Figure 7. Output signal vs. applied bias-voltage at various frequencies. (The signal level at 40V is taken as 0 dB.)

(C) "Active GaAs Ultrasonic Detector"
Supplement to the Journal of the Japan Society of Applied Physics Vol.39, 1970, p.57
Hisao Hayakawa, Takehiko Ishiguro, Susumu Takada, Nobuo Mikoshiba and Makoto Kikuchi
Electrotechnical Laboratory, Tanashi, Tokyo

In this case f_c = 280 MHz), the characteristic curve is separated into two branches. The signal levels take a maximum at about 3,600 V/cm and decrease sharply at about 4,000 V/cm. This feature, observed in the lower bias region, is tentatively referred to as 1st modes. As the bias voltage is further increased, the output levels again increase abruptly. Such a feature found in the higher bias voltage region is referred to as 2nd mode.

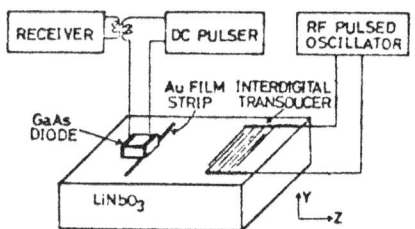

Figure 8. Experimental arrangement.

© "Detection of Surface Waves with GaAs Bulk Effect Diode" Applied Physics Letters Vol.18, No.1, 1971, p.31
Hisao Hayakawa, Susumu Takada, Takehiko Ishiguro and Nobuo Mikoshiba
Electrotechnical Laboratory, Tanashi, Tokyo

In Fig. 8, the experimental arrangement for the detection of the surface wave is shown.[4] In this case, the diode with length 1 mm, cross section 0.6x0.25 mm^2, resistivity 500 Ω-cm, and mobility 6,000 cm^2/V. sec was used. The acoustical surface wave propagated at Z direction on the Y cut LiNbo$_3$ and this frequency was varied from 105 MHz to 300 MHz. In Fig. 9, the dependence of the signal levels on the applied electric voltage is shown.

When the applied field exceeded a critical field of about 3 kV/cm, which was somewhat lower than that of the current saturation in the current-voltage characteristic, the signal levels increased abruptly with the applied electric field and tend to saturate in the higher bias region. In this case, the 1st mode which appeared in the bulk wave detector was not present.

The features of this type detector are, (1), small size; (2), high sensitivity of the diode, since the amplification of the signals simultaneously takes place in a diode; (3), broadband; and (4), that for the surface wave, the diode is easily movable on the surface of the substrate, if the metal strip is constructed on the diode.

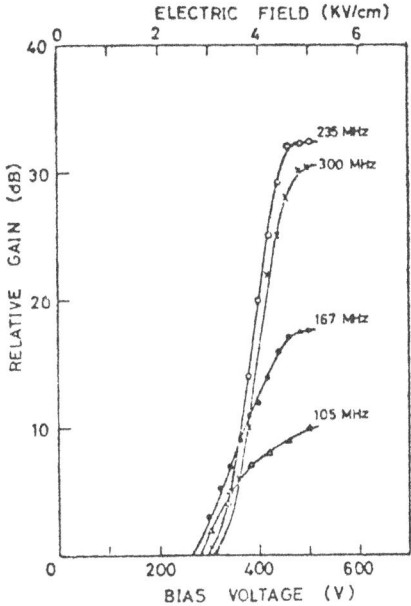

Figure 9. Bias voltage dependences of the detected signal levels for various frequencies of the surface waves.

© "Detection of Surface Waves with GaAs Bulk Effect Diode" Applied Physics Letters Vol.18, No.1, 1971, p.31
Hisao Hayakawa, Susumu Takada, Takehiko Ishiguro and Nobuo Mikoshiba
Electrotechnical Laboratory, Tanashi, Tokyo

II. APPLICATIONS OF ULTRASOUND DIFFRACTION

As the application of the interaction of light with the acoustical surface wave, Bragg deflection of optical guided waves in thin film and optical guided wave mode conversion have been recently demonstrated. In ETL there are two typical applications concerning ultrasound diffraction; one is the active AM mode locking by using acoustic surface wave, and the other is ultrasonic imaging using light diffraction by low frequency ultrasound.

1. <u>Application of the Acoustical Surface Wave</u> - In general, the bulk ultrasonic standing wave [12] has been used as an active

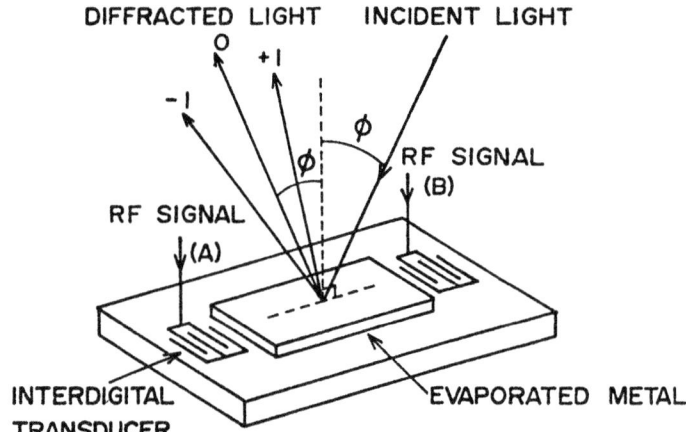

Figure 10 (A). Model of standing wave modulator.

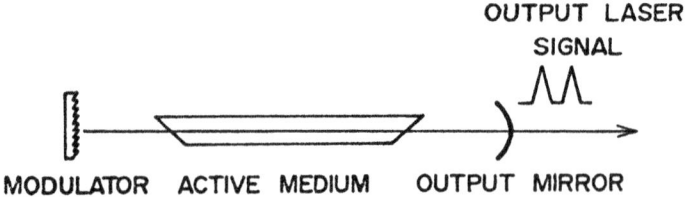

Figure 10 (B). Arrangement of mode locking laser.

AM mode-locking device. But in that case, it is necessary that the crystal for the modulator has high optical properties and the surface of the crystal is antireflection coated to reduce reflection loss. Moreover, the crystal must be exchanged against the frequency of light used because of the absorption characteristic of the crystal. To overcome these defects, one method is to make use of the reflection at the surface of the crystal.

In this section, we shall discuss the capability of the acoustical surface wave as a mode-locking device.

In Fig. 10 (A), the model of this type of the mode-locking device is shown. The rf signal is fed from the A and B interdigital transducers and the standing wave is constructed at the central region at angle ϕ against the incident face which is Raman-Nath diffracted. The diffracted light is amplitude modulated at the same frequency as that of the standing acoustic wave.

According to R.J. Hallermeier,[13] the intensity of the 0th diffracted light is following as the incident angle which is $\phi = 0$ and the incident light intensity is 1.

$$I_o = 1 - \frac{W^2}{4} - \frac{W^2}{4S^2} - \frac{W^2}{2S} \cos 2 W{*}t \qquad (2)$$

Here, $W = 2ak$
a : amplitude of the acoustical wave (i.e. displacement of the surface)

k : wave number of incident light

W* : angle-frequency of acoustic fundamental wave[15]

S : standing wave ratio

From this, the depth of modulation m follows

$$m = \frac{I_{omax} - I_{omin}}{I_{omax}} = \frac{\frac{W^2}{S}}{1 - \frac{W^2}{4} - \frac{W^2}{4S^2} + \frac{W^2}{2S}} \qquad (3)$$

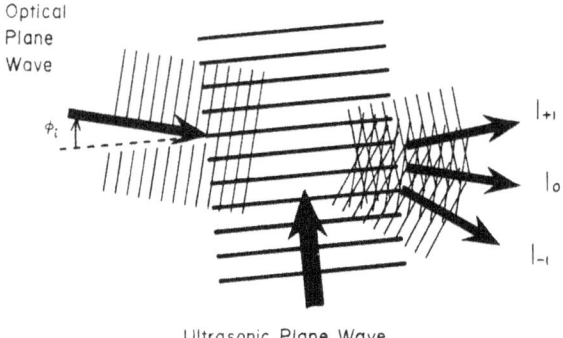

Figure 11. Diffraction of optical plane wave by ultrasonic plane wave.

Table 2. Absorption of ultrasound in living body.

Tissue	Frequency (MHz)	Absorption Coefficient (Neper/cm)
blood of man	0.7	0.013
	1	0.017
	2	0.046
	4	0.115
	7	0.230
	10	0.345
skull bone of man	0.3	0.17
	0.6	0.52
	0.8	0.92
	1.2	1.7
	1.6	3.2
	1.8	4.3
	2.25	5.3
	3.5	7.9
fat of pig	1.6	0.07
	2.5	0.2
	4.0	0.4
	6.0	0.6
	7.0	0.8
liver of cow	0.8	0.07
	1.5	0.2
	2.4	0.2
	4.5	0.35
	10	1.37
	23	3.0

At $W = 0.2$, $S = 1$ (this value is corresponding 200 Å peak-to-peak surface corrugation at $\lambda = 6328$ Å), the depth of modulation $m = 0.04$.

This value of the depth of modulation is sufficient to AM-mode-locking, though the threshold depth of modulation for mode-locking is very dependent on the gain of the active medium.

As shown in Fig. 10 (B), when cavity is constructed so that the diffracted 0th light may be feed-backed to the laser medium, the mode-locking laser pulse will be obtained. The greatest feature of this method is applicable to any wavelength of light.

2. <u>Ultrasonic Imaging Using Light Diffraction by Low-Frequency Ultrasound</u> - A new method of ultrasonic imaging called "Bragg Imaging" because it applied acousto-optical Bragg diffraction was first described by A. Korpel[7,8] and has been studied by various authors from different points of view.[9-11] When we can put this ultrasonic imaging method to practical use and obtain ultrasonic real-time pattern information, it will have wide application in the various fields such as biological metrology, medical diagnoses, underwater imaging and nondestructive testing. The absorption of ultrasound increases in proportion to $(\text{frequency})^{1-2}$; hence, imaging methods as in the lowest possible frequency region is in great demand. For instance, absorption coefficients of ultrasound in several tissues of living body are tabulated in Table 2.[15] When high-frequency ultrasound is applied to the interior of the body, it must apply high-power ultrasound in a living body and there is the risk that tissues may be affected or damaged.

On the other hand, in application to underwater or underground exploitation, it is impractical to use ultrasound, the frequency of which is over 10 Megahertz. In consideration of these conditions, we shall try to expand Bragg imaging to the lower frequency range of ultrasound.

2.1 <u>Expansion of Bragg Imaging to Lower Frequency Range</u>[16]

In considering the interaction between complicated ultrasound and optical fields, we expand each field into plane waves and investigate a simple model as shown in Figure 11. In Figure 11, ϕ_i is an incident angle of optical plane wave to ultrasonic plane wave.

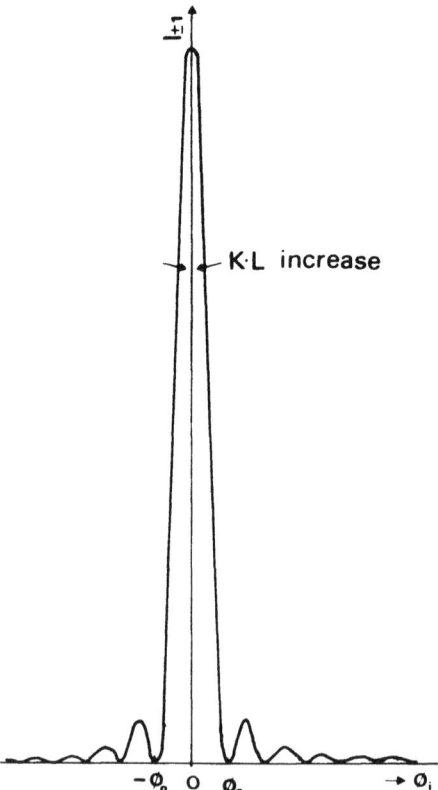

Figure 12. Dependence of diffracted light intensity $I_{\pm 1}$ on incident angle i.

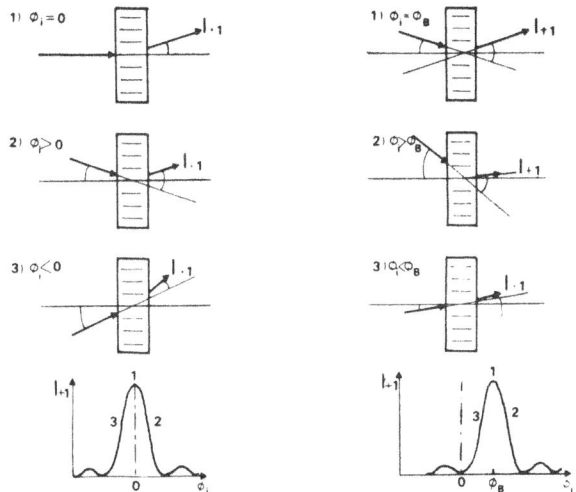

Figure 13. Incident angle dependence: a) Raman-Nath diffraction; b) Bragg diffraction.

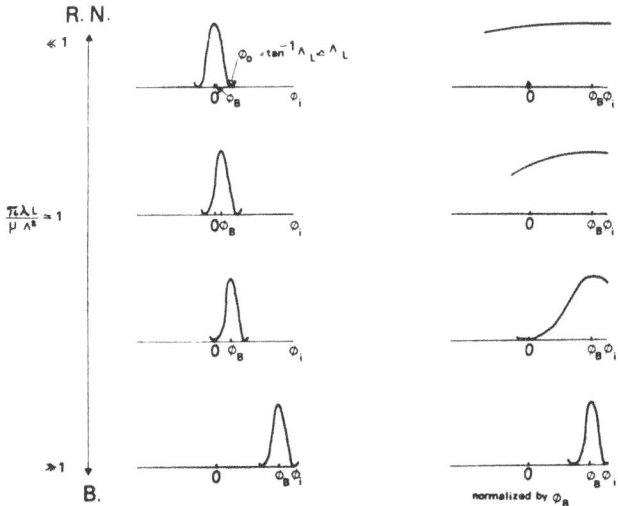

Figure 14. Relation between R.N. imaging and B. imaging.

In the following analysis, we shall deal with the range in which Raman-Nath's approximation of phase diffraction grating, where the frequency of ultrasound is lower than that in the range in which typical Bragg diffraction is observed and the inequality

$$\pi \lambda L / \mu \Lambda^2 \ll 1, \tag{4}$$

holds, where λ and Λ are the wavelength of light and ultrasound respectively, L is the effective width of the ultrasound in which light is diffracted and μ is the mean refractive index of the medium. In this case, the intensity of the plus or minus first order diffraction light may be written as

$$I_{\pm 1}(\phi_i) = J_1^2(k\Delta\mu L \sec \phi_i \, \text{sinc}(\frac{KL}{2} \tan \phi_i)), \tag{5}$$

where $J_1(x)$ is the first order Bessel function, $k=2\pi/\lambda$, $K=2\pi/\Lambda$, are wave number of light and ultrasound respectively, μ is the maximum variation of refractive index caused by ultrasound and $\text{sinc}(x) = \sin x/x$. As the value K.L becomes large, $I_{\pm 1}(\phi_i)$ tends to be characteristic of delta-function (Figure 12).

This may allow us to assume that diffraction occurs only when $\phi_i = 0$; that is, wavefront of ultrasound is perpendicular to that of light. The ultrasonic imaging in the range where the above assumption holds will be called "Raman-Nath Imaging". Of course, Raman-Nath imaging and Bragg imaging merge with each other according to certain conditions (such as the frequency of ultrasound). The relation between Raman-Nath imaging and Bragg imaging is shown in Figures 13 and 14. Two-dimensional imaging rules can be proved in almost the same way as that of Korpel. For instance, in Figure 15, a proof of two-dimensional imaging of an ultrasound point source (using ray optics) is depicted. S' and O are the point sources of ultrasound and light, respectively. According to the above assumption, ultrasound from S' diffracts light from O at limited points such as X', X'', X''' ...... where the condition $\phi = 0$ holds. Because diffraction angles at every point, X', X'', X''' ..... are the same, it is easily shown by using elementary geometry that every diffracted light seems to leave Point S. Although S is an optical virtual point image of O, we can regard S as an image of S' (and imaging is proved).

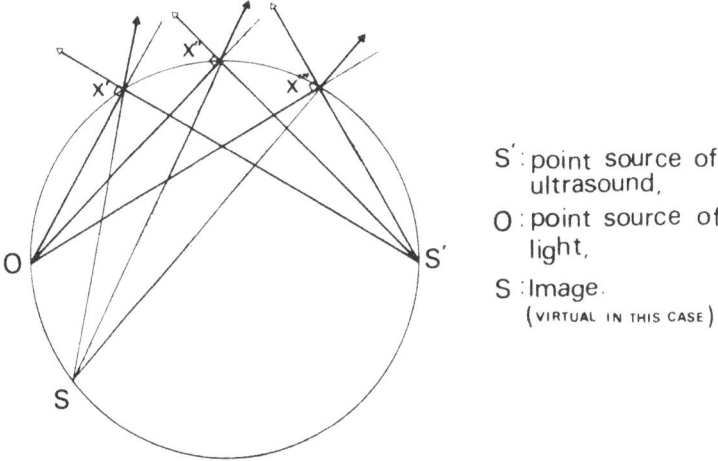

Figure 15. Imaging of ultrasound point source.

The experimental arrangement is very similar to that of the Bragg imaging, Figure 16. A monochromatic light wave is converged to a line image which is perpendicular to a paper by a cylindrical lens, L_1. The object to be imaged is illuminated by ultrasound. The diffracted, reflected or transmitted ultrasound diffracts light to make an optical image of the object. Cylindrical lens, L_2, plays the role of projecting the image onto a screen. One example of preliminary experimental results is shown in Figure 17. The object to be imaged is a rubber-made O-ring whose internal diameter is 4.5 mm, illuminated from the backside with an ultrasound plane wave whose frequency is 2.5 MHz, corresponding to a wavelength of 0.6 mm in water. The lens used for imaging is a very inexpensive one whose numerical aperture is about 0.1, and whose aberrations are not compensated. Other parameters are:

$$\lambda = 630 \text{ nm}, \quad L = 18 \text{ mm}, \quad \text{and} \quad \Delta \mu \simeq 3 \times 10^{-6}$$

Since in this case the demagnification factor is very small, $\lambda/\Lambda \simeq 10^{-3}$, and we cannot compensate vertical-lateral distortion ratio of the image by a lens system, 1:1 vertical-lateral ratio is oblique to the optical axis; hence, deterioration of image quality is introduced. Vertical fringes superimposed on the image are caused by diffraction at the edge of a slit for stopping zero-order light.

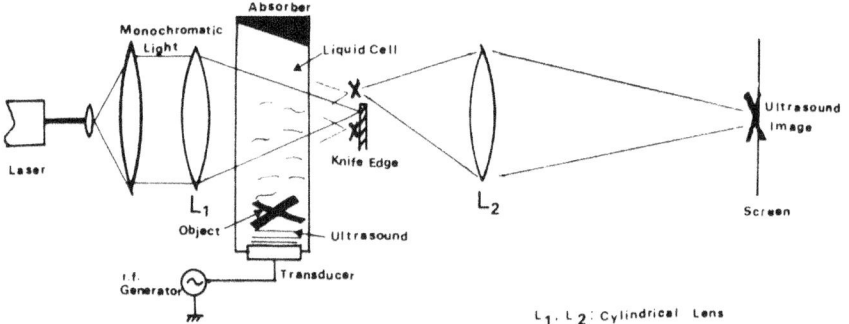

Figure 16. Experimental arrangement.

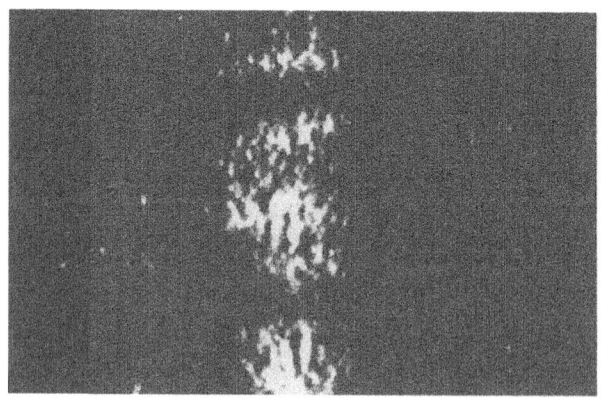

4.5 mm

Figure 17. Ultrasonic image of rubber-made O ring.

2.2 Consideration of Resolution Power[17] - Since in this imaging method, pattern information of an object is obtained by illuminating ultrasound "wave" to the object, the resolvable distance is limited by the wavelength of the "wave". In this section we shall consider a few parameters which determine the resolution power in the direction parallel to that of light propagation[18] and will discuss practical problems.

2.2.1 Angular Resolution Power Limit - When ϕ_i, which gives the first minimum value of Equation (5), is called ϕ_o, as shown in Figure 12,

$$\phi_o = \arctan \Lambda/L \tag{6}$$

This gives the width of a kind of impulse response of this imaging[19] and limits the angular resolution as shown in Figure 18. When the ultrasound, the width of which is 18 mm and the frequency, 2.5×10^6 Hz, and water ($23°C$) as a medium is used, the wavelength of ultrasound becomes 0.6 mm and the theoretical value of ϕ_o is calculated from Equation (4) to be $1.91°$.

2.2.2 Resolution Limit Due to Quality of Cylindrical Lens
The deviation of a cylindrical lens used for producing an optical line image from an ideal lens makes it impossible to image a line image of infinitesimal width, and makes worse the resolution power of a total image (Figure 19).

When Δx is the deviation of an actual line image from an ideal line image, Δy corresponds to the deviation distance in the object space:

$$\Delta y = \frac{\Lambda}{\lambda} \Delta x \tag{7}$$

When a wavelength of light is 600 nm and other conditions are the same as in the above section, in order to make the deviation in object space smaller than one millimeter, the deviation of the actual line image must be smaller than one micrometer. The limit due to the numerical aperture of a lens is discussed in the same manner.

If a numerical aperture (N.A.) of a cylindrical lens is small, a light wave component which should be diffracted by ultrasound does not exist. The diffracted light by a lower component of spatial frequency of object, other than a certain component, is imaged (Figure 20), and minimum resolvable distance, d_{min}, will be given as follows:

$$d_{min} = \begin{cases} = \dfrac{\mu^2 \Lambda}{2\sin\theta\sqrt{\mu^2-\sin^2\theta}} & (\sin\theta \leq \dfrac{\mu}{\sqrt{2}}) \\ = \Lambda & (\sin\theta > \dfrac{\mu}{\sqrt{2}}) \end{cases} \quad (8)$$

The case of $\mu = 1.33$ (water) is shown in Figure 21. This figure shows that in order to make $d_{min} = 3$, we must use a lens of f/2.

It finally became evident that in order to improve the resolution power when the frequency of ultrasound is restricted within lower range, one should take a width of ultrasound as large as possible. Another important conclusion is that qualification of a cylindrical lens for producing an optical line image is very severe.

2.3 A Proposal for Optical Holographic Improvements of Signal-to-Noise Ratio[20] - In the ultrasonic imaging method aforementioned, the frequency of light which is a carrier of image information shifts from the frequency of incident light, $W/2\pi$ by the frequency of ultrasound, $\Omega/2\pi$.

On the other hand, most of the background light which acts as a noise against the desired image is due to so called Rayleigh scattering (the Tyndall phenomenon in ultrasound medium, scattering on optical components, etc.) which does not change the frequency of light from $W/2\pi$. It has an important effect on the image especially when low frequency ultrasound is used. Scattering which shifts the frequency of light, such as Brillouin scattering, also exists.

The typical example of optical spectra of the image is illustrated in Fig. 22.

When the SN ratio is defined as the ratio of the image component to all other components, the image obtained by the ultrasonic imaging should be passed through a narrow bandpass filter for improvement of the SN ratio.

OPTICAL INFORMATION PROCESSING AND ACOUSTO-OPTICS

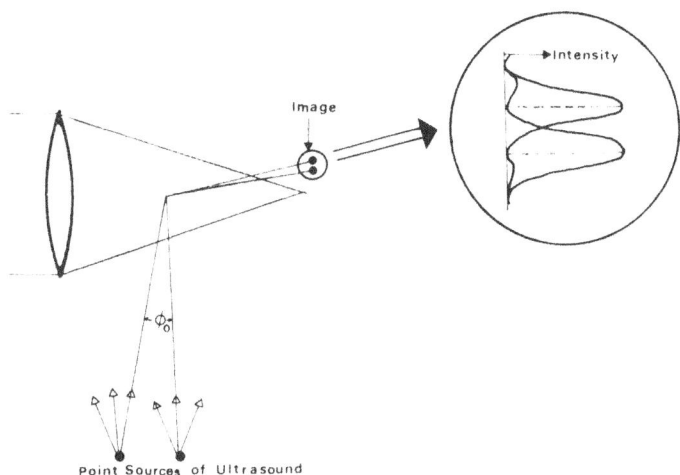

Figure 18. Angular resolution power limit.

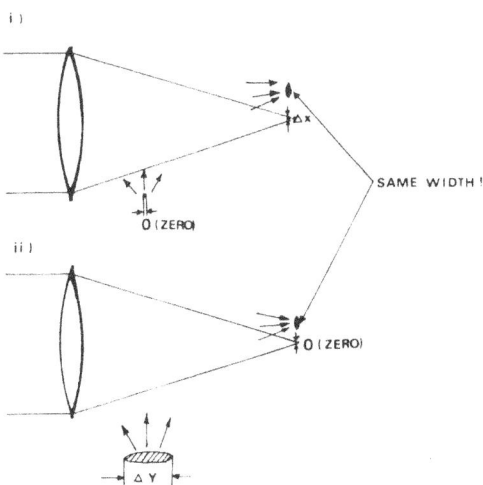

Figure 19. Resolution limit due to the quality of a lens.

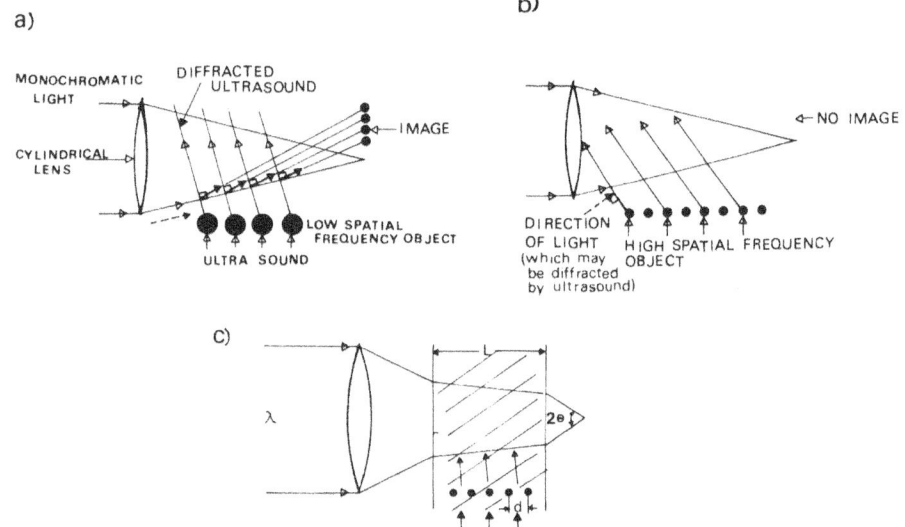

Figure 20. Resolution limit due to numerical aperture of a lens.

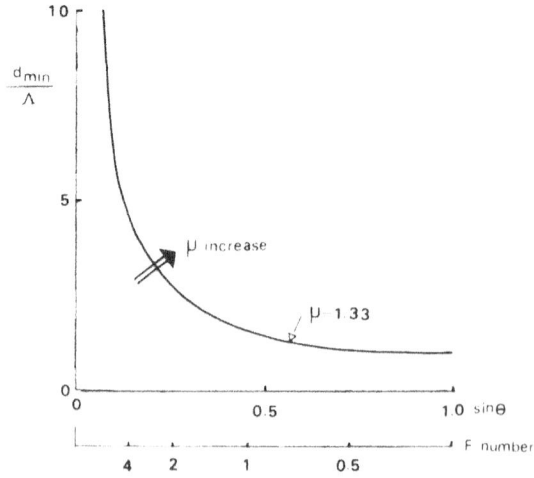

Figure 21. Resolution limit due to numerical aperture of a lens.

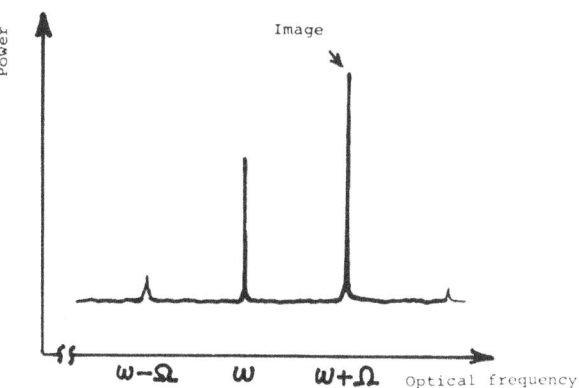

Figure 22. Optical spectra of ultrasound image.

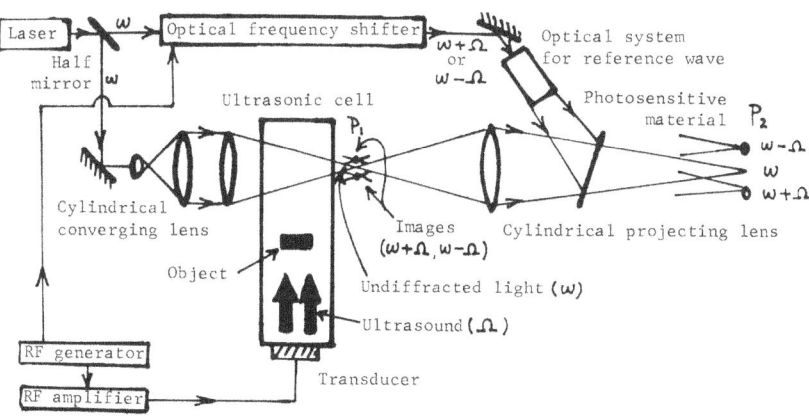

Figure 23. Optical holographic improvements of SN ratio.

As a filtering method, we propose the optical holographic method because a subject of filtering is an image which extends _spatially_. In this method at first a hologram is made from the object light wave which carries image information with noise component and the reference light wave whose optical frequency is the same as that of image component which we want to take out by filtering; next, the hologram is illuminated by the light wave whose spatial form is the same as that of the reference wave and re-constructs an image with improved SN ratio. In Fig. 23, the experimental arrangement of the method proposed above is shown.

A light beam from a laser is split into two beams, one of which is used in an ultrasonic imaging system and makes an image. The other beam is passed through an optical frequency shifter and is exposed on a photosensitive material with another beam which comes out from the imaging system. This method has the advantage of a heterodyne method proposed by J. Landry et al.[21], since our method does not need mechanical scanning, nor electric shield. Of course the reference wave with broad spectrum reduces the SN ratio of the reconstructed image.

III. CONCLUSION AND REMARKS

The present situation of the generation and detection of acoustical surface wave and the applications of ultrasound is reviewed and discussed.

The generation of bulk ultrasonic wave, 0.4 W/cm^2 at 355 MHz, and the generation of an acoustical surface wave at 105 MHz, 0.3-1 μsec. pulse width, and 30-100Hz repetition frequency, are obtainable.

A subcritically doped GaAs diode has differential negative conductivity, and it is proposed that this diode apply to ultrasonic detectors.

On the other hand, it is proposed that an acoustical surface wave has the capability of being a mode-locking device.

In the area of ultrasonic imaging technique, a few problems arising when the so-called Bragg imaging method is expanded into the lower frequency region of ultrasound were discussed. The conclusion from our preliminary experiment is that it is natural in the low frequency region that resolution power of ultrasound imaging becomes worse, and it is technically very difficult to manufacture a cylindrical lens with high precision (aberration-free and with large numerical aperture) so it might compensate for deteriorated resolution power. However, since there is great need in the various fields of application for expansion into the lower frequency of ultrasound, as stated at the beginning, an effective method of improvement of signal-to-noise ratio is proposed. We should proceed with research for the purpose of overcoming the difficulties.

REFERENCES

1. H. Hayakawa, T. Ishiguro, S. Takada, N. Mikoshiba, and M. Kikuchi, J. A. P. $\underline{41}$, 4755 (1970)

2. H. Hayakawa, S. Takada, T. Ishiguro and N. Mikoshiba, The 2nd Conference of Solid State Devices, Tokyo, 1970 J. J. A. P. $\underline{40}$, 143 (1970)

3. H. Hayakawa, T. Ishiguro, S. Takada, N. Mikoshiba, and M. Kikuchi The 1st Conference of Solid State Devices, Tokyo, 1969, J. J. A. P. $\underline{39}$, 56 (1970)

4. H. Hayakawa, S. Takada, T. Ishiguro, and N. Mikoshiba, A. P. L. $\underline{18}$, 31 (1970)

5. L. Kuhn, M. L. Dakss, P. F. Heidrich, and B. A. Scott, A. P. L. $\underline{17}$, 265 (1970)

6. L. Kuhn, P. F. Heidrich, and E. G. Lean, A. P. L. $\underline{19}$, 428 (1971)

7. A. Korpel, A. P. L. $\underline{9}$, 425 (1966)

8. A. Korpel, IEEE Trans. $\underline{SU-15}$, 153 (1968)

9. H. V. Hance et al., J. A. P. $\underline{38}$, 1981 (1967)

10. J. Landry et al., A. P. L. $\underline{15}$, 186 (1969)

11. R. G. Buckles et al., Nature 22, 771 (1969)

12. L. E. Hargnove, R. L. Fork, and M. A. Pollack,
    A. P. L. 5, 4 (1964)

13. R. J. Hallermeier and W. G. Mayer,
    J. Acoust. Soc. Amer. 47, 1236 (1970)

14. O. P. McDuff and S. E. Harris,
    Q. E. 3, 101 (1967)

15. Handbook of Ultrasound Technique, ed. J. Saneyoshi,
    Y. Kikuchi and O. Nomoto (Nikkan Kogyo Shimbunsha,
    Tokyo, 1966) 1372

16. S. Ishihara, Y. Mitsuhashi, K. Sakurai and S. Tanaka,
    Annual Meeting of Jap. Soc. of Appl. Phys. (1971)
    2P-B-9

17. S. Ishihara, Y. Mitsuhashi and K. Sakurai,
    Fall Meeting of Jap. Soc. of Appl. Phys. (1971)
    2P-D-11

18. Analysis for the resolution power in the direction
    perpendicular to the direction discussed in the text
    is omitted here because it is the same as the usual
    Bragg imaging.

19. This value also holds when Bragg diffraction is used.

20. S. Ishihara and K. Sakurai, Fall Meeting of Jap. Soc.
    of Appl. Phys. (1972) IP-B-9

21. J. Landry, R. Smith, and G. Wade, Acoustical Holography
    3, (Plenum Press, 1971) 47.

PRESENT ASPECTS OF "ULTRASONOTOMOGRAPHY" FOR MEDICAL DIAGNOSTICS

Yoshimitsu Kikuchi

Research Institute of Electrical Communication

Tohoku University, Sendai

I. HISTORICAL REVIEW OF ULTRASONIC DIAGNOSTICS

Penetration Methods - The beginning of research on the use of ultrasound for medical diagnostics was not so old as that on the use of ultrasonic energy for medical purpose. In 1942,[1] Dussik in Austria reported the possibility of diagnostic use, and in 1947,[2,a] he made public a sort of ultrasonic image of some brain tumors, which he named "hyperphonogram" (Figure 1). The upper picture is for normal, and the lower is for Astrocytoma, a sort of brain tumor. Although the images are obscure in comparison to current ultrasonotomograms, the historical significance is sharply dominant.[2,b] Dussik obtained these images by using ultrasonic beams of 1.2 to 1.5 MHz, which penetrated the human brain and showed varied attenuation, route-by-route, to reach the receiving transducer that is placed at the corresponding opposite side of the head. The dark patterns are for good penetration, and a simple pattern in the normal brain is the ultrasonic projection of the intracranial ventricle. In 1950,[3] Ballantine and Bolt, in the United States, made public a similar experiment with knowledge of ultrasonic attenuation in biological tissues. In 1951,[4] they made the instrumentation of this penetration method and named it "ultrasonic ventriculography."

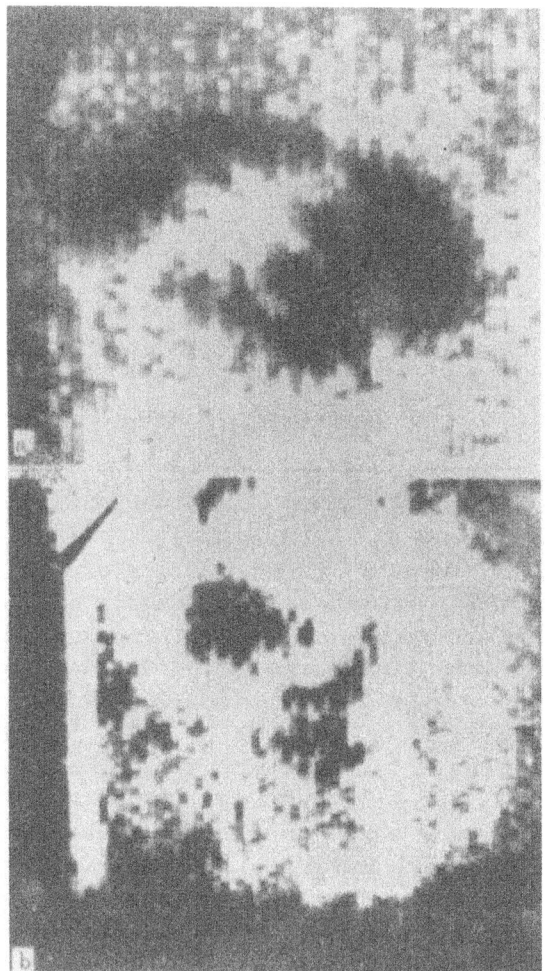

Figure 1. "Hyperphonograms" of the Brain (Dussik)

ⓒ Dussik, K. T.* : Ultraschallanwendung in der Diagnostik und Therapie der Erkrankungen des zentralen Nervensystems, Ultraschall in der Medizin, p. 283, Zurich, Hirzel, 1949. Related figure appears in the last part of the book. *Ref. Dr D. N. White, Queen's University, Kingston, Ontario, Canada; he quoted the same figure in his book entitled "Ultrasonic Encephalography,"p. 2, Figure 1-1.

This method, however, has not yet been used in medical clinics because the localization of brain tumors is too indirect to be used. The main reason is that ultrasonic attenuation in tumor tissues is not so different from the attenuation in normal brain tissues, so the tumor boundaries are not clearly shown on the hyperphonograms. In addition, ultrasonic attenuation at the skull varies place-by-place so much that the variation is generally dominant over the useful attenuation difference between the pathological and normal tissues. This fact results in masking of the objective pattern. Moreover, as Güttner[5] in Germany pointed out, the large attenuation in the skull, as large as 40 to 50 dB/cm, might be an additional barrier in the ultrasonic technique relevant to the method.

Echo Methods; A-Scope - In 1950, French and Wild[6] in the United States, tried an ultrasonic pulse-echo method and succeeded in diagnosing brain tumors. Figure 2 depicts the first A-scope representation of ultrasonic echoes from the inside of a brain tumor.[6] The echo pattern is quite different from that of normal tissues. In 1951, Kikuchi, et al.[7] reported the ultrasonic detection of hematoma in the brain. They used equipment for ultrasonic flaw detection which had been developed for industrial use at that time. Through some improvements, they were soon successful in detecting various brain tumors through the skull. As it had been said that ultrasonic waves in general have biological effects on living tissue, they made various animal tests to determine if the ultrasonic pulse used for a diagnostic purpose would affect the animal's brain.[8] After physiological confirmation that there were no effects, the application of ultrasonic examination has become clinical routine for the first time in Japan. Through hundreds of further investigations and development, clinical methods have been established.[9]

In the United States, however, the U. S. Atomic Energy Commission[10] made public a report in which it was stated that ultrasonic reflection methods as well as penetration methods were unsuitable for detection of brain diseases.

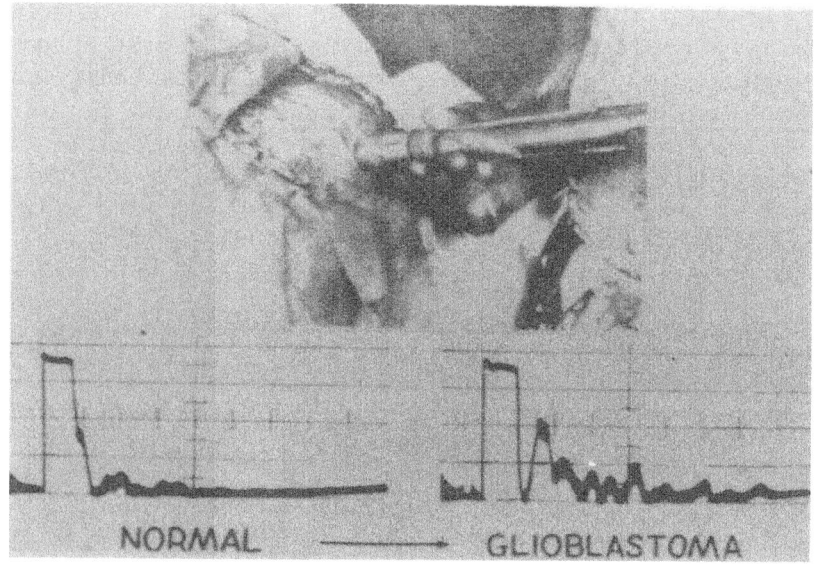

Figure 2. A-mode Echo Patterns Obtained for the
First Time by French and Wild.

Ⓒ Wild, J.J.* and Reid, J.: The effects of biological
tissues on 15 mc pulsed ultrasound, J.A.S.A., Vol.25,
No.2, March, 1953, p.270, p.276, Figure 7.
* J.J. Wild, Department of Electrical Engineering,
University of Minnesota, Minneapolis, Minnesota, USA.

It might not be denied that the Commission's report would not have delayed the general development of ultrasonic diagnostics in the United States, although the skull is really a big barrier for ultrasonic penetration. White in Canada[11] also stated the difficulty through such a statement, "If it were possible to make clinically useful tomograms (of the brain) as easily as they could be made of the pelvic organs, neurological investigation would be enormously helped and simplified." Though we[12] are at present able to obtain clinical tomograms of the brain in various tumor cases by means of a contact sector scanning from the outside of the skull, the tomograms cannot be so clear as those of other organs. With regard to brain mapping, Fry, et al.[13] reported in 1965 that ultrasonic echoes were obtained from a part of living tissue just after the part was otherwise irradiated by stronger ultrasound, and suggested that the echo sources were the acoustic impedance change of the local part due to a slight temperature difference induced by ultrasonic absorption.

In the field of cardiology, Edler and Hertz[14,15] in Lund University, Sweden, inaugurated the "Ultrasonic Cardiography." This is a sort of time-position-indication of ultrasonic echoes which come back from the inside of the pulsating heart.

The clinical application to other human organs was developed in a swift step dramatically after the pulse technique was employed in medical examinations, not only by A-mode representation, but also by B-mode, PPI, and other tomographical representations.

B-Scope; Tomography - Physicians in clinics were generally requiring a more instructive representation of the echo sources in the organs than that of the A-scope representation. Plan-position-indication type of representation was accordingly considered among the research workers in the field. In 1952, Wild[16] developed for the first time, the two-dimensional visualization of a section of the human body.

Howry, in Denver, Colorado, introduced an ingenious technique of a compound scanning into Wild's simple scanning and obtained clear tomograms during 1954 to 1957.[17] He named the technique "Somascope". Figure 3 shows an example of the tomograms, of a horizontal section of the neck of a patient. Scanning was as follows:

> The neck was immersed in a water vessel where an ultrasonic transducer was made traveling around the neck while the transducer was quickly swinging its directional beam.

The idea of this compound beam motion is provided to ultrasonically illuminate dark target planes which usually exist in the human tissues in relation to the incidental and reflected ultrasonic waves when the waves are to be managed by a single transducer.[18] Howry later improved the apparatus by introducing a converging beam transducer and some appropriate functions in the electronic circuits so that clearer tomograms were obtained.[17]

Another contemporary approach was going on in our laboratories where we were engaged in obtaining apparatus for clinical use. Figure 4 shows our first tomographic scanner, moving horizontally in a water vessel that is attached to the breast. Ash shown in Figure 5, a section of the breast is displayed on a cathode ray screen when the transducer makes one linear trip slowly while the ultrasonic pulses of a high repetition rate are making the echo-sounding through the mammary structure. By taking the picture of the screen with a slow exposure, or when the cathode ray screen is a sort of memory scope type, the section of the living organ is visualized as if observers were looking at the slice of the part. Such slice representation of human organs had been called "tomograms" in the X-ray examination; thus, this technique was named "ultrasonotomography".[19,20] Figure 6 shows a pair of examples of the actual tomograms for a normal breast and a malignant one. The cancer tissue is indicated by the gathered mass of bright spots in accordance with acoustically more inhomogeneous and reflective structure of the malignant tissue

"ULTRASONOTOMOGRAPHY" FOR MEDICAL DIAGNOSTICS

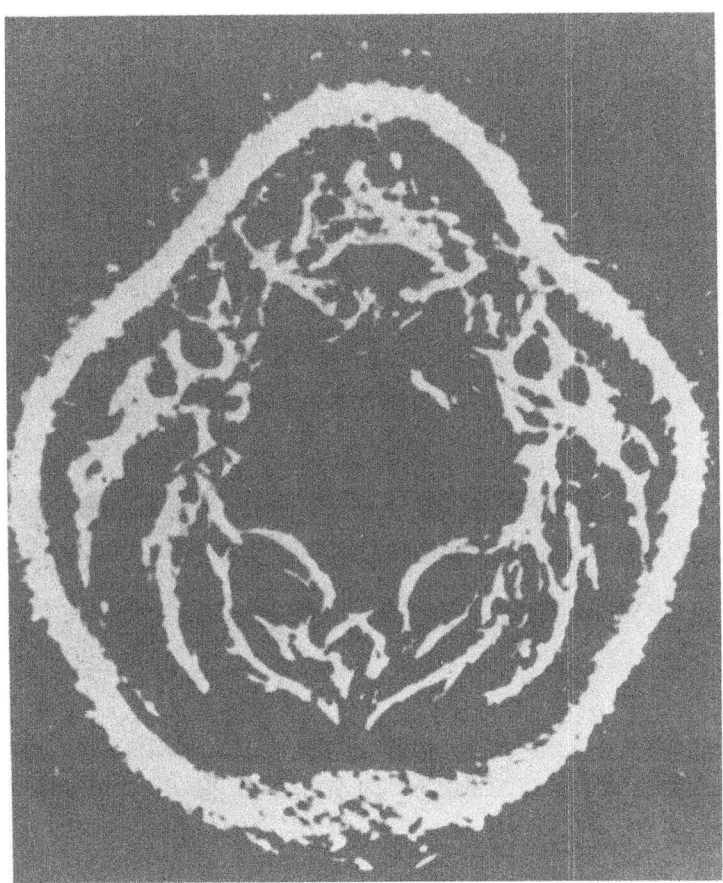

Figure 3. A Tomogram of the Neck (Howry)

© Howry, D. H.*: Techniques used in ultrasonic visualization of soft tissues; Ultrasound in Biology and Medicine (Ed., E. Kelly), p. 49, American Institute of Biological Science, Washington, D. C., 1957. P. 62, Figure 10.
* D. H. Howry, Ultrasonic Research Unit, University of Colorado School of Medicine, Denver, Colorado.

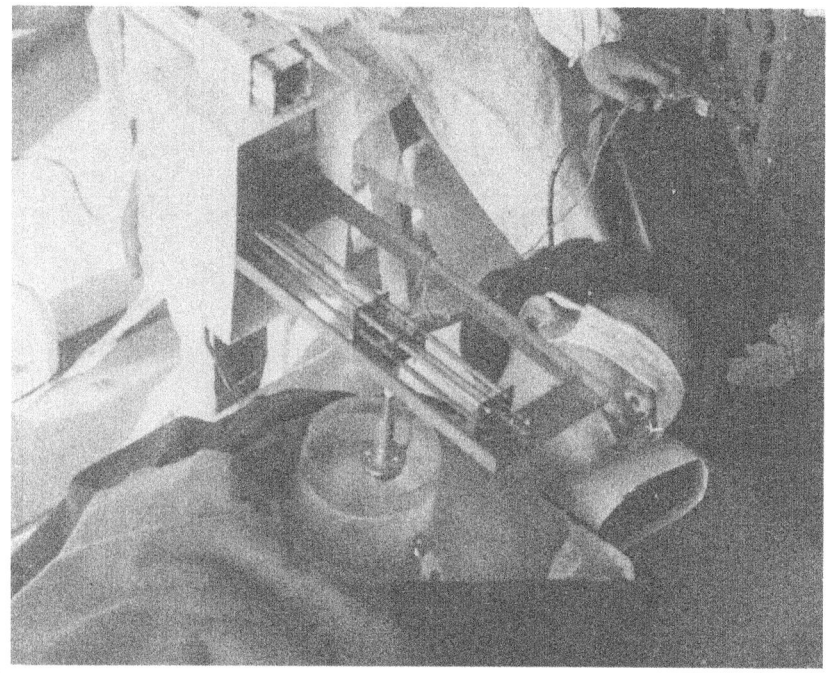

Figure 4. First Clinical Experiment of Ultrasonotomography for Breast Tumors. (Kikuchi, et al.)

© Kikuchi, Y., * Tanaka, K., Wagai, T. and Uchida, R.: Early cancer diagnosis through ultrasonics, Journal of Acoustical Society of America, 29(7), P. 824, 1957, p. 831, Figure 16. * Professor of Research Institute of Electrical Communication, Tohoku University, Sendai, Japan.

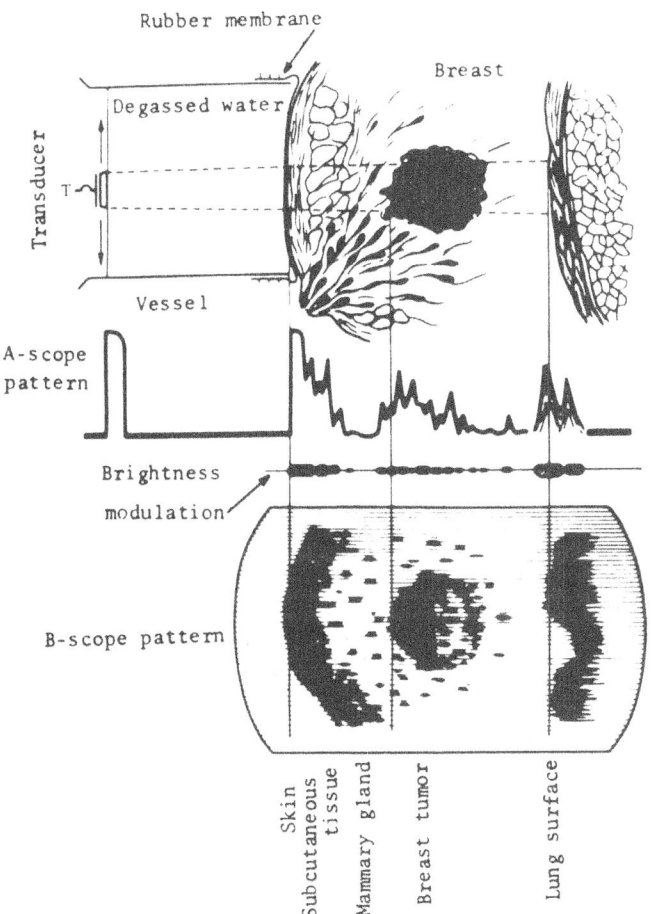

Figure 5. Illustration for Ultrasonotomography.

© Yamakawa, K.*, Wagai, T.*, Kikuchi, Y. and Uchida, R.: Application of Ultrasound in Diagnostic Fields (Tentative translation from Japanese), Inst. of Elect. Comm. Eng. Japan, Report of Professional Group in Electrical Apparatus for Medicine, Feb. 12, 1959, p. 14, figure 11.
* Juntendo U. School Med., 1-1, 2-Chome, Hongo, Bunkyo-Ku, Tokyo, Japan.

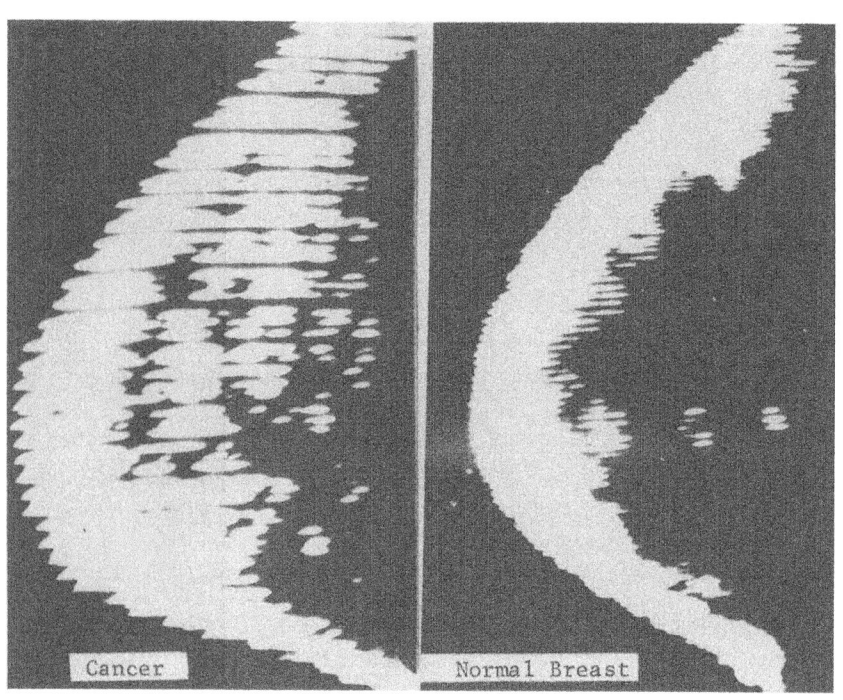

Figure 6. Ultrasonotomograms of the Breast.
(Kikuchi)

ⓒ Kikuchi, Y*.: Recent results of research and development in the field of ultrasonics in Japan, Proceedings Third International Congress on Acoustics, 1959(Ed. L. Cremer), Vol. II, P. 1193, Elsevier Pub. Company, Amsterdam, 1961, p. 1191, Figure 1. * Y. Kikuchi, Professor of Research, Institute of Electrical Communication, Tohoku University, Sendai, Japan.

Improved apparatus are widely used at present by which clearer tomograms are obtained. One of the tomographic apparatus currently used in Japan is shown in Figure 7.

Such ultrasonotomographic techniques have been extended to various fields of clinical medicine. In 1962, Kossoff and others[21] reported its application to obstetrics, and showed the tomograms of the fetus living in the uterus. In Figure 8 an example is shown of the tomograms presently obtainable

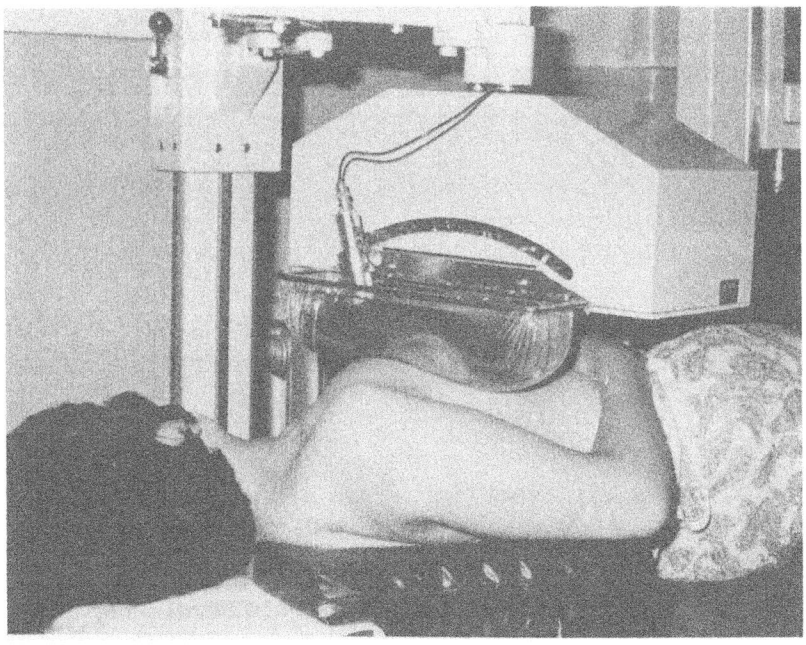

Figure 7. A Tomographic Apparatus Currently used in Japan.

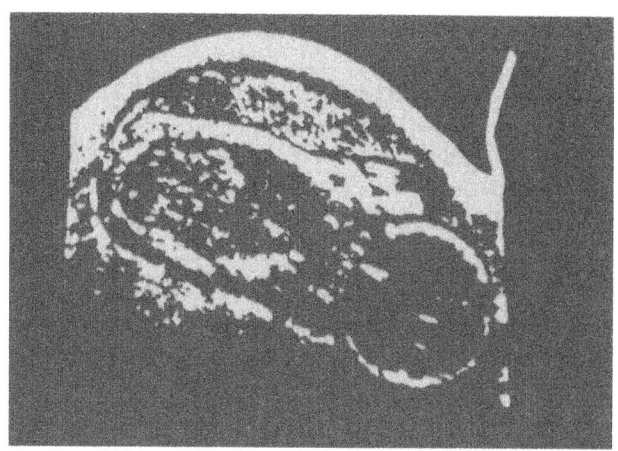

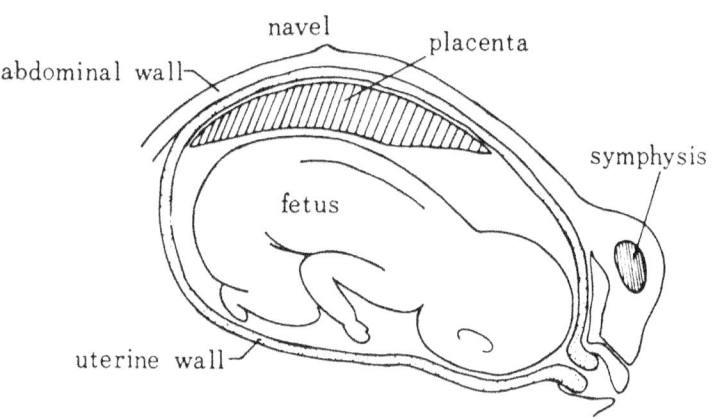

Figure 8. Ultrasonotomogram of the Uterus with 38 Week Pregnancy (Takeuchi)

© Kusano, R.*, and Takeuchi, H.*: Application to Obstetrical Field, Medical Apparatus Culture, 11, 64 (1970-8), P. 65, Figure 8. * Ryoichi Kusano and Hisaya Takeuchi, Juntendo University School of Medicine, 1-1, 2-Chome, Hongo, Bunkyo-Ku, Tokyo, Japan.

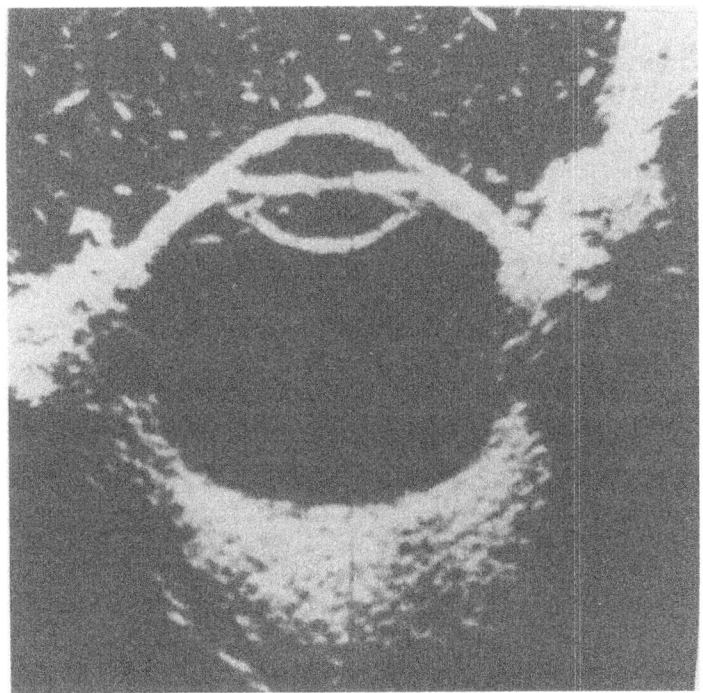

Figure 9. Ultrasonotomogram of the Human Eye
(Baum)

© Baum, G.*: A Comparison of the merits of scanned intensity modulated ultrasonography versus unscanned A-scope ultrasonography, Diagnostic Ultrasound (Ed: C.C. Grossman), P. 59, Plenum Press, New York, 1966, P. 62, Figure 1. * Gilbert Baum, M.D., Albert Einstein College of Medicine, Yeshiva University, Bronx, New York, USA.

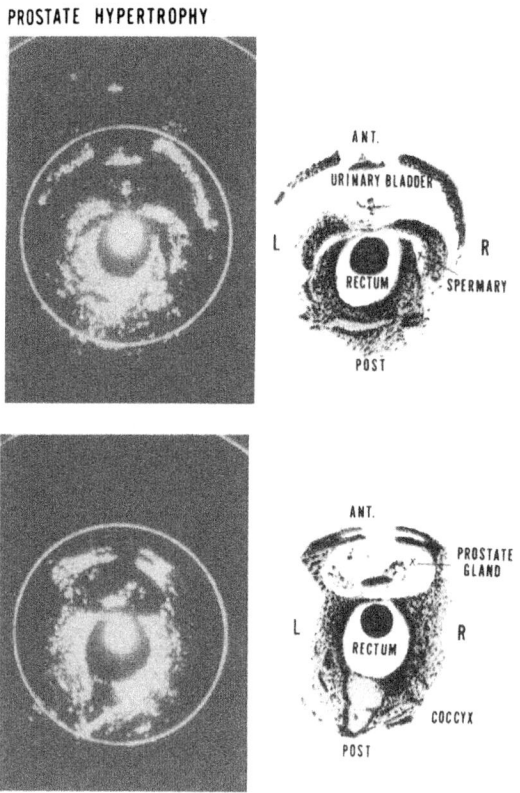

Figure 10. PPI Tomograms of the Urinary Bladder and Prostate Gland (Tanaka, et al.)

© Tanaka, M.*, Oka, S., Watanabe, H., Kato, T., Shishido, S., Ebina, T., Kosaka, S., Terasawa, Y., Unno, K., Nitta, K., Kikuchi, Y., Okuyama, D., Uchida, R.: Ultrasono-tomography of the intrapelvic organs (a tentative title translated from the Japanese), Proc. 13th Meeting Japan Society of Ultrasonic Med., 79, 1968, p. 80, Figure 2.

* Motonao Tanaka, M.D., Res. Inst. Tuberculosis, Leprosy and Cancer, Tohoku U., Sendai, Japan.

in any clinic. In 1958 through 1965, Baum[22,23] was engaged in the ultrasonic diagnostics in the field of ophthalmology, and obtained tomograms of the eye as shown in Figure 9, as a typical example. The cornea, iris, retina, etc., are clearly visualized in a living human eye. In 1957, Wild and Reid[16] reported an equipment for obtaining three-dimensional information from the lower bowel. The transducer revolves inside the rectum to obtain plan-position-indication of a section. The sectional plane is then shifted to the next successively. This technique is useful at the present time for the examination of the urinary bladder and prostate gland.[24,25] In Figure 10, examples are shown, which were obtained in our clinic, of a case of prostate hypertrophy with stones. By a similar technique, we have developed intraesophageal[26] and intratracheal[27] methods of ultrasonic examination. The height of ultrasonic frequency for these methods of examining tubular organs should be 10 to 15 MHz, as the transducers must be smaller in size; and in addition, we recommend concave transducers so as to use a narrower and nearly focusing ultrasonic beam. We have already proposed a suitable design procedure for the concave transducer. The outline will be described later in this paper.

Through many contributions given by research workers in the world, ultrasonic examination for medical diagnostics has been established in almost all fields of clinical medicine; the heart is the only organ that did not allow anyone to take its ultrasonotomogram because every part of a living heart is always pulsating with faster velocity than to be assumed stationary in any case of usual scanning.

Kikuchi and others[28] have been engaged in this problem since 1966, and introduced a technique named "ultrasonocardio-tomography" which is based on a method of display in synchronization of the heart movement. They have further developed "Kineto-Ultrasonotomography"[29] which is a sort of animation of the individual cardiotomograms. The details of these methods are described in a separate paper.

II. ACOUSTICAL PROPERTIES OF BIOLOGICAL TISSUES

In any case of ultrasonic visualization of biological tissues, the acoustical properties of the tissues such as the propagation velocity and attenuation of ultrasonic waves in the tissues and the acoustic impedance (ρc) of the tissues are important factors. Especially, in the pulse-echo method, the acoustical impedance difference among the relevant tissues determines the echo intensities. As to the propagation attenuation, not only does it have close relation with the echo intensities, but also is one of the important factors in the medical diagnostics for brain edema, hematoma, cancer, epilepsy of certain cases, breast tumors, etc.[8a,30-33]

Many research workers have been engaged in the measurement of the acoustic properties of biological tissues. In the following, some summarized data are shown: Table 1 shows the data obtained by Kikuchi and others[7,14,32] in 1952, giving sound velocity, c, specific gravity, ρ, acoustic impedance, ρc, of various tissues, and also the estimated percentage reflection of a plane wave when it is incident normally to the interface between any two tissues. Note that the reflection is usually very small.

With regard to the ultrasonic absorption, there are usually appreciable differences between the normal and tumor tissues. Figure 11 is an example of those in the brain.[34] Wells[35] has published a similar table of interface reflection, including more biological materials. There are, however, no data concerning pathological tissues. In 1956, Goldman and Heuter[36] reported a condensed data available at that time on the velocity and absorption of high frequency sound in mammalian tissues. In Tables 2 and 3, some essence relevant to human bodies is shown; in Table 2, the sound velocities of various human tissues; and in Table 3, the sound absorptions in the blood, plasma and in normal and various tumor tissues. The authors said in the report that the absorption coefficient of most tissues lies in the range from 0.5 to two dB/cm/MHz, and with other data which are not shown here, the authors showed an estimated frequency dependence of it as shown in Figure 12.

Table 1
Acoustical properties of human tissues

Tissue	Sound Velocity (c) cm/s	Density (ρ) gr/cc	Acoustic Impedance (ρc) ab·ohm	Power Reflection Coefficient in % (Normal incidences of plane waves)						
				air	steel	water	fresh cerebrum (cow)	glioma	meningioma	cerebellum	cerebrum
cerebrum	1.53×10^5	1.038	0.159×10^6	99.9	86.9	0.27	0.00001	0.046	0.0023	0.048	0
cerebellum	1.47	1.034	0.152	99.9	87.4	0.093	0.049	0.00001	0.029	0	
meningioma	1.49	1.056	0.157	99.9	87.0	0.23	0.0026	0.028	0		
glioma	1.46	1.042	0.152	99.9	87.4	0.095	0.048	0			
fresh cerebrum (cow)	1.54	1.032	0.159	99.9	86.9	0.28	0				
water	1.43	1.00	0.143	99.9	88.1	0					
steel	5.88	7.7	4.53	100	0						
air	0.33	0.0012	0.00004	0							

Power Reflection Coefficient: $R_o = \left(\dfrac{\rho_1 c_1 - \rho_2 c_2}{\rho_1 c_1 + \rho_2 c_2}\right)^2$

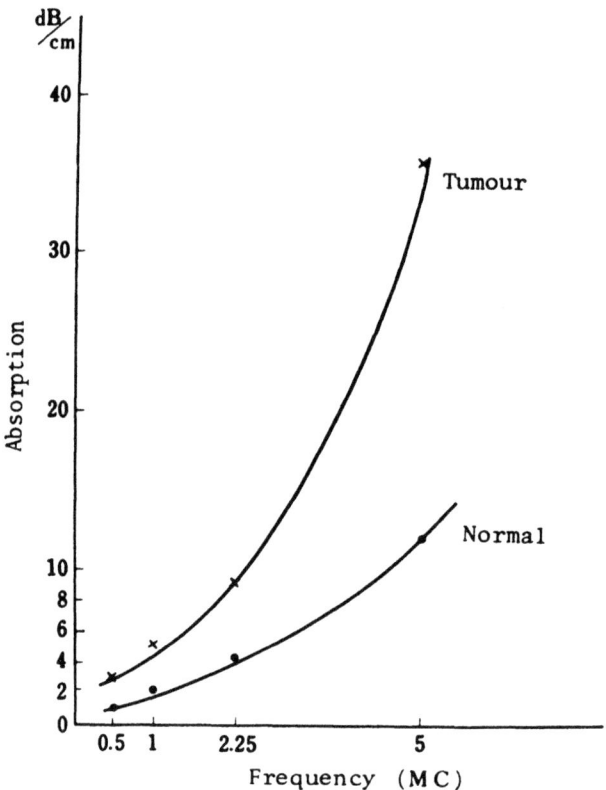

Figure 11. Sound Absorption in Brain Tissues
Upper: Brain Tumour Tissue
Lower: Normal Brain Tissue

© Tanaka, K., * Kikuchi, Y. and Uchida, R.: Ultrasonic Detection of anatomical abnormalities in the human cranium, Part 2. Reports of 1953 Spring Meeting, Acoustical Society of Japan, No. 2-9, May, 1953 (in Japanese) Page 54, Figure 3. * Kenji Tanaka, Professor Emeritus of Juntendo University School of Medicine, 2-1-1, Hongo, Bunkyo-Ku, Tokyo, Japan.

Table 2
Velocity of H. F. Sound in Human Tissues

Tissue	Temp. °C	Frequency MHz	Velocity m/s
muscle	24	1.8	1568
liver	24	1.8	1570
fat	24	1.8	1476
limb	body	2.5	1540
meningioma	body	2.26	1540
skull bone	body	0.8	3360
breast carcinoma	24	1.8	1573

© Goldman, D. E.* and Hueter, T. F.** Tabular data of the velocity and absorption of high-frequency sound in mammalian tissues, Journal Acoustical Society America, Volume 28, No. 1, P. 35, 1956. P. 35, Table I.
* D. E. Goldman, Naval Medical Research Institute, Bethesda, Maryland.
** T. F. Hueter, Massachusetts Institute of Technology, Cambridge, Massachusetts.

Table 3
Abosrption of h. f. sound in human tissues

Tissue	Condition	Frequency MHz	Absorption* cm^{-1}
plasma	refrig.	1.7	0.04
plasma	---	1.0	0.007
blood	---	1.0	0.02
muscle	---	0.80	0.1
brain	fixed	1.7	0.18
brain	fixed	3.4	0.37
medulla oblongata	---	1.7	0.14
medulla oblongata	---	3.4	0.34
liver	20 hr post mortem	1.	0.15
liver	do	3.	0.23
liver	do	5.	0.35
fat	melted	0.87	0.045
fat	melted	1.7	0.09
fat	melted	3.4	0.16
skull bone	fresh or fixed	0.6	4.5
skull bone	---	1.2	17.
skull bone	---	2.25	53.
skull bone	---	3.5	80.
sciatic nerve	----	3.4	0.35

* Absorption coefficient α in $A = A_o e^{-\alpha x}$ where A_o is an amplitude of sound.

ⓒ Goldman, D.E.* and Hueter, T.F.** : Tabular data of the velocity and absorption of high-frequency sound in mammalian tissues, Jour. Acoust. Soc. Am., Vol.28, No.1, p.35, 1956. p.36, Tab.II.

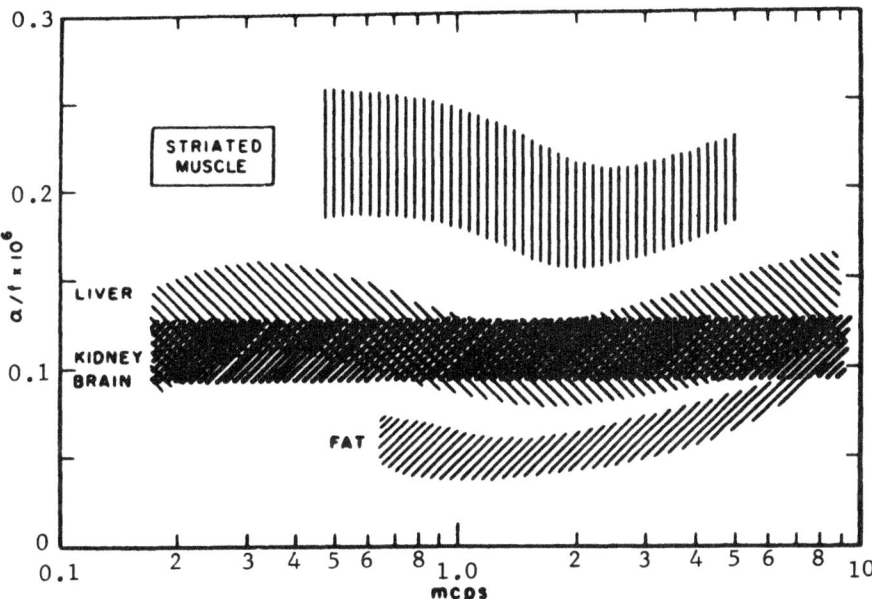

Figure 12. Estimated Relation Between Sound Absorption and Frequency for Several Mammalian Tissues. (Goldman and Hueter)

Ⓒ Goldman, D. E.* and Hueter, T. F.** Tabular data of the velocity and absorption of high-frequency sound in mammalian tissues, Journal Acoustical Society America, Volume 28, No. 1, Page 35, 1956. P. 37, Figure 1.
* D. E. Goldman, Naval Medical Res. Inst., Bethesda, Md.
** T. F. Hueter, Massachusetts Institute of Technology.

In 1964, Ishikawa[37] reported that the absorption in the brain tissues is different between the sagittal direction and the horizontal direction toward the forehead. At five Megahertz they are 7.2 dB/cm and 10.7 dB/cm, respectively; and at 10 MHz they are 16.6 dB/cm and 28.4 dB/cm, respectively. In the meantime, Nomoto[38] summarized available data to a comprehensive table on the acoustic properties of various tissues of man, cat, dog, horse, cow, calf, pig, rabbit, rat, mouse and guinea pig.

The methods of measurement for the acoustical properties of biological tissues involve many specific problems and difficulties arising from the fact that the subjects are changing under biological and physiological conditions. In the case of sample measurement, it is considerably difficult to keep the samples under a certain biological condition during the period of time required for a series of measurements. There are also problems such as individual deviation among subjects, acoustical inhomogeneity of a tissue, etc. Although it cannot be said that the methods of measurement have been established, some typical methods are as follows:

After Pohlman,[39] Hueter and others,[40] in 1948, placed an excised biological sample in a cylindrical glass tube filled with water and measured acoustical absorption of the biological tissue from the change of acoustical power reaching the other end of the tube. In later years (1952 and 1953), Kikuchi and others[34] used pulsed ultrasonic waves. Yoshioka and others[41] recently have made an instrumentation of the method with specific care. Ishikawa[37] used data on the temperature rise of tissues for the observation of ultrasonic absorption. The temperature rise with regard to the lapse of time is a suitable measure to indicate the relative absorption of biological tissues. Carstensen[42] has proposed a suitable method to measure liquid absorption.

With regard to the measurement of ultrasonic velocity in biological tissues, most research workers[7,43,44,32] employed some sort of ultrasonic interferometer and/or method of multiple reflection principle, placing a tissue or a biological liquid sample between two parallel planes, one of which or both of which were transducer planes. An ultrasonic thickness gauge was also used as a variation of interferometer principle. Venrooij[45] recently reported an accurate measuring system for the velocity in human tissues with some measured results. In Figure 13, the measuring system is shown. This is a sort of "sing-around" method.[46] A test sample is placed between an ultrasonic transmitter and receiver and the transmitter is struck by a pulse generator which is to be excited by every received pulse so the electronic system sings around with a definite frequency. This singing frequency is determined in accordance with the sound propagation velocity in the sample.

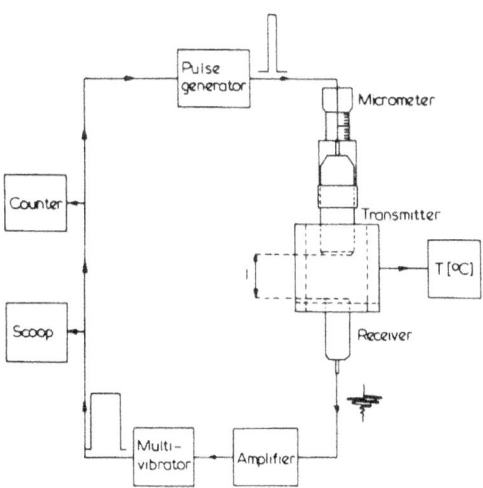

Figure 13. Velocity Measuring System Along the Sing-Around Principle (Venrooij)

© Van Venrooij, G. E. P. M. *: Measurement of ultrasound velocity in human tissue, Ultrasonics, Page 240, October, 1971. P. 240, Figure 1.
* Van Venrooij, G. E. P. M., Department of Neurosurgery, Medical Physics Department, State University, Utrecht, The Netherlands.

III. RECENT RESEARCH RESULTS ON THE GENERATION AND RECEPTION OF PULSED ULTRASOUND

1. <u>Distance Resolution</u> - In general pulse echod methods, distance resolution depends directly on the length of the wave train; that is, the pulse length. There were many reports in the past on this problem, and convenient circuits[47],[48] for the generation of pulsed electrical oscillation have been suggested. On the side of ultrasonic transducers, a means in which a piezoelectric plate is backed by a high damping material has been recommended.[49] For the backing material, a mixture of artificial resin and fine metal powder such as tungsten has been used. Available highest acoustic impedance is around 10×10^6 Kg/m^2.s, and is nearly sufficient for damping the free oscillation of a quartz and/or LH transducer. For other transducers made of piezo-active material of higher acoustic impedance such as $BaTiO_3$-ceramic, and/or PZT, Kikuchi and others[50] have proposed a compound backing as shown in Figure 14. The piezo-active ceramic plate is first backed by an inverted cone or wedge made of brass, and the tapered surface of the cone or wedge is then backed by tungsten-araldite mixture which absorbs ultrasonic energy traveling backward; thus, the piezo-active plate can be acoustically backed by such a high impedance as brass itself since no stationary waves nor echoes yield in the brass. In Figure 15, an example of our experiment is shown. The oscillograph, a, is an echo of a reflective plane in water obtained by a single pulse excitation given on a 2.25 MHz circular disk of 20 mm in diameter, made of PZT, with the aforementioned compound backing. The oscillograph, b, shows the front and rear surface echoes of a rubber plate, the thickness of which is as thin as 1.5mm.

In recent years we[51] also proposed an active damping which is based on a doubled electrical excitation. A piezo-active disk is struck by two electrical pulses, one of which is delayed by an electrical network so the natural mechanical ringing of the disk is forced to diminish immediately after the first wave.

Recently, we[52] have established another method of radiation and reception of very short ultrasonic pulses by using a quarter wave-length layer. This idea was first proposed by McSkimmin[53] and developed by Kossoff[54] to some extent. We[55,56] then developed an engineering procedure to determine the optimum impedance of the quarter wave length layer material. The procedure enables us to easily calculate the train of ultrasonic pulses, the resultant pulse-train width and the sound intensity in relation to the layer materials. A material of any desired acoustic impedance is not easily available among markets so we recommend use of a suitable mixture of artificial resin and metal powder. Though the mixture is originally absorbing material, the propagation loss in a thin layer, as thin as a quarter wavelength, is usually very small. In any choice of material,

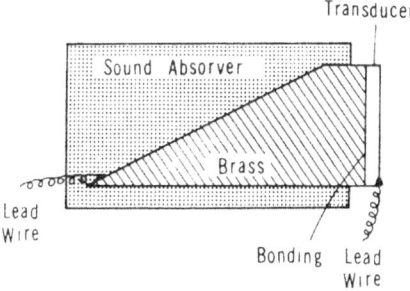

Figure 14. Compound Acoustic Backing
(Kikuchi, et al.)

ⓒ Kikuchi, Y.,* Okuyama, D., Kasai, C., and Tatebayashi, K.: Generation of extremely short ultrasonic pulses into liquid by acoustic backing transducer. Reports of Japan Society of Ultrasonics in Medicine, 17th Meeting, Page 53, 1970(in Japanese), P. 53, Figure 1.

* Y. Kikuchi, Professor of Research Institute of Electrical Communication, Tohoku University, Sendai, Japan.

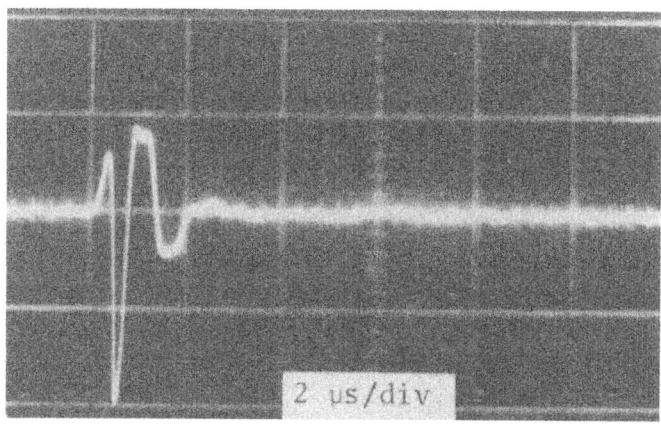

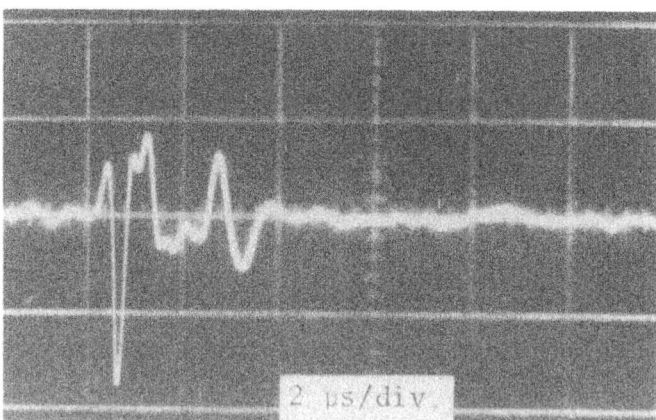

Figure 15. Echoes of High Resolution

© Kikuchi, Y., Okuyama, D., Kasai, C. and Tatebayashi, K.: Generation of extremely short ultrasonic pulses into liquid by acoustic backing transducer. Reports of Japan Society of Ultrasonics in Medicine, 17th Meeting, P.53, 1970 (in Japanese). P. 54, Figures 2c and 3b.

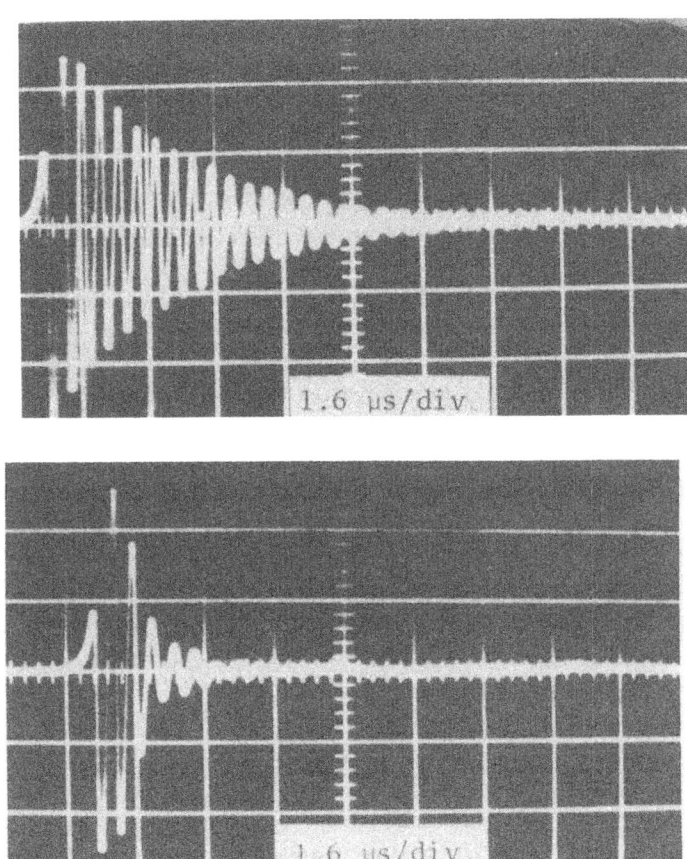

Figure 16. Effect of a Quarter-Wavelength Layer

© Kikuchi, Y., Okuyama, D., and Kasai, C.: Generation of short ultrasonic pulses by means of ultrasonic transducer with an intermediate layer of a quarter wavelength. Reports of Professional Group on Ultrasonics, Institute of Electronics Communications Engineers, Japan, US-71-13, 1971 (in Japanese). P.6, Figure 9.

the sensitivity of the transducer as a whole is much higher than that of the aforementioned transducers with the acoustical backings. In the acoustical backing, acoustical power is largely absorbed, and only a small portion of the total power is sent forward, to be used as acoustic waves. On the contrary, the quarter wavelength device does not dissipate acoustic power but does transfer the power completely to the acoustic medium, or does receive the acoustic power from the medium through an appropriate transforming action of acoustic impedances. In Figure 16, an example of the experimental results is shown. The ringing of a PZT disk such as is shown in Oscillograph a can be damped as shown in b with a favorable increase of amplitude.

2. <u>Azimuthal Resolution</u>.- Another important factor that determines the image resolution is the directional characteristics of the ultrasonic transducer. This factor is usually called azimuthal or lateral resolution. The diagnostic application of ultrasound is generally confined within the near field zone of the transducer. This zone is often called Fresnel's interference zone, and the axial length of the zone for a circular disk transducer[57],[58] of Radius R is about R^2/λ, where λ is the wavelength of the radiated sound. So far as pulse-echo method is concerned, the field pattern of ultrasonic waves within this region can be usually approximated as a beam of 2R in diameter, with parallel waves of constant intensity.

With regard to concave disks for focusing transducers, the near-field patterns are theoretically derived by O'Neil,[59] Torikai[60] and others. In Figure 17, an example is shown in which a dimensional parameter, $\underline{D}$, is chosen as 10. And $D = R^2/a\lambda$, where $\underline{a}$ is the focal length of the concave disk. In Figure 18, the sound field along the beam axis is shown with $\underline{D}$ as parameters. The horizontal axis $\underline{Z}$ is the distance along the sound beam which is normalized by the focal length, $\underline{a}$. Kikuchi and others utilized Torikai's result and have recommended that the dimensional parameter D should be around four, so as to obtain optimum transducers for diagnostic use.[61] In this case, "optimum" means that the lateral three dB width of

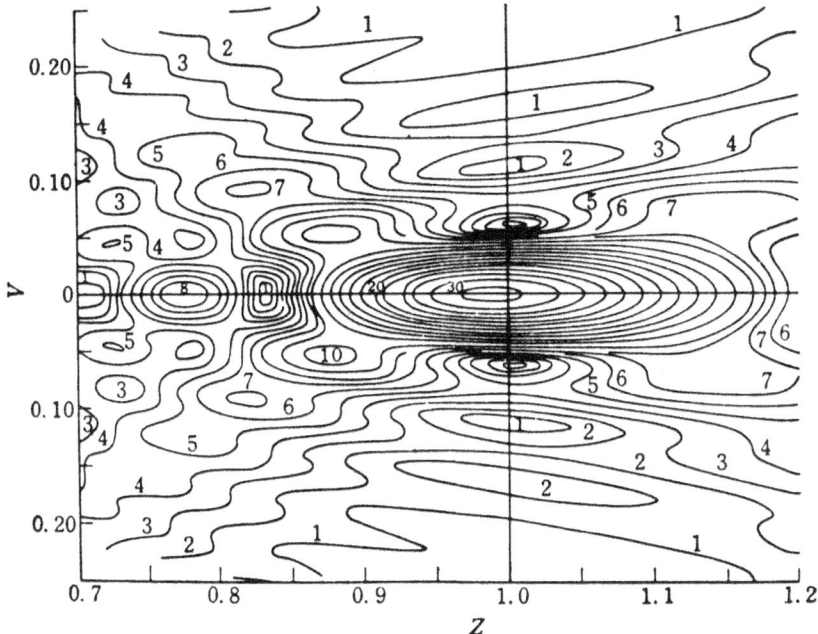

Figure 17. Near-field Pattern of a Focusing Transducer when $D = R^2/a\lambda = 10$ (Torikai)

© Torikai, Y.*: Cho-onpa Gijutsu Binran (A Handbook of Ultrasonic Engineering). Edited by J. Saneyoshi, Y. Kikuchi and O. Nomoto. Page 1418, Fig. 1-27, Nikkan Kogyo Press Ltd.** Tokyo, Japan.

* Yasuo Torikai, Professor of Institute of Industrial Science, University of Tokyo, Roppongi, Tokyo.
** Nikkan Kogyo Press, Ltd, 8-10, Kita 1-Chome, Kudan, Chiyoda-Ku, Tokyo, Japan.

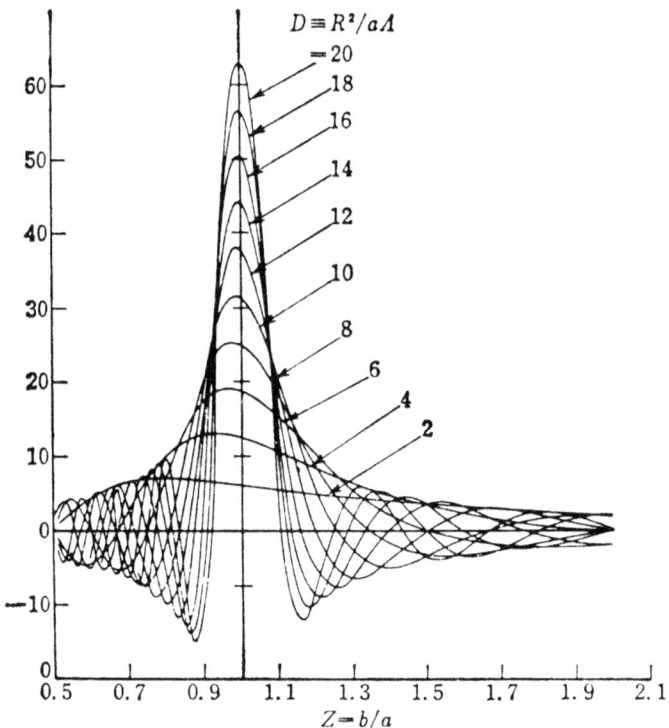

Figure 18. Sound Field on the Beam Axis of Concave Disks, $D = R^2/a\lambda$ being Dimensional Parameters. (Torikai)

Ⓒ Torikai, Y.: Calculations of near sound field (tentative translation from Japanese), The Institute of Electronics and Communication Engineers of Japan. Report of Professional Group in Ultrasonics, November, 1962, PP 17-18, Figure 7.

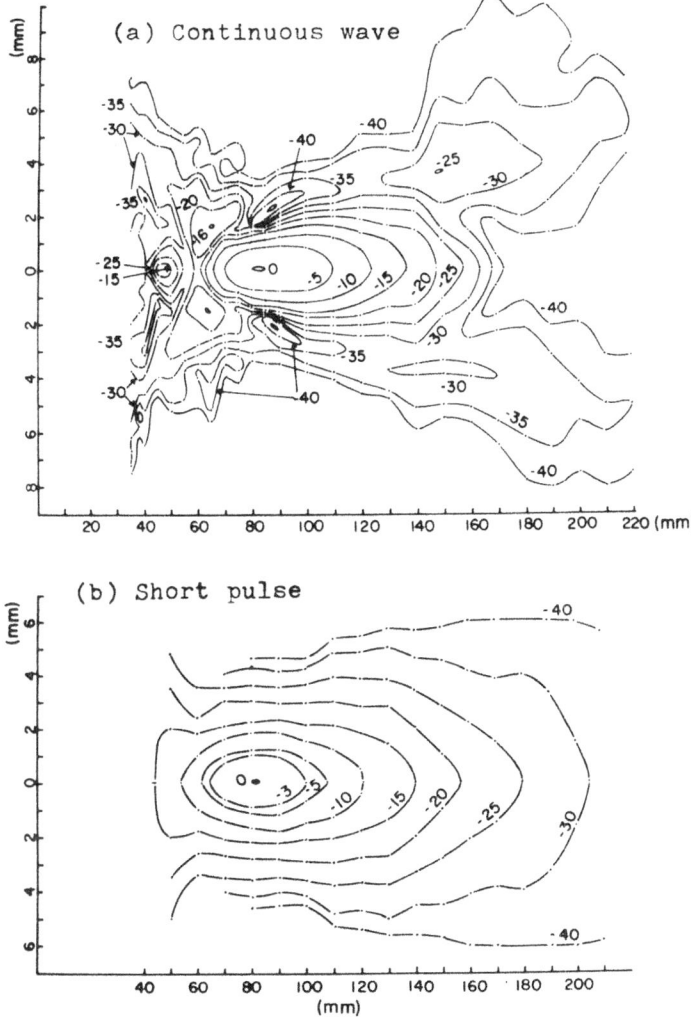

Figure 19. Near-field Patterns of a Concave Disk (a) Continuous Waves;(b) Extremely Short Pulse. (Kikuchi, et al.)

© Kikuchi, Y., Okuyama, D. and Kasai, C.: Sound field characteristics of extremely short ultrasonic pulse radiated by a concave transducer. Reports of Japan Soc. Ultrasonics in Med., 19th Meeting, 19-31, p. 61, 1971 (in Japanese). P. 62, Figure 2.

the sound beam in the focal region is to be as narrow as
possible, provided that the axial field of the beam is to be
as smooth as possible.

This design recommendation has been good so far,
provided that the pulse length of the ultrasonic waves is
not extremely short. In the case of extremely short pulse,
however, the order of wave interference decreases and the
field pattern becomes different. In Figure 19, one of our
experimental results[62] is shown. The patterns are the
equal echo-intensity contours when the target is a three mm
steel ball. The pattern a is for the usual pulse length and
the pattern b is for an extremely short pulse, as short as
one wavelength. The transducer under observation is a
concave disk, 25 millimeters in diameter and 90 millimeters in focal length. It is used at 2.25 Megahertz. Comparing both patterns, it is found that Pattern b is much simpler but has rather good resolving power. The beam width
in the focal region really becomes a little wider than that
for the longer pulse; no submaxima appear which do appear
on both sides of the beam axis when the pulse is longer, as
in Pattern a. This is of a favorable nature.

Although not yet employed in medical applications, side
lobe suppression has been suggested[63,64,65] by means of
electrode patterns applied on a piezo-active disk.

Uniformity control of the sound field has also been
suggested[66] by means of a concentric arrangement of transducers, each of which is to be excited by different voltage.
Olofsson[67] has proposed an acoustic reflector system in
place of concave disks of large size. The azimuthal resolving power in the focal region is extraordinarily excellent, but the axial field distribution is too rough to be used
in the usual way. Fry and others[68] then used this mirror
system through a computer control in the three-dimensional
sector-scanning mode. An idea of composite echograms
was suggested with a time-gated display and other relevant
techniques.

IV. SYSTEMS FOR ULTRASONOTOMOGRAPHY

1. For Clinical Equipment:

Ultrasonic imaging used in medical diagnostics at the present time is based mainly on a pulse-echo method. As this method was originally developed in RADAR practice, electronic circuits of the ultrasonic apparatus are similar to those of Radar equipment. There are two modes of operation: A mode and B mode. The B scope in Radar practice, however, is to display the targets on a coordinate system in which the abscissa is azimuthal angle and the ordinate is distance; whereas the B mode display in the ultrasonic apparatus means any mode of display in which the echo intensity is represented by the brightness of cathode ray spots.

The scanning device of the ultrasonic transducer consists of (1), linear scanner; (2), sector scanner; (3), PPI scanner; and (4), compound scanner, consisting of any combination of the other three.[69] In Figure 20 is seen one of the actual systems[70] of its block diagram in the case of a linear sector compound scanning. In a suitable water vessel, the ultrasonic probe; i.e. a transducer, repeats the sector swing which is driven by a motor (1), and at the same time, the probe makes a horizontal displacement driven by the other motor (2). The position and azimuthal direction of the probe are given to the cathode ray tube through a signal generated by a linear potentiometer and a three phase resolver respectively, as is easily seen in the diagram.

In recent years, a scanning mode called contact compound scanning has appeared in medical practice and is useful, especially in the examinations of the head and abdominal organs including the use for obstetrics. The transducer is manually made to swing and displace while it keeps a sliding contact on the skin of the patient.

There are two systems for positioning the ultrasonic probe. One is as shown in Figure 21a. The position of the

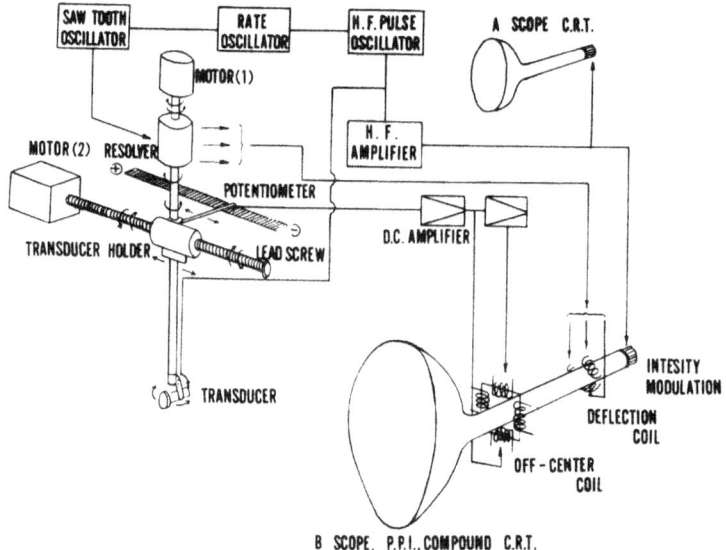

Figure 20. Linear Sector Compound Scanning
(Kikuchi, et al.)

© Ebina, T.,* Kikuchi, Y., Oka, S., Tanaka, M., Kosaka, S., Uchida, R., and Hagiwara, Y.: The diagnostic application of ultrasound to the diseases in mediastinal organs--Ultrasono-Tomography for the Heart and Great Vessels(First Report), Science Reports of the Research Institutes Tohoku University Series C (Medicine), Reports of the Res. Inst. for Tuberculosis Leprosy and Cancer, Vol. 12, No. 1, March, 1965, P. 58. P. 60, Figure 3. * Toshiaki Ebina, Prof. Emeritus, Res. Inst. for Tuberculosis and Leprosy, Tohoku University, Sendai, Japan.

ultrasonic probe with reference to a fixed point, 0, is represented by a pair of electric voltages that are produced by a potentiometer device so as to be proportional to x and y, respectively. The direction of the ultrasonic beam is represented by another voltage proportional to the angle, θ. The voltages x and y determine the origin of the time sweep on a cathode ray screen, and the voltage θ gives the direction of the sweep. The other system of positioning is as shown in Figure 21b. The voltages x and y in this case are constructed in such a way that:

$$x = \overline{AB} \sin \alpha + \overline{BC} \sin \beta$$
and
$$y = -\overline{AB} \cos \alpha + \overline{BC} \cos \beta$$

where $\overline{AB}$ and $\overline{BC}$ are the proportional constants representing the lengths of the two mechanical arms which hang the probe from the fixed point, A. The voltages proportional to $\sin \alpha$, $\sin \beta$, etc., are produced by certain resolvers or special potentiometer devices. The beam direction θ is the same as described above. In Figure 22, an actual system of the mechanical arms and resolvers are shown, for example. At the right tip is a cylindrical ultrasonic probe with which a physician makes manual contact onto the patient's skin. Figure 23 shows the entire equipment with another type of manual compound scanner.

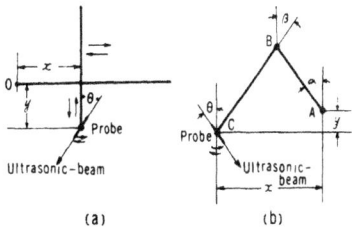

Figure 21. Two Kinds of Positioning Coordinates for Contact Compound Scanning.

© Kikuchi, Y., and Okuyama, D.: Ultrasonic Diagnostics in Medicine, II, Journal of Acoustical Society of Japan, Vol. 27, No. 11, P. 579, 1971, P. 580, Figure 11.

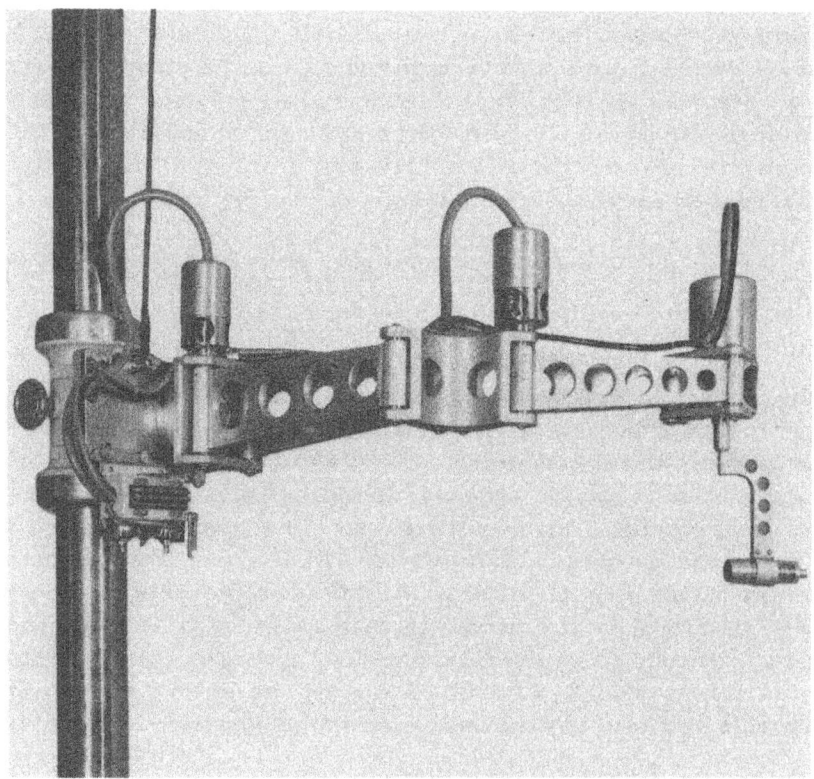

Figure 22. A Contact Compound Scanner
(Uchida, et al.)

ⓒ Uchida, R., * Wagai, T., Ito, K., and Ohashi, H.:
Ultrasonic Diagnostic Apparatus for Multiple Usage,
Proc., 4th Meeting of Japan Society of Ultrasonics in
Medicine, P. 31, Nov. 1963, P. 32, Figure 5.
* Rokuro Uchida, Japan Radiation and Medical Electronics, Inc., 6-22-1, Murei, Mitaka, Tokyo, Japan.

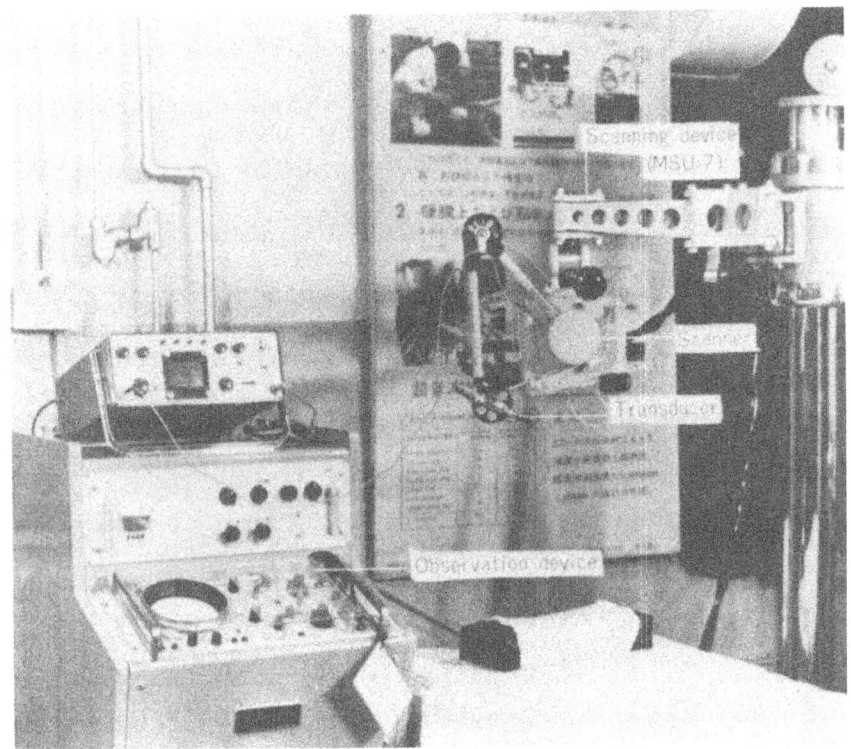

Figure 23. A Clinical Tomograph Setup of a
Contact Compound Scanning Type. (Tanaka)

© Tanaka, K.*: Diagnosis of Brain Disease by
Ultrasound, Shindan-to-Chiryo Sha, Ltd, Tokyo,
1969, P. 120, Figure 134.
* Kenji Tanaka, Professor Emeritus, Juntendo U.
School of Medicine, 2-1-1, Hongo, Bunkyo-Ku, Tokyo.

Actual circuits which treat the azimuthal angle θ are not simple in any scanning system because θ must control the cathod ray sweep in combination with the lapse of time after each pulse transmission. The system already shown in Figure 20 employs a three phase sawtooth voltage system for this function.

Systems for the synchronized cardiotomography are to be described in a separate paper in this book.

2. For Quantitative Examination in Clinics

As has been described in this paper, no one doubts nowadays that ultrasonic examination has become one of the important diagnostic methods in nearly every field of clinical medicine. Insofar as qualitative examination is concerned, it can be said that the method has been well established. As to quantitative description of the examination, however, many research workers are still making efforts toward that establishment. As one of the proposals applicable to tomographic description, the author has introduced a method of quantitative comparison of patterns, and has tentatively named the method "sensitivity graded tomogram pair method."[71] A set of several numbers of tomograms are first taken for a single part of an organ under examination by changing the sensitivity of the apparatus in several steps. Visual comparison is then performed by physicians with any pair among the several tomograms. Targets with higher reflection can appear in both tomograms of the pair, but those with lower reflection cannot; thus, the pair contains much important information for diagnostic judgment. Another pair obtained at another pair of sensitivity settings will do the same. It has been recommended that a set of four tomograms be taken for a single pathological part in six dB steps which may be considered optimum in most of the current diagnostic apparatus. At present, this method is spreading widely among various clinics. Examples of the successful results are shown by Dr Wagai in a separate paper.

V. SOME NEW TECHNIQUES PROPOSED FOR SOLVING PROBLEMS IN ULTRASONOTOMOGRAPHY

There are many requirements and problems that arise from clinics for the improvement of ultrasonotomography toward its wider applicability, its compact instrumentation, its possibility for clinical usage with reduced time, etc. As most of the problems relate to transducers and/or display systems, new technical proposals concerning them are reviewed briefly in the following:

1. High Speed Scanning of Ultrasonic Beams:

Methods for high speed scanning of ultrasonic beams have been considered in accordance with some clinical needs. Asberg[72] has been proposing a high speed sector scanning of a focusing mirror system receiver for obtaining an ultrasonic cinematogram of the living heart. Pätzold and others[73] have developed another high speed scanning system such as is shown in Figure 24. A focusing transducer is made to make a rapid circular motion around the focal line of a paraboloid reflector. The whole system is covered by a plastic sheet and filled with a certain liquid medium. The surface of the sheet is attached to the patient's skin. The reflected focused beam then linearly scans the subject. It is reported that one tomogram of a subject's section of 14 cm x 16 cm in size can be displayed within 60 ms.

As an alternative, Uchida and others[74] have recently proposed a completely stationary system. The idea is as follows: 200 strip transducers are arranged on a plane to form an arrayed probe as shown in Figure 25; and any 20 neighboring transducers are chosen to make an ultrasonic beam. By switching this choice successively, the beam linearly scans the subject on whose skin the arrayed probe is attached, in a stationary manner. An ultrasonotomogram of a sponge-test piece is shown in Figure 26 with its illustration. For a cross-section of 16 cm x 25 cm, it takes about 60 ms to make a tomogram; for a smaller section, however, the required time can be easily shortened.

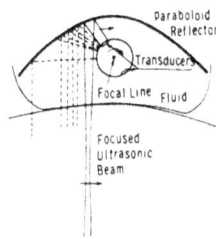

Figure 24. A High Speed Scanner by Means of a
Paraboloid Mirror System

ⓒ Pätzold, J., * Kranse, W., Kresse, H. and Soldner, R.:
Present State of an Ultrasonic Cross-section Procedure
with Rapid Image Rate, IEEE Trans. on Bio-Medical
Engineering, P. 263, 1970, P. 263, Figure 1.
* J. Pätzold, Siemens Aktiengesellschaft, Bereich
Medizinesche Technik, Erlangen, Germany.

Figure 25. A High Speed Scanner of a Stationary
Transducer System. (Uchida, et al.)

ⓒ Uchida, R., * Hagiwara, Y., and Irie, K.: Electronic
Scanning of Ultrasonic beam for Medical Diagnostic
Apparatus (a tentative English translation from the
Japanese title), Report of Japan Society Ultrasonics in
Medicine, 19th Meeting, No. 19-33, 1971 (in Japanese),
P. 65, Figure 2.
* R. Uchida, Japan Radiation and Medical Electronics,
Inc., 6-22-1, Murei, Mitaka, Tokyo, Japan.

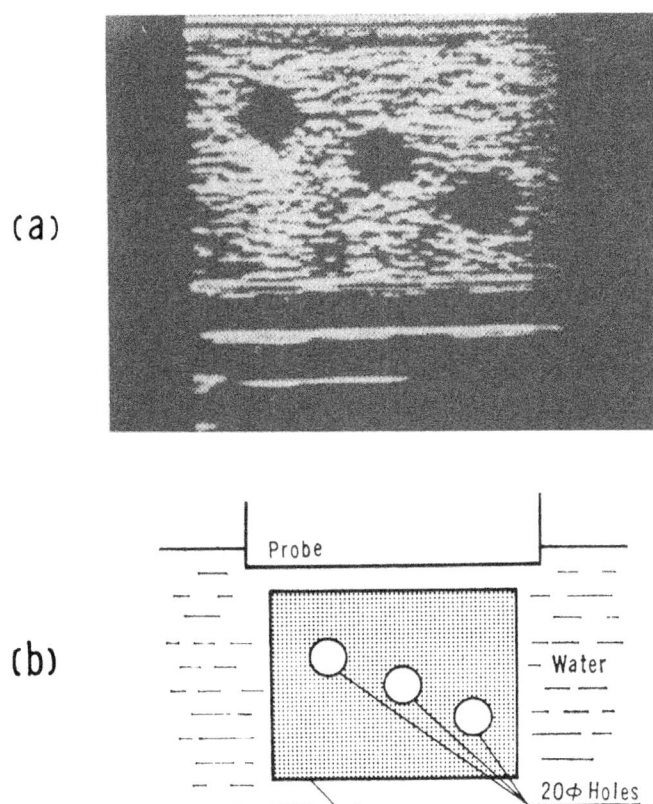

Figure 26. Test Tomogram Obtained by High Speed Scanning. (Uchida)

ⓒ Uchida, R., Hagiwara, Y., and Irie, K.: Electronic Scanning of Ultrasonic Beams for Medical Diagnostic Apparatus (a tentative English translation from the Japanese title), Report of Japan Society Ultrasonics in Medicine, 19th meeting, No. 19-33, 1971, P.66(a)-Figure 10, (b), Figure 8.

Methods of electronic scanning have also been considered for use in medical ultrasonics. Somer[75] reported a prototype experiment in which every unit transducer in an array made up by 21 units is each excited by individual local oscillators which are to make pulsed oscillation in a proper time sequence so the ultrasonic beam of the array can make a sector scanning. Okujima[76] recently reported an idea of a high speed scanning technique in a polarity correlation principle in which a fewer number of unit transducers are required for sublobe suppression.

2. Tomographic Display, Recording and Reproduction

Recently, several new techniques have been published on tomographic display. On one hand, efforts have been made toward color display by such a means as color-photography itself,[77] a certain electronic method,[78] or a certain digital computer method[79] The authors of this computer method, Yokoi and Ito have employed a method of converting analog signals of ultrasonic echoes into digital signals through an A-D converter to be memorized by a certain memory device. On memorization, echo information is processed so that every echo intensity is represented by three bit signals according to certain preset intensity levels of eight steps. On reproduction, the three-bit signals are applied on R (red), G (green) and B (blue) electrodes of a color kinescope tube, while the other memorized data are being reproduced on the kinescope screen. An example of the black and white reproduction of the color tomogram is shown in Figure 27. The authors have claimed that a single tomogram thus obtained in color presentation has information on intensity level simultaneously and that the sensitivity-graded pair method will be therefore needless in the clinic.

On the other hand, Baum[80] recently disclosed a quantized ultrasonography in which the usual tomogram of extended dynamic range is processed by a certain image quantizer that prints out each isodensitometric area in a code form. Then an acoustic contour map can be prepared for differential diagnosis. In Figure 28 is shown an example of the quantized image of the human orbit, clearly showing tumor tissues by black masses at Arrows A and B.

Robinson[81] at MIT, recently published a three-dimensional display of ultrasonic data for medical diagnosis. The method is based on a picture digitization and the use of a digital computer. The paper demonstrates the formation of a section at any arbitrary angle through the three-dimensional information. The author has indicated that this technique is to be incorporated in acoustical holography or other means than the current pulse-echo methods used in clinical routine.

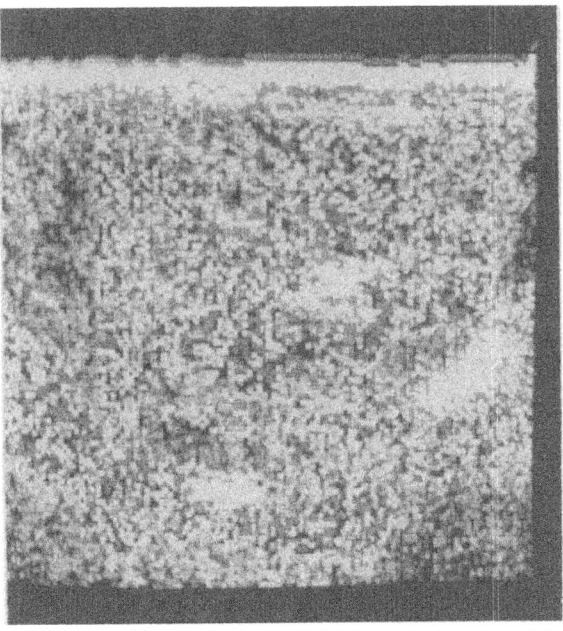

Figure 27. A Tomogram of the Liver and Gall Bladder (Yokoi) (The original picture was in color)[79]

© Hiromu Yokoi, M.D. Nissei Hospital, Itachi-Bori, Nishi-ku, Osaka, Japan.

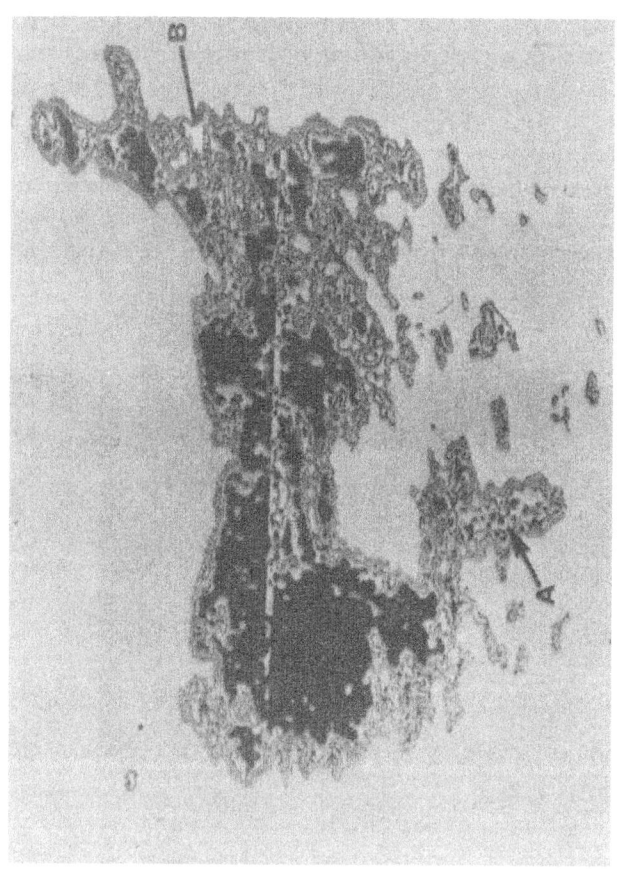

Figure 28. A Quantized Image of the Human Orbit (Baum)

© Baum, G.: Quantized ultrasonography, ultrasonics, P. 14, January, 1972. P. 15, Figure 2.

As to video recording techniques for ultrasonotomography, Ide and others [82] have been developing a system for clinical use. Kikuchi, Okuyama and others [83] are engaged in developing a system of multiple information recording and reproduction which is applicable to the synchronized tomography for the living heart. Details of the above two systems will be described in separate papers.

VI. SUMMARY

In this paper, the historical background of ultrasonic diagnostics was first reviewed briefly with some descriptions of A-mode display used in the early stage of the echo-method development. In the next place, the development of ultrasonic tomography was outlined. In Chapter II, the acoustical proportions of biological tissues were briefly summarized in such a way that anyone can reach the outlines of the concept easily. In Chapter III, recent findings concerning the ultrasonic pulse technique were described. In Chapter IV, present systems for ultrasonotomography were explained, together with the concept of sensitivity-graded tomograms, one of the useful kinds of software in medical diagnostics. In the last chapter, some new techniques were shown that are relevant to ultrasonic tomography. These techniques are being developed in this field toward the advancement of ultrasonic diagnostics.

The author hopes that this paper give some knowledge concerning the technique of ultrasonic imaging of human organs, currently used in clinics, so the knowledge would be of use in future development of holographic techniques that will be applicable to clinical medicine.

REFERENCES

Chapter I

1. Dussik, K. T., Über die Möglichkeit, hochfrequente mechanische Schwingungen als diagnostisches Hilfsmittel zu verwerten, Ztschr. ges. Neurolg. Psychiatry, 174, 153, 1942.

2a. Dussik, K. T., Dussik, F., and Wyt, L., Auf dem Wege zur Hyperphonographie des Gehirnes, Wien. med. Wchnschr., 97, 425, 1947.

2b. Dussik, K. T., Ultraschallanwendung in der Diagnostik und Therapie der Erkrankungen des zentralen Nervensystems, Ultraschall in der Medizin, P. 283, Zurich, Hirzel, 1949.

3. Ballantine, H. T., Bolt, R. H., Hueter, T. F. and Ludwig, G. D., On the Detection of Intracranial Pathology by Ultrasound, Science 112, 525, 1950.

4. Hueter, T. F. and Bolt, R. H., Ultrasonic Method for Outlining the Cerebral Ventricles, J. Acoustical Soc. America, 23, 160, 1951.

5. Güttner, W., Fiedler, G. and Pätzold, J., Über Ultrasschallabbildungen am Menschlichen Schädel, 2, 148, 1952.

6. Three references:
    a. French, L. A., Wild, J. J. and Neal, D., Detection of Cerebral Tumors by Ultrasonic Pulses, Cancer, 3, 705, 1950.

    b. French, L. A., Wild, J. J. and Neal, D., The Experimental Application of Ultrasonics to the Localization of Brain Tumors, J. Neurosurgery, 8, 198, 1951.

c. Wild, J.J. and Reid, J., The Effects of Biological Tissues on 15 mc Pulsed Ultrasound, J.A.S.A., Vol. 25, No. 2, March, 1953, Page 270.

7. Two references:
   a. Kikuchi, Y., Tanaka, K. and Uchida, R., Medical Application of Ultrasonic Flaw Detector, Report of Scientific Research of Ministration of Education, Japan, 1952. (Japanese)

   b. Kikuchi, Y., Tanaka, K. and Uchida, R., Ultrasonic Diagnosis of Intracranial Disease, Journal Acoustical Society Japan, 8 (2), 111, 1952. (Japanese)

8. Two references:
   a. Kikuchi, Y., Tanaka, K., Wagai, T. and Uchida, R., Early Cancer Diagnosis through Ultrasonics, J.A.S.A., 29 (7), 824, 1957.

   b. Kikuchi, Y., Recent Results of Research and Development in the Field of Ultrasonics in Japan, Proceedings III ICA, II, 1193, Stuttgart, 1959, Elsevier Pub. Co., Amsterdam.

9. Wagai, T., Miyazawa, R., Ito, K., and Kikuchi, Y., Ultrasonic Diagnosis of Intracranial Disease, Breast Tumors and Abdominal Diseases, P. 347, Ultrasonic Energy, ed., E. Kelly, University of Illinois Press, Urbana, 1965.

10. U.S. Atomic Energy Commission, Studies in Methods in Instruments to Improve the Localization of Radioactive Materials in the Body with Special Reference to the Diagnosis of Brain Tumors and the Use of Ultrasonic Techniques, AECU-3012, Minneapolis, 1955.

11. White, D.N., Ultrasonic Encephalography, Queen's University, Canada, 1970.

12. Tanaka, K., Wagai, T., Kikuchi, Y., Uchida, R., and Uematsu, S., <u>Ultrasonic Diagnosis in Japan</u>, Diagnostic Ultrasound, (Ed, C.C. Grossman, et al.) Page 27, Plenum Press, New York, 1966.

13. Two references:
   a. Fry, W.J., <u>New Approaches to the Study and Modification of Biological Systems by Ultrasound</u>, Ultrasonic Energy, (Ed., E. Kelly) Page 242, American Institute Biological Sciences, Wash., D.C., 1965.

   b. Fry, W.J., Fry, F.J., Kelly, E., Fry, T.A. and Leichner, G.H., <u>Ultrasound Transmission in Tissue Visualization</u>, Diagnostic Ultrasound, (Ed. C.C. Grossman, et al.) Page 13, Plenum Press, New York, 1966.

14. Two references:
   a. Edler, I., <u>The Diagnostic Use of Ultrasound in Heart Disease</u>, Ultrasonic Energy, (Ed. E. Kelly) P. 303, University of Illinois Press, Urbana, 1965.

   b. Hertz, C.H., <u>The Continuous Registration of the Movement of Heart Structure by the Reflectoscope Techniques-Method of Registration</u>, Ultrasonic Energy, (Ed. E. Kelly) University of Illinois Press, Urbana, 1965.

15. Edler, I., <u>Ultrasound Cardiography</u>, Ultrasound as a Diagnostic and Surgical Tool, (Ed. D. Gordon) Page 124, E & S Livingstone, Ltd, Edinburgh & London, 1964.

16. Wild, J.J. and Reid, J.M., <u>Progress in Techniques of Soft Tissue Examination by 15 MC Pulsed Ultrasound</u>, Ultrasound in Biology and Medicine, (Ed. E. Kelly) Page 30, American Institute of Biological Science, Washington, D.C., 1957.

17. Howry, D. H., Techniques Used in Ultrasonic Visualization of Soft Tissues, Ultrasound in Biology and Medicine, (Ed. E. Kelly), Page 49, American Institute Biological Science, Washington, D. C., 1957.

18. Gordon, D., Ultrasound as a Diagnostic and Surgical Tool, P. 100, E & S Livingstone Ltd, Edinburgh and London, 1964.

19. Ibid, Ref. 8a.

20. Kikuchi, Y., Recent Results of Research and Development in the Field of Ultrasonics in Japan, Proc. Third International Congress on Acoustics, 1959, (Ed. L. Cremer) Vol. II, P. 1193, Elsevier Pub. Co., Amsterdam, 1961.

21. Kossoff, G., Garrett, W. J. and Robinson, D. E., An Ultrasonic Echoscope for Visualizing the Pregnant Uterus, Ultrasonic Energy, (Ed., E. Kelly) P. 365, University of Illinois Press, Urbana, 1965.

22. Baum, G. and Greenwood, I. A., Current Status of Ophthalmic Ultrasonography, Ultrasonic Energy, (Ed. E. Kelly) P. 260, l.c., 1965.

23. Baum, G., A Comparison of the Merits of Scanned Intensity Modulated Ultrasonography Vs Unscanned A-scope Ultrasonography, Diagnostic Ultrasound, (Ed., C. C. Grossman), P. 59, Plenum Press, New York, 1966.

24. Tanaka, M., Oka, S., Watanabe, H., Kato, T., Shishido, S., Ebina, T., Kosaka, S., Terasawa, Y., Unno, K., Nitta, K., Kikuchi, Y., Okuyama, D. and Uchida, R., Ultrasono-Tomography of the Intrapelvic Organs (a tentative title translated from the Japanese), Proc. of 13th Meeting of Japan Society of Ultrasonic Medicine, 79, 1968. (In Japanese)

25. Kikuchi, Y., Ultrasonic Approach to Internal Medicine--Application to Diagnostic Examination, Japan Journal of Medicine, Vol. 2, No. 2, PP 70-80, 1970.

26. Tanaka, M., Oka, S., Ebina, T., Kosaka, S., Terasawa, Y., Unno, K., Kikuchi, Y. and Uchida, R., Ultrasonic Diagnosis of the Mediastinal Organs Including the Heart and Great Vessels (5th report)-Ultrasono-Tomography of the Heart, Great Vessels and Other Mediastinal Organs by Means of the Plan Position Indication from the Esophagus, (a tentative title translated from the Japanese) Kokenshi 19, P. 425, 1967. (In Japanese)

27. Haneda, Y., Tanaka, M., Naritomi, T., Oshiba, M., Kosaka, S., Terasawa, Y., Unno, K., Yamaura, G. and Nitta, K., Ultrasonic Visualization of Mediastinal Tumor-Especially in Regard to Transtrachael Method, Proc. 12th Meeting, Japan Society Ultrasonics in Medicine, P. 29, 1967.

28. Kikuchi, Y., Ebina, T. and Tanaka, M., Some Improvements in Ultrasonotomograph for the Heart and Great Vessels, 2. Tomography Synchronized with any Cardiac Phase, IEEE Ultrasonic Symposium, J3, Cleveland, Ohio, 1966.

29. Kikuchi, Y., Okuyama, D., Ebina, T., Oka, S. and Tanaka, M., Cardiac Kineto-Ultrasonotomography, Reports of 6th I.C.A., Tokyo, M-1-8, P.M-25, 1968.

Chapter II

30. Tanaka, K., Kikuchi, Y. and Uchida, R., Ultrasonic Diagnosis of Brain Tumor, Proc. 3rd ICA, Stuttgart, 1959, Vol. 2, P. 1291, Elsevier Pub. Co., Amsterdam, 1961.

31. Tanaka, K., <u>Diagnosis of Brain Disease by Ultrasound,</u> Shindan-to-chiryo Sha, Ltd, Tokyo, 1969.

32. Kikuchi, Y. (Chief Editor), Cho-onpa Igaku (Ultrasonics in Medicine), Igaku Shoin, Ltd, Tokyo, 1966. (in Japanese)

33. White, D.N., <u>Ultrasonic Encephalography,</u> 1c, Ref. 11.

34. Tanaka, K., Kikuchi, Y. and Uchida, R., <u>Ultrasonic Detection of Anatomical Abnormalities in the Human Cranium</u>, Part 2, Reports of the 1953 Spring Meeting, Acoustical Society of Japan, No. 2-9, May, 1953. (in Japanese)

35. Wells, P.N.T., <u>Physical Principle of Ultrasonic Diagnosis,</u> P. 13, Academic Press, London, 1969.

36. Goldman, D.E. and Hueter, T.F., <u>Tabular Data of the Velocity and Absorption of High-Frequency Sound in Mammalian Tissues,</u> Journal Acoustical Society of America, Vol. 28, No. 1, P. 35, 1956.

37. Ishikawa, S., <u>On the Measurement of Ultrasonic Attenuation in Biological Tissues, Especially Brain Tissue</u> (a tentative title translated from the Japanese) Nihon Geka Hokan, Vol. 33, P. 923, 1964. (Japanese)

38. Nomoto, O., PP 566-571, Ref. 32.

39. Pohlman, R., <u>Über die Absorption des Ultraschalls im menschlichen Gewebe und ihre Abhängigkeit von der Frequenz,</u> Phys. Zeitschr., Vol. 40, P. 159, 1939.

40. Heuter, T.F., <u>Messung der Ultraschallabsorption in Tierischen Geweben und Ihre Abhängigkeit von der Frequenz,</u> Naturwissenschaften, Vol. 35, P. 285, 1948.

41. Yoshioka, K., On the Ultrasonic Absorption of the Brain-A Measurement in a Living Tissue (a tentative English translation of the Japanese title), Reports of Japan Society of Ultrasonics in Medicine, No 48, the 11th meeting, Tokyo, 1967. (in Japanese)

42. Carstensen, E. L., Measurement of Dispersion of Velocity of Sound in Liquids, Journal Acoustical Society of America, Vol. 26, P. 858, 1954.

43. Two references:
    a. Goldman, D. E. and Richards, J. R., Measurement of High-Frequency Sound Velocity in Mammalian Soft Tissues, Journal Acoustical Society of America, Vol. 26, P. 981, 1954.

    b. Begui, Z. E., Acoustic Properties of the Refractive Media of the Eye, Journal Acoustical Society America, Vol. 26, P. 365, 1954.

44. Two references:
    a. Kikuchi, Y., Okuyama, D., Kasai, C., and Yoshida, Y., Measurements on the Sound Velocity and the Absorption of Human Blood in 1-10 MHz Frequency Range, Record of Electronics and Communications Engineering Conversazione, Tohoku University, Vol. 41, P. 152, 1972. (Japanese)

    b. Welkowitz, W., Ultrasonics in Medicine and Dentistry, Proceedings IRE, P. 1059, Aug. 1957.

45. Van Venrooij, G. E. P. M., Measurement of Ultrasound Velocity in Human Tissue, Ultrasonics, P. 240, October, 1971.

46. Two references:
    a. Cedrone, N. P., and Curran, D. R., Electronic Pulse Method for Measuring the Velocity of Sound in Liquids and Solids, J. Acoustical Society of America, Vol. 26, P. 963, 1953.

b. Holbrook, R. D., Pulse Method for Measuring Small Change in Ultrasonic Velocity in Solids with Temperature, J. Acoustical Soc. America, Vol. 20, P. 590, 1948.

Chapter III

47. Firestone, F. A., See Heuter, T. F. and Bolt, R. H., Sonics, John Wiley & Sons, Inc., New York, P. 393, 1966.

48. Kossoff, G., Robinson, D. E. and Garrett, W. J., Ultrasonic Two-Dimensional Visualization Techniques, IEEE Transactions, Vol. SU-12, P. 31, 1965.

49. Wells, P. N. T., Physical Principles of Ultrasonic Diagnosis, Academic Press, London, P. 38, 1969.

50. Kikuchi, Y., Okuyama, D., Kasai, C. and Tatebayashi, K., Generation of Extremely Short Ultrasonic Pulses into Liquid by Acoustic Backing Transducer, Reports of the Japan Society of Ultrasonics in Medicine, 17th meeting, P. 53, 1970. (in Japanese)

51. Kasai, C., Okuyama, D. and Kikuchi, Y., Active Damping of Piezo-electric Transducer to Generate Extremely Short Ultrasonic Pulse, Medical Ultrasonics, J. of the Japan Society of Ultrasonics in Medicine, Vol. 8, P. 40, 1970.

52. Kasai, C., Okuyama, D. and Kikuchi, Y., Transient Response of Piezo-electric Composite Transducer, Reports of Professional Group on Ultrasonics, Inst. Electronics Communications Engineering, Japan, US-70-13, 1970. (in Japanese)

53. McSkimmin, H. J., Transducer Design for Ultrasonic Delay Lines, J. A. S. A., Vol. 27, P. 302, 1955.

54. Kossoff, G., The Effect of Backing and Matching on the Performance of Piezo-electric Ceramic Transducers, IEEE Transactions, Vol. SU-13, P. 20, 1966.

55. Kikuchi, Y., Okuyama, D. and Kasai, C., Ultrasonic Transducers with a Quarter Wavelength Layer for the Generation of Short Ultrasonic Pulses, Reports of the 1971 Spring Meeting, Acoustical Society of Japan, P. 189. (In Japanese)

56. Kikuchi, Y., Okuyama, D. and Kasai, C., Generation of Short Ultrasonic Pulses by Means of an Ultrasonic Transducer with an Intermediate Layer of a Quarter Wavelength, Reports of Professional Group on Ultrasonics, Inst. of Electronics Communications Engineers, Japan, US-71-13, 1971. (In Japanese)

57. Backhans, H., and Trendelenburg, F., Z. Tech. Phys., Vol. 7, P. 630, 1926. See also Osterhammel, K., Akust. Z., Vol. 6, P. 73, 1941.

58. Lommel, E., Die Beugungserscheinungen einer kreisrunden Öffnung und eines kreisrunden Schirmchens, Abh. der K. Bayer. Akad. der Wissen., Vol. 15, P. 233, 1886.

59. O'Neil, H. T., Theory of Focusing Radiators, J. A. S. A., Vol. 21, P. 516, 1949.

60. Torikai, Y., Sound Field of a Spherically Concave Transducer, Report of Professional Group on Ultrasonics, Inst. of Electronics Communications Eng. Japan, October, 1957. (In Japanese)

61. Kikuchi, Y. and Tanaka, M., Some Improvements in Ultrasonotomography for the Heart and Great Vessels - Azimuthal Resolution of Concave Transducers, Reports of Professional Group on Ultrasonics, Institute Elect. Communications Engineers, Japan, October 26, 1966. (In Japanese)

62. Kikuchi, Y., Okuyama, D. and Kasai, C., Sound Field Characteristics of Extremely Short Ultrasonic Pulse Radiated by a Concave Transducer, Reports of Japan Society Ultrasonics in Medicine, 19th meeting, 19-31, P. 61, 1971. (In Japanese)

63. Kikuchi, Y. (Chief Editor), Ultrasonic Material Testing, Japan Society for Promotion of Science, Tokyo, 1964. (In Japanese)

64. Krautkrämer, J., Determination of the Size of Defects by the Ultrasonic Impulse Echo Method, British Journal Applied Physics, Vol. 10, P. 240, 1959.

65. Kimura, K. and Suzuki, T., On Experimental Models of Ultrasonic Transducers for Inspection in Near-Field Region (a tentative translation of the Japanese title), J. NDI, Vol. 9, P. 270, 1960. (In Japanese)

66. Kossoff, G., A Transducer with Uniform Intensity Distribution, Ultrasonics, P. 196, October, 1971.

67. Olofsson, S., An Ultrasonic Optical Mirror System, Acustica, Vol. 13, PP 361-367, 1963.

68. Fry, W.J., Leichner, G.H., Okuyama, D., Fry, F.J. and Fry, E.K., Ultrasonic Visualization System Employing New Scanning and Presentation Methods, Journal Acoustical Society America, Vol. 44, No. 5, PP 1324-1338, 1968.

Chapter IV

69. Howry, D.H. and Gordon, D., Ultrasonic Tomography, Ultrasound as a Diagnostic and Surgical Tool. (Ed. D. Gordon) P. 104, E & S Livingstone, Ltd, Edinburgh and London, 1964.

70. Ebina, T., Kikuchi, Y., Oka, S., Tanaka, M., Kosaka, S., Uchida, R. and Hagiwara, Y., <u>The Diagnostic Application of Ultrasound to the Diseases in Mediastinal Organs-Ultrasono-Tomography for the Heart and Great Vessels</u> (the first report), Science Reports of the Research Institutes Tohoku University Series C (Medicine), Vol. 12, No. 1, March, 1965, P. 58.

71. Two references:
    a. Kikuchi, Y., <u>Way to Quantitative Examination in Ultrasonic Diagnosis</u>, Medical Ultrasonics, Vol. 6, No. 1, PP 1-8, 1968.

    b. Kikuchi, Y., <u>On the Significance and Problems of the Sensitivity-graded Tomograms in Ultrasonic Diagnostics in Medicine,</u> Record of Japan Institute Ultrasonics in Medicine, 18th meeting, Sendai, 1970. (In Japanese)

<u>Chapter V</u>

72. Asberg, A., <u>Ultrasonic Cinematography of the Living Heart</u>, Ultrasonics, Vol. 5, P. 113, 1967.

73. Pätzold, J., Kranse, W., Kresse, H. and Soldner, R., <u>Present State of an Ultrasonic Cross-section Procedure with Rapid Image Rate</u>, IEEE Trans. on Biomedical Engineering, P. 263, 1970.

74. Uchida, R., Hagiwara, Y. and Irie, K., <u>Electronic Scanning of Ultrasonic Beam for Medical Diagnostic Apparatus</u> (a tentative English translation from the Japanese title), Report of Japan Society of Ultrasonics in Medicine, 19th meeting, No. 19-33, 1971. (In Japanese)

75. Somer, J.C., <u>Electronic Sector Scanning for Ultrasonic Diagnostics</u>, Ultrasonics, P. 153, 1968.

76. Okujima, M., Endoh, N. and Nishimura, M., Basic Research of High Speed Scan in Ultrasonic Tomograph for Diagnosis, Reports of 1970 Autumn meeting, Acoustical Society Japan, No. 3-2-5, 1970. (In Japanese)

77. Yokoi, H., Yoshitatsu, M. and Tatsumi, T., A Trial of Photograph by Different Sensitivities, Report of Japan Society of Ultrasonics in Medicine, 19th meeting, No. 19-5, 1971. (In Japanese)

78. Ide, M. and Masuzawa, N., Electronic Color Tomograph, Report of Japan Society of Ultrasonics in Medicine, 21st meeting, P. 123, 1972. (In Japanese)

79. Yokoi, H. and Ito, K., Ultrasonic Diagnostic Equipment for Simultaneous Tomogram Method, Toshiba Review, Vol. 27, No. 7, PP 661-665, 1972. (Japanese)

80. Baum, G., Quantized Ultrasonography, Ultrasonics, P. 14, January, 1972.

81. Robinson, D. E., Display of Three-dimensional Ultrasonic Data for Medical Diagnosis, Journal Acoustical Society of America, Vol. 52, No. 2 (Part 2), P. 673, 1972.

82. Ide, M., Masuzawa, N. and Wagai, T., Pattern Processing in the Reproduction of VTR Tomograms, Report of Japan Society of Ultrasonics in Medicine, 21st meeting, P. 111, 1972. (In Japanese)

83. Two references:
    a. Kikuchi, Y. and Okuyama, D., Ultrasono-Cardio-Tomography, Japan Electronic Eng. No. 47, PP 53-60, October, 1970.

b. Kikuchi, Y., Okuyama, D., Ebina, T., Tanaka, M., Terasawa, Y. and Uchida, R., Multi-Information Recording and Reproduction in the Ultrasono-Cardio-Tomography, Acoustical Holography, (Ed., G. Wade) PP 113-126, Plenum Press, New York, 1972.

NEW FORMS OF ULTRASONIC AND RADAR IMAGING

Winston E. Kock

Visiting Professor, The University
of Cincinnati, and Consultant
The Bendix Corporation

ABSTRACT

Recent developments in synthetic aperture techniques to be described, which are applicable to both ultrasonic and radar imaging, include: a solution to the ambiguity problem, creating end-fire gain and synthetic gain against moving targets for both monostatic and bistatic systems. Other subjects to be discussed include acoustic kinoforms, real-time holographic detection of concealed weapons, synthetic aperture hologram interferometry, and new developments in (1) real-time imaging with sonar arrays comprising thousands of elements, and (2) seismic imaging.

EARLY ACOUSTIC HOLOGRAMS

If one defines an acoustic hologram as the photographically recorded interference pattern between a set of coherent sound waves of interest and a coherent reference wave generated by the same source, then one can say that the first acoustic holograms were made at the Bell Telephone Laboratories in 1950.[1] In that work, only the holograms (the interference patterns) were of interest and no wave reconstruction was performed; nevertheless, the method employed in recording the interference patterns is still useful today.

The interference patterns were made visible by photographic scanning, using a scanning microphone. To generate an interference pattern, two wave sets from a common source of coherent sound waves were used. Because sound wave signals can be conducted through wire lines directly to the receiver without passing through the microphone, the usual spatial form of reference wave addition is not required in making acoustic holograms. The sound recordings could have been made in the same way in which several microwave holograms were made at that time, with a spatial reference wave as shown in Fig. 1. Here a single coherent source feeds energy both to the microwave lens at the left, and to the horn at the right, which provides the reference waves. The interference pattern is formed in the rectangle outlined by the dotted lines. A scanning microwave pickup (a microphone, if an acoustic hologram is being recorded) has its signal rectified and amplified, and then fed to a small neon lamp affixed to it. The intensity of the light varies in accordance with the microwave field (or the sound level) encountered by the pickup. A camera, set at time exposure, records the variation of light intensity. The desired interference pattern is thus built up by scanning the wave field somewhat after the manner in which a television image is formed.

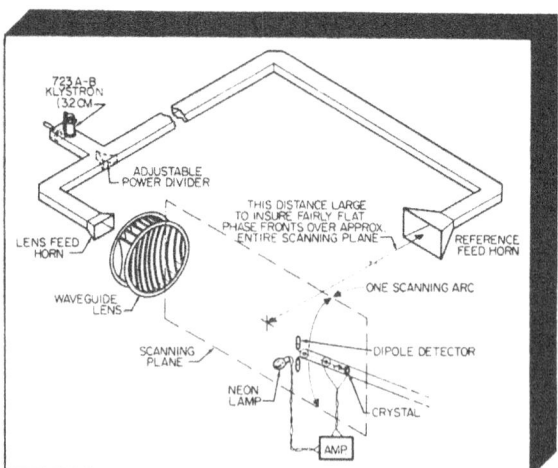

Fig. 1. In making a microwave hologram, reference waves from the feed horn interfere with waves emanating from the lens. Interference patterns are detected in the scanning plane by the dipole, amplified to light the lamp and photograph.

NEW FORMS OF ULTRASONIC AND RADAR IMAGING

For the sound wave case, when only the amplitude of the wave field of interest was desired, the method of Fig. 2 was employed. The mechanism caused the microphone to follow circular vertical paths while the whole scanning mechanism moved slowly backwards (to the right of the figure). Fig. 3 is an example of an amplitude-only pattern of sound waves issuing from a horn showing the directional effect of the horn as a sound wave radiator. Fig. 4 is a photo of the actual mechanism used in forming these early acoustic holograms.

As noted above, the reference wave need not be introduced in spatial form for an acoustic interference pattern. Fig. 5 shows how the spatial reference waves of Fig. 1 can be bypassed in generating an interference pattern of sound emanating from a telephone receiver. A coherent oscillator signal is fed both to the receiver (acting as the sound source) and to the neon tube (which is simultaneously being energized) by the (amplified) microphone signal. At integral wavelength circles (having the receiver as a center), the two signals will add (constructive interference)

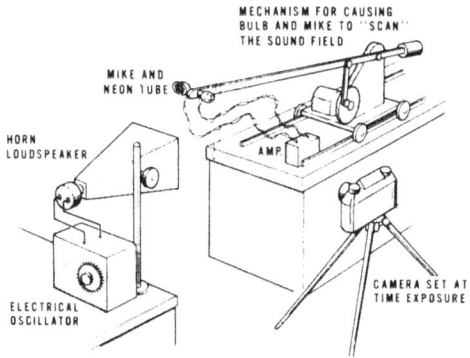

Fig. 2. A microphone-light combination can scan an area of interest and thus record on film the space pattern of sound radiated by an acoustic horn.

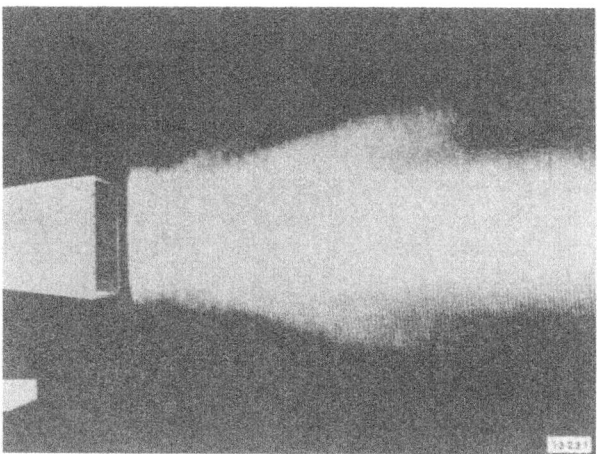

Fig. 3. The sound pattern of a pyramidal horn, having a six-inch square aperture and radiating sound waves having a frequency of 9 Khz, as recorded by the arrangement of Fig. 2.

Fig. 4. A photograph of the scanning mechanism used to generate Fig. 3.

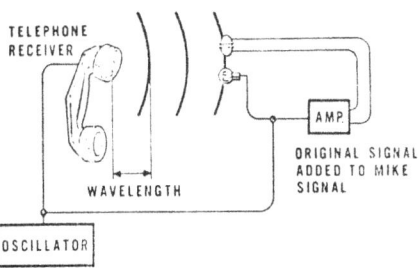

Fig. 5. By adding electronically, a signal from the coherent oscillator to the signal picked up by the scanning microphone, the resulting constructive and destructive interference effects produce a pattern identical to one generated with a spatial reference wave.

whereas at half wavelength points they will cancel (destructive interference). The record of this interference pattern is shown in Fig. 6. As can be seen, this unreconstructed "hologram," through its interference *fringe* pattern, makes visible not only the progress of the sound waves, but also the fact that a small acoustic transducer such as a telephone receiver does not provide the directionality of the horn of Fig. 3.

The microwave hologram of Fig. 7, which is similar to the acoustic hologram of Fig. 6, could disclose, with only the last few fringes at the far right (because of their curvature), the existence of the microwave source (the waveguide at the left), and its approximate location. Such unreconstructed holograms have therefore been suggested as a method of detecting concealed weapons.[2] Forward scatter procedures, in which the concealed object passes between a source and a hologram-generating receiver, might also be useful for this purpose. Fig. 8 is such an unreconstructed *acoustic* hologram,[2] showing the definite forward-scatter pattern located behind a disk. The use of such unreconstructed, forward-scatter acoustic holograms might

Fig. 6. The electronically injected reference signal of Fig. 5 shows, by the hologram thus generated, the progress of the sound waves emerging from the receiver of a telephone handset.

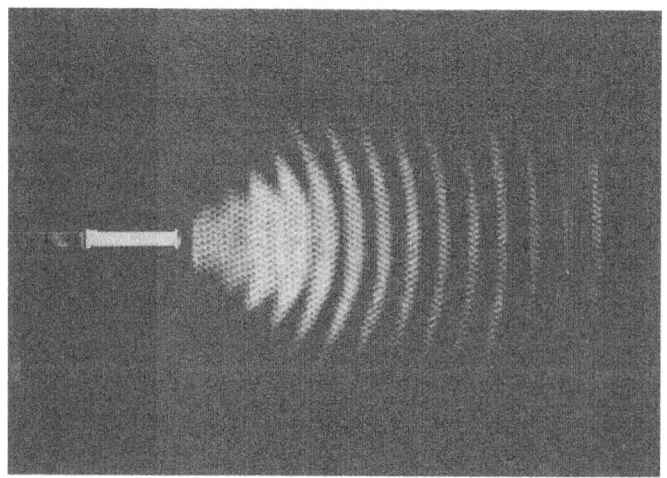

Fig. 7. Very short electromagnetic waves are here portrayed emerging from a microwave wave-guide (the arrangement of Fig. 1 was used in making this photo).

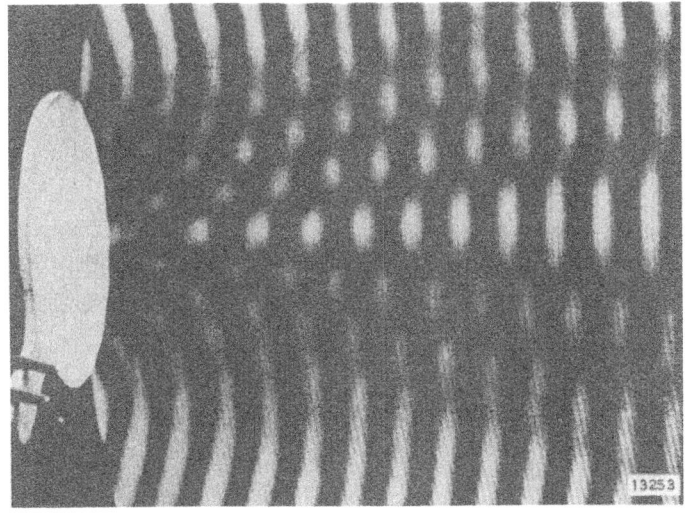

Fig. 8. The forward-scatter fringe pattern (hologram) formed by a 90° reference wave and the waves diffracted into the shadow region of a disk.

also be useful in the nondestructive testing of structures immersible in a liquid, whereby undesired enclosed air pockets could be detected.

Because the electronically introduced reference signal of Fig. 5 corresponds to a plane wave spatial reference whose wave fronts are parallel to the scanning plane of the microphone, an in-line hologram can be produced by orienting the scanning plane so as to be perpendicular to the waves of interest. This was done for Fig. 9. The arrangement is similar to that of Fig. 10 with the electronically injected reference corresponding to the plane reference waves in the upper path of the figure, and the waves from the acoustic lens of Fig. 9 corresponding to the spherical waves diverging from the right beam splitter in Fig. 10. The hologram of Fig. 9 resembles a zone plate pattern so that if the white rings of Fig. 9 were to be transformed into acoustically opaque structure, the result would resemble the somewhat similar acoustic zone plate described by Lord Rayleigh.[3]

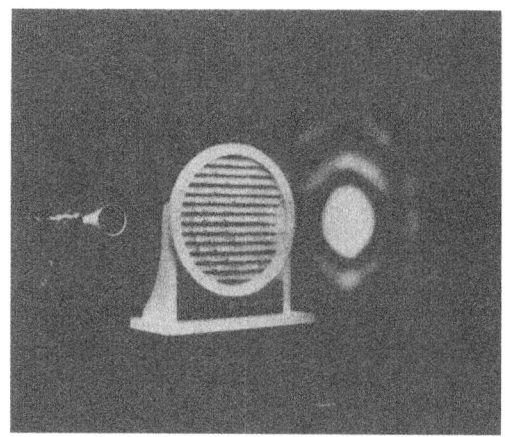

Fig. 9. By causing the scanner of Fig. 4 to scan a plane which is perpendicular to the direction of wave propagation, a cross-section of the zone structure of the acoustic radiator pattern is portrayed (corresponding to the original in-line holograms of Gabor).

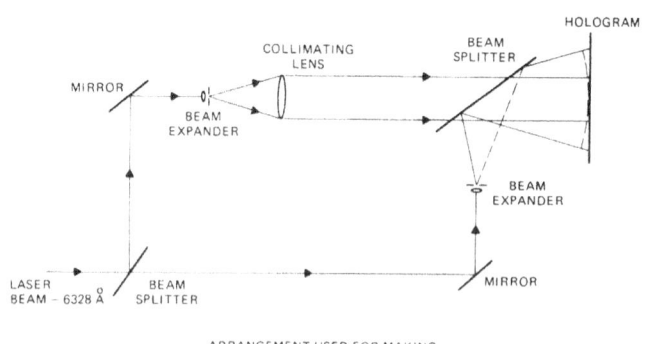

Fig. 10. A procedure for generating, very accurately, a photographic zone plate, for possible later use as a rather high quality optical "lens."

Because the directional properties of certain microphones (nonreversible ones, such as the carbon microphone, extensively used in telephone transmitters) cannot be determined by reciprocity (i.e., by using them as acoustic radiators), their directional patterns were obtained by using a scanning sound *source*. Thus, in Fig. 5, the oscillator signal which energized the telephone receiver was fed instead to the scanning (radiating) unit, and the signal received by the telephone microphone was fed to the amplifier. Fig. 11 was the result. It is quite similar to the pattern of Fig. 6, since both units are small compared to a wavelength and therefore exhibit negligible directivity.

An interesting mixup occurred when the photo of Fig. 11 was first published.[4] The journal involved had learned of the experiments and had requested permission to publish a brief report on some of the results sound portrayed. Numerous photos were accordingly provided, with six being used, including Fig. 11. The article's title was "Photographing Sound Waves" and the caption for Fig. 11 was "Pattern of Sound Waves from a Telephone Handset." Observant readers noted that the carbon microphone was at the

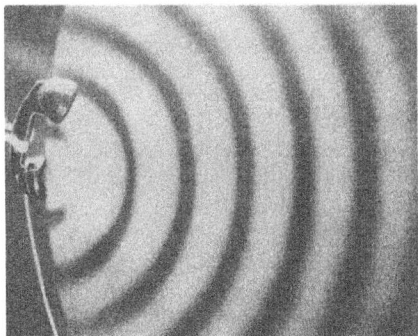

Fig. 11. By causing the pick-up microphone of the scanner of Fig. 4 to *radiate* sound, and connecting the microphone output of the telephone hand set to the scanning neon tube, the directional *receiving* properties of the carbon telephone microphone are made visible. This technique of scanning the sound *source* has recently been revived for acoustic holograms.

center of the circles and accordingly wrote the journal, pointing out that sound does not issue from the carbon transmitter of a handset. As a result, a second article had to be printed, showing both Figs. 6 and 11, and explaining the use of a scanning sound source in the case of the microphone pattern. Fig. 11 is thus of interest in modern acoustic holography as it is the forerunner of the scanned-source procedure proposed 17 years later.[5,6]

In the article[4] which included Fig. 11, experiments were also described in which a succession of holograms such as Fig. 12 were recorded, with the relative phase of the reference wave changed slightly for each one; up to a maximum value of 360°. A motion picture was then made, with these recorded in succession, "the result being a motion picture presentation of waves moving outward, . . . converging to a focus, and then diverging again" (from Reference 4).

The experiments thus demonstrated several holography firsts, including the first use of a scanning source, an early movie using holograms (true hologram movies[7] were

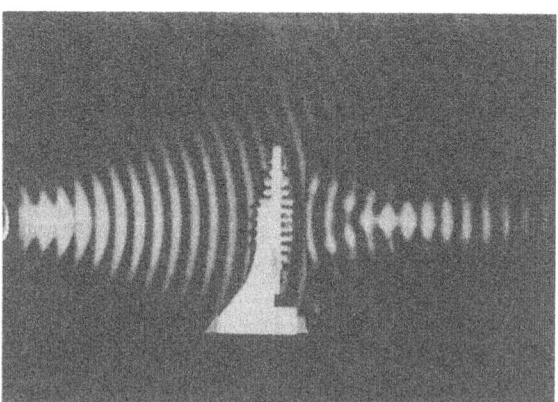

Fig. 12. The fringes in this (static) pattern of sound waves being focused by an acoustic lens can be shifted slightly by modifying the phase of the reference wave. Successive such photos, when repeated rapidly, provide the viewer with a motion picture of sound waves moving toward the lens and continuing on through the focal point.

first reported much later) and the first recorded use of
an off-axis reference wave in holography as noted in the
well-known German text on holography:[8] "Before Leith and
Upatneiks, Kock and Harvey in 1951, had already used off-
axis reference waves for providing unambiguous phase de-
terminations in diffracted wave fields."

THE SYNTHETIC APERTURE CONCEPT

One of the most extensive uses of nonoptical holograms
has occurred in the microwave radar field in the form of
synthetic aperture radar. Because of the outstanding success of such hologram radars, similar hologram concepts are
beginning to be examined in the sonar field. In 1971,
shortly before the announcement of his award of the 1971
Nobel Prize in Physics for holography, Dennis Gabor commented:[9]

"Unknown to me, a most interesting branch of holography was developing from 1955 onwards at the Willow Run
Laboratory attached to the University of Michigan. It was
holography with electromagnetic waves, and reconstruction
by light, which was called 'Side Looking Radar' or 'Synthetic Aerials.' It was classified work; the first publication by Cutrona, Leith, Palermo and Porcello occurred in
1960. Reconstructions of the object plane by illumination
with a monochromatic mercury lamp were of impressive perfection. So, curiously, in the first 12 years, the aim of
holography was the reconstruction with light of electron or
X-ray records, with wavelengths about 100,000 times shorter
than light, and reconstruction with light of electromagnetic holograms with wavelengths 100,000 times longer."
Because Emmett Leith was one of the principal contributors
to synthetic aperture development, and because he also was
with Upatneiks the first to use the laser in holography,
Gabor has commented:[10] "Emmett Leith arrived at holography
by a path just as adventurous as mine was. I came to it
through the electron microscope, he through side-looking,
coherent radar."

Thus, whereas Gabor's brilliant conception of optical
holography in 1947 lay almost dormant until 1963, the ele-
tromagnetic form received extensive attention all through
the 1950's. As noted by Gabor (above) the airborne form
of synthetic aperture radar was pioneered by Dr. L. J.
Cutrona and his group at the University of Michigan[11,12]

and sizable support was given it in classified projects in numerous laboratories across the U.S. Indeed, it is likely that this early classified program was in large part responsible for the extreme rapidity of growth in the development of true optical holography from 1963 on, particularly at the University of Michigan.

As University of Michigan's Professor William G. Dow has noted, the involvement of numerous University of Michigan professors in this classified work later enabled them to teach their students the subject of optical processing and holography in a much more timely and effective way.[13] Actually, the relation of coherent radar to holography was not appreciated until rather recently[14,15,16] but Emmett Leith, a key contributor to this radar development, instinctively applied the side-looking radar technique of offset beams,[11] i.e., the two-beam recording technique, in his first experiments in the optical holography. The success of synthetic aperture radar led this author to consult with Dr. Cutrona in 1957 on the possibility of applying this radar concept to sonar devices. Unfortunately, when the original synthetic aperture technique is applied to sonar, the low propagation velocity of sound somewhat limits its usefulness; however, as we shall see, new procedures have recently been described which can materially extend the use of synthetic aperture procedures in sonar applications.[17]

For simplicity we define hologram radar or sonar systems as those which employ waves having extremely high temporal coherence, and which utilize that coherence with an equally coherent reference wave to generate (e.g., by synchronous detection), a recorded interference pattern (a hologram) for later processing (reconstruction) of the reflections from objects. As noted, many of the early synthetic aperture developments took place in the radar field during the middle 1950's, with the acoustic versions following some years later. In that early work and until quite recently, the synthetic aperture concept was considered as a Doppler detection process. Even in the classic 1966 paper of Cutrona, Leith, Porcello, and Vivian,[12] no mention of holograms was made. In 1967, the relation of this technique to holography was discussed at length,[14] and since then numerous publications have appeared[15,16,18,19,20,21,22,23] discussing this connection, with one[16] stating that "the holographic viewpoint appears more flexible than the communications theory or cross-correlation viewpoint and has led to designs which are not easily explicable from the latter viewpoint.

The earliest *published* paper on hologram radar (as defined above) was one authored by G. L. Rogers.[24] In those 1955 experiments, the reflector was moving, and the radar was stationary. Hologram concepts were employed, including the use of a "signal taken from the transmitter's master oscillator, suitably attenuated, and fed into the receiver, where it combines with the downcoming (reflected) signal in accordance with the well-known 'coherent demodulator' technique" (quoted verbatim from Reference 24). Rogers, in these experiments, generated a one-dimensional image of a Dakota aircraft, which had purposely been sent up, so as to permit him to recover, "by diffraction from one of the aerial records" such as image. In 1966, G. L. Tyler[25] discussed a bistatic form of hologram radar, noting that his bistatic radar mapping scheme corresponded to the process of making a (radar) hologram and "playing it back." The widest use, presently, of hologram (synthetic aperture) systems involves airborne radar, although synthetic aperture acoustic (ultrasonic) systems are now being examined.[26]

In a synthetic aperture radar system,[12] an aircraft moving along a very straight path continually emits successive microwave pulses. The frequency of the microwave signal is very constant (the signal remains coherent with itself for very long periods). During these periods the aircraft may have traveled several thousand feet, but because the signals are coherent, all the many echoes which return during this period can be processed as though a single antenna as long as the flight path had been used. The effective antenna length is thus quite large and this large "synthetic" aperture provides records having extremely fine detail.

The photographic record of the echoes received by such a coherent radar is a form of hologram with the microwave generator which provides the illuminating signal also providing a reference wave. The reflected signals received along the flight path are made to interfere with this reference signal (by synchronous detection), and the complex interference pattern thereby generated and photographed is a form of hologram.

The method of operation is sketched in Fig. 13. For simplicity, only one reflecting point is shown. Waves returning from this point have spherical wave fronts whereas the oscillator reference signal acts like a set of plane waves perpendicular to the path of the airplane. The received signal is combined with the coherent reference signal and amplified to intensity modulate a cathode-ray tube trace as shown in Fig. 14. Each vertical line in that figure thus plots signals received from all range points, with the points at greater range being recorded near the top of the vertical trace. As the airplane moves along and new pulses are emitted, the film is indexed to record a new set of returns on a new vertical line.

For the case of only one reflecting point at fixed range, the upward moving cathode ray beam would, for every pulse, be modulated only at that one point in range, and the result would be a single horizontal line of recorded echoes. But this line is not continuous. The returning waves are circular and as the slant azimuth range from the aircraft to the reflecting point changes, the combination of return and reference waves produces successive

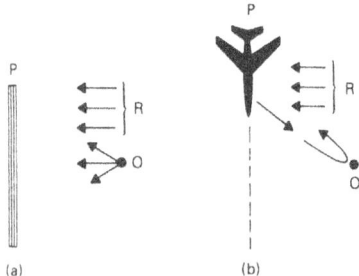

Fig. 13. In a synthetic aperture radar, echoes from the reflecting point are received over a considerable interval and assembled by a holographic method that creates interference patterns between the return signal and a reference sampled from the transmitted signal. The result is a synthetic-aperture, high-resolution antenna whose length is that of the flight path.

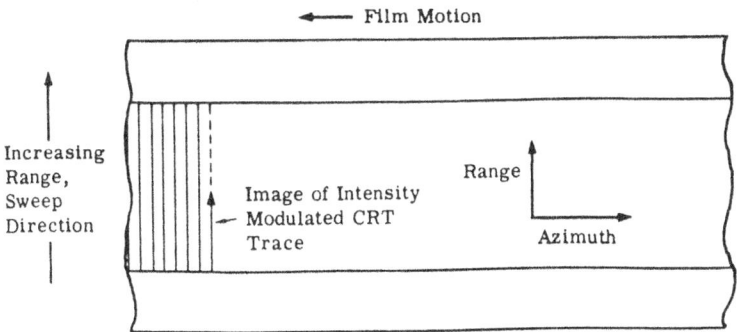

Fig. 14. A hologram record in a synthetic aperture radar is formed by photographing repeated intensity-modulated cathode ray tube traces.

constructive and destructive interference. At the greater slant angles, this succession of in-phase and out-of-phase conditions occurs rapidly, whereas when the aircraft is practically abreast of the point, it occurs slowly. The resulting record is a one-dimensional zone plate hologram, as sketched in Fig. 15. When illuminated by laser light, as shown in Fig. 16, it would reconstruct the reflecting point just as a hologram does. Indicated in the figure are two reflecting points which are displaced appreciably in range and slightly in azimuth. All reconstructed images fall on a tilted plane as shown, with the tilt of this plane being determined by the amount of radar vertical tilt.

A typical microwave hologram as generated by a side-looking radar is shown in Fig. 17, and an enlarged portion of this radar hologram is shown in Fig. 18. In this latter figure, a prominent one-dimensional zone plate is seen near the bottom of the central blank area. As in optical holograms, this assemblage of microwave holograms does not identify the object or area in view. Nevertheless, the *processed* hologram yields photos of extremely good detail. One of the early examples, the city of Washington, D.C. and its environs, is shown in Fig. 19.

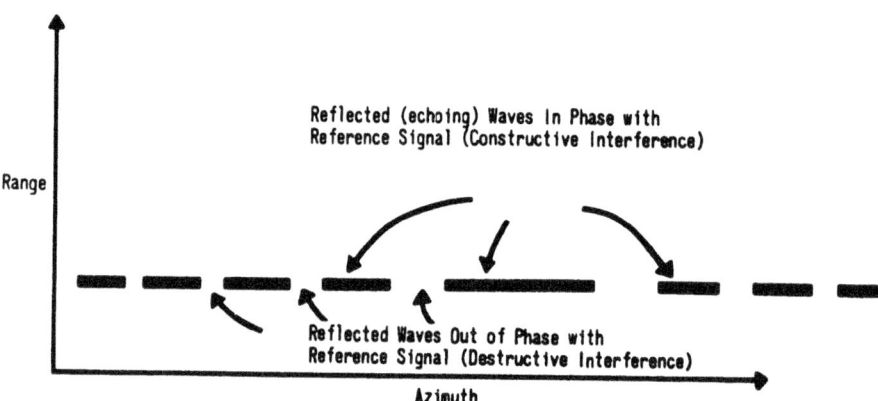

Fig. 15. When a photographically stored hologram of a single point is made, the recorded signals form a one-dimensional zone plate.

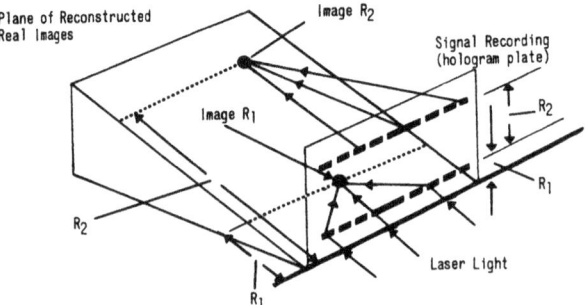

Fig. 16. Reconstruction by laser light of a side-looking radar hologram formed by two reflecting points.

Fig. 17. A synthetic aperture (hologram) record.

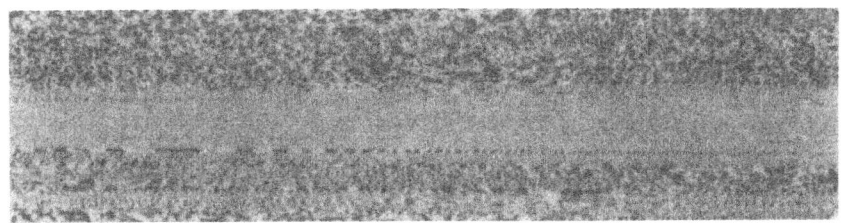

Fig. 18. A portion of the hologram record of Fig. 17. The one dimensional zone plate pattern of a particularly distinct reflecting object is seen at the lower portion of the blank area.

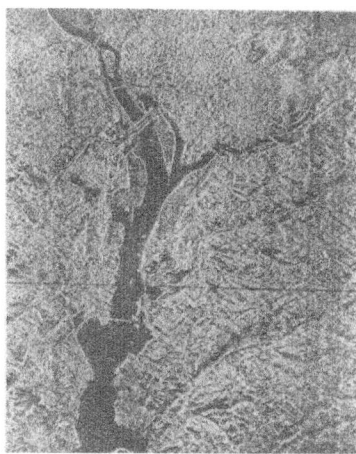

Fig. 19. Proper optical processing of synthetic aperture radar holograms provide high resolution detail of the terrain flown over by the aircraft. This is a record of Washington, D. C. (Courtesy L. J. Porcello).

During 1972, the entire Amazon River Basin was surveyed by radar;[27] Fig. 20 is an example of a (reconstructed) record made during this mapping project.

This synthetic aperture radar survey of the Amazon Basin (3 million square kilometers) was conducted by Aero Service Corporation of Litton Industries under the supervision of Homer Jensen. Fig. 20 is a print of the area near the town of Esmerelda in the State of Amazonas in Venezuela. The river in the lower left hand corner is the upper Orinoco. The radar employed was an X-band (10 GHz) radar made by the Goodyear Aerospace Corporation, and the Venezuelan title for such a record is "Mosaico de Radar." The object of this survey was to provide accurate maps of the area, not just a reconnaissance or surveillance of it. For this special requirement, Aero Service made several additions to the processing equipment, including an anamorphic printer to correct for scale variations, and a special optical mask to correct for tonal effects.

Even in fairly recent discussions of synthetic aperture radar, including several which note the relation to

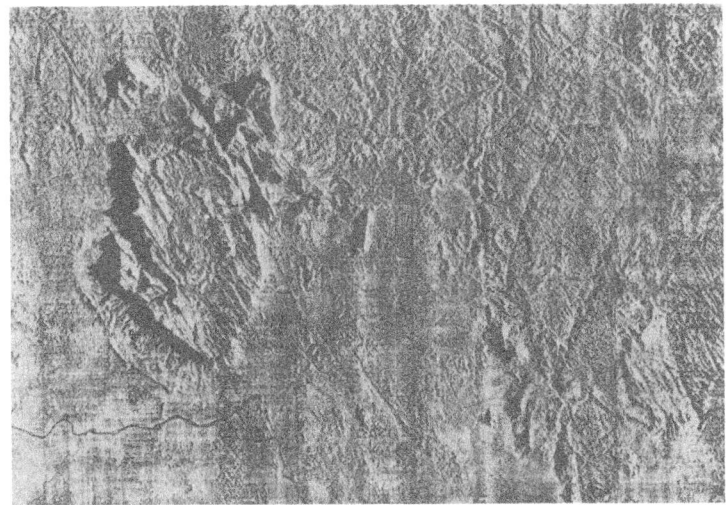

Fig. 20. A reconstructed synthetic aperture radar record taken during a 1971-72 survey of the Amazon River Basin (Courtesy of Homer Jensen, Aero-Service).

holography, the very early Doppler concepts are still stressed.[19] In true holography, there is no Doppler, so that once the holographic viewpoint is fully embraced, there is no need to consider Doppler effects.[18] The way is then open for consideration of other forms of coherent systems, such as stationary (Doppler-free) radars (and sonars) and others.[15,18,20,21,22]

Thus, until recently, the usual form of synthetic aperture (hologram) system involved a moving transmitter and receiver, but, as in true holography, uses of the same concepts can be made in stationary arrays including large aperture arrays. Such sonar or radar arrays could take the form of long linear arrays of independent receivers, crossed arrays (Mills Crosses), or square arrays. In holography, even a small portion of the hologram is able to reconstruct the full image. Similarly, in stationary coherent sonars or radars, retaining only the end sections of long linear arrays, or the four corner sections of square arrays (thereby maintaining maximum resolution) should be satisfactory in many situations. Such a use in a linear array is shown in Fig. 21; in this figure the transmitter is separate.

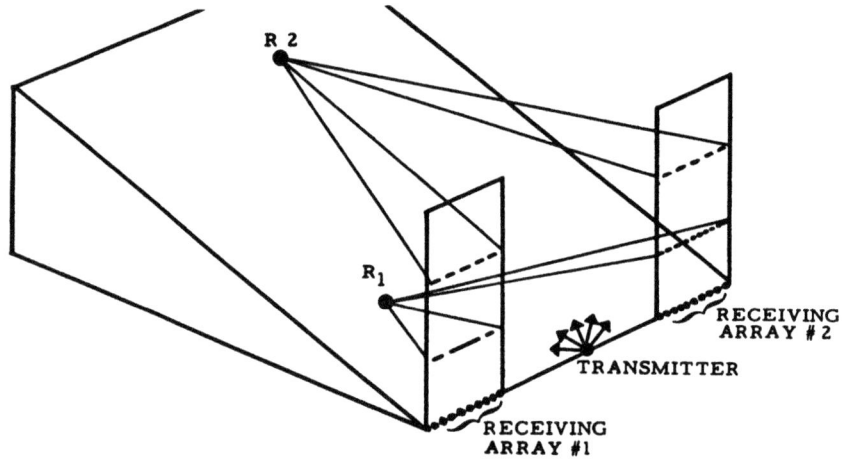

Fig. 21. A stationary, line-array, hologram sonar can also be effective when only a portion (e.g., the two ends) is employed.

Continuous wave bistatic systems (radars or sonars) are often employed to provide information on moving targets. Because such targets generate signals at the receiver which closely approximate a zone plate pattern, hologram procedures offer a signficant improvement in the signal-to-noise ratio in such radars.[28] When an aircraft flies across the line joining a transmitting station and a receiving station; the signal reflected from the aircraft combines, at the receiver, with the direct transmitter signal and this generates interference effects which cause the (combined) received signal to vary periodically, rapidly at first, then slowly, then rapidly again. This effect is often observed on home television sets. A time-frequency plot of such changing-frequency signals is shown in Fig. 22 for a number of passing aircraft. Comparable records would be generated by targets moving between a continuous wave underwater sound radiator and a similarly placed acoustic receiver. To indicate the similarity between such transmission path interference signals and synthetic aperture records, two situations are sketched in Fig. 23. The sketch at the left portrays the process of recording a single reflecting object O on a synthetic aperture radar

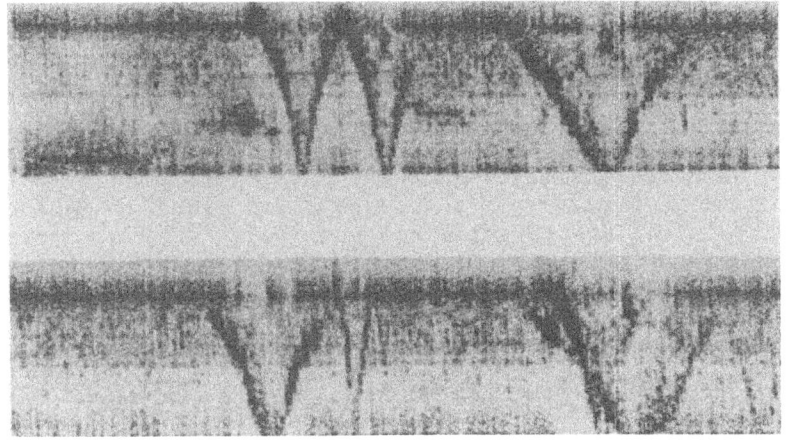

Fig. 22. Frequency versus time records of the interference effect caused by aircraft flying over a radio transmission link.

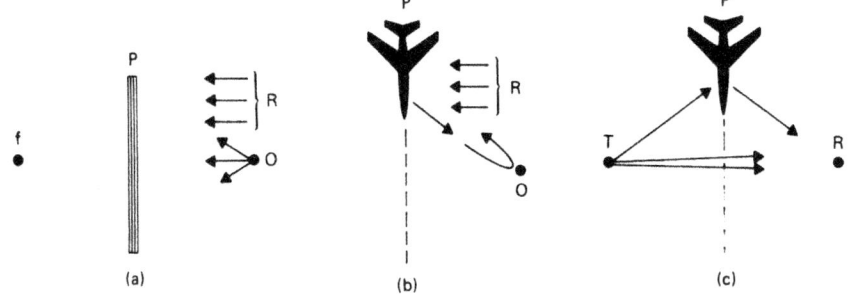

Fig. 23. The recording, by a synthetic aperture radar, of echoes from a single reflecting point (left), is similar to the recording of the signal received in a radio transmission link, when it is flown over by an aircraft (right). The direct signal from T to R (right) corresponds to the reference wave R (left).

record. The right hand sketch indicates the arrangement involved in generating the signals of Fig. 22. Here the continuous wave signal from the transmitter acts (at the receiver) as the reference signal, and now there is a moving reflector P rather than a moving transmitter and receiver. It is thus evident that the individual V curves of Fig. 22 could have been recorded photographically, as one-dimensional zone plates. For the targets traveling at uniform velocity along a straight line, such hologram procedures would provide a gain equivalent to the signal-to-noise improvement obtained in moving synthetic aperture systems.

The usual form of synthetic aperture system is the side-looking version (Fig. 13), which generates a synthetic aperture corresponding to a linear broadside array. Recently, the concept was extended to a form in which the synthetic aperture resembles a linear *end-fire* array.[29] When a series of radiators are energized by a single source and a delay is introduced between the individual radiating elements, the radiated waves can be made to add along the line of radiators, i.e., an end-fire pattern, as shown in Fig. 24. The photographically recorded interference

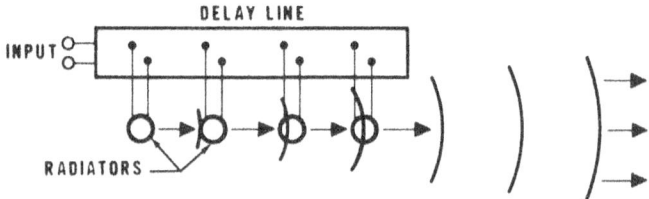

Fig. 24. An end-fire array results when radiators are placed in a row and energized, not in phase, but in such a way that proper phase addition occurs in a given direction. In the illustration, this direction is to the right.

pattern that is produced by combining the signals echoing from a target which is ahead, (e.g., on the forward course of an aircraft), with a coherent reference wave, is a uniformly spaced varying density pattern, rather than the nonuniformly spaced zone-plate pattern of the side-looking case. The recorded pattern is a one-dimensional grating, and to reconstruct the object (the target), a lens is employed to convert plane laser light waves diffracted by this photographic grating into circular waves converging at a focal point. The light concentration thereby obtained is equivalent to that obtained with a one-dimensional zone plate, and a correspondingly high array gain (with its accompanying high signal-to-noise ratio) is realized. As in the side-looking case, pulsed transmissions would normally be employed to provide range information on the reflecting objects of interest.

Whereas the usual synthetic aperture system operates from a moving platform, the same end-fire concept can be applied to a stationary active radar or sonar, whereby a small transmit-receive transducer would effectively be provided with a large synthetic gain against targets moving toward it or past it at uniform speeds.[30] Fig. 25 indicates the procedure for imparting synthetic gain to a stationary radar antenna. At the right is sketched a moving reflector P, rather than the moving transmitter and receiver of an airborne synthetic aperture radar shown at the left. When a reference signal is introduced, the combined signal is, for identical aircraft motion and with a similar moving film technique, the same for both cases. The concept is of course also applicable to sonars. For underwater targets traveling at uniform velocity along a straight line, the signal-to-noise gain thereby achieved at the stationary sonar would be equivalent to the signal-to-noise improvement of the moving synthetic aperture sonar.

We noted above that end-fire gain could be imparted to a moving synthetic aperture system. This technique is also possible for a stationary system when inform velocity targets are moving directly toward it. Since the synthetic gain thereby achieved would be generated only by moving targets, reverberation effects would be minimized for sonars and similarly ground clutter effects minimized for radars. The range capability of the system should accordingly be markedly improved.

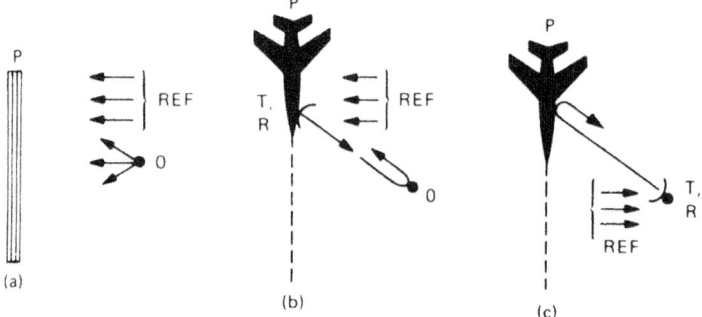

Fig. 25. The synthetic antenna gain achieved in an airborne synthetic aperture radar (left) can be duplicated for a stationary radar against an aircraft target moving in a straight line at uniform speed (right).

ACOUSTIC SYNTHETIC APERTURE SYSTEMS

We noted that synthetic aperture hologram techniques have been investigated in acoustic applications. One of the earlier proposals for such uses was described in a U.S. patent applied for in March 1967, entitled "Synthetic Aperture Ultrasonic Imaging Systems."[26] Two procedures were described, both aimed at medical ultrasonic applications for the "examination of the interior of living bodies." In the first, a transmit-receive transducer was moved (scanned) along a straight line. It was submerged in a liquid reservoir, with the liquid supported by, and in contact with, the body surface under examination. The second procedure utilizes a large number of fixed transducers positioned along the line of motion of the moving transducer of the other method, with each energized in succession so as to simulate the moving transducer.

More recently, experiments using the ultrasonic synthetic aperture procedure for true sonar use were described, with very excellent records resulting.[31] Fig. 26 shows the arrangement of equipment used in these tests; the

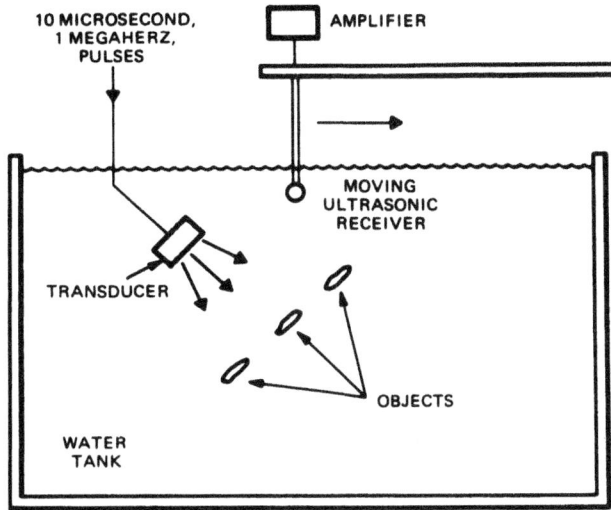

Fig. 26. Arrangement for synthetic aperture (hologram) sonar experiments (after Pekau and Diehl).

transmitter radiated 10 microsecond pulses of 1 MHz sound waves. In these experiments only the receiving transducer was moved. The round trip distance to each of the three reflecting objects from transmitter and receiver was made sufficiently different so as to permit the one-dimensional zone plates generated by the reflections from each of the three objects to be adequately separated in the holographic record. For these tests, the reference signal was supplied electronically from the stabilized oscillator directly to the receiver, which also supplied energy (amplified) to the transmitting transducer (as in Fig. 5). In Fig. 27, the three one-dimensional zone plates (one for each reflecting object) are seen, all adequately separated to permit reconstruction, from the zone plates, of the three individual objects. The similarity of these zone-plates to the radar one of Fig. 18 is evident. It is seen that the left-hand zone plate has its central portion slightly above center of the photo, whereas the middle zone plate has its slightly below, and the right hand one has its below the bottom of the photo. The right hand portion of Fig. 27 shows the reconstruction (by laser light) of the three reflecting objects (the light areas at the lower left are artifacts). One of the two co-investigators in this work

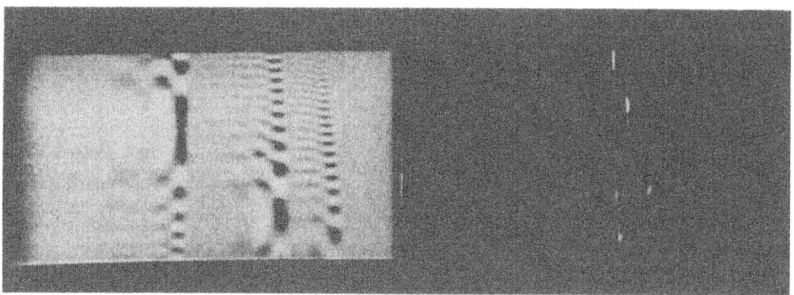

Fig. 27. Zone-plates generated by equipment of Fig. 26 (left). The three dots at the right are the reconstructions of the objects of Fig. 26 (after Pekau and Diehl).

(D. F. Pekau), had earlier been a colleague of R. K. Mueller at the Bendix Research Laboratories.

One of the limiting factors in the synthetic aperture process results from the need to delineate accurately the finer fringes of the zone plates. As the platform carrying the transducer moves, the distance to the reflecting object changes, and a round-trip change of one half-wavelength causes the interference between the fixed reference wave and the varying echo wave to change from a constructive interference case to a destructive one. The former corresponds to the black portions of the zone plates of Fig. 27 and the latter to the white or blank portions. It is obvious that if two few pulses return during the period when a fringe (adjacent white and dark areas) is being generated, the zone-plate cannot be delineated properly.

This need to send out closely spaced pulses causes the maximum useful (unambiguous) range to be limited, since the time interval between pulses corresponds to the round-trip time to the most distant targets. While this is somewhat annoying the radar case, it is far more serious in the sonar case because of the much lower velocity of sound waves versus electromagnetic waves. Thus, in a recently published report of a National Academy of Sciences Summer Study, it was noted that for a ship traveling at a speed of 6 knots and carrying a 1 kc, synthetic aperture, sonar, the maximum unambiguous range would be only 1.33 nautical miles.[17]

The procedure which was recently described[43] and which overcomes the ambiguity problem is sketched in Fig. 28. The usual single transmit and receive transducer case, with its need to transmit pulses whenever the sonar moves a distance equal to one half the aperture dimension D,[5] is shown at the top left. At the bottom left, a receive-only transducer is added ahead of the transmit-receive one. It is evident that the signal received by this second unit for pulse number 1 is the same as that received for pulse number 2 in the usual (upper left) case, since for that case, the outward path is shorter, and the receive path is longer. The pulse repetition rate can therefore be halved. If instead of one, three receive-only transducers are added, the pulse repetition rate can be reduced by a factor of four (with a consequent increase in the maximum unambiguous range by that same factor). If seven receive units are added the increase becomes eight, etc. The cathode ray tube patterns (comparable to those of Fig. 14) for the two cases are shown at the right; for the second case, the tube is equipped with two adjacent upward-moving beams, with the second one amplitude-modulated by the receive-only transducer signal (properly combined with the reference wave, i.e., synchronously demodulated). This recent development will materially extend the usefulness of holographic (synthetic aperture) sonars (and also satellite-borne synthetic aperture radars).

Recent experiments[32] involving the long range transmission of fairly low frequency (367 Hz) sound waves showed that such sound waves exhibit fairly high coherence over a rather long path (700 nautical miles) and over a resonably long period of time (105 seconds). The conclusions drawn from these experiments suggest that hologram techniques might be useful in underwater applications at these lower frequencies. Thus, the statement was made[32] that the results were "taken to be evidence that the sound velocity structure is frozen in the volume of the ocean, or at least partially so, over this period" (105 seconds). These experiments were conducted in the Atlantic ocean, using a bottom

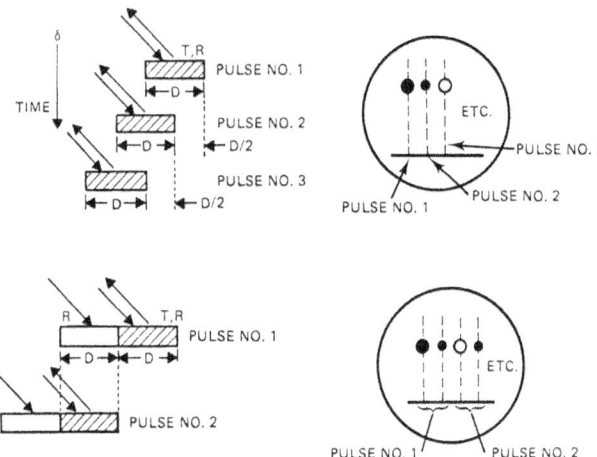

Fig. 28. In a standard synthetic aperture (hologram) sonar (top left) a new pulse must be transmitted each time the moving platform advances one-half the aperture dimension D. Top right indicates how a target generates, with successive pulse echoes, a horizontal one-dimensional zone plate on the cathode ray tube of Fig. 14. The heavy black dot corresponds to a constructive interference situation and the white circle to a destructive one. The lower sketches show how the pulse spacing (and maximum range) can be doubled through the addition of a receive-only transducer whose signal is fed to a second cathode ray tube beam moving up with the original one.

Fig. 29. An experimental deep water, low frequency, acoustic sound source (Bendix Research Laboratories).

mounted source located off the island of Eleuthera in the Bahamas, and a receiving array off Bermuda.

In the holographic utilization of the high coherence of low frequency sound in the sea, the stationary forms of hologram sonar, as described earlier, would appear to be the most likely candidates. Thus, use might be made of a line array of receivers (or the partial line array of Fig. 21), in monostatic systems[15,16] or in bistatic systems. The latter could use a transmitter radiating to a receiving *array* for detecting stationary targets, or to a *single* receiving transducer[28] whereby the received zone plate signals (the V curves of Fig. 22), as generated by a moving target (Fig. 23) would be processed so as to utilize the gain inherent in such signals. Stationary monostatic hologram systems (Fig. 25) could also be of interest, again against moving targets, in either the side-looking arrangement[30] or the end-fire one.[29]

Fig. 29 shows an early low-frequency transmitting array; designed by the Bendix Research Laboratories, and also used in acoustic transmission experiments between Bermuda and Eleuthera, about to be lowered into the Atlantic from a ship located just south of Bermuda. Pulses of low frequency (250 Hz) sound transmitted from this ship supported unit were received both at Bermuda and at Eleuthera (Fig. 30). Those received at the nearby Bermuda receiving point (top) exhibit reverberation effects which invariably accompany a strong energy pulse. At the much more distant Eleuthera receiver, the local Bermuda reverberation artifact is too weak to be received and the pulses are seen to be very clear and sharp (bottom). These pulses were 30 seconds in length, and the reverberation in Bermuda is seen to last for several seconds.[71] Long range coherence measurements were also made by cross correlating the transmitted and Eleuthera-received pulses and a quite definite indication of coherence was observed, providing further support today for possible holography uses. Fig. 31 shows the transmitting array of Fig. 29 mounted on a supporting structure as it was later being lowered to the bottom off Eleuthera, to act as a bottom-mounted transmitter for further transmission tests from Eleuthera to Bermuda. Results of such tests using this bottomed unit were later reported by Bell Laboratories scientists.[33]

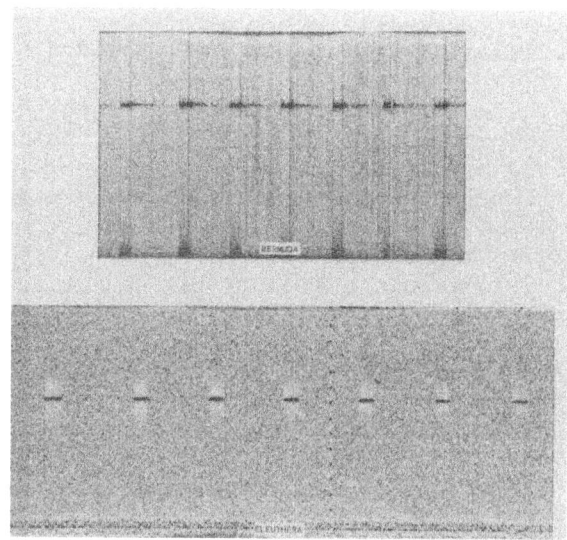

Fig. 30. Transmissions (top) from the sound source of Fig. 29 as received 700 miles away (bottom).

Fig. 31. The transducer array of Fig. 29 arranged for deep bottom mounting.

ACOUSTIC KINOFORMS

It was recognized rather early that cathode ray tube signals, originating from a properly programmed computer, could become, when photographically recorded, computer-generated holograms. Numerous such examples and their reconstructions have been reported[34,35,36] by A. W. Lohmann who is probably recognized as the pioneer in this field. More recently, a computer-generated hologram called the kinoform has been described[37,38], interesting because it accomplishes what normal (planar) holograms cannot do: it eliminates both the straight-through (zero order) component, and one of the two first-order diffracted hologram components (either the virtual or the real). Although it thus constitutes one of the most efficient holograms, its application possibilities are not as yet fully determined. We discuss it here because it has some interesting counterparts and possible applications in acoustics.

The kinoform, in a general sense, is a computer-generated wavefront reconstruction device, which, like the hologram, provides a display of a three-dimensional image. When illuminated, however, it yields only a single diffraction order, so that, ideally, all the incident light is used to reconstruct one image. The kinoform may be thought of as a complex lens which transforms the reference wave incident upon it into the wavefront needed to form the desired image. Although first conceived as an optical focusing device, the kinoform can be used as a focusing element for any physical waveform including sound waves. The useful information in a kinoform is, like that in a hologram, a coded description of the wavefront of light scattered from a particular object of interest. Upon illumination with a reconstruction beam, the kinoform provides the display of an image.

The kinoform concept can perhaps be best explained by a comparison of Figs. 32 and 33. In the first, the slit variety of a grating and the photographic variety are sketched. Both diffract several orders, but the latter diffracts only the zero order and the two first orders. In the second, a periodic array of prisms is sketched; this too can be considered as a form of grating, but because the diffracting elements are dielectric prisms rather than slits

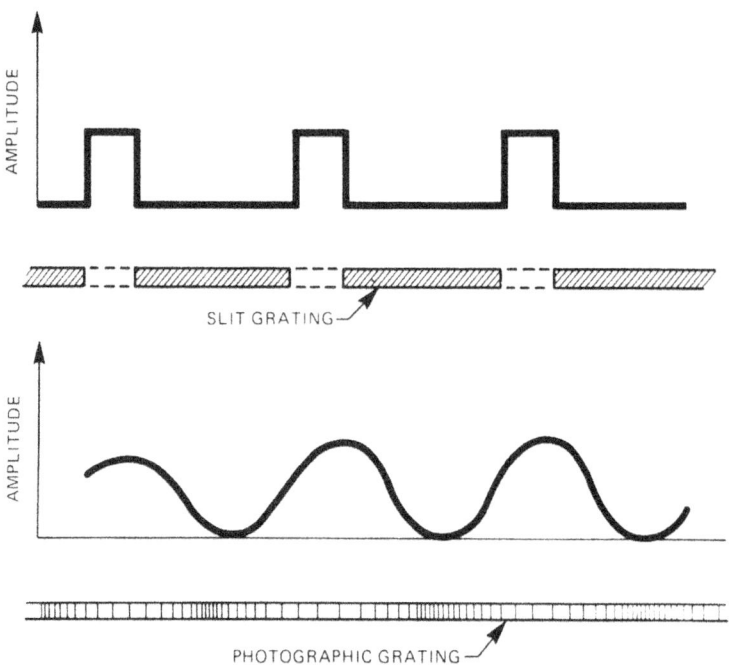

Fig. 32. A slit grating (above) generates abrupt changes in light intensity, whereas a photographically made grating (or zone plate) does not (below).

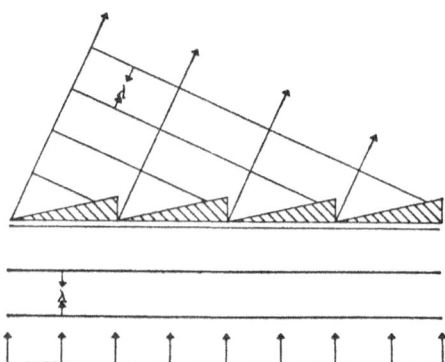

Fig. 33. A properly shaped array of prisms can act as a highly efficient grating.

or sinusoids, only one diffracted order results (even the zero order, straight-through, waves are eliminated). For this latter form of grating, the prisms must be designed rather carefully, so as to cause all elemental, prism-refracted, waves, to combine properly with all other such elemental waves. Now it has been known for some time[39] that if the positions where the prisms are stepped back are points of 2π radians phase delay, then all of the elemental diffracted waves add. This is illustrated in the figure by the parallel wavefront lines being spaced one wavelength apart.

This same relation holds when the refracting grating shown is altered so as to become a lens. Thus whereas a circular zone plate generates a zero-order wave and two sets of first-order diffracted waves, a zoned (stepped) *lens*, patterned after the stepped prism of Fig. 33, generates only one wave set. If the lens is designed to be a converging lens, the waves converge upon a real focal point; if it is a diverging lens, the waves diffracted and refracted by it emerge as waves diverging from a virtual source.

A microwave lens, fabricated in 1944, for use in a parallel plate antenna[40] is shown in Fig. 34. The maximum step thickness corresponds to a phase change in the dielectric of 2π radians as outlined above, and because the dielectric has a refractive index which is greater than unity the lens shown produces converging waves. It is of interest to compare this lens with the figure of the Collier *et. al.* book Optical Holography which was used to explain kinoforms. The lends of Fig. 35 is also a microwave kinoform lens; the refractive material has an index less than unity (it is a parallel plate wave-guide design) so that for the lens to be converging (convex), the contours must be concave.[39] Again the steps are placed at those points where the thickness reaches a value equal to a wavelength in the refractive material (2π radians). The similiarity of this microwave lens to a kinoform was recently noted.[41]

In the more general form of optical kinoform, a complicated object is reconstructed. The hologram for this object is expressed analytically so that a computer can generate the light pattern. Each point of the reconstruction can be looked upon as being generated from a simple kinoform lens (a circular form of the one in Fig. 34, but

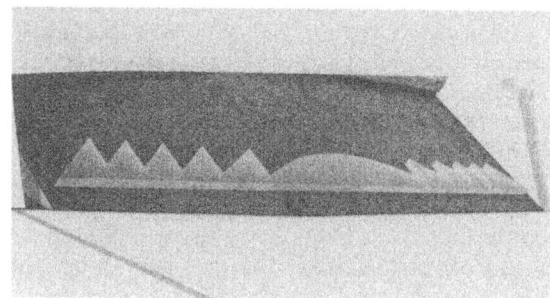

Fig. 34. A dielectric microwave lens designed in 1944 for use in a parallel plate antenna. Becuase it is stepped at one-wavelength thickness, it can be considered a Kinoform lens.

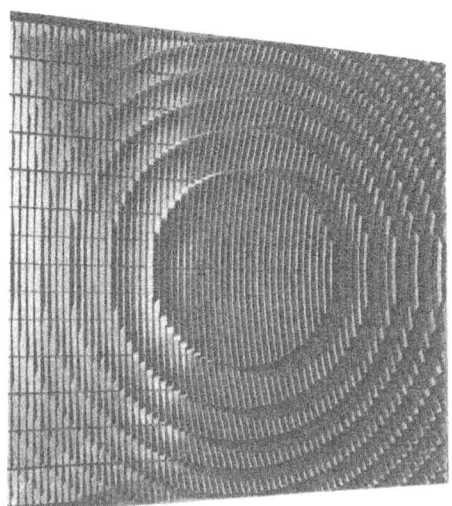

Fig. 35. A microwave lens, designed in 1946 and employed in the Bell System's New York to Boston Microwave radio relay circuit, has a circular "step" design which matches that of a zone plate. This lens is related to a computer generated hologram called a kinoform.

drastically reduced in size, of course, so as to be functional at optical wavelengths). The superposition of all such lenses causes it to become the complex lens which then transforms the reference wave incident upon it into the wavefront needed to form the desired image. In an optical kinoform, the varying thickness and the steps at 2π radians are obtained by bleaching the film in a very careful way, so that the hologram becomes a combination diffracting and refracting device (like the prism array of Fig. 33) instead of a purely diffracting one.

Kinoform lenses have been considered for use in underwater acoustic systems because they can be thin and lightweight, thus minimizing attenuation and mounting considerations. Recently an ultrasonic technique was described in which a kinoform mirror was used as a focusing element.[42] The mirror was a reflective variant of the simple kinoform lens and was constructed so as to transform an incident plane wave into a converging spherical one. The limitation, in optics, which exists for kinoform lenses or reflectors, namely that they must be used with nearly monochromatic radiation, is not so serious in ultrasonic applications, where single frequency sound waves are often employed.

The kinoform mirror was called a stepped paraboloidal reflector when it was employed in the microwave field during World War II (Fig. 36). Fig. 37 shows the recently described stepped paraboloidal reflector.[42] On the other hand, the cross-section of the lens of Fig. 35 is that of a stepped ellipse. Fig. 38 shows how the 2π radian steps must be formed to achieve proper phase correction. The top equation is that of the cross-section of the basic ellipsoid of revolution (corresponding to the paraboloid of revolution of the reflector of Fig. 37). When a lens thickness corresponding to a phase shift of 2π radians or one wavelength is reached, the lens is stepped back, and a new elliptical contour is followed corresponding to a focal length increase of one wavelength in the refracting medium (for a reflector the increase is a half wavelength). The process is then continued. The stepped shaped and curved sections are required to insure that the wave energy is refracted into a single diffraction order with maximum efficiency. Obtaining such an intricate relief at optical frequencies (in hologram kinoforms or kinoform lenses) through the highly controlled shrinkage of the emulsion poses certain difficulties. However, at the wavelengths considered useful

Fig. 36. An experimental zoned (stepped) parabolic reflector designed as wide-angle scanning microwave antenna (Bell Telephone Laboratories).

Fig. 37. A flat, two dimensional stepped parabolic reflector, i.e., a kinoform reflector. (I.B.M.)

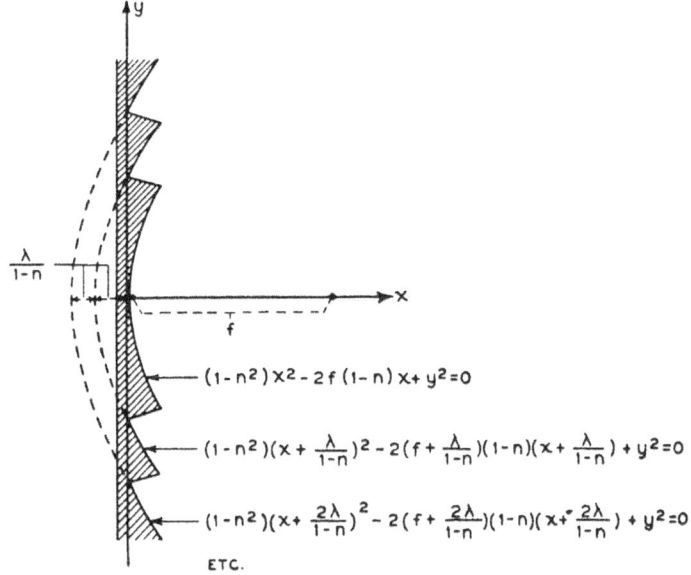

Fig. 38. Step design for the elliptical contours of a wave guide microwave lens (a microwave kinoform).

for ultrasonic applications, machining procedures were considered adequate in fabricating the reflector of Fig. 37. It was designed as an f/1 kinoform mirror for use with 5 MHz ultrasound in water (λ = 0.29 mm). The mirror is 50 mm in diameter and was made on a lathe. As noted earlier, the steps here are at $\lambda/2$ thickness points because it is a reflector rather than a lens. A comparable, circular, microwave lens, with index of refraction less than unity, is shown in Fig. 39. Both the flat reflector of Fig. 37 and the flat lens of Fig. 39 could have been made to exhibit less aberration if they had been made spherical instead of flat. The reason for which the microwave parabolic reflector enclosed in the World War II antenna of Fig. 36 was stepped was to permit it to be given a circular shape and hence a wider field of view (less coma aberration).

Acoustic kinoforms (variable thickness reflectors or refractors) might be useful in ultrasonic nondestructive testing procedures. Thus, for example, a kinoform could be designed to convert a plane acoustic wave into a multiplicity of narrow beams, all aimed in predetermined directions, so that when used for testing a particular structure which

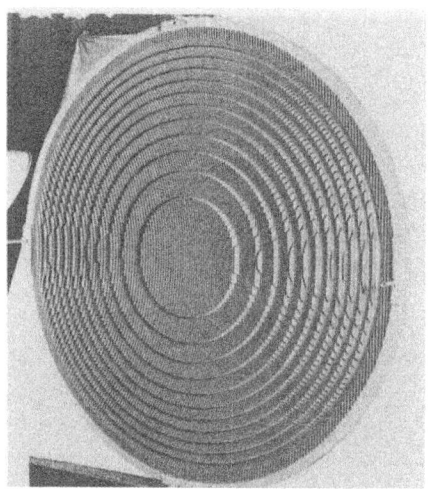

Fig. 39. A circular microwave kinoform lens. The resemblance to the kinoform reflector of Fig. 37 is interesting.

is opaque to light waves but transparent to sound waves and which should have a number of properly positioned hollow areas (voids), reflections caused by the voids could be ascertained by sensors placed in the paths of the acoustic kinoform sound beams.

ACOUSTIC HOLOGRAPHIC INTERFEROMETRY

Holographic interferometry has found important applications in nondestructive testing procedures.[43,44,45,46,47] Because of the longer coherence length of sonic and ultrasonic waves, acoustic uses of such techniques could permit the testing of very large structures (including the use of synthetic aperture methods).

The hologram interferometry process involves recording two holographic exposures of the structure under test on the same photographic plate, with the structure stressed in some way for the second exposure. When this double hologram is developed and re-illuminated with the reference wave, two superimposed images are reconstructed, and where the stress has caused a deformation of the structure, optical interference fringes are observed, with the number of fringes per square area being a direct measure of the topographical change.

Recently, some microwave experiments were reported[48,49] in which the photographic plate was replaced by a large plane area on which the two interference patterns were formed (a first and second microwave hologram). Fig. 40 is a sketch of the scanning procedure employed. (The wavelength employed was 3.2 cm). Fig. 41 shows the object used in the tests. It is a square sheet of aluminum with its surface made diffusing by providing it with extended relief. For the deformed condition the sheet was bent as shown, with a deformation depth of 7 cm. Microwave holograms were made of both structures, with both holograms superimposed on the same photographic plate (a Polaroid 107) through the use of an oscilloscope camera.

The plane (hologram) area was 110 cm square, and the object under test was 100 cm by 100 cm. The doubly exposed hologram was greatly scaled down and illuminated with laser light (Fig. 42). The reconstruction shows only two, roughly circular (microwave) fringes, which were not too

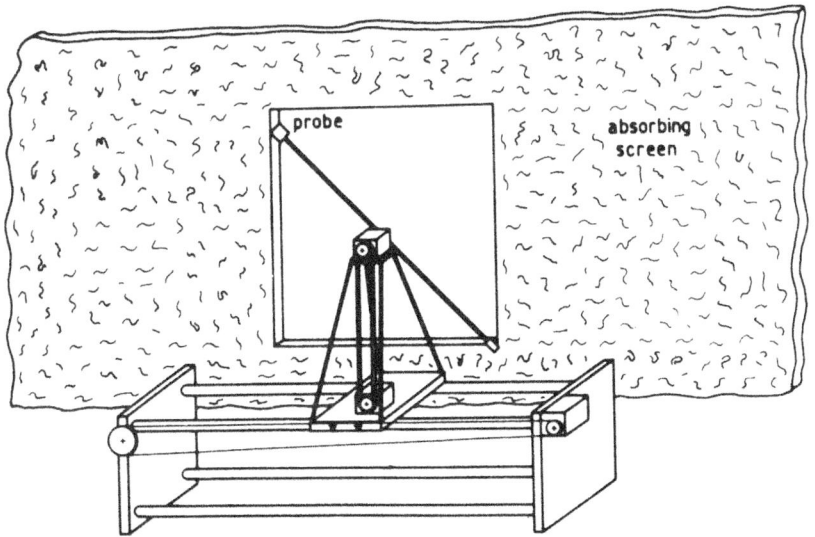

Fig. 40. Arrangement for scanning a square microwave hologram area for achieving microwave hologram interferometry records.

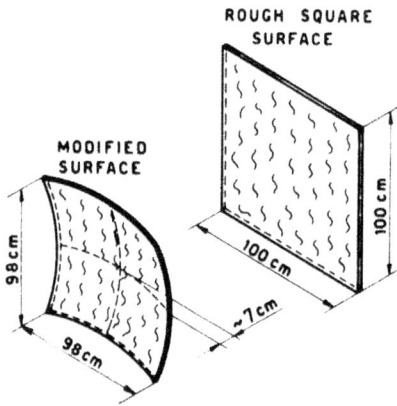

Fig. 41. A test object used for microwave hologram interferometry experiments: right, without deformation, and left, deformed (Reference 48).

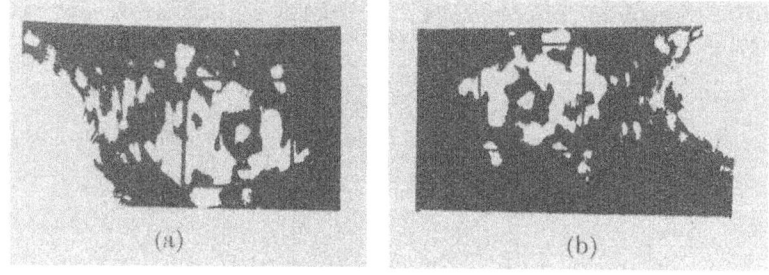

Fig. 42. Microwave interferometric hologram fringes generated by the deforming of the test object of Fig. 41 (from Reference 48).

recognizable. Thus, it is seen that for the wavelengths employed (3.2 cm), and even with rather large hologram areas (110 cm by 110 cm), only modest fringe information was available. The situation could have been improved with higher microwave frequencies and larger hologram areas. Acoustic techniques would have the advantage of the relative ease of generating much shorter wavelengths than the 3 cm microwaves.

A recently suggested extension of microwave hologram interferometry involves the use of synthetic aperture techniques.[50] In such a procedure, a small pulsed transmitting and receiving transducer would be moved at constant speed along a long, straight, track, and a photographic record made of the returning acoustic (or microwave) signals. The first synthetic aperture (sonar or radar) record would be made with the structure under examination unstressed, and the second made (on the same photographic film) with the structure stressed. As in the regular holographic interferometry process, deformations would cause changes in the phase of the second set of reflected waves, causing visible fringes densities would, as usual, pinpoint areas of high displacements. However, because synthetic aperture holograms are superpositions of one-dimensional zone plates, interference fringes between the stressed and unstressed hologram patterns will only occur along the direction of the moving transducer, and, accordingly, only those profile deformations which occurred along that direction will be portrayed. Deformations at right angles to that direction can be ascertained by generating a second pair of holograms for which the track of the transducer is at right angles to the first direction.

Fig. 43 shows one conceivable arrangement; an aircraft wing is rigidly supported at its fuselage end, and, for the second hologram, it is stressed by applying the force indicated. The synthetic aperture transducer track is positioned vertically, causing the resulting interference pattern to permit detection of possible construction faults. In general, since the wing-tip end in this arrangement moves the farthest, it will represent the point of largest fringe difference, with the number of fringes continually decreasing toward the fuselage end. Should this fringe density variation be too rapid at some point, or if the displacement does not follow the expected design equation, corrective measures would be called for. As shown in the insert, the first record would be comparable to one obtained by an airborne synthetic aperture radar (Fig. 13) of a high wall extending along a line perpendicular to the path of the aircraft, while the superimposed record would correspond to the wall modified as shown by the dotted line.

In some synthetic aperture sonars procedures, for example, one described in Reference 26, each element of a fixed, linear array of transmit-receive units is energized

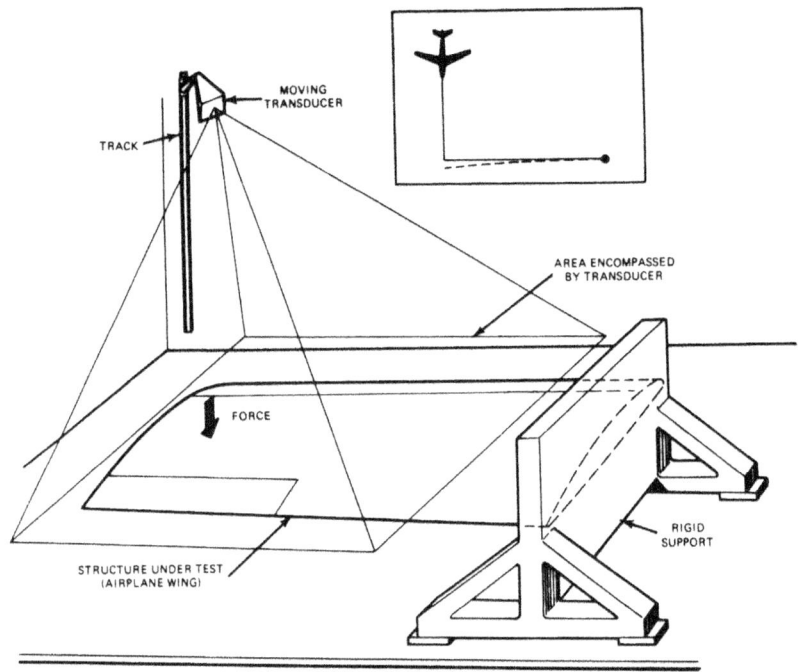

Fig. 43. Use of synthetic aperture techniques in microwave hologram interferometry: The high coherence length of microwaves compared to laser light permits much larger structures to be tested, and the synthetic aperture process provides, for the same aperture, twice the resolution and twice the number of fringes.

in turn so as to duplicate the effect of the moving antenna. The holographic interferometry procedure outlined here could also be utilized in such arrangements. Thus, instead of one transducer moving along the vertical track of Fig. 43, a vertical *array* could be used, with each array element activated in succession. Two significant advantages gained by using the synthetic aperture procedure are: (1) the linear hologram aperture (the track length or linear array length) provides a resolution twice that of a normal acoustic hologram of equal aperture, and (2) the number of recorded interferometric fringes is double the number generated in a normal interferometric hologram for the same displacement of the test object.[50]

UNDERWATER APPLICATIONS OF ACOUSTIC HOLOGRAPHY

Because sound waves are the only form of waves which are able to propagate to any signficant distance in sea water (or fresh water), the use of acoustic devices for detecting underwater objects (such as submarines) has been explored and used for many decades; in fact sound wave echo-location systems (later referred to as sonar systems) were used during World War I. In more recent time, sonar design has developed into a highly sophisticated technology, so that some who are involved in the design of modern sonars often tend to dismiss the thought of applying holography to sonar because of their conviction that the later techniques are surely already being employed. Also, because the concept of phase has been associated with steered beam sonars for quite some time, the stress often put on the phase recording ability of a hologram has again led some sonar experts to feel that the technique provides nothing new.

Now it is true that as yet, acoustic holography has not been able to make much use of the three-dimensional properties which are so striking in optical holograms. We know that in the usual hologram, a small portion of it can be used to reconstruct the original scene. However, this smaller portion is like a smaller window, so that for it the three-dimensional parallax effect is much less useful. The viewer is required to position himself more carefully in order to see objects through this smaller window, and accordingly, the three-dimensional realism he would obtain in a full hologram through the parallax effect is appreciably reduced. His ability to see around objects in the foreground is obviously greatly limited when the window size is reduced to a significantly smaller area.

Because of the much longer wavelengths of sound waves as compared to light waves, even fairly large acoustic holograms act like small windows, so that when they are optically reconstructed very little parallax or three-dimensional effects are available. Another way of stating the above is to say that the average acoustic hologram portrays very little of the near (Fresnel) field. Because many of the advantages of holography and synthetic aperture systems relate to the ability to provide detailed information concerning objects located in the near field, some are tempted to dismiss consideration of the far field performance of hologram sonar as unimportant, since ordinary sonar performs this task well. On the other hand, hologram techniques often provide simpler ways of achieving the performance requirements of the usual, far field, sonars.

Consider, for example, a multi-element, square array, sonar, having 100 elements on a side and hence 100 x 100 or 10,000 total elements. If the elements are positioned $\lambda/2$ apart, the aperture dimension is 50λ on a side. We now suppose further that the array itself is used only for receiving, and that the transmitted signal, radiated by another, high power transducer, illuminates a pyramidal volume of sizable angular extent. The array is to acquire, to the best of its ability, information regarding reflecting objects located in the far field and within the radiation pyramid of the transmitting transducer.

One way of accomplishing the desired function of the array would be to establish thousands of "pre-formed" beams, whereby, for any single beam direction, phase shifts would be provided as necessary for each of the 10,000 array elements, so that all elements would contribute properly to the receiving beam for that direction. Because the array is 50 wavelengths on a side, the width of each of the pre-formed beams will be approximately 1 degree ($51\lambda/a$), so if a 90° pyramid of coverage (±45°) is desired, approximately 90 x 90 or 8,100 preformed beams would be required. Whether the required phase shifts (for 10,000 elements and 8,100 beams, the total is approximately 80,000,000) are provided individually (by analog methods) or by digital computer techniques the task is obviously not a simple one.

Consider now providing these same far field beams by holographic sonar procedures. The coherent transmitted signal would be provided by a stable oscillator, and a small amount of this oscillator signal would be used as a reference signal. It would be supplied continuously to each element of the receiving array (possibly by direct line to the element) with the phase adjusted so as to simulate a plane wave of this frequency impinging (preferably at some angle) on the entire array. Reflecting objects located within the cone of the transmitter will reflect some of the transmitted signal back to the array, thereby generating, in combination with the reference signal, a holographic interference pattern which is "sampled" by all the elements. The resultant values at all elements are then recorded photographically, and, **after development of the photographic** record, the far field reflecting objects would be reconstructed by coherent light. We shall see shortly that real-time reconstruction methods are now under development.

At first glance it would appear that the first procedure, which combines all 10,000 elements into a preformed beam, thereby achieving an extremely high directivity gain (40 decibels or so), *must* be superior to the procedure in which each single isolated unit of the 10,000 elements is made to affect, by itself, the exposure of the photographic plate at that point. In the 10,000 element preformed beam, a reflected target signal arrives in phase at all elemental receivers (after proper phase shifts are taken into account) and the summed output is therefore very large; on the other hand for the same elements, the sea noise coming from all directions adds in a *random* way, and the noise signal build-up is thus far less. In the hologram case it would appear as though each elemental receiver with its own amplifier treats noise and signal equally, thereby losing the array gain of the preformed beam. Actually, the coherence of the signals generating the interference pattern on the array and the similar coherence of the laser reconstructing beam, causes each recorded point to combine coherently (during the reconstruction process) with all the rest of the points, thus providing, holographically, a comparable array gain for each of the 8,100 directions.

An alternative standard design for the 100 x 100 element receiving array would employ *one* receiving beam, made steerable to all of the 8,100 directions through the use of a much smaller number of *variable* phase-shifts mechanisms. This technique, however, loses the important time integration advantages accrued through individual storage of the signals in each of the outputs of the 8,100 preformed beams, or equivalently, the storage of each of the receiver element outputs in the hologram version, since the scanning beam looks in any one direction for only 1/8100th of the time.

This consideration also brings out an interesting point regarding the ability of the medium to maintain the extraordinary coherence which is usually assumed to be necessary in hologram sonar. Thus, it is known that the stable oscillator signal of acoustic hologram systems can have extremely long coherence lengths in a perfect medium. But because sea water introduces velocity variations due to

currents, thermal fluctuations etc., many assume that a
holographic sonar will automatically suffer far more than
an ordinary sonar from these effects. However, for the
same size array (such as the 100 x 100 wavelength, one just
discussed) the performance of the standard form of sonar
will also be degraded by velocity variations. A reflected
signal will only arrive at the array with a perfectly flat
wavefront if the medium is perfect, and hence any changes
in *spatial* coherence caused by the medium will degrade the
performance of a standard sonar array designed to receive
flat wavefronts. Also, the phasing methods used in forming
or steering sonar array beams are often highly frequency
sensitive, so that an impairment, by the medium, of the
frequency coherence of the signal will also affect performance. It is because of these effects that beam output
integration is useful, since the magnitude of the variations in spatial and temporal coherence as caused by the
medium is variable with time, and a steered beam might be
aimed in the target direction exactly at a time when the
coherence impairment is a maximum, with the result that
the received target signal may at that instant be undetectable. This need for beam output integration tends to force
the designers to the very complicated, preformed beam, version, for which the hologram form offers such an attractive,
simpler, alternative.

Recently, in a program funded by the U.S. Navy Office
of Naval Research, a holographic imaging system was constructed.[51] It included a 400 element hydrophone array
(Fig. 44) with all the necessary electronics. Zone plate
patterns generated by groups of hydrophones are shown in
Fig. 45. A real-time reconstructor tube, permitting the
successful reconstruction of acoustic targets was also
constructed.

This program is now being followed by the design and
construction of a large-scale deep-operating holographic
imaging system using a 9,216 element hydrophone array. It
is a square array (96 elements on a side), and, based on
the successful results of the earlier 400 element array, it
is expected to provide a resolution of 0.00487 radians
(between three tenths and four tenths of a degree).

One of the most important components of this system is
a real-time reconstructor tube[52,53] shown in Figure 46. In
the complete sonar system, an ultrasonic signal is sent out

Fig. 44. A 400 element receiver array for underwater acoustic holography.

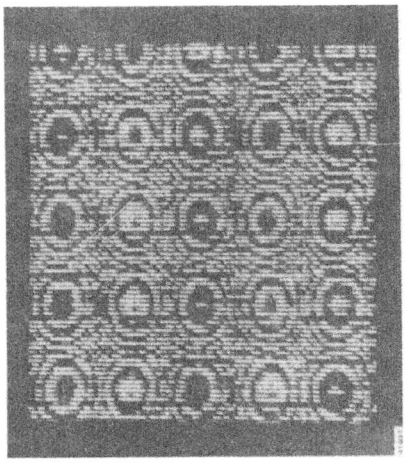

Fig. 45. Zone plate records generated by groups of the transducers in the array of Fig. 44.

Fig. 46. An experimental Model of a DKDP reconstructor tube for real-time viewing of acoustic holograms (Bendix Research Laboratories).

by the transmitter and reflected by objects (which can be *moving*). The reflected waves are detected by the large receiving array, and the resulting electrical signals are converted into holographic information by an electronically injected reference wave. The reconstructor tube converts the holographic signals into real-time visible images of the object scene, and these are then displayed for the operator. This tube consists of a thick (DKDP) crystal, an off-axis scanning electron gun, and associated optics. In operation, the scanning electron beam is modulated with the holographic information, so that a hologram is written on the crystal in the form of a positive charge pattern. The electric field within the DKDP varies over the crystal according to the holographic signal, thus modulating its refractive index by the electro-optic effect, so that coherent-light transmitted *through* the DKDP crystal becomes modulated with (that is, it reconstructs) the holographic information. The hologram is periodically erased by flooding the crystal with electrons with an appropriate potential on a near-by grid. Moving objects generate varying holograms and the real-time reconstruction follows the variation which occurred in the object illuminated. The real-time capabilities are now of the order of 16 frames per second. Fig. 47 shows a hologram and Fig. 48 the reconstruction of it as accomplished by this tube. As an example of a possible configuration for using this underwater viewing system, Fig. 49 shows an artist's concept of an experimental search vehicle.

SEISMIC HOLOGRAPHY

In sonic geophysical prospecting, low-frequency sound waves are sent into the earth, and the returning echoes are analyzed to appraise the likelihood of oil or gas being present in the substructure. Earlier, this analysis was made by inspecting visually the recorded, multiple-receiver, traces that resulted. More recently, optical processing techniques have been introduced in such forms as spatial filtering in order to suppress artifacts that would otherwise obscure the desired information. It is interesting to note that this particular development in seismology benefited from the optical processing concepts that originated with the early synthetic aperture radar work. Thus, several authors of early papers on optical seismic-data processing

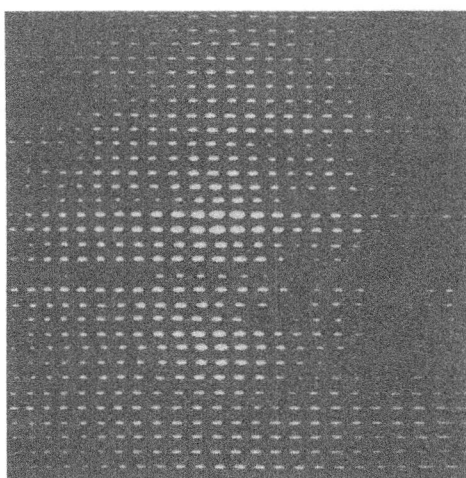

Fig. 47. Acoustic Hologram of the Letter E, as recorded by the real-time viewing tube of Fig. 46.

Fig. 48. Real-time reconstruction of the hologram of Fig. 47.

Fig. 49. Underwater viewing using acoustic holography.

had been contributors to the development of such radar.[54,16] Now the cycle has run full course, and the concept of seismic holography has been patented.[55] When this form of holography is fully developed, it should provide a better knowledge of the subterranean geologic structure and afford a higher probability of success to those in search of oil and minerals.

In seismic work sonar methods are used to propagate sound pulses into the earth's surface by especially designed sound sources. Records of the echoes of these sounds as they are reflected from various underground layers of rock or sediment produce profiles which often show numerous sedimentary layers and their shape can give information regarding the possibility of oil and mineral deposits being present. Thus, a rounded or dome shaped layer quite often indicates an oil and gas deposit.

Recent field tests suggest that the technique of holography may improve the acoustic seismic process by more definitely indicating underground formations. An arrangement as shown in Fig. 50 can be used to generate the

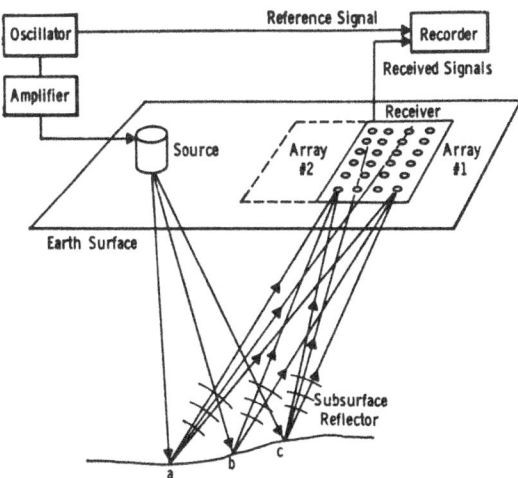

Fig. 50. Hologram array arrangement for seismic exploration (Courtesy D. Silverman).

hologram records as in the case of the array sonar described above. So far, the application of holographic techniques to large-scale underground viewing is still in the initial development stages. The frequency range employed (10 to 100 Hz) is so low (and wavelengths so great) that detector arrays must be extraordinarily long. However, scanning techniques, such as those used in synthetic aperture sonar, should be useful since the objects under study are immobile.

For offshore seismic exploration, a system, as depicted in Fig. 51, might involve a cable, which in practice would be 100 wavelengths or more in length, and be towed behind a ship equipped with a high-power transmitter capable of emitting low-frequency coherent acoustic energy into the ocean depths. Signals reflected or scattered from the ocean bottom, or from the geophysical layers below, would be picked up by the cable array. Holographic processing of seismic data obtained from large arrays of hydrophones should provide a maximum of information retrieval from such acoustic signals. Recently Lerwill,[56] using mechanical vibrators operating at 90 Hz, and using a single

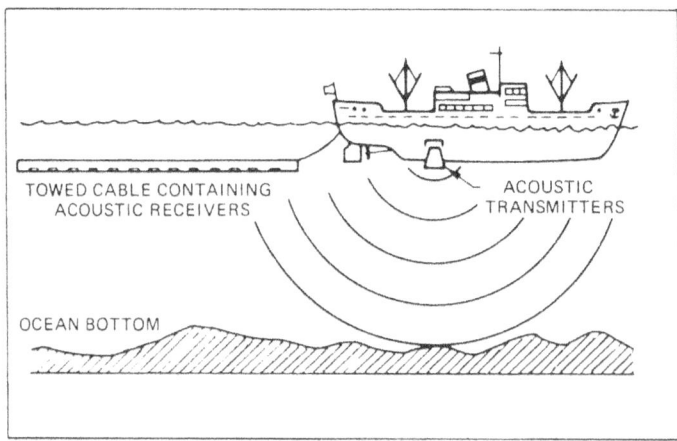

Fig. 51. Acoustic holography offers certain advantages in off-shore geological exploration.

4,350 foot line of geophones at 50 foot spacing, was able to portray, holographically, a known, 2,500 foot deep, reflecting layer.

ACKNOWLEDGEMENTS

The author wishes to thank L. J. Porcello of the University of Michigan for Figs. 17, 18, and 19; Homer Jensen of Aero-Service Corporation for Fig. 20; Messrs. Papi, Russo, and Sottini for Figs. 40, 41, and 42; D. F. Pekau and R. Diehl for Figs. 26 and 27; J. A. Jordan, Jr., of I.B.M. for Fig. 37; and to the Bendix Research Laboratories for Figs. 29,30,31,44,46,47, and 48. I am indebted to F. K. Harvey of the Bell Telephone Laboratories who made the photos of Figs. 3, 6, 7, 8, 9, 11, and 12 (See Reference 1).

REFERENCES

1. W. E. Kock and F. K. Harvey, "Sound Wave and Microwave Space Patterns," Bell System Technical Journal, Vol. 20, No. 7, pp. 564-587, (July, 1951).

2. W. E. Kock, "Real Time Detection of Metallic Objects Using Liquid Crystal Microwave Holograms," Proc. IEEE, (Lett), November, 1972.

3. Lord Rayleigh, "Theory of Sound," Dover, Vol. 2, p. 142, (1945).

4. "Photographing Sound Waves," Bell Laboratories Record, July 1950, pp. 304-306.

5. A. F. Metherell and S. Spinak, "Acoustical Holography of Non-Existent Wavefronts Detected at a Single Point in Space," App. Phys. Lett. 13-22, (1968).

6. V. L. Neeley, "Source Scanning Holography," Phys. Lett. 28A (7); pp. 475-476, (1968).

7. "Holographic Movies," Laser Focus, 1, No. 17, 1965, pp. 8-9.

8. H. Kiemle and D. Ross, "Einfauhrung in die Technik der Holographie," Akademische Verlagsgesellschaft, 1969, p. 17.

9. E. Camatini, Editor, Optical and Acoustical Holography, Plenum Press, 1972, p. 11.

10. D. Gabor, "The Hologram," Fri. Evening Discourse, Feb. 7, 1969, Proc. Royal Institution Great Britain, 43, No. 200.

11. L. J. Cutrona, E. N. Leith, C. J. Palermo, and L. J. Porcello, "Optical Data Processing and Filtering," IRE Trans. Prof. Group on Infor. Theory, Vol. IT6, No. 3, pp. 386-400, 1960.

12. L. J. Cutrona, E. N. Leith, L. J. Porcello, and W. E. Vivian, "On the Application of Coherent Optical Processing Techniques to Synthetic-Aperture Radar," Proc. IEEE, Vol. 54, pp. 1026-1032, Aug., 1966.

13. W. E. Kock, "Acoustics and Optics," (Introductory guest-editor paper), Applied Optics, Vol. 8, No. 8, August 1969, pp. 1525-1530.

14. W. E. Kock, "Holography and Microwaves," presented at the U.S.-Japan Seminar on Holography, Tokyo, Japan, Oct. 5, 1967.

15. W. E. Kock, "Side-Looking Radar, Holography, and Doppler-Free Coherent Radar," Proc. IEEE, Vol. 56, pp. 238-239, Feb., 1968.

16. E. N. Leith and A. L. Ingalls, "Synthetic Antenna Data Processing by Wavefront Reconstruction," Appl. Opt., Vol. 7, Mar. 1968, pp. 539-544.

17. W. E. Kock, "Extending the Maximum Range of Synthetic Aperture (Hologram) Systems," to appear in the Proc. IEEE (Lett.).

18. W. E. Kock, "Stationary Coherent (Hologram) Radar and Sonar," Proc. IEEE, Vol. 56, No. 12, December 1968, p. 2180.

19. W. M. Brown and L. J. Porcello, "An Introduction to Synthetic Aperture Radar," IEEE Spectrum, September 1969, pp. 52-62.

20. W. E. Kock, "Holography, A New Dimension for Radar," Electronics, Vol. 43, No. 21, Oct. 12, 1970, pp. 80-88.

21. W. E. Kock, "Radar and Microwave Applications of Holography," pp. 323-356, in Applications of Holography, New York, Plenum Press, 1971.

22. E. N. Leith, "Quasi-Holographic Techniques in the Microwave Region," Proc. IEEE, Vol. 59, No. 9, September 1971, pp. 1305-1318.

23. W. E. Kock, "Holographic Computing in Radar and Ultrasonics," (Invited Paper), Proc. of the Optical Computing Symposium, Darien, Connecticut, April 12, 1972.

24. G. L. Rogers, "A New Method of Analyzing Ionospheric Movement Records," Nature, March 31, 1956, pp. 613-614.

25. G. L. Tyler, "The Bistatic, Continuous-Wave Radar Method for the Study of Planetary Surfaces," Journal of Geophysics Research, Vol. 71, No. 6, pp. 1559-1567, March 15, 1966.

26. J. J. Flaherty, K. R. Erikson, and Van Metre Lund, "Synthetic Aperture Ultrasonic Imaging Systems," U.S. Patent No. 3,548,642, December 22, 1970 (filed March 2, 1967).

27. W. E. Kock, "Nobel Prize for Physics: Gabor and Holography," Science, Vol. 174, Nov. 12, 1971, pp. 674-675.

28. W. E. Kock, "Holographic Techniques in Continuous-Wave Bistatic Radars," Proc. IEEE, (Lett.), Vol. 58, Nov. 1970, pp. 1863-1864.

29. W. E. Kock, "Synthetic End-Fire Hologram Radar," Proc. IEEE, (Lett.), Vol. 58, Nov. 1970, pp. 1858-1859.

30. W. E. Kock, "A Holographic (Synthetic Aperture) Method for Increasing the Gain of Ground-to-Air Radars," Proc. IEEE, (Lett.), Vol. 59, No. 3, March, 1970, pp. 426-427.

31. D. F. Pekau and R. Diehl, "Recording of One Dimensional Holograms as a Function of Object Range," presented at the Int. Symp. Applications of Holography, Besancon, France, July 6-11, 1970.

32. B. P. Parkins and G. R. Fox, "Measurement of the Coherence and Fading of Long-Range Signals," I.E.E.E. Transactions on Audio & Electro-Acoustics, Vol. AV-19, June 1971, pp. 158-165.

33. R. H. Nichols and H. J. Young, "Fluctuations in Low-Frequency Acoustic Propagation in the Ocean," Jour. Acout. SJC. Am., Vol. 43, April, 1968, pp. 716-723.

34. B. R. Brown and A. W. Lohmann, "Complex Spatial Filtering and Binary Masks," Applied Optics 5, p. 967, (1966).

35. A. W. Lohmann and D. Paris, "Binary Fraunhofer Holograms Generated by a Computer," Applied Optics 6, p. 1739, (1967).

36. B. R. Brown and A. W. Lohmann, "Computer-Generated Binary Holograms," I.B.M. Jour. Res. and Dev., Vol. 13, p. 130, (1969).

37. L. B. Lesem, P. M. Hirsch, and J. A. Jordan, Jr., "The Kinoform: A New Wavefront Reconstruction Device," IBM J. Research & Dev., Vol. 13, 1969, pp. 150-155.

38. J. A. Jordan, P. M. Hirsch, L. B. Lesem, and D. L. Van Rooy, Applied Optics, Vol. 9, Aug., 1970.

39. W. E. Kock, "Metal Lens Antennas, Proc. of IRE, Vol. 34, p. 828, (1946).

40. W. E. Kock, Internal Bell Telephone Laboratories Memorandum, M.M. 44-160-67, "Experiments with Metal Plate Lenses for Microwaves," March 27, 1944, (Secret; since declassified).

41. W. E. Kock, Applied Optics, July 1972, pp. 1653-1654.

42. A. L. Boyer, P. M. Hirsch, J. A. Jordan, Jr., L. B. Lesem, and D. L. Van Rooy, "Kinoform Mirror for Acoustic Imaging, I.B.M. Publication No. 2220-6100, June 18, 1970.

43. B. P. Hildebrand and K. A. Haines, "Interferometric Measurements Using the Wavefront Reconstruction Technique," Applied Optics, Vol. 5, pp. 172-173, January 1966.

44. L. O. Heflinger, R. F. Wuerker, and R. E. Brooks, "Holographic Interferometry," J. Applied Physics, Vol. 37, pp. 642-649, February, 1966.

45. K. A. Stetson and R. L. Powell, "Hologram Interferometry," Journal Opt. Soc. Am., Vol. 56, No. 9, Sept. 1966.

46. J. A. Haines and B. P. Hildebrand, Jour. Opt. Soc. Am., Vol. 57, 1967, p. 55.

47. G. M. Brown, R. M. Grant, and G. W. Stroke, Jour. Opt. Soc. Am., May 1969, pp. 1166-1179.

48. G. Papi, V. Russo, and S. Sottini, "Microwave Holographic Interferometry," IEEE Transactions on Antennas and Propagation, Vol. AP-19, No. 6, November 1971, pp. 740-746.

49. G. Papi, V. Russo, and S. Sottini, "Two-Frequency Microwave Holographic Interferometry, Proc. IEEE (Lett.), Vol. 60, No. 8, Aug., 1972, pp. 1004-1005.

50. W. E. Kock, "Holographic Synthetic Aperture Interferometry," to appear shortly in the Proc. IEEE (Lett.).

51. E. Marom, R. K. Mueller, R. F. Koppelmann, and G. Zilinskas, "Design and Preliminary Test of an Underwater Viewing System Using Sound Holography," Acoustical Holography, Vol. 3, New York, Plenum, 1971.

52. G. G. Goetz, "Real-Time Holographic Reconstruction by Electro-Optic Modulation," Applied Optics Letters, Vol. 17, No. 2, July 15, 1970, pp. 63-66.

53. G. G. Goetz, R. F. Koppelmann and R. K. Mueller, "Real-Time Reconstruction and Display of Acoustic Holograms for an Underwater Viewing System," Proceedings of the Electro-Optic Systems Design Conference, East, pp. 202-208, September, 1971.

54. M. B. Dobrin, A. L. Ingalls, and J. A. Long, "Velocity and Frequency Filtering of Seismic Data Using Laser Light," Geophysics 30 (No. 6), pp. 1144-1178, December, 1965.

55. D. Silverman, "Wavelet Reconstruction Process for Sonic, Seismic, and Radar Explorations," U.S. Patent 3,400,363 (September 3, 1968).

56. W. E. Lerwill, "Holography at Seismic Frequencies," presented at the European Association Exploration Geophysical Conference, Venice, Italy, 1969.

ACOUSTIC MICROSCOPY

A. Korpel

Zenith Radio Corporation

6001 W. Dickens Ave., Chicago, Ill. 60639

INTRODUCTION

> "I mean, for example, is there any truth for men in their sight and hearing?"
>
> Plato "Phaedo"

The development of new microscopy techniques has historically always added a new dimension of information to the existing knowledge. Thus the phase contrast microscope opened up the study of transparent objects, the electron microscope gave access to superfine structure, the scanning electron beam microscope added a three-dimensional aspect, etc. In view of these historical precedents it is tempting to predict that a similar dimensional breakthrough will be initiated by the acoustic microscope. It is not easy, however, to specify exactly what this new dimension will be. The most one can say is that it will have to do with revealing mechanical structure, (e.g. bonding, polymerization, elasticity, density, viscosity) rather than the electronic structure (band transitions, free carriers, excitons, etc.) indirectly observed by the optical microscope. Yet many optical and mechanical properties are intimately coupled and thus it may well be that in many cases the new dimension of acoustic microscopy will be more quantitative than qualitative in nature. Frequently this quantitative difference between optical and acoustic properties is very pronounced. For instance, whereas the optical reflection off a water-air interface amounts to no more than 2%, the acoustic reflection is virtually total.

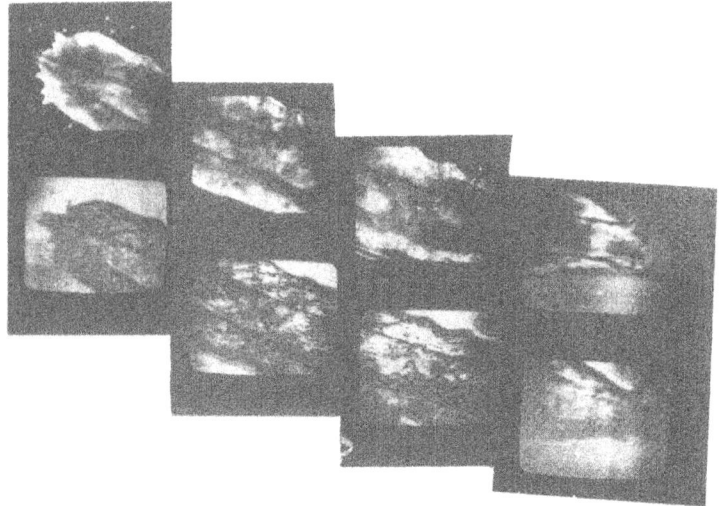

Fig. 1. Composite optical (top) and acoustical (bottom) image of the fruit fly larva (Drosophila Melanogaster). The acoustic frequency is 100 MHz. Each frame is approximately 1.7 mm wide, successive frames overlap by 40%. (After Kessler et al).

Figure 1 shows an example of this in the air-breathing structure (tracheas) of the fruit fly larva. The top picture is an optical dark field image taken in through-transmission with the same scanning laser system that produced the acoustic picture on the bottom.[1] (This system will be discussed later.) The acoustic frequency is 100 MHz, corresponding to a wavelength of 15 μm in water. The resolution in both pictures is of the order of 20-25 μm. Note the fine tracheal network in the acoustic picture; the little air tubes present a perfect mismatch to the

sound and in consequence show up very clearly. To view
the tracheas optically, special Schlieren or phase contrast
techniques have to be used and even then these structures
may be masked by general opacity. Thus to the extent
that the object is optically but not acoustically opaque or
that acoustic boundaries are clearly very different from
optical ones, acoustic microscopy does not have to prove
its credentials.

A second example is shown in Fig. 2. The object here
is the wing of a house-fly.[2] The overall acoustic image

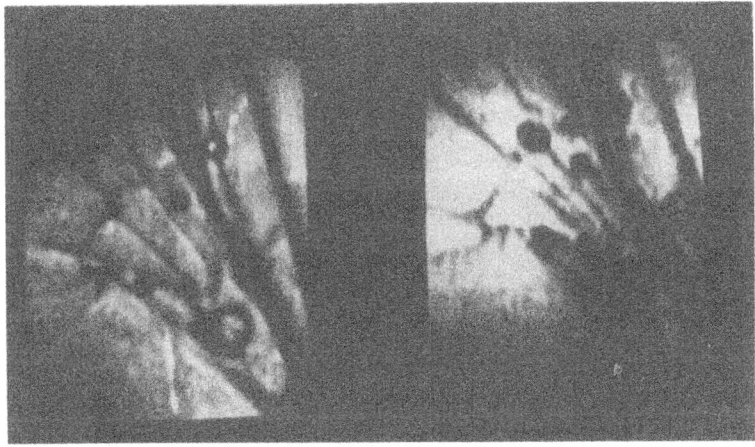

Fig. 2. Optical (left) and acoustical (right) image of
house-fly wing. The acoustic frequency is 100 MHz, the
field of view is 1.5 mm. (Courtesy Zenith Radio Corp.).

(right) is not greatly different from the optical one (left)
apart from some small accidental air bubbles. Note, however, in the acoustic picture, an additional black line in
between the two structural membranes in the center. This
may be a break in the wing which is either filled with air
or severely scatters the acoustic radiation. In any case
it does not show up in the optical image.

Where neither optical opacity nor boundaries are involved the acoustic microscope's advantage would have to

lie in the superior contrast of differential acoustic absorption versus say, optical differential phase contrast. There is no a priori reason to rule this out; note that the acoustic absorption in water is of the order of 200 db/mm at 1000 MHz.

If there is no objection to damaging biological material, acoustic staining, yet to be developed, may offer advantages over optical staining. Conceivably, thermal staining techniques could be developed which would induce structural change at certain critical temperatures.

At 100 MHz the <u>absorption</u> in small tissue structures, e.g. 15 μm diameter, is insufficient to cause observable contrast. An example is seen in Fig. 3 which shows the optical (left) and acoustical (right) images of onion cells.[2]

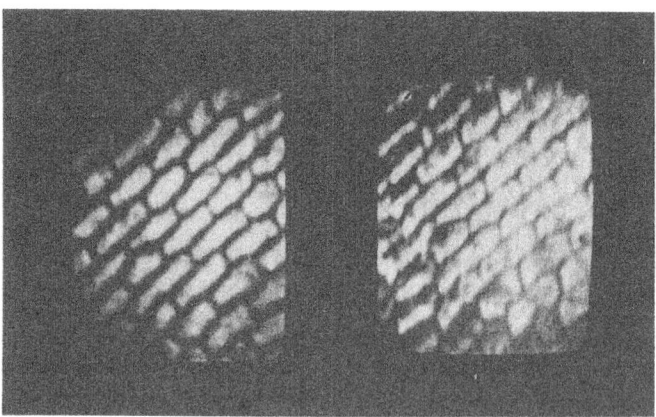

Fig. 3. Optical (left) and acoustical (right) image of onion skin cells. The acoustic frequency is 100 MHz, the field of view 1.5 mm. (After Kessler et al).

Little structure is seen inside the cell, only the cell walls are visible. At higher frequencies the attentuation increases very quickly (in water with the square of the frequency) and it would be fascinating to look at the onion cells of Fig. 3 at a frequency of 2 GHz with a resolution of 1μm. It is not unreasonable to assume that such pictures will be a reality within a period of a few years. Whether the final results will be useful in the field of medicine is a question that cannot be answered yet in any but an emotional sense. One thing is certain, without building the instrument we will never know.

HISTORICAL BACKGROUND

Acoustic microscopy is an entirely new field that began to expand when the feasibility of constructing a practical instrument was demonstrated. Acoustic visualization methods already employed in the fields of medical diagnosis[3] and non-destructive testing[4] had sought to reveal macroscopic size structures, and thus only long wavelength, low frequency radiation was necessary. Acoustic microscopy, on the other hand,[5] aims at the visualization of structural detail of 0.5 to 25 μm, resolution never before achieved with acoustic imaging. In this region high frequency illumination (viz., 100 MHz to 3000 MHz) is required since the wavelength of sound must be of the same order of magnitude as the desired image detail.

The concept of acoustic microscopy dates back to 1959 when it was demonstrated[6] that a resolution of 75 μm could be achieved by mechanically scanning a single 12 μm thermocouple junction across a sound field of interest and continuously recording its output signal. A real time or even quasi real time instrument based on this principle is difficult to realize physically because it would require the fabrication and addressing of a matrix of subminiature acoustic detectors for which the technology is not yet available. Renewed interest in acoustic imaging, largely due to implications of holography,[7] produced subsequent technological achievements. With these achievements, the feasibility of fabricating a practical ultrasonic microscope became apparent.[8]

There are four projects currently underway to develop a practical device. The first project,[9] under Professor Quate at Stanford University, is based on a variation of the technique of Bragg diffraction imaging. (Bragg diffraction imaging was first described by Zenith researchers in 1966.[10]) In the Stanford variation, a sound field which has passed through the specimen is first Fourier transformed by means of an acoustic lens and then made to interact with a laser beam. This is shown in Fig. 4.

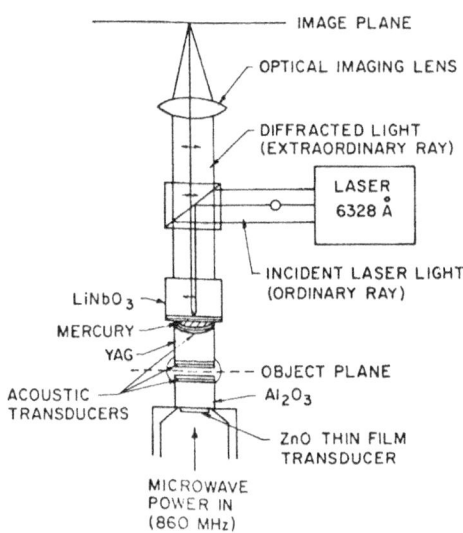

Fig. 4. Collinear Bragg diffraction acoustic microscope (After Havelice and Quate).

The interaction is of the collinear kind, made possible by using an anisotropic material ($LiNbO_3$) as the interaction medium.[11] Each plane wave of sound gives rise to a backscattered plane wave of light with polarization orthogonal to that of the incident light. Thus the far field (Fourier transform) of the sound image is translated into the optical domain. Subsequent focusing by a lens then produces an optical image of the sound field itself. The sound frequency used is 860 MHz; it is determined by the anisotropic properties of the $LiNbO_3$ medium. The resolution obtained with the system is of the order of 20 μm, i.e. approximately 12 wavelengths of sound in water at the operating frequency.

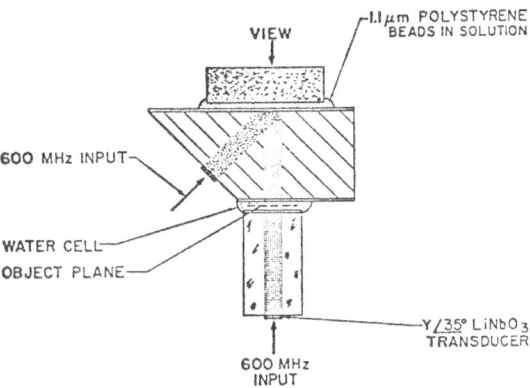

Fig. 5. Radiation pressure holographic acoustic microscope. (After Cunningham and Quate).

More recently Quate has been experimenting with a second technique[12] in which small polystyrene beads in solution are lined up by the radiation pressure pattern caused by two intersecting plane waves. One arrangement is shown in Fig. 5. In operation, the polystyrene beads arrange to form a hologram of the object. This is then photographed and reconstructed. Results with this configuration so far have shown a resolution of 300 μm at 600 MHz with a sensitivity on the order of 10^{-4} watts/cm^2 to 10^{-3} watts/cm^2.

A more recent non-holographic arrangement[13] is shown in Fig. 6.

Here, the object is placed in intimate contact with the solution of beads, and is traversed by both beams at once. The resulting image is a non-holographic shadow picture of the object. The resolution at 260 MHz is better than 60 μm (about 10 wavelengths in water).

Quate and Cunningham have also experimented[13] with the one-beam configuration shown in Fig. 7. The acoustic frequency as used was 1.1 GHz and the achieved resolution better than 9 μm (about 7 wavelengths in water). Quoted sensitivity is 5×10^{-2} W/cm^2.

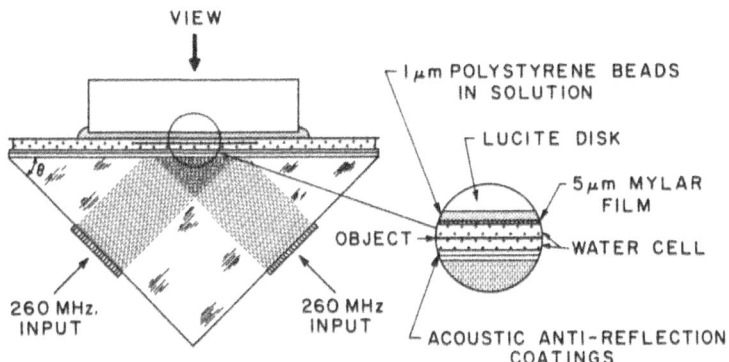

Fig. 6. Double beam radiation pressure acoustic microscope (after Cunningham and Quate).

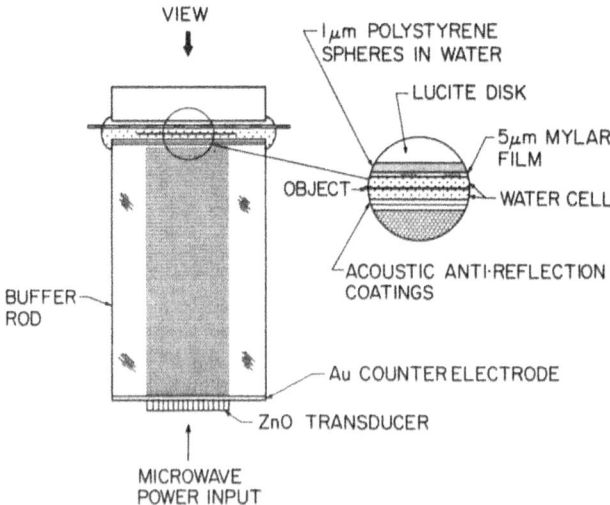

Fig. 7. Single beam radiation pressure acoustic microscope (after Cunningham and Quate).

The third technique, by Auld et al. of Stanford University[14,15] is based upon focusing a sound beam to a small diameter upon the sample to be examined, then locally measuring the transmitted amplitude as a function of position of the spot and subsequently displaying the pattern on a TV screen. The implementation of this technique[14] depends upon using a single material which is both piezoelectric and photoconductive. With such a material used as a transducer, it should be possible to optically control both the acoustic beam shape and position.

In another mode of operation illustrated in Fig. 8, a similar transducer is used as a receiver by monitoring the change in electric field developed across it while locally quenching its piezoelectric activity by exposure to a focused spot of light. Thus far, difficulty has been experienced in producing a suitable transducer. A potential material such as CdS can be used efficiently as either a photoconductor or as a piezoelectric. The use envisaged here, however, requires a difficult-to-achieve combination of both properties. The resolution thus far observed (of the order of $\approx 25 \mu m$ at a frequency of 1000 MHz[15] is restricted by the intrinsic transducer material parameters which set a power limit to the size of the region that can be quenched.

The fourth method of acoustic microscopy, the one we have used at Zenith, is based on direct optical scanning and recording of the dynamic displacement pattern caused by an acoustic wave impinging upon a suitable optical mirror.[5] This technique has demonstrated resolution better than 1.5 wavelengths of sound.[16] That is, at an operating frequency of 230 MHz (the highest used so far) the observed resolution is of the order of $10 \mu m$. The technique will be described more fully in the next section.

Prior to the construction of this instrument, we conducted a study to compare various possible methods of acoustic microscopy.[5] This study did not include Quate's and Auld's methods mentioned before. The result of this study indicated that the laser beam sampling technique

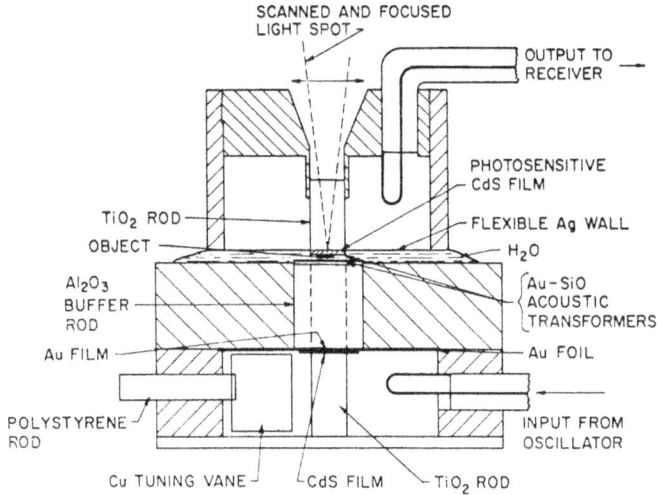

Fig. 8. Photoconductive-piezoelectric acoustic microscope (after Auld et al.).

not only could be made practical, but that the sensitivity was sufficient to warrant pseudo real time operation, i.e. <1 sec. per frame. On the basis of this, a prototype device was constructed, operating at 100 MHz, at television rates. At present the frequency has been raised to 230 MHz, and the frame time to 1/5 sec. The resolution is currently 10-12 µm. Future plans are to increase the frequency by a factor 10 with a corresponding increase in resolution. Concurrently with these developments we have achieved simultaneous display of acoustical and optical images with, as a refinement, the superposition of these two images in different colors and in real time.

THE SCANNING LASER ACOUSTIC MICROSCOPE

The basic principle of operation of the acoustic microscope is similar to that employed earlier by Zenith in an experiment with acoustic holography conducted at lower frequencies[17] and is based on a technique used by Adler, et al.[18], for the visualization of acoustic surface waves. A detailed analysis may be found in Ref. 19. A simplified diagram of a very early configuration is shown in Fig. 9.

ACOUSTIC MICROSCOPY

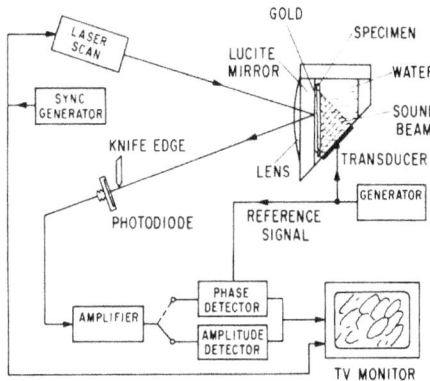

Fig. 9. Scanning laser acoustic microscope (after Korpel et al.).

An ultrasonic transducer launches a 100 MHz plane wave through a specimen which is suspended in water. The transmitted sound which carries spatial information about the specimen, strikes a plastic mirror oriented at 45° with respect to the nominal direction of sound travel and causes a dynamic ripple pattern on the mirrored surface. The ripples cause a periodic angular deflection of a focused laser beam which is reflected off the mirror. This angularly modulated, reflected light beam gives rise to an electrical signal in the photodiode when half of the beam is intercepted by the knife edge. The electrical signal is directly proportional to the angular displacement of the mirrored surface and hence to the local sound pressure. To achieve real time operation, the focused light beam is made to scan rapidly over a selected area of the mirror by means of two acousto-optic light deflector cells.[20] The scan of the light beam is synchronized with that of a conventional TV monitor. The lens shown in Fig. 6 serves to keep the exit pupil of the laser scanner focused on the knife edge during the scanning of the beam. If the 100 MHz signal from the photodiode is rectified and applied to the TV monitor, a magnified picture of the sound field appears on the screen. In an alternative mode of operation, the signal from the photodiode is first applied to the phase detector where it is compared to an electronic reference signal. In this case, the TV screen shows an acoustic hologram rather than a conventional sound picture. This hologram may be recorded on film and reconstructed to obtain additional depth information.

In thin samples placed close to the faceplate, a good shadow image will be made. If the sample is thin, that is, of the order of a few wavelengths of sound, holographic processing for additional depth information will not be required. However, for thick objects or for objects spaced far from the faceplate, holographic techniques will be required to form an image. In addition, holographic techniques reveal the existence of object regions which only affect the local sound velocity (phase objects). An example of the holographic capabilities of the instrument is seen in Fig. 10. This shows the hologram of the Drosophila larva of which optical and acoustic images have been presented earlier in Fig. 1.

A diagram of the present version of the object holder used for simultaneous optical and acoustic imaging is shown in Fig. 11. Note that part of the focused beam now goes through the object to be detected by an auxiliary photodiode. The output of this diode is a measure of the local optical transmissivity of the object and is displayed on a second television monitor as shown in Fig. 12. For

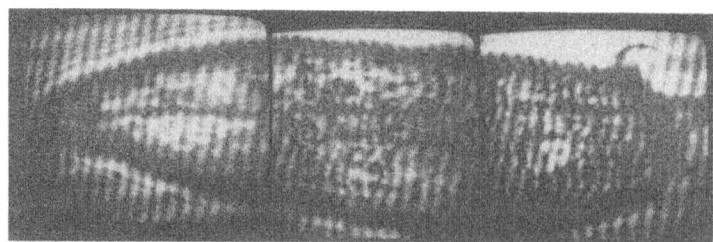

Fig. 10. Composite acoustic hologram of fruit fly larva (Drosophila Melanogaster). The acoustic frequency is 100 MHz. Total field of view is about 3 mm (after Kessler et al.).

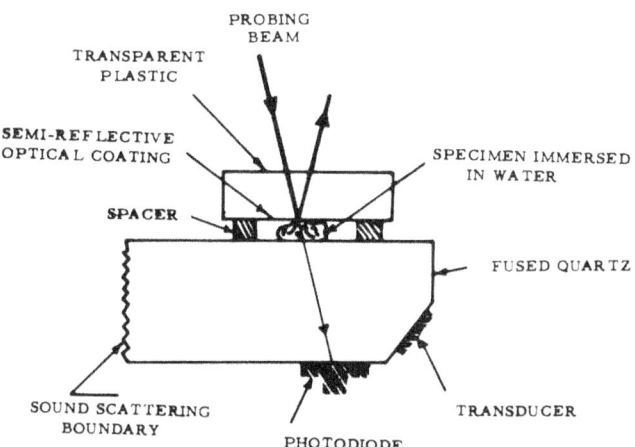

Fig. 11. Object holder for simultaneous acoustic and optical imaging (after Kessler et al.).

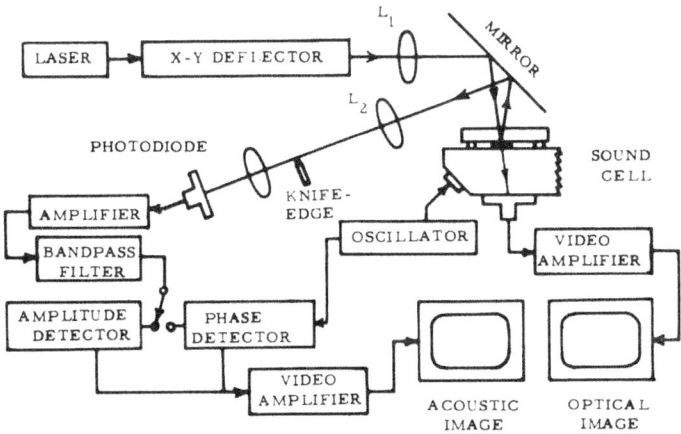

Fig. 12. Diagram of acoustic scanning laser microscope with simultaneous optical imaging (after Kessler et al.).

convenience the single lens that images the scanner exit pupil onto the knife edge in Fig. 6 has been replaced by the two lenses shown in Fig. 9.

Figure 13 demonstrates the resolution capabilities of

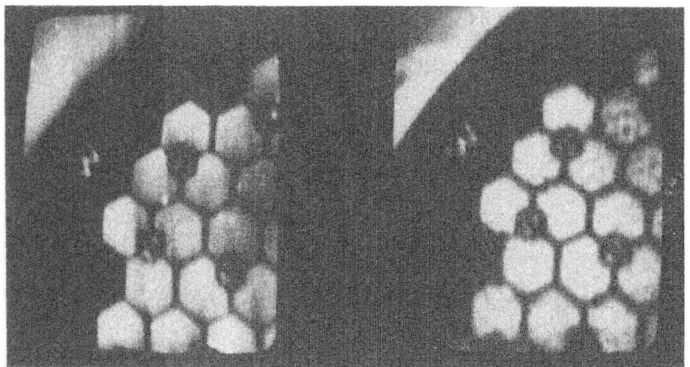

Fig. 13. Optical (left) and acoustical (right) picture of microscope reference grid. The small circles have a diameter of 100 μm. The acoustic frequency is 100 MHz (after Kessler et al.).

the 100 MHz instrument. The object is a metallic reference grid with dimensions of 250 μm between the sides of the hexagons. The small circles which carry the reference letters have a diameter of about 100 μm. The optical picture is on the left, the acoustical picture on the right. It is seen that the resolution is comparable and of the order of 20-25 μm. For this particular test object the visual character of the pictures is quite similar; the metal stops sound as well as light. In general, however, as shown before in Fig. 1, the simultaneous mode of operation greatly enhances the facility of interpretation of the acoustic picture, by immediately showing up the difference between the two images.

The sensitivity of the instrument at 100 MHz has been measured to be about $10^{-3} W/cm^2$, when operating at television rates, i.e. 30 frames/sec. Scaling up to 1 GHz will require slowing down the frame rate to about 1 frame per 3 sec. in order to achieve comparable sensitivities.

For completeness, a photograph of the current instrument is shown in Fig. 14. The laser (50 mW, He-Ne) is under the table; the beam is brought up by a mirror which

Fig. 14. Photograph of acoustic scanning laser microscope. An optical picture is displayed on the top monitor, an acoustical picture on the bottom one (after Kessler et al.).

is to the bottom right, outside the picture. The object holder is at the extreme left of the optical bench in front, near the two television monitors.

CONCLUSION

It will be evident from this short review that acoustic microscopy has almost reached the stage where its potential applications can be evaluated. Technologically, resolution has to be increased by an order of magnitude to reach the resolving power of the optical microscope.

This does not appear to require a new invention; proper development of the existing techniques should achieve this goal.

Once it has been reached, a proper assessment of potential should be made by comparising optical and acoustic micrographs of a wide range of structures. This will require intensive cooperation between researchers in widely varying disciplines. An example taken at random might be a group consisting of pathologists, electrical engineers and holographers.

All this will take time, effort and, above all, a belief in the promise of this new technique. It has become fashionable (and to a certain extent justified) to grade scientific research in terms of relevance to contemporary problems. It is my belief that the potential of acoustic microscopy warrants its further development, when judged by this criterion.

REFERENCES

1. L. W. Kessler, A. Korpel and P. R. Palermo "Simultaneous Acoustic and Optical Microscopy of Biological Specimens", Nature, 239, 111 (1972).

2. L. W. Kessler, P. R. Palermo and A. Korpel "Practical High Resolution Acoustic Microscopy", to be published in Acoustical Holography Vol. 4, Plenum Press, December 1972.

3. A good review of medical applications of ultrasound can be found in E. Kelly, "Ultrasonic Energy: Biological Investigations and Medical Applications", U. of Illinois Press, Urbana, Ill. (1965).

4. P. S. Green, International Journal of Nondestructive Testing, 1, 1 (1969).

5. A. Korpel and L. W. Kessler in Acoustical Holography Vol. 3, ed. by A. F. Metherell, Plenum Press, New York, N.Y. 1971, p. 23.

6. F. Dunn and W. J. Fry, J. Acous. Soc. Amer. 31, 632 (1959).

7. A. Korpel, IEEE Spectrum, May 1969, p. 42.

8. R. K. Mueller, Proc. IEEE 59, (1971) p.13.

9. A description of the work of Quate and Havelice may be found in Ref. 7.

10. A. Korpel, Appl. Phys. Lett. 9, 425 (1966).

11. R. W. Dixon, IEEE J. Quantum Electron, QE-3, 85 (1967).

12. J. A. Cunningham and C. F. Quate "Acoustic Interference in Solids and Holographic Imaging", M. L. Report No. 2086, Hansen Laboratories of Physics, Stanford University, Stanford, Calif.; also to be published in Acoustical Holography, Vol. 4, Plenum Press.

13. J. A. Cunningham and C. F. Quate "High-Resolution, High-Contrast Acoustic Imaging", M. L. Report No. 2094, Hansen Laborarories of Physics, Stanford University, Stanford, Calif.

14. B. A. Auld, D. C. Webb and D. K. Winslow, Proc. IEEE 57, 713 (1969).

15. B. A. Auld, R. C. Addison and D. C. Webb in Acoustic Holography, Vol. 2, A. F. Metherell and L. Larmore (ed.), Plenum Press, New York, N.Y. 1970, p. 117.

16. A. Korpel, L. W. Kessler and P. R. Palermo, Nature, 232, 110 (1971).

17. A. Korpel and P. Desmares, J. Acoust. Soc. Amer. 45, 881 (1969).

18. R. Adler, A. Korpel and P. Desmares, IEEE Trans. on Sonics and Ultrasonics, SU-15, 157 (1968).

19. R. L. Whitman and A. Korpel, Applied Optics 8, 1567 (1969).

20. A. Korpel, R. Adler, P. Desmares and W. Watson, Applied Optics 5, 1667 (1966).

SOME ASPECTS OF OPTICAL HOLOGRAPHY THAT MIGHT BE OF INTEREST FOR ACOUSTICAL IMAGING

A. W. Lohmann

Dept. of Applied Physics & Information Science
University of California at San Diego
La Jolla, California 92037

ABSTRACT

Some concepts of optical holography might be useful for acoustical imaging. First we compare the current state of these two fields. Then we summarize some data economy procedures such as sampling and quantization, which are important in optical computer holography and probably also for acoustical imaging. Finally we describe some specific holographic tricks that can perhaps be translated into the realm of acoustics.

1. MY ENCOUNTERS WITH ACOUSTICAL HOLOGRAPHY

Acoustic imaging may be categorized according to the frequency range into

Seismology	< 100 Hz
Oceanography	< 1 MHz
Medical Applications	$\geq$ 1 MHz.

Frequency ranges correspond to ranges of object thickness since the acoustic attenuation is generally proportional to the square of the frequency.

I tried twice to bring the blessings of holography to the acoustics community, once for the benefit of seismology, since my brother is a geophysicist, and once to my colleagues at the Scripps Institution of Oceanography. The replies from the seismological and from the oceanographic communities were similar to what is stated in articles by R. A. Peterson [1] and F. N. Spiess [2]. Basically I am now convinced that a simple-minded translation of the general holographic concept from optics to acoustics is not necessarily a step forward for acoustical imaging, but may sometimes be even a step backward. I think the basic reason for this becomes obvious when considering which of the three wave types are used in the various fields.

Wave Type	Field	Data Parameters
stochastic polychromatic	incoherent optics	only amplitude (stationary)
deterministic monochromatic	holography	amplitude and phase (stationary)
deterministic broadband	acoustical imaging	varying amplitude and phase

When optics progressed from amplitude-only data to amplitude-and-phase data as a consequence of the invention of holography it was essential that a monochromatic wave be utilized. The term "deterministic" implies that essentially no random fluctuations of the wave can be tolerated. In the field of physical acoustics several image formation experiments with monochromatic waves and with phase measurements had been performed already before Gabor invented holography. In other words, the holographic concept of phase measurement is not needed in acoustics since in that field it is not difficult to measure the phase directly.*

---

* This last statement is not entirely valid for very high sound frequencies which are used for small objects as they occur in medical applications. In that situation the liquid interface is a suitable recording medium that has much in common with the photographic plate, the standard receiver in optical holography. Both receivers are phase-blind. Hence the holographic principle of indirect phase measurement is essential. But other sound receivers such as the Bragg cell are phase sensitive even at high frequencies. Hence it is still true that the acoustics community is less dependent on the holographic principle.

Based on these comments it looks as if optics and acoustics were at the same level of development. But actually acoustics is even a step ahead of optics, since it is possible to generate and measure continuously the amplitude and the phase of an acoustical *broadband* wave. One way of judging this skill is to say that in acoustical imaging we may explore simultaneously the amplitudes and phases of different monochromatic waves by means of frequency multiplexing. Another way of appraising the capabilities of deterministic broadband waves is to consider how depth information is obtained. Remember, with the help of an optical lens we can measure the depth location as well as we can measure the depth of focus δz_0. It depends on the wavelength λ and on the F-number kappa, which is the inverse of the aperture angle. Typically $\delta z_0 = \lambda \kappa$ is several wavelengths large (λ = optical wavelength, κ = kappa). On the other hand with acoustical waves of a bandwidth $\Delta \nu_t$ we can measure the time of the arrival of a pulse with accuracy $\delta t = 1/\Delta \nu_t$ and hence the depth location of a reflecting object with an accuracy $\delta z_A = v \delta t/2 = \Lambda \nu_t / 2 \Delta \nu_t$ (v = wave velocity, Λ = acoustical wavelength, ν_t = mean frequency). Assuming a relative bandwidth $\Delta \nu_t / \nu_t$ of 1/5, the range accuracy is one tenth of a wavelength, which is superior by an order of magnitude over the case of monochromatic image formation. In terms of angular resolution both methods depend in the same manner on the wavelength and on the width of the array or lens.

In spite of these fundamental considerations it is still sensible for acoustical and optical people to talk to each other. Firstly, once the acoustical image data are available they have to be communicated to the human brain. The best way of bringing three-dimensional information into the brain is via the eyes. In other words, visual *display* technology is an area where we optics people can contribute to acoustical imaging. Considerable effort on display systems (holography, integral photography, carrier frequency photography) and on real-time display materials (photoplastics, liquid crystals, electro-optical crystals) is reported in almost every issue of journals such as "Applied Optics".

Secondly, the acoustics and optics people share problems under the heading of "data economy". For example in computer holography, which I am particularly interested in, a hologram consists typically of an array of 64 × 64 data points. Bigger computer holograms are certainly possible, but costly. Some comments on data economy will follow in the second chapter.

Thirdly, there exist some rather specialized holographic methods which can measure certain parameters exceptionally well. Other methods are surprisingly insensitive to some disturbances that have their counterpart in acoustics. It might be worthwhile to bring some of these methods to the attention of the acoustics community. This will be attempted in the third section under the heading "The Virtues of Some Holographic Illnesses".

2. DATA ECONOMY IN HOLOGRAPHY

A continuous function such as the intensity $I(x)$ may in principle contain infinitely many data. However in practice $I(x)$ will not vary wildly without constraint. Hence we get all the inherent information if we measure only at discrete sampling points, such getting the sampled intensity $I_S(x)$ (see figure 1). And since we cannot measure the intensity accurately anyway we might cut $I(x)$ into horizontal slices, which leads to the quantized signal $I_Q(x)$. If the signal $I(x)$ is N samples long and up to M quantization levels high the simplified signal (sampled and quantized) can have only a finite number of different shapes. The logarithm N log M of this number indicates the amount of information in bits that is contained in such a signal. This basic piece of information theory is of course well known [3].

Well known also is how sampling can be performed in the context of Fraunhofer or far field diffraction [3]. An object $u_0(x)$ that is illuminated by a monochromatic plane wave (figure 2) produces as diffraction pattern at infinity* the Fourier transform $\tilde{u}_0(x/\lambda f)$ of that object. If the object is confined to $|x| < \Delta x_0/2$ the Fourier transform $\tilde{u}_0(\nu)$ may be sampled without loss of information at steps $\delta\nu_1 = 1/\Delta x_0$ according to the sampling theorem [4]. Remembering the connection between the spatial frequency ν of the object and the coordinate $x = \lambda f \nu$ in the rear focal plane of the lens (see figure 2) we get the following set of equations pertinent to sampling and Fraunhofer diffraction.

$$u_0(x) = \int \tilde{u}_0(\nu) \exp(2\pi i \nu x) \, d\nu$$

spatial frequency analysis in $z = 0$

---
* "Infinity is where parallel rays meet. That happens in the rear focal plane of a lens.

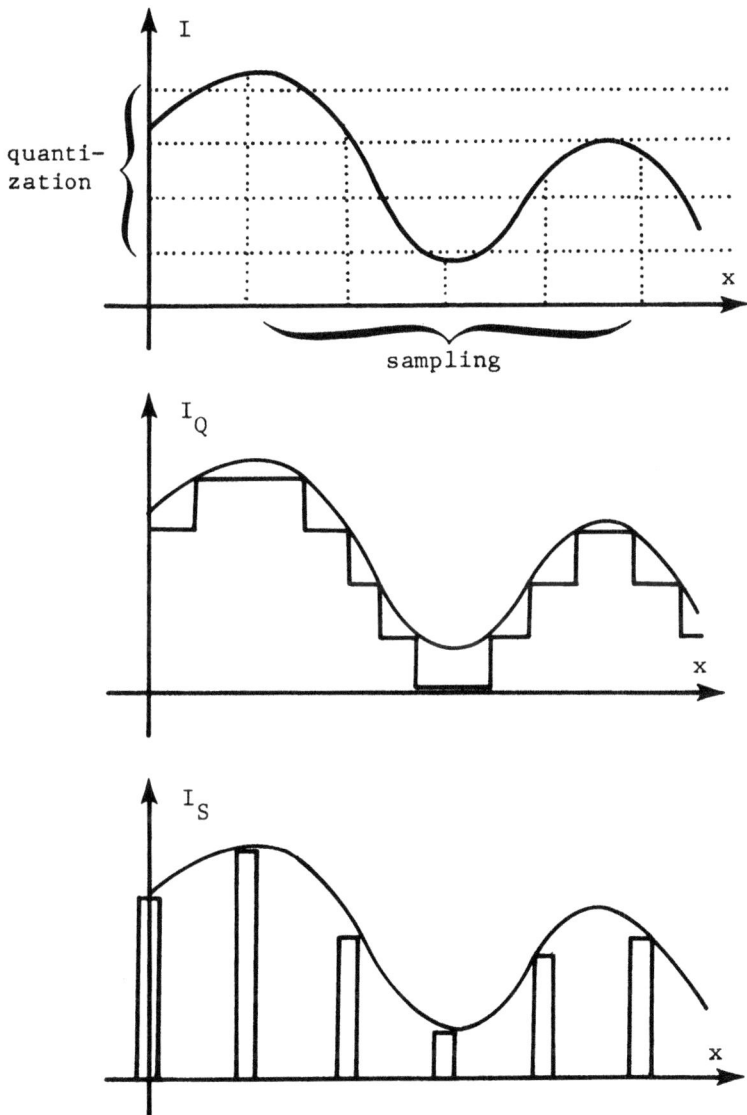

Figure 1. Simplification of signal I(x). Horizontal slicing I_Q is called "quantization"; vertical slicing I_S is called "sampling".

$$\tilde{u}_0(\nu) = \sum \tilde{u}_0(n/\Delta x_0) \, \text{sinc}(\nu \Delta x_0 - n)$$

sampling theorem; $\text{sinc}(y) = \sin(\pi y)/(\pi y)$

$$\tilde{u}_0(x/\lambda f) = \sum \tilde{u}_0(n/\Delta x_0) \, \text{sinc}(x\Delta x_0/\lambda f - n)$$

in $z = 2f$. (1)

Sampling step in $z = 2f$ is $\delta x_1 = \lambda f \delta \nu_1 = \lambda f/\Delta x_0$. A Fourier transformation by means of wave propagation occurs also without any lens involved. The wave amplitude on a sphere with a radius R that is centered around the object $u_0(x)$ is $\tilde{u}_0(x/\lambda R)$ if R is very large and if the spatial frequencies of the object are well below $1/\lambda$. If the object size is Δx_0 the wave amplitude at the sphere may be sampled at intervals $\lambda R/\Delta x_0$.

More frequently occurs the case of Fresnel diffraction (figure 3), which is most commonly described by the Huygens-Fresnel-Kirchhoff integral. That equation can be consider-

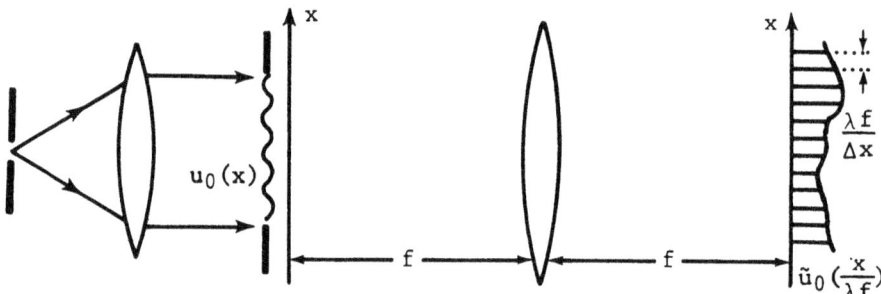

Figure 2. Sampling in the Fraunhofer diffraction plane.

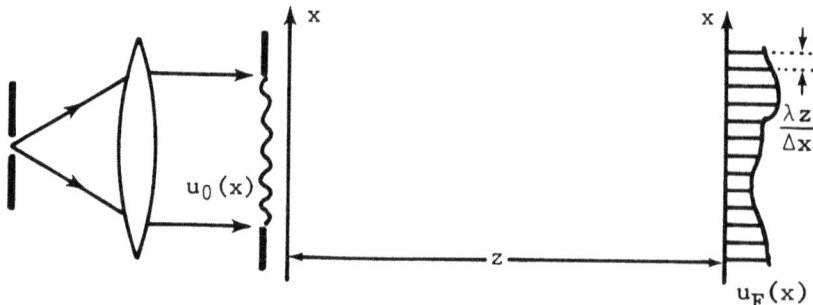

Figure 3. Sampling in a Fresnel diffraction plane.

ably simplified if only moderate diffraction angles occur, say less than 20°. The resulting approximation is sometimes called shadow transformation [5] or Fresnel transformation [6] (equation (2)).

$$u(x,z) \propto \int u(x',0) \exp[i\pi(x-x')^2/\lambda z] \, dx'$$

$$u(x,0) \propto \int u(x',z) \exp[-i\pi(x-x')^2/\lambda z] \, dx' \qquad (2)$$

The similarity of this transformation and the Fourier transformation had been recognized already by Gabor [5]. Based on this similarity it is possible to derive a sampling theorem for the Fresnel transformation as outlined in equ. (3).

$$\hat{u}(x) = \int u(x',0) \exp[i\pi(x-x')^2/\lambda z] \, dx'$$

$$= \exp(i\pi x^2/\lambda z) \int \{u(x',0) \exp(i\pi x'^2/\lambda z)\}$$
$$\exp(-2\pi i x x'/\lambda z) \, dx';$$

if $u(x',0) = 0$ for $|x'| > \Delta x_0/2$ then also

$$\{u(x',0) \exp(i\pi x'^2/\lambda z)\} = v(x');$$

hence

$$\int v(x') \exp(-2\pi i x x'/\lambda z) \, dx' = \tilde{v}(x/\lambda z)$$

$$= \sum \tilde{v}(n/\Delta x_0) \, \text{sinc}(x\Delta x_0/\lambda z - n);$$

now $\hat{u}(x) = \exp(i\pi x^2/\lambda z) \sum \tilde{v}(n/\Delta x_0) \, \text{sinc}(x\Delta x_0/\lambda z - n);$

find $\tilde{v}(n/\Delta x_0)$ from $\hat{u}(n\lambda z/\Delta x_0) = \exp[i\pi(n/\Delta x_0)^2/\lambda z] \, \tilde{v}(n/\Delta x_0);$

finally insert

$$u(x,z) = \sum u(n\lambda z/\Delta x_0, z) \exp[(i\pi/\lambda z)\{x^2 - (n\lambda z/\Delta x_0)^2\}]$$
$$\text{sinc}(x\Delta x_0/\lambda f - n) \qquad (3)$$

Apparently the sampling step $\delta x_2 = \lambda z/\Delta x_0$ increases with the distance from the object in $z = 0$.

The sampling theorem enters into optics in yet another way. Suppose we decompose the wavefield behind the object not into *spherical* waves but into *plane* waves, as Lord Rayleigh did about 90 years ago. This plane wave representation [3] has always been much less popular in optics than is deserved. Now we assume that the spatial frequency spectrum $\tilde{u}_0(\nu)$ of the object $u_0(x)$ in $z = 0$ is limited to frequencies $|\nu| < \Delta \nu_0/2$. On that basis we may sample the object at intervals $\delta x_3 = 1/\Delta \nu_0$. But not only the object at $z = 0$ but also the wavefield in planes z = constant may be sampled at intervals δx_3 because the bandwidth $\Delta \nu$ remains the same as the wave propagates from $z = 0$ to plane z. The relevant formulas are shown in equ. (4).

$$u_0(x) = \int \tilde{u}_0(\nu) \exp(2\pi i \nu x) \, d\nu \quad \text{in } z = 0$$

$$u_0(x) = \sum u_0(n/\Delta \nu_0) \, \text{sinc}(x\Delta \nu_0 - n)$$

$$\text{due to } \tilde{u}_0(\nu) = 0 \text{ for } |\nu| > \Delta \nu_0/2$$

$$u(x,z) = \int \tilde{u}_0(\nu) \exp[2\pi i (\nu x + \sqrt{1 - \lambda^2 \nu^2} \, z/\lambda)] \, d\nu$$

plane wave representation

$$\tilde{u}(\nu,z) = \int u(x,z) \exp(-2\pi i \nu x) \, dx$$

$$= \tilde{u}_0(\nu) \exp[2\pi i \sqrt{1 - \lambda^2 \nu^2} \, z/\lambda]$$

$$u(x,z) = \sum u(n/\Delta \nu_0, z) \, \text{sinc}(x\Delta \nu_0 - n). \qquad (4)$$

It remains to compare the two near field applications of the sampling theorem (Fresnel, Rayleigh). In the very near field the Fresnel approximation is not valid. Hence we have to use the rigorous plane wave representation (equ. (4)) with a sampling step $\delta x_3 = 1/\Delta \nu_0$. But at intermediate distances z we may select the larger of $\delta x_2 = \lambda z/\Delta x_0$ and $\delta x_3 = 1/\Delta \nu_0$, because a larger sampling step means fewer data for a complete description of $u(x,z)$. From a mathematically rigorous point of view we will never have to choose between δx_2 and δx_3 because the two underlying conditions $u_0(x) = 0$ for $|x| > \Delta x_0/2$ and $\tilde{u}_0(\nu) = 0$ for $|\nu| > \Delta \nu_0/2$ are incompatible. However these two conditions become compatible if one or both of the = signs are replaced by $\approx$ signs. That is good enough for our considerations to be valid with experimentally reasonable accuracy, particularly when the space-bandwidth product [7] $\Delta x_0 \Delta \nu_0 = \Delta x_0/\delta x_3 = \Delta \nu_0/\delta \nu_1$ is larger than say ten or twenty. This space-bandwidth product indicates the necessary number of elements of the receiver array. Rarely will that number be less than ten. Hence as experimentalists we do not worry about the question of mathematical rigor. We estimate qualitatively the object size Δx_0, the object bandwidth $\Delta \nu_0$ and the object distance z. On that basis we select the larger of the two possible sampling steps $\delta x_2 = \lambda z/\Delta x_0$ and $\delta x_3 = 1/\Delta \nu_0$ because coarser sampling is more economical.

A last remark on sampling has to do with "dilute apertures" or "dilute sampling arrays". That is an old art in radio astronomy [8] which is certainly not unknown in acoustics [9], for example due to the efforts of Metherell and Hildebrand, who also worked on related problems in optical holography [9]. Goodman [10] wrote a review article on "synthetic aperture optics". More recent contributions are due to Stroke [11] and Russel and Goodman [12]. Such tricks as frequency multiplexing, time sharing, or exploitation of *a priori* information about the object are utilized in order to overcome the shortcomings of a receiver with holes in it. Most of these tricks require a low noise environment to be useful.

Now we want to comment briefly on "quantization" in optical holography. When synthesizing a hologram by means of a digital computer the amplitude and/or the phase might be quantized in order to simplify the job of plotting the hologram [13]. In other instances we might want to quantize the *intensity* that is recorded on a hologram [14]. For example, a hologram made on Mars will be scanned. The scanned intensity profile has to be simplified perhaps by quantization since the telemetry channel can transmit data only at a very low rate.

An important term in quantization theory is "hard clipping", which means that only two different amplitude levels can be distinguished. One-level-amplitude quantization is also called amplitude equalization. The resulting holograms are called "phase-only holograms" [15]. A hologram recorded close to the object or close to the image plane can tolerate a fairly coarse phase quantization. But for far field and Fourier holograms it is more important to maintain reasonably good phase information. The amplitude information is especially unimportant in Fourier holograms of diffuse objects. The most brutal quantization procedure that can be applied to a Fourier hologram sets all amplitudes equal to one and all phases to either zero or π, whatever is closer to the true phase. Surprisingly, decent images can be reconstructed from those one-amplitude-level and two-phase-level holograms if certain precautions are taken [13].

A quantization procedure is a special case of a nonlinear process. Hence a short comment on the influence of nonlinearities is in order here. In optical holography it is mainly the nonlinear response of the photographic emulsion which is

of concern. Basically one can avoid nonlinear distortions
when using only weak signals. But that approach is not
always desirable since the signal-to-noise ratio might be
poor. Hence one tolerates moderate amounts of nonlinear distortions. This subject has been thoroughly studied and it
is well covered in the newer books on optical holography.
But very little known is an amazing study by Landau et al.
[16] on the reversibility of nonlinear distortions. Although
a nonlinear process will enlarge the bandwidth one can afford
to throw away these newly created high frequencies. The
original information can be recovered perfectly if the nonlinearity was strictly monotonic.

3. THE VIRTUES OF SOME HOLOGRAPHIC ILLNESSES

Recent Navy survival tests revealed that sickly children
are later as adults more resistant to psychic pressures than
formerly always-healthy children. A recurrent theme in Thomas
Mann's prose is the superior creativity and sensitivity of
people who are or have been afflicted by illnesses. Holography fits into this pattern since it was a very feeble child.
Even Gabor around 1960 [17] apparently was not entirely happy
with the developmental difficulties of the adolescent holography. But then came the laser, which was for holography
what penicillin is for medicine. Yet penicillin does not cure
everything, for example not allergic hypersensitivities.
Holography is allergic, among others, to the slightest vibrations. The obvious counteraction is to carefully avoid
vibrations. Not so obvious was Powell and Stetson's reaction
in 1965, when they converted the allergic affliction into a
virtue. The holographic sensitivity to vibrations is now the
basis for one of the most valuable applications of holography,
the study of small vibrations.

Holographic vibration analysis is only one out of several
benefits derived from the various holographic illnesses. A
list of other holographic illnesses together with their cures
or exploitations follows below. Could not perhaps some of
the optical tricks listed in this table be useful for acoustical imaging as well? I will now describe some acoustical
situations for which I suspect some optical tricks might be
applicable.

Optical holographic vibration analysis has been applied
successfully to airplane bodies. Could one investigate a

Some Holographic Illnesses

Problem	*Cure or Exploitation*
vibrations	Powell and Stetson
motion blur	shutter modulation [18]
spatial incoherence	incoherent holography
spectral incoherence	achromatic holography
turbulence	Goodman, Labeyrie [19]
fog	Spitz, Stetson et al. [20]
not real time	new materials, or TV, or [20]
not in historical times (before 1948)	see [21]
not without real object	computer holography [22]

ship's body similarly with acoustical waves? Or some medical objects have periodic pulsations whose amplitude, phase and location might be of interest.

The problems encountered in optics due to the turbulent atmosphere have no counterpart in seismology but to some degree they do in underwater acoustics. Large hydrophone arrays moved at high speeds through the ocean will probably generate turbulence in their neighborhood. Or a sound wave that is reflected off the wavy ocean surface while travelling from the object to the receiver will be distorted in much the same way as an optical wave in the turbulent atmosphere. Goodman and recently Labeyrie [19] invented some tricks for overcoming the deteriorating influence of turbulence.

Air bubbles and dirt in the water cause scattering of the sound waves, and hence a blurred low contrast image. If the scattering bodies move, the scattered waves will be shifted in frequency due to the Doppler effect. The situation is likewise when light propagates through fog. It has been demonstrated that optical holography through fog produces considerably better images than optical photography [20]. Hence acoustical imaging might benefit as well from holography, which acts like a heterodyne system that rejects the Doppler-shifted waves.

One of the reasons for developing optical computer holography was to produce wavefields that seem to come from a non-existing object. More specifically, when a digital computer has produced a result that is three dimensional we may display the computer output as the reconstructed image from a computer hologram, even in color [22].

REFERENCES

In this list we do not always cite for acknowledging priorities. Instead we hope to provide a convenient access to the optical literature for the acoustics community.

1. R.A. Peterson, "Seismography 1970: The Writing of the Earth Waves", copies on request from the author at Bendix-United Geophysical Company, Pasadena, California.
2. F.N. Spiess, in "Underwater Photo-Optical Instrumentation Applications", Vol. 24 of SPIE Seminar Proceedings, Redondo Beach, California, 1971.
3. D. Gabor, "Light and Information", in "Progress in Optics" (E. Wolf, ed.), Vol. 1, p. 109-153, Amsterdam, 1961.
4. For example E.L. O'Neill, "Introduction to Statistical Optics", Reading, Mass., 1963.
5. D. Gabor, Proc. Roy. Soc. $\underline{A197}$, 462 (1949).
6. L. Mertz, "Transformations in Optics", New York, 1965.
7. A.W. Lohmann, IBM Technical Report RC 438, San Jose, California 1967.
8. R.N. Bracewell, Handbuch der Physik (S. Flügge, ed.), Vol. LIV, p. 42-129, Springer Verlag, Berlin, 1962.
9. B.P. Hildebrand, J. Opt. Soc. Am. $\underline{59}$, 1; $\underline{60}$, 259, 1511 (1970); A.F. Metherell, in "Developments in Holography", SPIE Seminar Proc., Vol. 25, p. 137, Redondo Beach, California, 1971.
10. J.W. Goodman, in "Progress in Optics" (E. Wolf, ed.), Vol. 8, p. 1-50, Amsterdam, 1970.
11. G.W. Stroke, Opt. Comm. $\underline{1}$, 283-287 (1970).
12. J.W. Goodman, F.D. Russell, J. Opt. Soc. Am. $\underline{61}$, 182 (1971).
13. W.J. Dallas, A.W. Lohmann, Appl. Opt. $\underline{11}$, 192 (1972).
14. W.J. Dallas, A.W. Lohmann, Opt. Comm. $\underline{5}$, 78 (1972).

15. A.W. Lohmann, in "Acoustical Holography" (A.F. Metherell et al., ed.), Vol. 2, p. 203, New York, 1970.

16. H.J. Landau et al., Bell System Tech. J. $\underline{39}$, 351 (1960); J. Math. Anal. & Appl. $\underline{2}$, 97 (1961).

17. D. Gabor, in "Developments in Holography", SPIE Seminar Proc., Vol. 25, p. 129, Redondo Beach, California, 1971.

18. O. Bryngdahl, A. Lohmann, J. Opt. Soc. Am. $\underline{59}$, 1175 (1969); O. Bryngdahl, J. Opt. Soc. Am. $\underline{60}$, 510 (1970).

19. J.W. Goodman et al., Appl. Phys. Lett. $\underline{8}$, 311 (1966); J. Opt. Soc. Am. $\underline{60}$, 506 (1970); A. Labeyrie et al., Astrophys. J. $\underline{173}$, L1-L5 (1972).

20. A.W. Lohmann, C.A. Shuman, Opt. Comm. $\underline{3}$, 73 (1971); $\underline{6}$, (1973).

21. O. Bryngdahl, A. Lohmann, J. Opt. Soc. Am. $\underline{58}$, 141 (1968).

22. A.W. Lohmann, in "Developments in Holography", SPIE Seminar Proc., Vol. 25, p. 43, Redondo Beach, California, 1971; W.J. Dallas, Y. Ichioka, A.W. Lohmann, J. Opt. Soc. Am. $\underline{62}$, 739A (1972).

SOME AFTERTHOUGHTS

After listening to all of the speakers I became more optimistic about the usefulness of holography in acoustics. Three reasons for this optimism will be given below, after comparing the types of objects which one encounters in optical and in acoustical image formation.

<u>Comparison of the Object Types</u> - The most impressive optical holograms are made with objects that are solid bodies (toy trains, etc.) surrounded by a homogeneous medium (air). Not so impressive are the reconstructions of inhomogeneous three-dimensional objects as they occur in biological microscopy. In that case, the practical image resolution is often an order of magnitude worse than the theoretical resolution due to the finite aperture. The comparatively poor image quality is caused by speckling or "out-of-focus noise".

If one wishes to make impressive acoustical holograms, one ought to select solid opaque objects surrounded by a homogeneous medium. The image quality will be as good as the array size permits. However, in acoustical image formation, one encounters more often objects that are three-dimensional and inhomogeneous in attenuation and dispersion, especially in seismological and medical application. Hence, one must expect the same difficulties as in optical holography and in coherent optical image formation.

Several approaches for improving the signal-to-noise ratio have in common that more than one image is recorded and incoherently superposed, sequentially or simultaneously. The signal intensities are the same in each image, but the noise structure should be different in the N images; thereby, the signal-to-noise ratio is improved by the factor $\sqrt{N}$. The noise itself might be time-varying, or the recording geometry is changed between exposures such that the image of the object remains the same, while the

noise structure is modified or shifted. Furthermore,
different wavelengths create different noise structures.
Hence, polychromatic or temporal broadband radiation also
improves the signal-to-noise ratio.

Three Reasons for Optimism

1. The depth accuracy obtained by holography, δ_{Z_O}, and
   by the pulse echo method, δ_{Z_A}, are roughly equal to
   the near field. The two accuracies are exactly alike
   if the relative bandwidth, $\Delta\nu_t/\nu_t$, equals $1/2\, k^2$,
   where k is the F-number; that is, the ratio of object-
   to-array distance versus array diameter. (Notice,
   in the middle of the third page are two minor errors.
   A square should be attached to kappa, and "one-tenth"
   should be replaced by "five halves".)

2. The depth accuracy of a far field hologram can be
   improved to the level of the pulse echo method if
   holography is properly modified as it was done, for
   example, by Gabor in his "Sonoradiography", which
   could also be called pulse-holographic tomography.
   When going to broadband radiation it may become
   difficult to record interference fringes. However,
   there exist ways like the "achromatic holography"
   Journal of Optical Society of America, $\underline{60}$, 281, 1970)
   which allow to record optical interference fringes in
   white light. Properly modified, this way should make
   it also possible to record polyphonic or broadband
   interferences.

3. Many ways have been invented that make optical
   holography much more versatile than optical imaging.
   When acousticians buy the holographic principle,
   they will get a whole bag of tricks as a gift, in addition.

SIGNAL PROCESSING METHOD IN ULTRASONIC SCANNING TECHNIQUE

Motoyoshi Okujima

Professor, Research Laboratory of Precision
Machinery and Electronics
Tokyo Institute of Technology
O-okayama, Meguro-ku, Tokyo, Japan

INTRODUCTION

In the field of sonics, it is practical to use a wideband signal. The author is investigating an imaging method with impulsive sound and is planning its application in the fields of medicine and civil engineering. On the other hand, Katakura et al. are investigating a two-dimensional viewing system with frequency modulated ultrasound, at Hitachi, Ltd.

This paper summarizes these investigations.

PART 1
SIGNAL PROCESSING TO OBTAIN VERY SHARP DIRECTIVITY IN NEAR-FIELD[1-8]

I. INTRODUCTION

The lateral resolution in the ultrasonic tomograph for diagnosis is nearly equal to the transducer diameter when we use a plane transducer. Therefore, we designed a ring transducer in order to improve the lateral resolution in the near-field. Directivity of the main lobe of the ring transducer is very sharp in the near-field, whereas, the amplitudes of its side lobes are large, so that suppression of the side lobes is required.

In order to suppress the side lobes, we employed two methods. One of them is to use an impulsive waveform as a transmitting signal, and another is to use two half-ring receivers and obtain a received output by means of the signal processing of individually amplified signals from the half-ring receivers.

II. METHOD OF SIGNAL PROCESSING AND SIMULATION EXPERIMENT

Our purpose is to improve the directivity characteristics in the y-z plane of the ring receiver laid on the x-y plane, which is split into two along x-axis, as shown in Figure 1-1a.

Pressure distribution on the ring by the radiated sound wave from the point source at P (O, Y, Z) shown in Figure 1-1a is equal to that of the incident plane wave from P' (O, Y, $\sqrt{Z^2+a^2}$) shown in Figure 1-1b to the center of the ring, if $Y \ll \sqrt{Z^2+a^2}$, where a is radius of the ring. Accordingly, the output signals of the two half-receiving elements are equal to the outputs of the simulator shown in Figure 1-2.

A waveform shown in Figure 1-3 was used as an input voltage, expecting that it will minimize the side lobes. Figure 1-4 shows waveforms of the output signals of the two elements, where

$$c\tau_s = \frac{2aY_1}{\sqrt{Z^2+a^2+Y_1^2}} \simeq \frac{2aY_1}{\sqrt{Z^2+a^2}}$$

Figure 1-5 shows the block diagram of the signal processor for the received signals. In this processor, a similar operation is carried out to get the product of the two input signals; that is, the upper block produces the product of the polarity of the two input signals. The output of this block is one of the following three values: + 1, 0, or -1. The output is equal to zero when the absolute value of at least one signal of the two inputs is smaller than the rejection level. The lower block produces the absolute value of the sum of the two input signals. Then, the product of the output signals of the two blocks is led to the RC circuit for short time averaging, and we get the output signal of this signal processor.

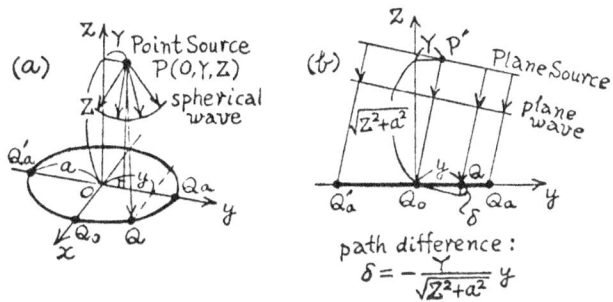

Figure 1-1. Incident Sound Wave to Ring Receiver

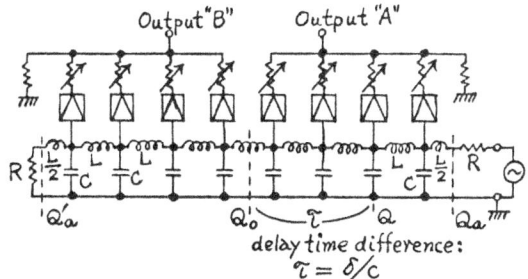

Figure 1-2. Simulator of Ring Receiver

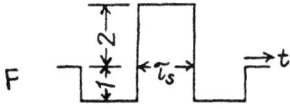

Figure 1-3. Transmitting Waveform

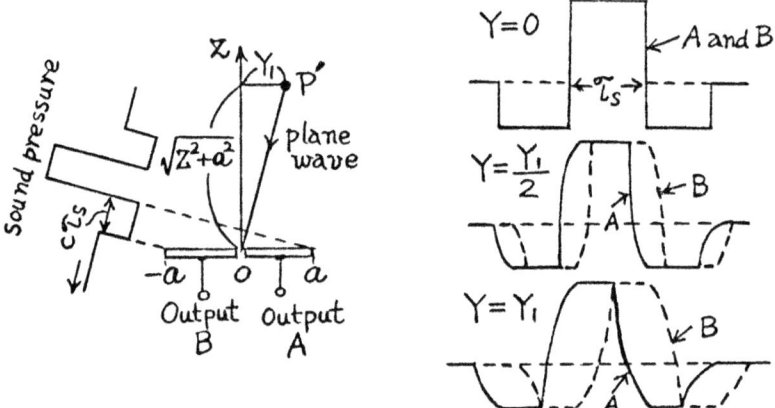

Figure 1-4. Output Waveforms of Two Elements of Ring Receiver

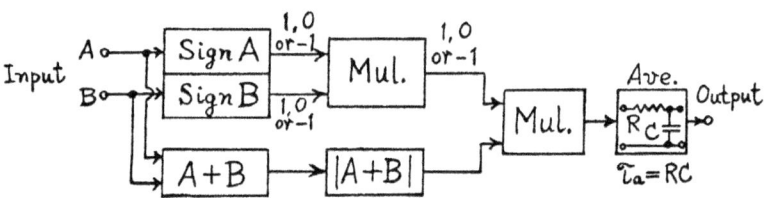

Figure 1-5. Block Diagram of Signal Processor

Figure 1-6 shows the waveforms of the input signals of the averaging. Figure 1-7 shows the directivity characteristics of the peak value of the signal processor output. In the figure, for comparison, the directivity of the ring receiver is shown when we use a sinusoidal continuous wave without signal processing. The figure illustrates that the use of the impulsive transmitting signal and our signal processing method give very sharp directivity to the ring receiver.

SIGNAL PROCESSING METHOD IN ULTRASONIC SCANNING 383

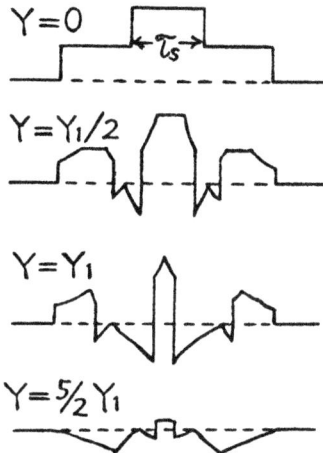

Figure 1-6. Waveforms of Input Signals of Averaging Circuit

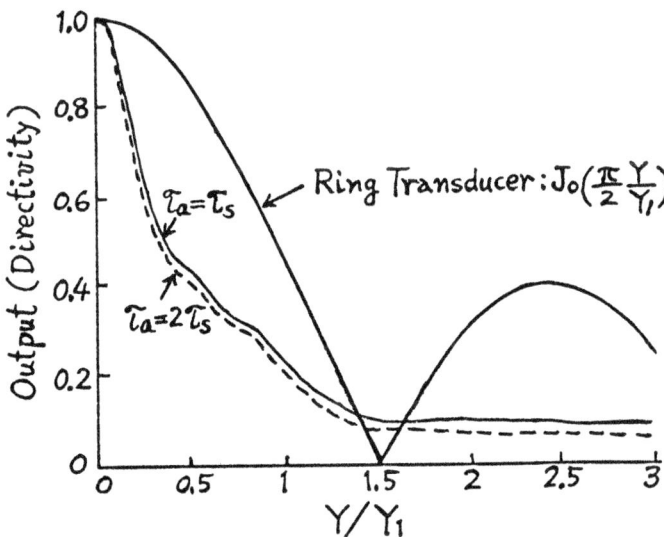

Figure 1-7. Directivity Characteristics of Peak Value of Signal Processor Output

III. TEST EQUIPMENT AND RESOLUTION MEASUREMENT

Figure 1-8 shows a transducer made for test. A PZT disk transmitter of one millimeter thickness and 21 mm diameter, and a PZT ring receiver split into two, of one millimeter thickness, 26 mm outer diameter and 23 mm inner diameter, are cemented onto a block of lead powder containing rubber. Figure 1-9 shows a block diagram of a test imaging equipment.

In the resolution test, two parallel hollow vinylchloride tubes of one millimeter diameter were used as a reflector. Their interval was varied from four to nine millimeters. The reflector was fixed at a distance of 30 cm from the transducer in parallel to the split line of the ring receiver, and the transducer was moved linearly along the perpendicular direction to the split line of the ring.

In Figure 1-10, the abscissa corresponds to the moving distance of the transducer, and the ordinate shows the video amplifier output of the echo signal. The figure illustrates that by the signal processing the two echo signals corresponding to the two tubes are observed if the tube separation is more than 5.5 mm.

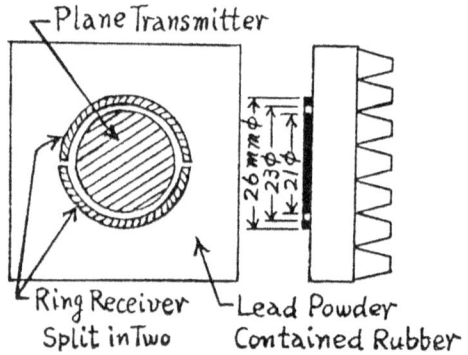

Figure 1-8. Transducer

SIGNAL PROCESSING METHOD IN ULTRASONIC SCANNING 385

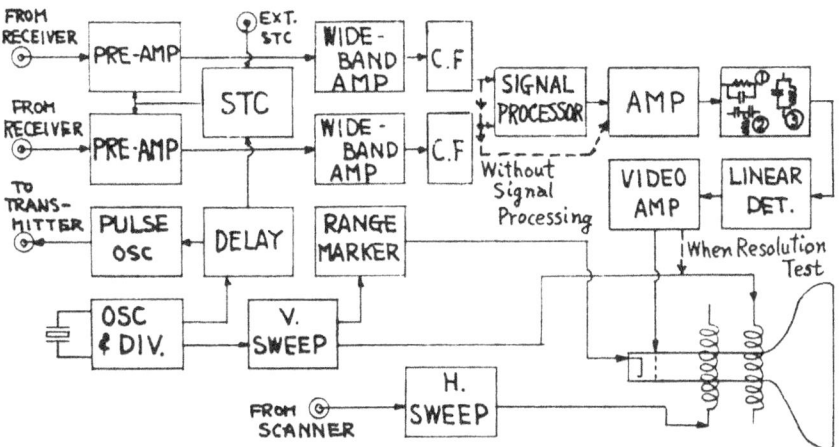

Figure 1-9. Block Diagram of Imaging Equipment.

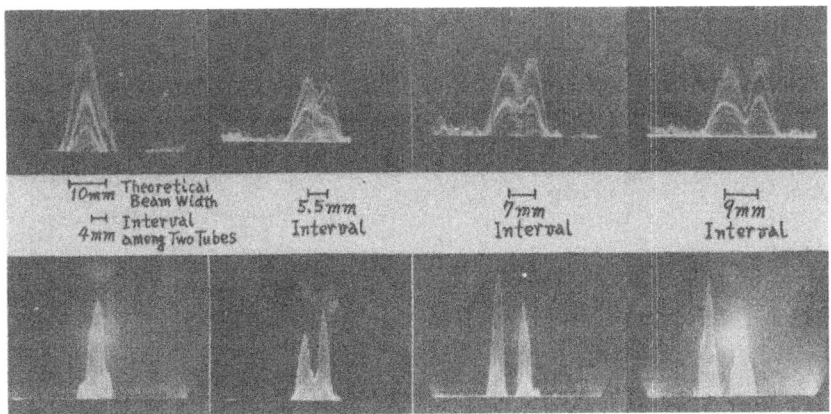

Figure 1-10. Lateral Resolution Test
              Upper: Without signal processing
              Lower: With signal processing

IV. EXAMPLES OF IMAGE PATTERNS

With the test equipment, image patterns of three objects immersed in water were obtained.

Figure 1-11 shows the image pattern of a sponge with a hole. Figure 1-11b shows an image pattern when a plane transducer of 21 mm diameter was used as a transmitter and a receiver; Figure 1-11c shows one when a transmitter and a receiver were a plane and a ring transducer respectively (without signal processing); and Figure 1-11d shows one when a split ring receiver was used with signal processing.

Figure 1-12 shows the image pattern of a wooden bar with a rectangular cross-section. Figure 1-12b shows an image pattern when a plane transducer of 15 mm diameter was used as a transmitter and a receiver, and Figure 1-12c shows one when a transmitter was a plane transducer and a receiver was a split ring type transducer. The signal processing was carried out.

Figure 1-13 shows the image pattern of a cross-section of a human forearm. Figure 1-13b, c and d show image patterns obtained by the same way as in Figures 1-11b, c, and d, respectively.

In every pattern of an object, a traditional method which employs a plane type transducer as a transmitter and a receiver gives laterally spread out spot patterns. On the other hand, the proposed method with a split ring type receiver, accompanied by signal processing, gives a fidelity image pattern of an object without laterally spread out spots.

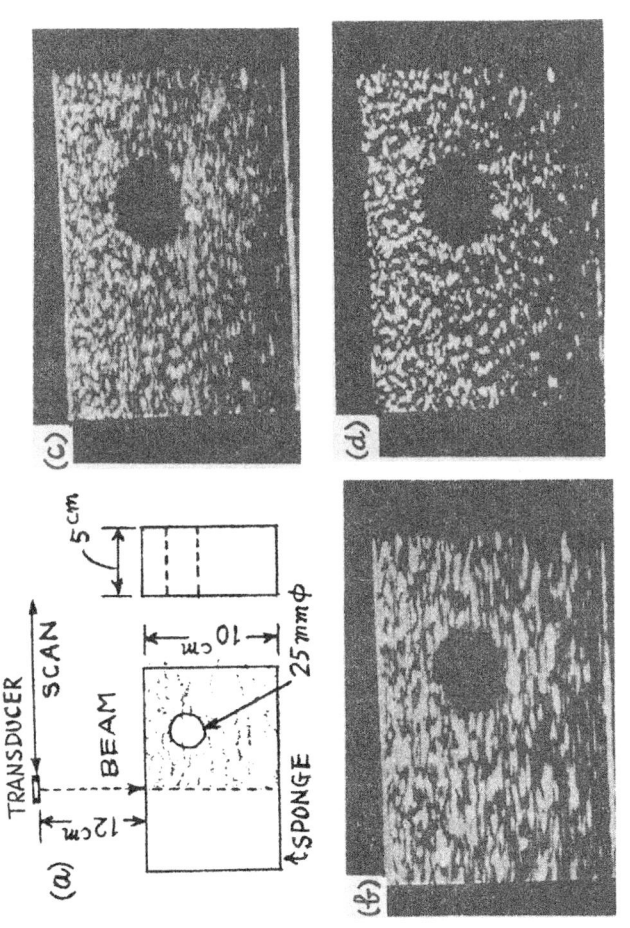

Figure 1-11. Image Patterns of Sponge with Hole
a. Schematic
b. Plane transmitter-receiver
c. Plane transmitter and ring receiver without signal processing
d. Plane transmitter and ring receiver with signal processing

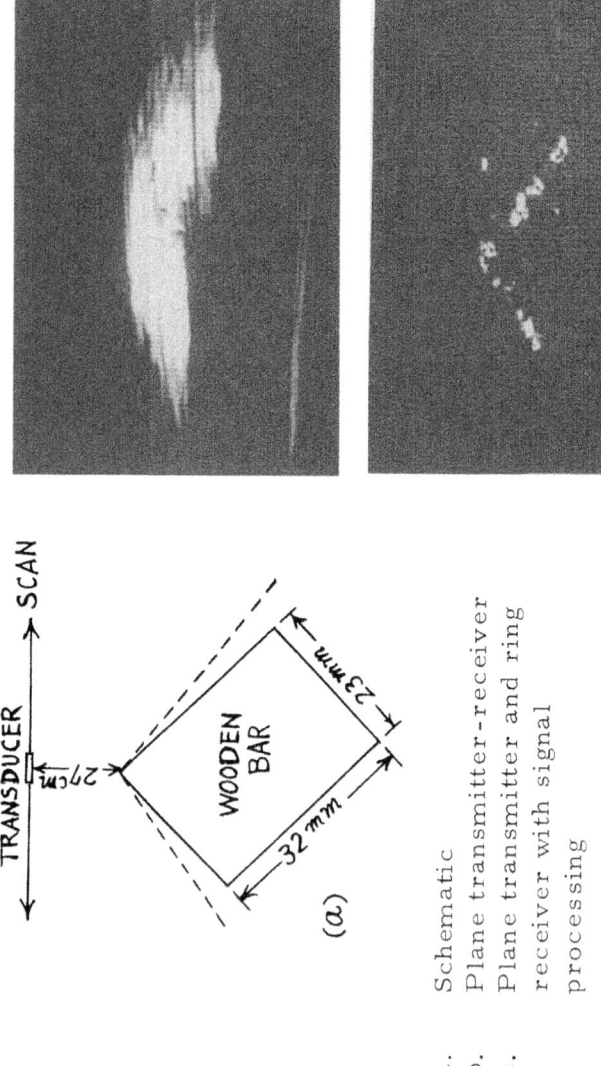

Figure 1-12. Image Patterns of Wooden Bar

a. Schematic
b. Plane transmitter-receiver
c. Plane transmitter and ring receiver with signal processing

SIGNAL PROCESSING METHOD IN ULTRASONIC SCANNING 389

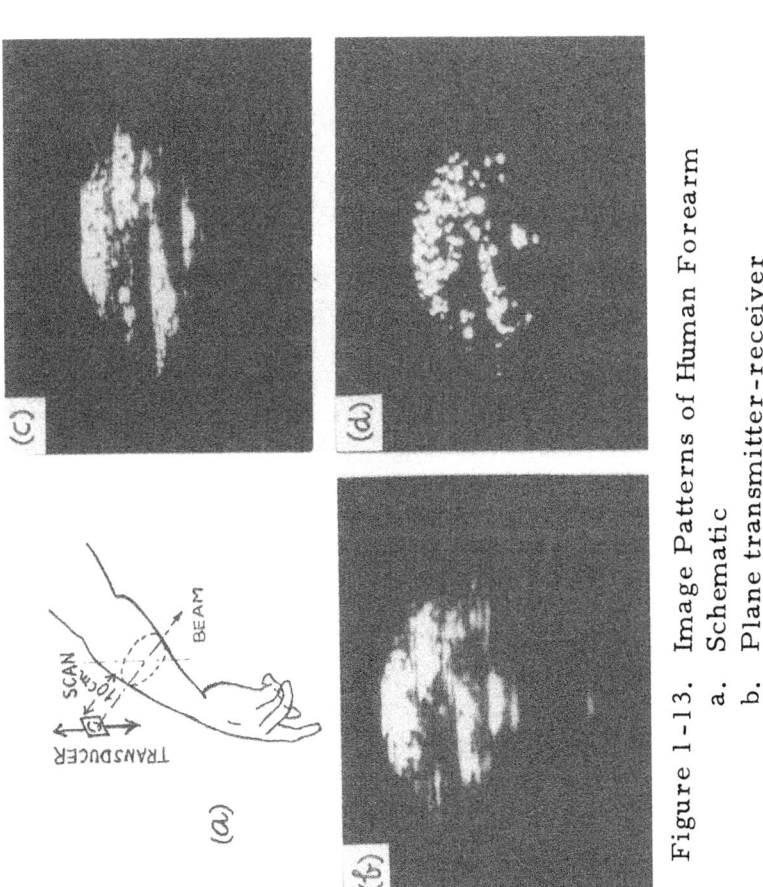

Figure 1-13. Image Patterns of Human Forearm
 a. Schematic
 b. Plane transmitter-receiver
 c. Plane transmitter and ring receiver without signal processing
 d. Plane transmitter and ring receiver with signal processing

V. CONCLUSION

In order to improve the lateral resolution in an ultrasono-diagnostic tomograph, we devised a method of a split ring receiver and a processing of received signals. In addition, we confirmed, by a simulation experiment, that an impulsive waveform of a transmitting signal gives a very sharp receiving directivity.

Furthermore, image patterns of three objects were obtained by test equipment, and they represented shapes of the objects clearly by our devised method.

PART 2
APPLICATION OF SIGNAL PROCESSING TO HIGH SPEED SCANNING DIAGNOSTIC EQUIPMENT[9-12]

I. INTRODUCTION

In order to obtain image patterns of moving human organs like a heart, it is required to scan an ultrasonic beam at high speed. For this purpose, J. C. Somer used a method in which a transmitted ultrasonic beam is deflected by supplying an electric pulse of different delay time to each element of an array transducer consisting of several ten bar elements.[13]

We had planned to decrease the number of transducer elements and delay circuits attached to each element. The decrease of transducer elements, however induces an enlargement of side lobes in the directivity. In order to suppress the side lobes, we devised a process using output signals of a receiving array transducer by the signal processing method presented in Part 1.

We are still at the fundamental research stage. Following are the results we have obtained so far.

II. CONSTRUCTION OF ULTRASONIC SCANNING EQUIPMENT

Figure 2-1 shows the schematic of a planned scanning equipment for directivity beams of a transmitter and a receiver. Both transmitter and receiver consist of an array of about four bar elements placed at suitable intervals.

A variable delay circuit for a trigger pulse and a pulse generator are connected to each element of the transmitter array, and a transmitted ultrasonic beam is scanned electronically by varying delay time of the variable delay circuits. A preamplifier and a variable delay line for a received waveform are connected to each element of the receiver array, and a receiving directivity beam is scanned electronically by varying delay time of the variable delay lines, synchronizing with the deflection of the transmitted beam.

The four outputs of these delay lines are grouped in pairs of two each. The sum of the two outputs of each pair is amplified individually. A received output is then obtained through the signal processing circuit presented in Part 1.

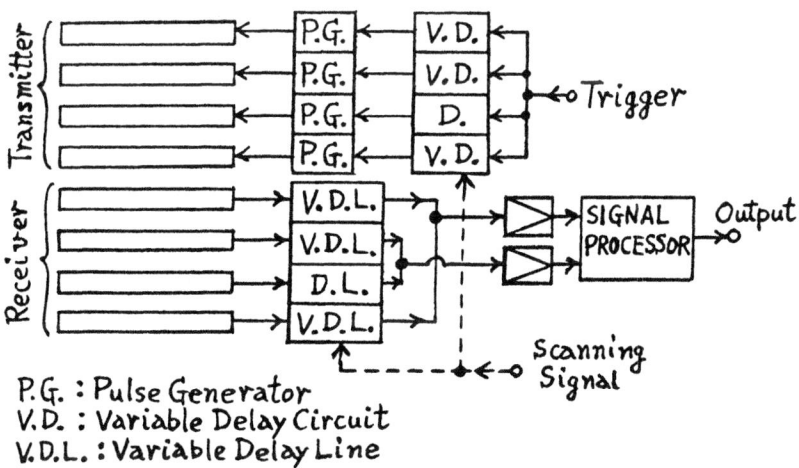

P.G. : Pulse Generator
V.D. : Variable Delay Circuit
V.D.L. : Variable Delay Line

Figure 2-1. Schematic of Scanning Equipment

III. FUNDAMENTAL EXPERIMENT BY TRANSDUCER OF ELEMENTS ARRAYED WITH EQUAL INTERVALS

Figure 2-2 shows a transducer used in this experiment. Eight elements of this transducer are PZT bars. Dimensions are: Thickness, 1 mm; Width, 1 mm; Length, 20 mm. They are placed at equal intervals of one millimeter. Three elements of A_1, A_2, and B were used as a receiver, and the other five elements were used as a transmitter.

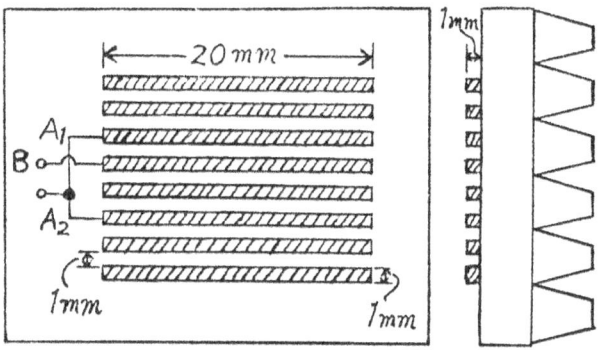

Figure 2-2. Array Transducer (First Made) With Equal Intervals

Experiments were made with the equipment presented in Part 1, Figure 1-9. In the experiments, a mechanical, instead of an electronic linear scanning method was employed, using the transducer to obtain image patterns.

Figure 2-3 shows the image pattern of a sponge with a hole. The figure shows a deformed pattern due to large side lobes. Too small a number of receiver elements might be the cause of the deformation. Accordingly, as shown in Figure 2-4, four elements to a receiver and the other four elements to a transmitter were allocated. A considerably clear image pattern of the sponge was thus obtained.

SIGNAL PROCESSING METHOD IN ULTRASONIC SCANNING

Then, as shown in Figure 2-5, the image pattern of a wooden bar with a rectangular cross-section was obtained. This pattern shows that the remaining effect of side lobes is still large.

IV. ELEMENT ARRAYS TO REDUCE SIDE LOBES AND IMAGE PATTERN OBTAINED BY DEFLECTING DIRECTION OF DIRECTIVITY BEAM

A suitable array of the elements should reduce the side lobes. Because of this, a transducer, as shown in Figure 2-6, was made.

Figure 2-7a shows the image pattern of a sponge obtained with this transducer and the same method as the previous experiment. The figure shows that a clear pattern was obtained.

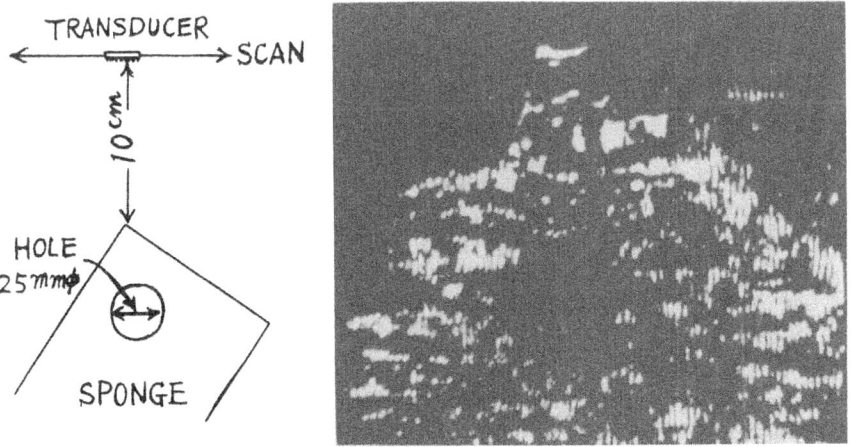

Figure 2-3. Image Pattern of Sponge with Hole, By Array Transducer Shown in Figure 2-2

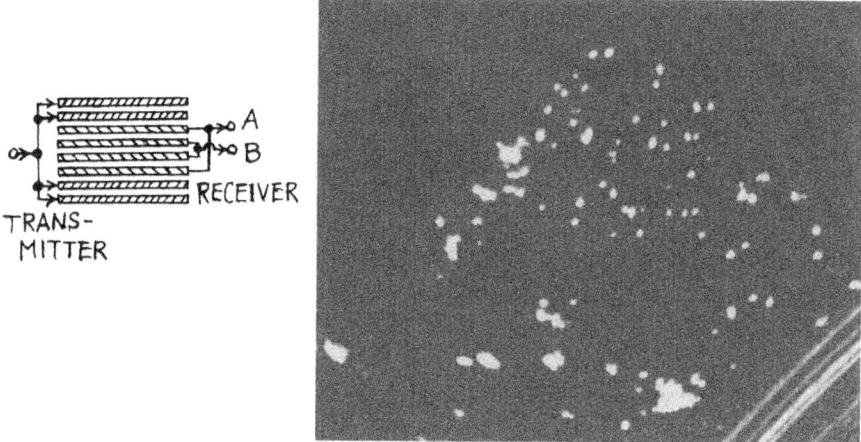

Figure 2-4. Image Pattern of Sponge with Hole, by Four Transmitter and Four Receiver Elements

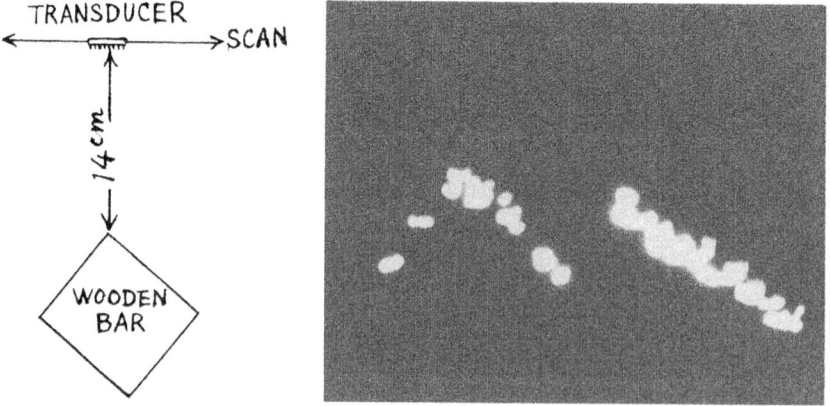

Figure 2-5. Image Pattern of Wooden Bar, By Array Transducer Shown in Figure 2-4

SIGNAL PROCESSING METHOD IN ULTRASONIC SCANNING

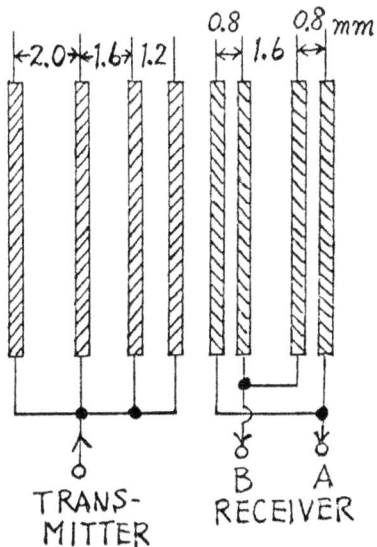

Figure 2-6. Array Transducer (Second Made)

Figure 2-7b shows the image pattern obtained by the following method: The directions of the transmitting and receiving directivity beams were deflected by 45° with fixed delay circuits. The transducer was scanned as presented previously. In this case, a clear pattern could not be obtained because a considerably large side lobe existed at the direction closer to the front axis of the transducer.

Because of this, another array of elements was designed to induce disappearance of large side lobes whenever the directivity beam is deflected. Figure 2-8 shows the designed array transducer and the image pattern of a sponge obtained by this transducer. The image pattern by deflecting the beam is not yet obtained.

CONCLUSION

In order to minimize the number of transducer elements, we designed the signal processing technique of four receiving element outputs, employing a high speed scanning method by varying delay time applied to the transmitting and receiving signals of the transducer element arrays.

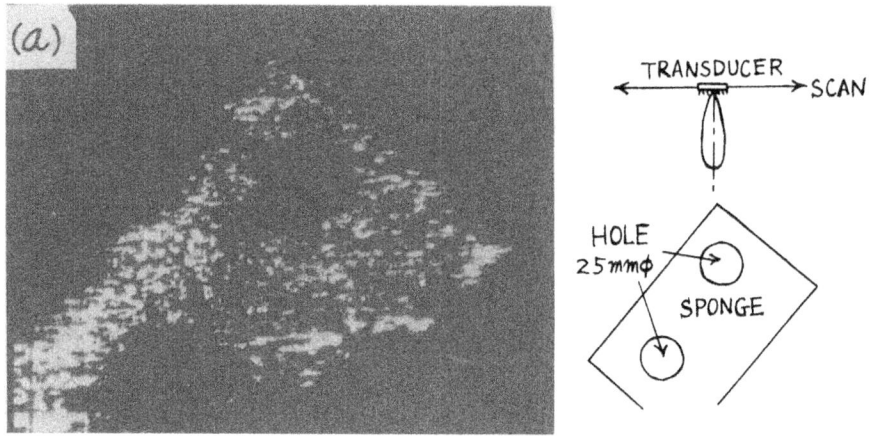

Figure 2-7a. Image Pattern of Sponge with Two Holes By Array Transducer Shown in Figure 2-6

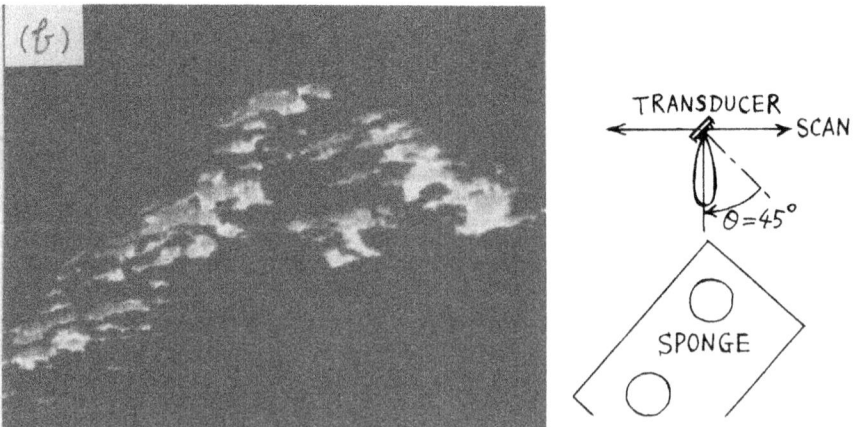

Figure 2-7b. Image Pattern of Sponge with Two Holes, By Array Transducer Shown In Figure 2-6 When Deflecting Direction of Beam

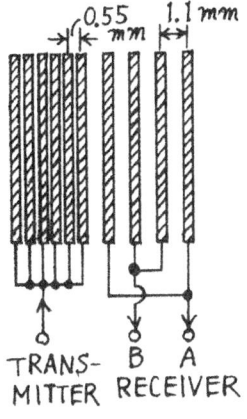 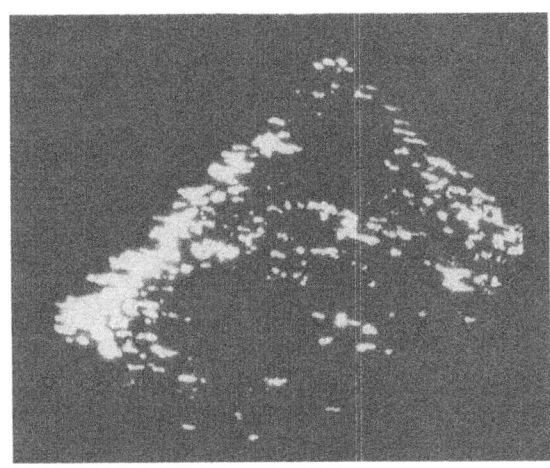

Figure 2-8. Array Transducer (Third Made) and Image Pattern of Sponge with Two Holes

According to the experiment, it was expected that considerably clear image patterns can be obtained by designing a proper array with only about four to six transmitter elements and four receiver elements, making a total of from eight to ten elements.

PART 3
APPLICATION OF SIGNAL PROCESSING TO ACOUSTIC IMAGING OF BODIES BURIED UNDERGROUNL[14-23]

I. INTRODUCTION

If we want to obtain image patterns of bodies buried underground by a pulse-echo method, completion of the patterns by continuous scanning of transmitters and receivers contacted to the ground surface is difficult and almost impossible.

It is necessary, therefore, to obtain an image pattern by transmitters and receivers fixed on the ground surface. In this case, it is required to remove transmitters and receivers many times to search a wide area because the covering area at one time is limited. To make it easier to remove, it is important to reduce the number of transmitters and receivers as much as possible. Therefore, we devised a method to obtain a clear image pattern of an object underground in which one electromagnetic induction type sound source to radiate an impulsive sound with wide directivity into the underground is used, and the signal processing of outputs from several receivers placed on the ground surface is carried out.

II. SIMULATION EXPERIMENT IN WATER FOR TESTING EFFECTIVENESS BY SIGNAL PROCESSING

In order to test the effectiveness to improve the resolution by the signal processing method presented in Part 1, a simulation experiment was carried out in water.

Figure 3-1 shows a block diagram of the experimental equipment. All of the transmitter and the four receivers comprised a PZT transducer of a cylindrical shell of about 100 kHz of resonant frequency. The transmitter was energized by an impulse voltage. The output signals of the four receivers were processed by the same method presented in Part 2, and the output waveform was observed by an oscilloscope.

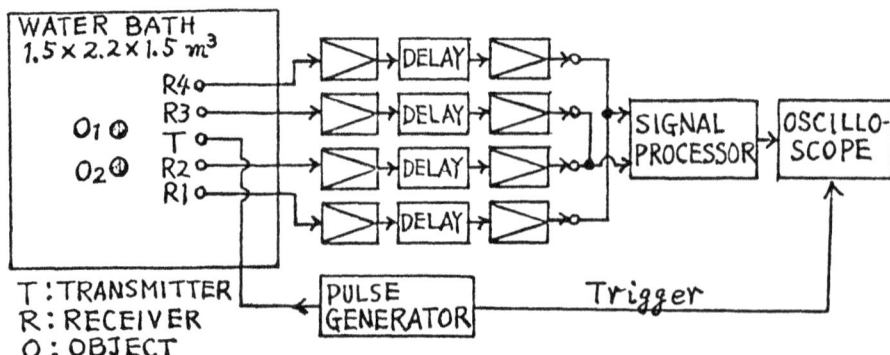

Figure 3-1. Block Diagram of Equipment for Simulation Experiment

SIGNAL PROCESSING METHOD IN ULTRASONIC SCANNING

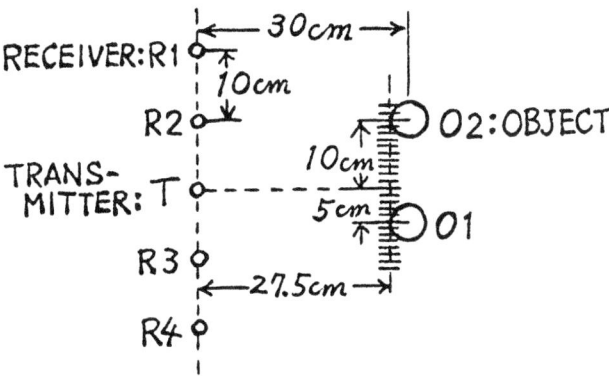

Figure 3-2. Arrangement of Transmitter, Receivers and Objects

Two air-filled cans, five centimeters in diameter and 14 centimeters in length, were the objects. Figure 3-2 shows the arrangement of the transmitter, receiver and objects.

Figure 3-3 shows output waveforms after the signal processing in case of scanned receiving directivity by varying delay time. Delay time given to each receiver output was set so as to compensate transmission time of sound of each receiver from the position of every interval of one centimeter on a line connecting each front surface of the two cans.

Moreover, the signal processing is carried out after each sum of two pairs introduced from the four receivers' delayed outputs, and the combinations are the following three cases: The output waveform of each combination is shown as (A), (B), and (C) in the figure. In addition, the waveform of the sum of the four receivers' delayed outputs is shown for comparison in the figure.

The figure illustrates that this signal processing method is effective in improving the resolution. The combination of the receivers' outputs of (C)-(R1 + R4, R2 + R3)- is the best.

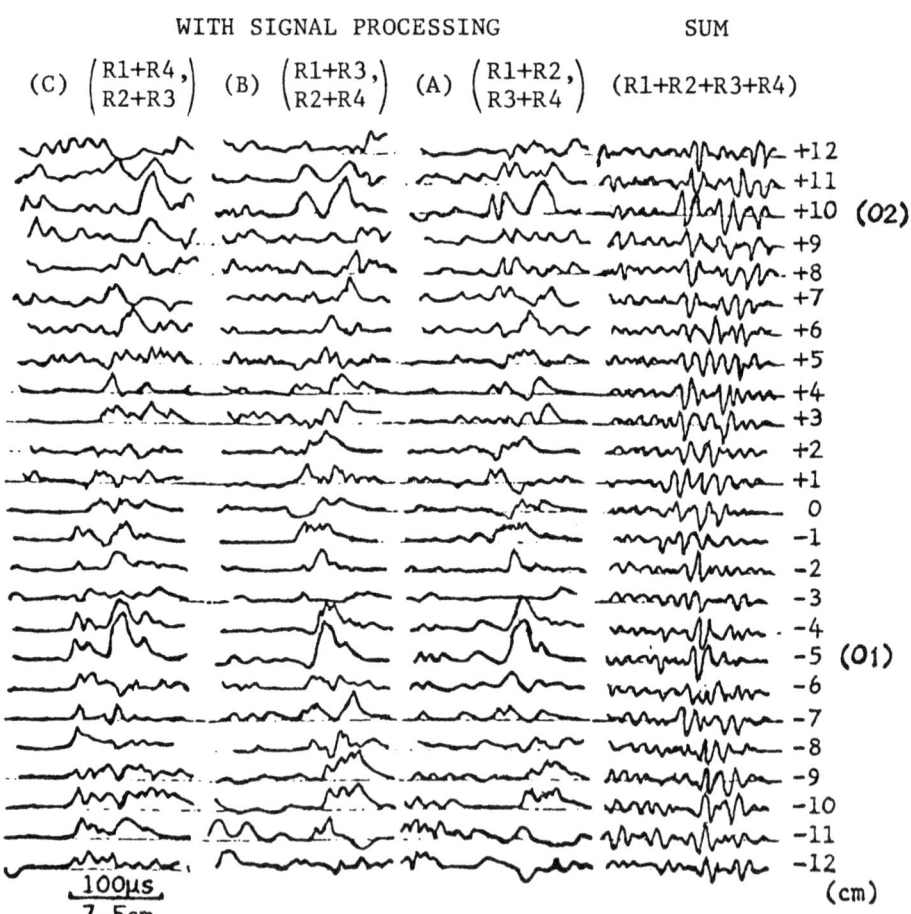

Figure 3-3. Output Waveforms by Two Objects

SIGNAL PROCESSING METHOD IN ULTRASONIC SCANNING 401

III. IMAGE PATTERNS OF PIPES BURIED IN THE SAND

Experiments for detecting pipes buried in sand were performed. Figure 3-4 shows photographs of an electromagnetic induction type sound source to radiate an impulsive sound wave and a piezoelectric receiver to receive echo signals from reflective bodies.

Figure 3-4. Electromagnetic Induction Type Sound Source (Upper); Piezoelectric Receiver (Lower)

Figure 3-5 shows the arrangement of the sound source and the receivers, and a block diagram of the measuring equipment. At the place where the experiment was performed, the outputs of the four receivers were recorded in a magnetic tape by a tape recorder. The signal processing and the image pattern display were made at our laboratory. The output of the signal processing circuit was recorded once in a tape and was reproduced at a speed of 1/20 by a pen recorder on paper.

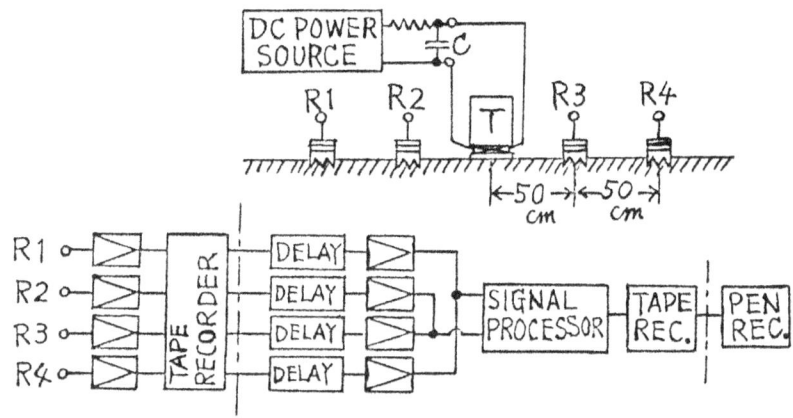

Figure 3-5. Arrangement of Sound Source and Receivers; Block Diagram of Equipment

Figure 3-6 shows the sand bath and the location of the buried pipes. In the sand bath, a small electromagnetic induction type sound source was also buried at a depth of one meter. With this sound source and a receiver placed on the ground surface, sound velocity was measured at about 153 m/s. The measurements were made at three points, A, B, and C; but this time, we shall show its result at Point C.

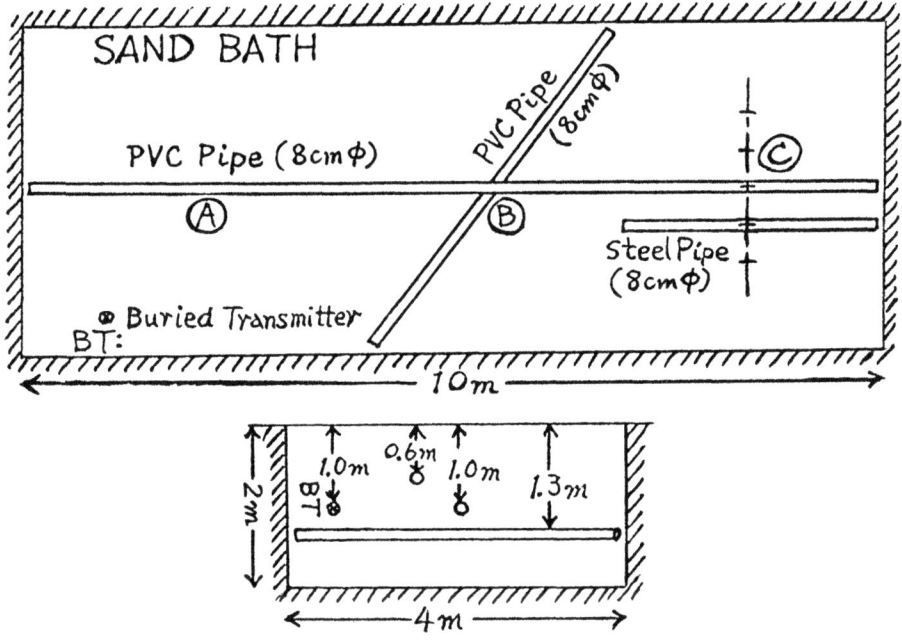

Figure 3-6. Sand Bath for Field Test and Location of Buried Pipes

As shown in Figure 3-7, a received output waveform obtained by setting delay time differences for the receivers' outputs so as to compensate transmission time differences of a reflective wave from a given point 0_1 to their receivers corresponds approximately to a reflective wave pattern of objects laid on a line connecting the sound source, T, and the point, 0_1. As an object is placed further from the point 0_1, the output from the object decreases because of the increase of difference between delay time difference and transmission time difference.

In Figure 3-8, 8b shows the output waveforms of four receivers which are arrayed perpendicular to the pipe axis. The sound source was placed just above the poly-vinyl chloride pipe of eight centimeter diameter, buried at a depth of one meter. Figure 3-8c shows received output waveforms which were signal processed. In this case, delay time difference corresponds to the positions of four different depths.

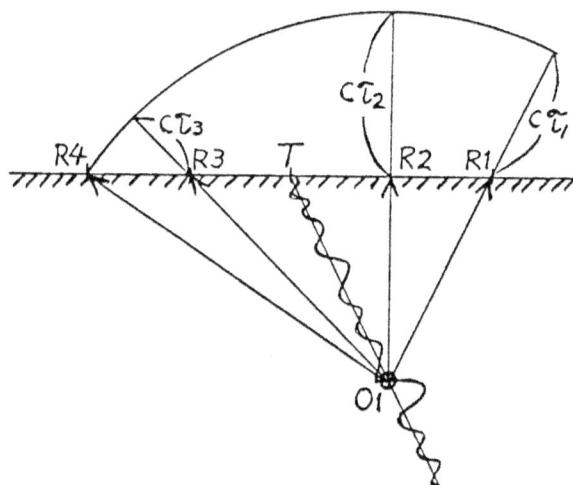

Figure 3-7. Delay Time Given to Each Receiver Output and Displayed Output Waveform

The figure illustrates that the amplitude of the output signal from the object laid at one meter depth is decreased only a little when we give delay time differences which correspond to the positions of 80 centimeters and 120 cm depths, while the amplitude is decreased by about one-half when it is set at a depth of 60 centimeters. This fact explains that when delay time difference corresponds to the position of one meter depth, the received output represents a reflective wave pattern of objects laid at about 80 to 120 cm depth. In other words, the effective range of the displayed patterns is about 80 to 120 cm depth. Accordingly, in order to obtain the entire image patterns of objects laid over a wide range of depth, it is necessary to synthesize some displayed patterns in each effective range. These were obtained by setting delay time differences which correspond to some different depths.

Figures 3-9a and 3-9b show image patterns obtained by setting delay time differences corresponding to a depth of one meter and 60 centimeters, respectively. Figure 3-10 shows the synthesized patterns of these two patterns in

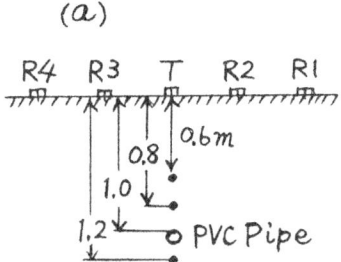

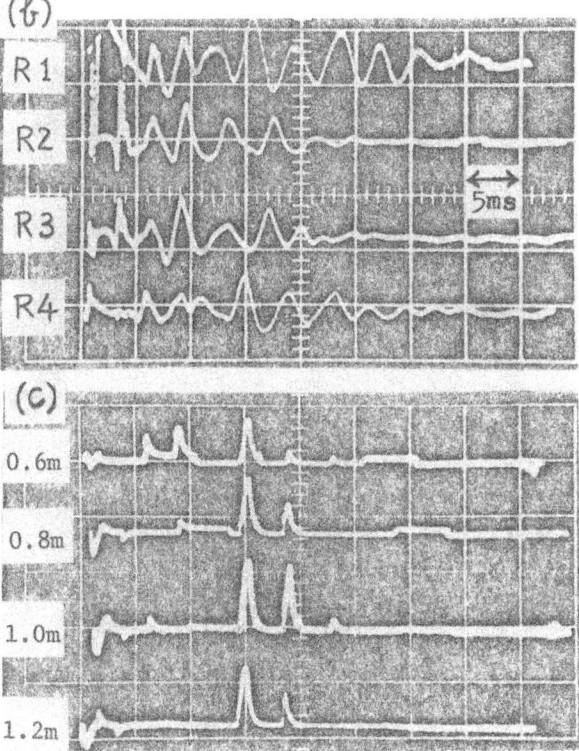

Figure 3-8. (a) Circuit Diagram; (b) Output Waveforms of Four Receivers; (c) Output Waveforms after Signal Processing when Four Different Delay Time Differences are Given

each effective range of 80 cm to 120 cm depth, and of 50 cm to 80 cm. The figure illustrates that the clear image patterns of a steel pipe and a poly-vinyl chloride pipe laid at depths of 60 cm and one meter, respectively, are obtained successfully.

CONCLUSION

Fundamental experiments were made in order to develop the technique to obtain underground profiles by an impulsive sound wave. As a result, a poly-vinyl chloride pipe and a steel pipe, eight centimeters in diameter, buried at less than one meter depth (approximate) were represented in a display profile successfully.

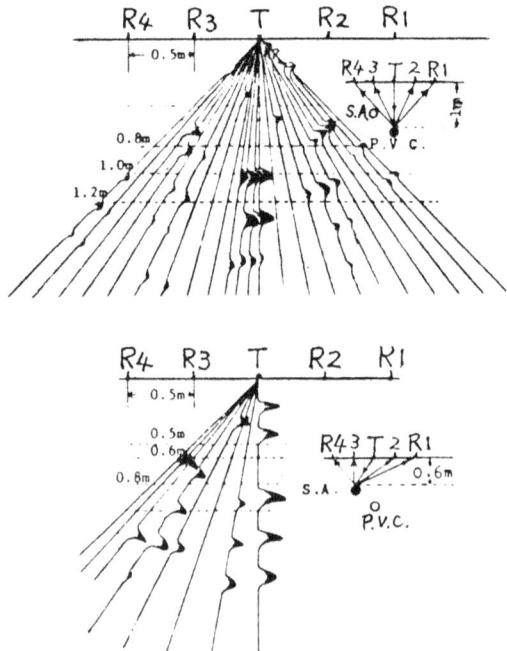

Figure 3-9. Cross-sectional Display of Underground Effective Range; (a), 80 cm to 120 cm Depth (b), 50 cm to 80 cm Depth

PART 4
HIGH SPEED ULTRASONIC IMAGING SYSTEM USING FAN BEAM SCANNING BY FREQUENCY MODULATION METHOD[24-27]

I. INTRODUCTION

In order to obtain a two-dimensional image of a body immersed in water at high speed, a high speed ultrasonic viewing system is being investigated in the Central Research Laboratory of Hitachi Ltd.

In this system, a multi-element transducer rod and a cylindrical acoustic lens are used as transmitting equipment to radiate a fan-shaped sound beam, and the direction of the beam is scanned by sweeping the frequency of the transmitting signal. An analog switch array and a large scale arrayed receiver plate are used as receiving equipment.

One type application of this system is the backscattered sound imaging. In this case by using a one to two MHz frequency swept sound wave, this system indicates a sound image of an object on the cathode ray tube. The frame rate of this system is up to 70 frames per second. Thirty degrees of viewing angle and a resolution of three centimeters at a distance of 10 meters is also realized.

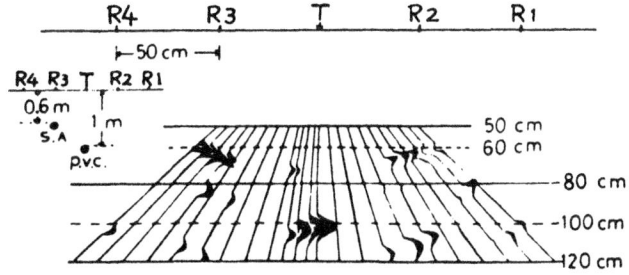

Figure 3-10. Combined Cross-sectional Display of Displays shown in Figures 3-9a, 3-9b

Another type application is the shadow imaging. In this case, 1.5 mm resolution and 10 cm by 10 cm viewing field are realized.

II. SCANNING OF TRANSMITTED FAN BEAM

A polarization-inverted piezoelectric arrayed transmitter is used to radiate an acoustic fan beam and to deflect its direction, depending upon signal frequency. Figure 4-1 shows the essential components of this system, especially its acoustic devices. Transmitter A in this figure is a linear array, constructed of small piezoelectric elements. The polarization axes of these elements are inverted on each adjacent element as shown in Figure 4-2. This transmitter radiates oblique plane waves in several directions. Any change in sound wave frequency varies its propagation direction, as shown in Figure 4-3. The far-field directivity pattern, $R(\theta)$ of this transmitter is given by:

$$R_{(\theta)} = \sin \frac{nH}{2} / \sin \frac{H}{2}$$
$$= \pi - \frac{2\pi d}{\lambda} \sin \theta \tag{1}$$

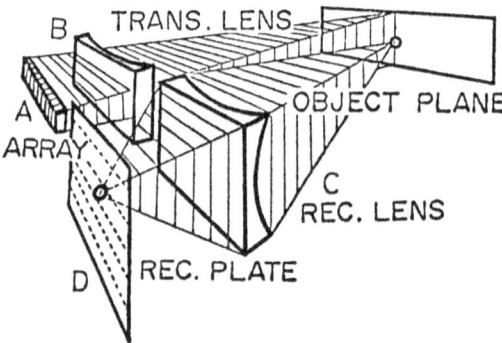

Figure 4-1. Arrangement of Acoustic Devices

SIGNAL PROCESSING METHOD IN ULTRASONIC SCANNING

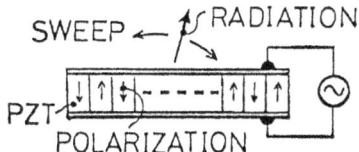

Figure 4-2.   Transmitter

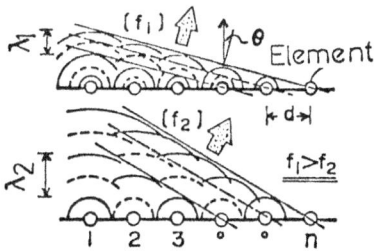

Figure 4-3.   Radiation of Oblique Plane Wave

where
   n is the total number of elements, λ is the wavelength in water, and θ is the bearing angle measured from the array front.

   A transmitter of 50 centimeters in length, constructed of 500 elements was made. The directivity pattern in the horizontal plane of this transmitter with a cylindrical acoustic lens, B, as shown in Figure 4-1, was measured at a distance of 10 meters, and is shown as a solid line in Figure 4-4. In the figure, the theoretical pattern is also shown as a broken line. This figure shows that the horizontal resolution is about three centimeters, at a distance of 10 meters.

   The field of view in the vertical direction is determined by width of transmitter and sound wavelength in water.

The measured value of this vertical field is shown in Figure 4-5. This figure shows 45 centimeters of the vertical field of view on a plane three meters away. Thus, an object 10 meters away can be illuminated by a thin sound beam, 1.5 m long and three cm wide.

The radiation angle against the signal frequency is obtained from Equation (1) as:

$$\theta = \sin^{-1}\left(\frac{\lambda}{2d}\right) \qquad (2)$$

Theoretical and experimental values of this angle are shown in Figure 4-6. The figure shows that the transmitting beam is scanned in a thirty degree horizontal angle of view by varying the signal frequency from 0.9 to 1.7 MHz.

III. BACK SCATTERED SOUND IMAGING

The object is illuminated by the acoustic fan beam. The frequency component of this beam corresponds to the

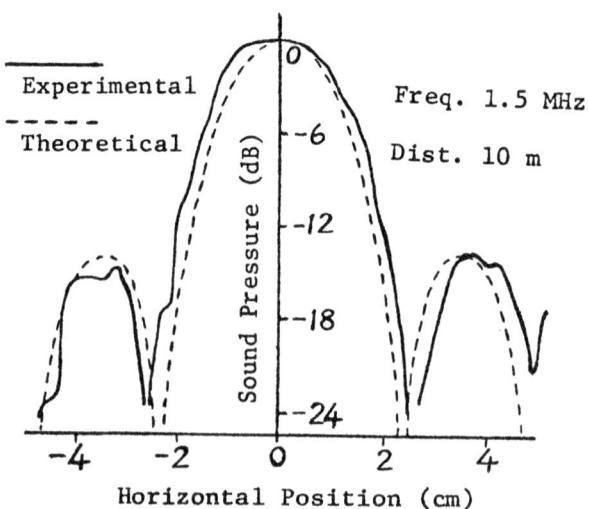

Figure 4-4. Horizontal Distribution of Sound Pressure

SIGNAL PROCESSING METHOD IN ULTRASONIC SCANNING

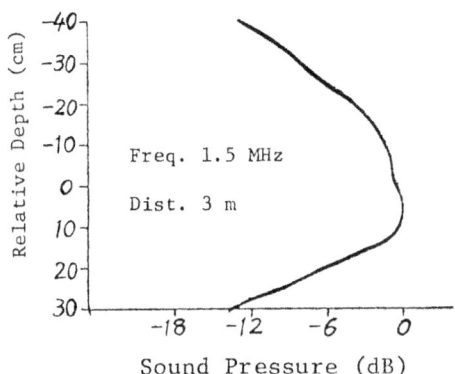

Figure 4-5. Vertical Distribution of Sound Pressure

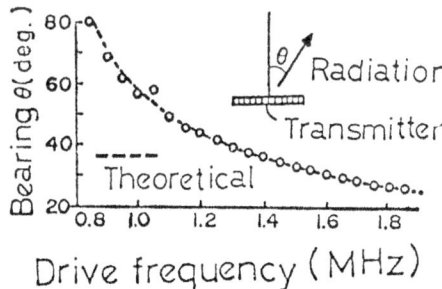

Figure 4-6. Beam Deflection

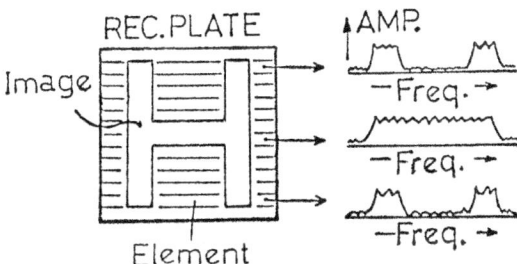

Figure 4-7. Receiver Plate

horizontal coordinates of the object. Back-scattered acoustic waves from the object are collected by a convergent lens, C, shown in Figure 4-1, and focused in the image plane. This lens is the solid-liquid compound lens with a focal length of 20 centimeters and a diameter of 26 centimeters. In this image plane, a 10 cm by 10 cm piezoelectric receiver plate, D, shown in Figure 4-1, is placed. The receiver plate is divided into 100 segments along its vertical direction as shown in Figure 4-7. From the mechanism of this transmission, the frequency component of the acoustic image varies according to its horizontal axis coordinates on the image plane as shown in Figure 4-7. Accordingly, the object horizontal position can be obtained as a frequency component of the receiver plate output. Its vertical coordinates can be also obtained as a position of the receiver element. Thus, a two-dimensional acoustic image can be displayed on the cathode ray tube, using brightness modulation by the amplitude of the frequency component.

In addition, the acoustic image appears on the receiving plate after a propagation delay time from the transmitter via the object to the receiver plate. Therefore, using a spectrum analyzer matched to the delay time, an acoustic image of the objection section laid at a desirable distance can be selected.

Figures 4-8 and 4-9 are photographs of the transmitter and the receiver plate, respectively.

IV. DIRECT SHADOW IMAGING

Figure 4-10 shows the compositions of the direct shadow imaging. Almost all components are the same as those shown in Figure 4-1, except that the focal length of the transmitter lens is 70 centimeters and the output cables of each pair of elements adjacent to the receiver plate are connected together.

SIGNAL PROCESSING METHOD IN ULTRASONIC SCANNING 413

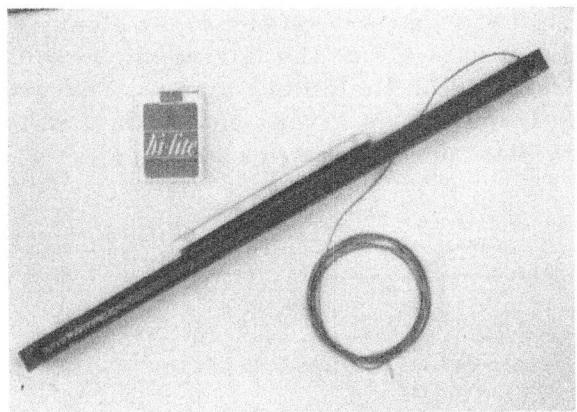

Figure 4-8. Transmitter

Figure 4-9. Receiver Plate

In this case, the horizontal position of the shadow is known from the transmitting frequency and is equivalent to the frequency shift period, with a little propagation delay. Therefore, the frequency analysis is not necessary for this type imaging.

A sound beam pattern on the focal plane is shown in Figure 4-11. It shows that the horizontal resolution of 1.5 mm is realized on the object plane. The range of the transmitting frequency is 1.4 to 1.6 MHz, and the frame rate is more than 100 frames per second.

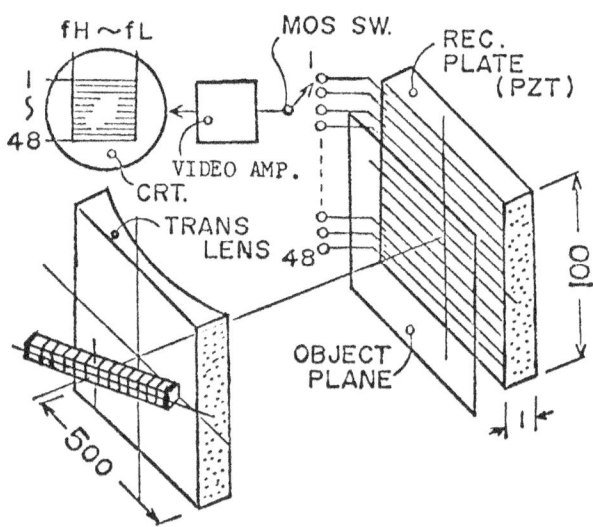

Figure 4-10. Shadow Imaging System

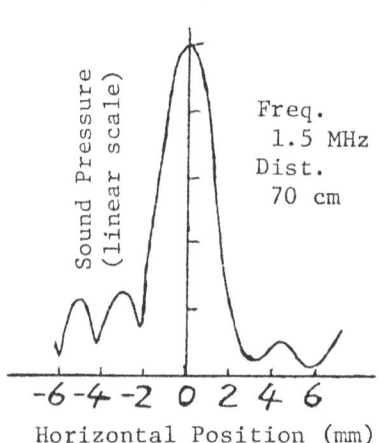

Figure 4-11. Sound Beam for Direct Shadow Imaging

V. APPARATUS

Figure 4-12 shows an apparatus for high-speed back-scatter imaging constructed with a transmitting section, a receiving section, a display section and a control section.

The transmitting section is composed of a sweep generator, a power amplifier, a transmitter and a cylindrical acoustic lens. In this case, the frequency range of the signal is 0.9 to 1.7 MHz. The output level of the power amplifier is 30 V, peak-to-peak. The receiving section is composed of a one hundred channel, parallel spectrum analyzer. Every spectrum analyzer has a single amplifier, a mixer, a local oscillator, an intermediate frequency amplifier, and a detector. Bandwidth of the intermediate frequency amplifier is six kHz. Cutoff frequency of the detector is six kHz. The display section consists of a switching circuit and a cathode ray tube display circuit.

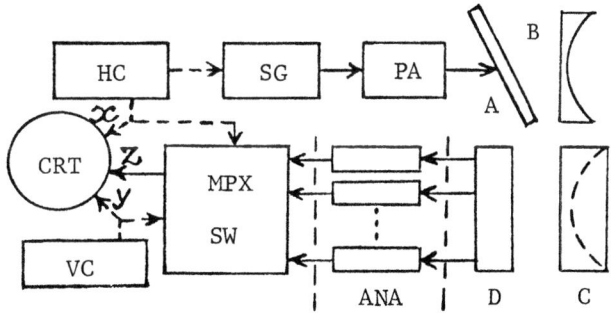

SG   : Sweep Generator
PA   : Power Amplifier
ANA   : Spectrum Analyzer
VC   : Vertical Controller
HC   : Horizontal Controller
MPX SW : Analog Multiplexer Switch

Figure 4-12. Composition of Back-Scattered Sound Imaging System

Figure 4-13 shows a composition of the shadow imaging system. Almost all equipment is the same as the back-scatter imaging system, except that a non-spherical Fresnel's lens of 69 centimeter focal length is used as a transmitting lens, and a video amplifier of 40 dB voltage gain is used instead of the spectrum analyzer.

VI. RESULTS

With the back-scatter imaging system, images of three typical objects were taken in a water bath and in a swimming pool, and they are shown in Figure 4-14. For a synchronizing analysis of the received signal with the transmitting signal, this system has substantially high-range resolution. For example, its value is about 13 cm under the following conditions: Field of view, $30°$; image-making speed, 70 frames per second; frequency sweeping width, 800 kHz.

With the direct shadow imaging system, images of three objects were also taken in a water bath; they are shown in Figure 4-15. Frequency range of these images is 1.4 to 1.6 MHz, and the image-making speed is 140, or 70 frames per second.

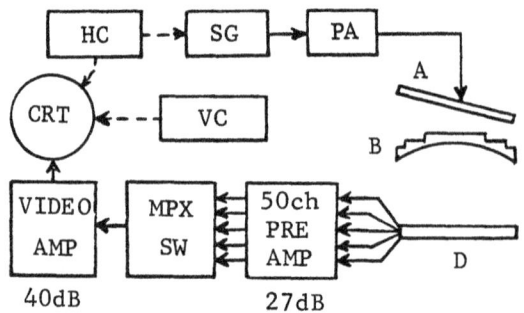

Figure 4-13. Composition of Direct Shadow Imaging System

SIGNAL PROCESSING METHOD IN ULTRASONIC SCANNING

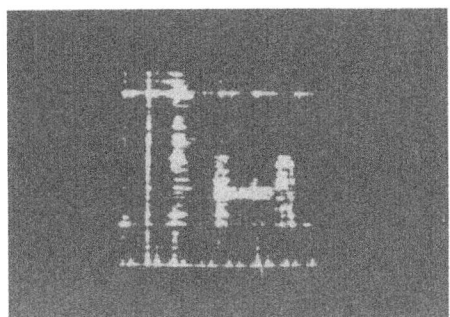

Figure 4-14a. Back-scattered Sound Images; No. 1 Distance, Three Meters

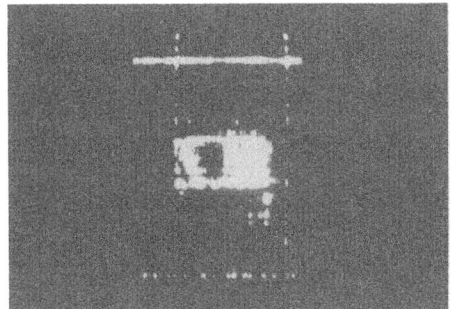

Figure 4-14a. Back-scattered Sound Image No. 2 Distance, 70 Centimeters

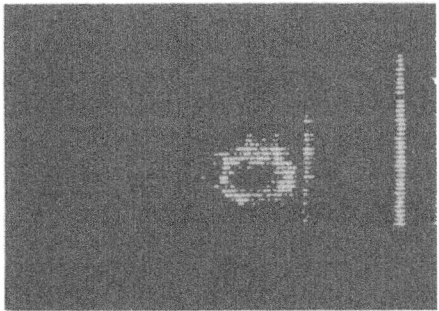

Figure 4-14a. Back-scattered Sound Image No. 3 Distance, Six Meters

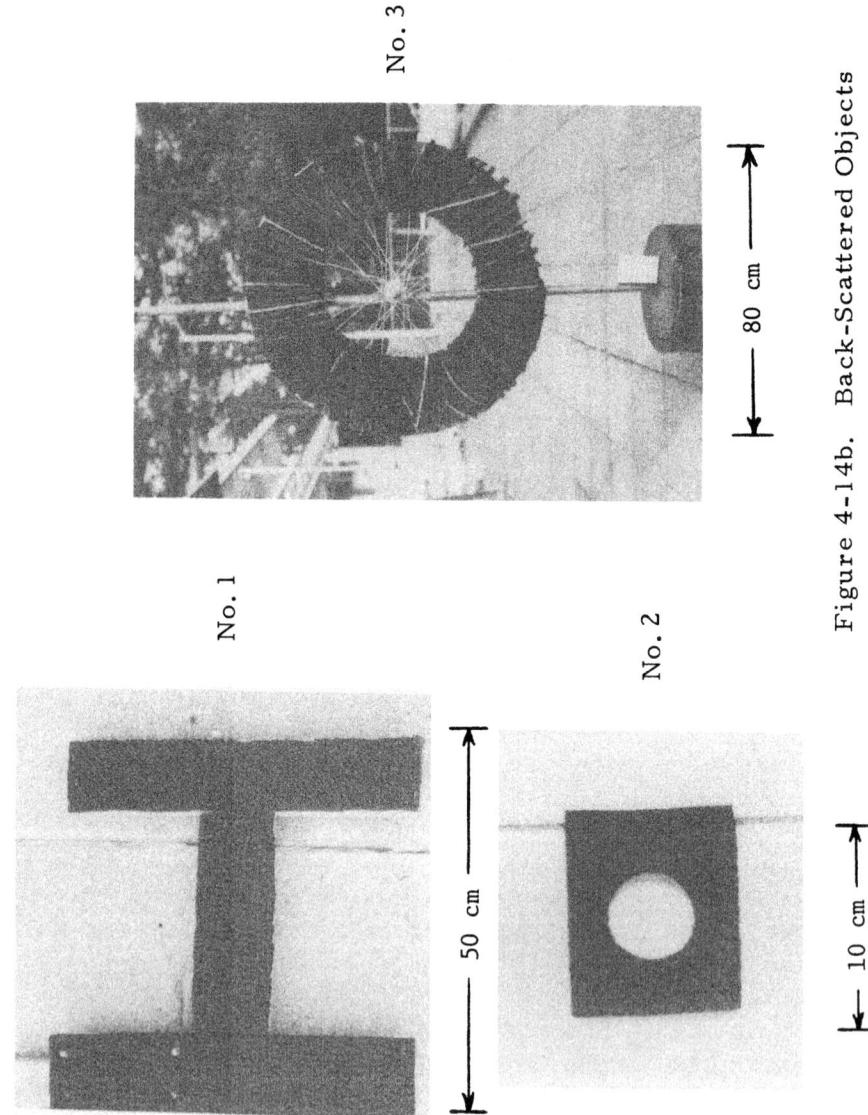

Figure 4-14b. Back-Scattered Objects

SIGNAL PROCESSING METHOD IN ULTRASONIC SCANNING

No. 1

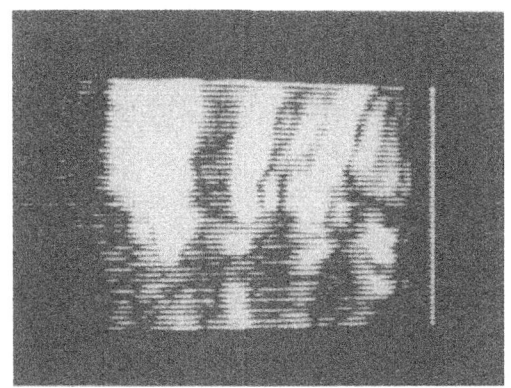

No. 2

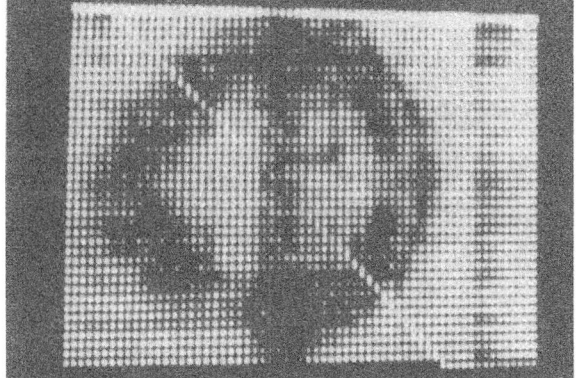

No. 3

Figure 4-15a. Direct Shadow Images

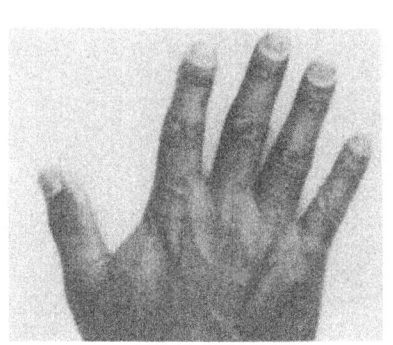

No. 1

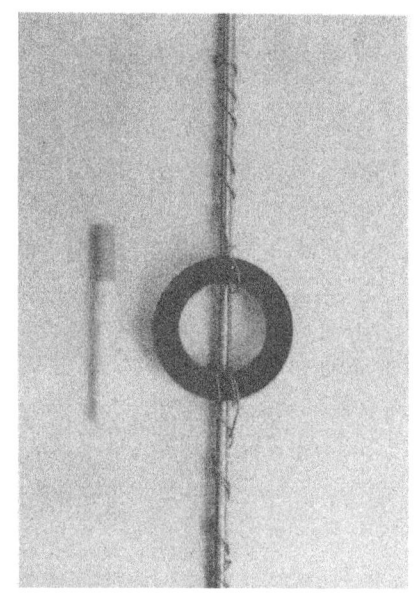

No. 2

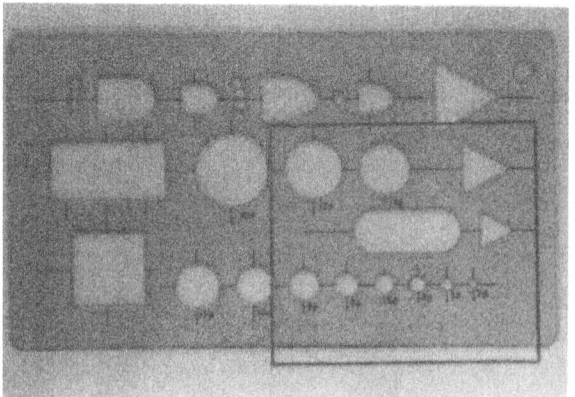

No. 3

Figure 4-15b. Direct Shadow Objects

REFERENCES

1. Okujima, M. and Isobe, M., Formation of Ultrasonic Fine Beam in Near-field by Ring Type Transducer, Reports 1967 Spring Meeting, Acoustical Society of Japan, No. 1-1-21, 1967. (In Japanese)

2. Awaya, K., Okujima, M., Imai, S., and Katakura, K., Simulation Experiment for Improvement in Receiving Directivity Characteristics of Ring Transducer, Reports of 1967 Autumn Meeting, Acoustical Society of Japan, No. 2-1-3, 1967. (in Japanese)

3. Awaya, K., Okujima, M., Imai, S. and Katakura, K., Simulation Experiment for Improvement in Receiving Directivity Characteristics of Ring Transducer (II), Reports 1968 Spring Meeting, Acoustical Society of Japan, No. 2-1-18, 1968. (in Japanese)

4. Okujima, M., Katakura, K., and Nakano, K., Experiment for Improvement in Receiving Directivity Characteristics of Ring Transducer, Report of Japan Society of Ultrasonics in Medicine, 13th Meeting, No. 13-18, 1968. (in Japanese)

5. Awaya, K., Okujima, M., Imai, S., and Katakura, K., Outline of our Trial-Made Ultrasonic Medical Inspector and Its Reception Patterns, Reports 1968 Autumn Meeting, Acoustical Society of Japan, No. 1-1-12, 1968. (in Japanese)

6. Awaya, K., Okujima, M., Imai, S., and Katakura, K., Resolving Power of the Correlation-Type Ultrasonic Medical Inspector, Reports 1969 Spring Meeting, Acoustical Society of Japan, No. 2-1-14, 1969. (in Japanese)

7. Awaya, K., Okujima, M., Imai, S., and Katakura, K., Ring Transducer with Very Sharp Receiving Directivity for Ultrasonic Tomograph, Reports of the 6th International Congress on Acoustics, No. M-1-4, 1968.

8. Okujima, M., Katakura, K., and Nakano, K., Ring Transducer with Very Sharp Receiving Directivity for Ultrasonic Tomograph, Medical Ultrasonics, Vol. 6, No. 2, P. 135, 1968.

9. Okujima, M., Endoh, N. and Nishimura, M., Basic Research of High Speed Scan in Ultrasonic Tomograph for Diagnosis, Reports 1970 Autumn Meeting, Acoustical Society of Japan, No. 3-2-5, 1970. (in Japanese)

10. Okujima and Endoh, N., Directional Pattern of Transducers for High Speed Scan Ultrasonic Tomograph for Diagnosis, Reports 1971 Autumn Meeting, Acoustical Society of Japan, No. 3-2-15, 1971. (in Japanese)

11. Okujima, M., Endoh, N., and Yoshida, H., Deflected Beam Pattern of Transducers for High Speed Scan Ultrasonic Tomograph for Diagnosis, Reports 1972 Spring Meeting, Acoustical Society of Japan, No. 3-1-16, 1972. (in Japanese)

12. Okujima, M., and Endoh, N., Design of Directional Array Transducer for Electronic Scan Ultrasonic Tomograph, Reports 1972 Autumn Meeting, Acoustical Society of Japan, No. 2-3-3, 1972 (in Japanese); Report of Japan Society of Ultrasonics in Medicine, the 22nd Meeting, No. 22-7, 1972. (in Japanese)

13. Somer, J.C., Electronic Sector Scanning for Ultrasonic Diagnostics, Ultrasonics, P. 153, July, 1968.

14. Okujima, M., Motooka, S., and Tsuchiya, I., Geological Survey on Land by Electromagnetic Induction Type Sound Source, Reports 1967 Autumn Meeting, Acoust. Society of Japan, No. 2-1-5, 1967. (in Japanese)

15. Okujima, M., Motooka, S., and Tsuchiya, I., Geological Survey on the Ground by Electromagnetic Induction Type Sound Source, Reports 1968 Spring Meeting, Acoustical Society of Japan, No. 2-1-16, 1968. (in Japanese)

16. Okujima, M., Motooka, S., and Endoh, N., Geological Survey from the Ground Surface by Electromagnetic Induction Type Sound Source (3), Reports 1969 Spring Meeting, Acoustical Society of Japan, No. 3-1-10, 1969. (in Japanese)

17. Okujima, M., and Motooka, S., Geological Survey from the Ground Surface by Electromagnetic Induction Type Sound Source (5), Reports 1969 Autumn Meeting, Acoustical Society of Japan, No. 1-1-17, 1969. (in Japanese)

18. Okujima, M., and Motooka, S., On the Ability of Geological Survey After Correlation Operation using Four Receivers Array, Reports 1970 Autumn Meeting, Acoustical Society of Japan, No. 3-2-6, 1970. (in Japanese)

19. Okujima, M., Motooka, S., and Kinoshita, N., Experiment of Geological Survey at the Fountain of Kakita River Head by Impulsive Sound, Reports of the 26th Annual Meeting, Japan Society of Civil Engineers, No. III-94, 1971. (in Japanese)

20. Okujima, M., Motooka, S., Nakamuta, T., and Nakao, S., Detection of Buried Pipes in Model Sand Bath by Using Impulsive Sound, Reports 1972 Spring Meeting, Acoustical Society of Japan, No. 3-1-12, 1972. (in Japanese)

21. Okujima, M., and Motooka, S., Detection of Buried Pipes by Using Impulsive Sound, Reports of the 27th Annual Meeting, Japan Society of Civil Engineers, No. III-159, 1972. (in Japanese)

22. Okujima, M., and Motooka, S., <u>Experiment in Scale Model for Geological Survey from Ground Surface,</u> Reports of the 6th International Congress on Acoustics, No. K-5-4, 1968.

23. Okujima, M., and Motooka, S., <u>Investigation of Geological Survey from Ground Surface by Impulsive Sound Wave,</u> Proceedings of the 7th International Congress on Acoustics, No. 23-U-2, 1971.

24. Katakura, K., Kobayashi, M., Tannaka, Y., and Koshikawa, T., <u>High Speed Ultrasonic Imaging System</u>, Report of Japan Society of Ultrasonics in Medicine, the 20th Meeting, No. 20-28, 1971. (in Japanese)

25. Katakura, K., Kobayashi, M., Tannaka, Y., and Koshikawa, T., <u>High Speed Ultrasonic Imaging System using Fan Beam Scanning,</u> Reports 1972 Spring Meeting, Acoustical Society of Japan, No. 3-1-13, 1972. (in Japanese)

26. Katakura, K., Kobayashi, M., Tannaka, Y., and Koshikawa, T., <u>High Speed Ultrasonic Imaging System Using Fan Beam Scanning,</u> Report of Japan Society of Ultrasonics in Medicine, the 22nd Meeting, No. 22-8, 1972. (in Japanese)

27. Katakura, K., Kobayashi, M., Tannaka, Y., and Koshikawa, T., <u>Underwater Viewing System,</u> Preprints of the 2nd International Ocean Development Conference, P. 427, 1972.

ULTRASONO-CARDIO-TOMOGRAPHY

Yoshimitsu Kikuchi*, Daitaro Okuyama*,
and Motonao Tanaka**

* Research Institute of Electrical Communication, Tohoku University, Sendai, Japan

** Research Institute for Tuberculosis, Leprosy and Cancer, Tohoku University, Sendai, Japan

I. INTRODUCTION

In obtaining ultrasonic tomograms of the heart and great vessels, there are, unlike other organs, some particular conditions: The pulsation of movement of all the echo sources, and the anatomical situation that the heart is covered by the lung and the costa. The positions of the echo sources repeatedly vary with time in the thoracic cavity. To obtain the stationary tomograms of the living heart at various cardiac phases, there have been proposed several methods. To overcome the pulsation movement, a synchronization method, a high speed scanning method and a tomo-kymographic method have been introduced. To solve the anatomical situation, an intra-cardiac, intra-esophagal, intra-tracheal and transthoracic methods have been developed. Those methods will be described in this paper, by placing some emphasis on the methods which the present authors have developed within the present clinical routine.

II. SYNCHRONIZED ULTRASONO-CARDIO-TOMOGRAPHY

Figure 1 shows the schematic diagram of the ultrasonic apparatus which operates in synchronization with the cardiac cycle.[1,2] For the synchronizing signal, the usual ECG meter is utilized. Although any wave of ECG can be used for the signal, the R-wave of ECG is used primarily for the purpose at present. As shown in the figure, the R-wave of ECG is first detected by the R-wave detector, and its waveform is shaped into a pulse which is used for the signal of time origin. Then, in accordance with the required phase for the stationary tomogram, a time delay is given to the signal of origin by the delay circuit. This signal is used as the synchronizing signal for operating the ultrasonic apparatus.

1. Intracavitary Methods

a. <u>Intra-Cardiac Method</u>[3] - Omoto, Atsumi and others have developed a method of detecting and measuring an atrial septal defect by using a catheter type ultrasonic probe which is inserted into the right atrium through the external jugular vein or the femoral vein. They tried various ultrasonic intravenous probes called "cardiac sonde". One cardiac sonde consists of a 18/8 stainless steel tube, 1.2 mm in diameter and 750 mm in length, with an ultrasonic transducer at the tip of the tube. The transducer is 3.2 mm in diameter and is used at five Megahertz. A modified C-scan indication system with ECG synchronizing circuit and a variable gate circuit has been employed as shown in Figure 2. An X-Y recorder is to be used in recording the image of the septal defect. The cardiac sonde was inserted into the superior or inferior vena cava as shown in Figure 3A, and it scanned the interatrial septum as shown in Figure 3b, at a number of different levels. An example of the obtained pattern is as shown in Figure 4. In this case, the authors reported that the estimated size of the defect from the pattern was 2.5 cm x 1.5 cm, and the measured size under surgery was 2.7 cm x 1.6 cm.

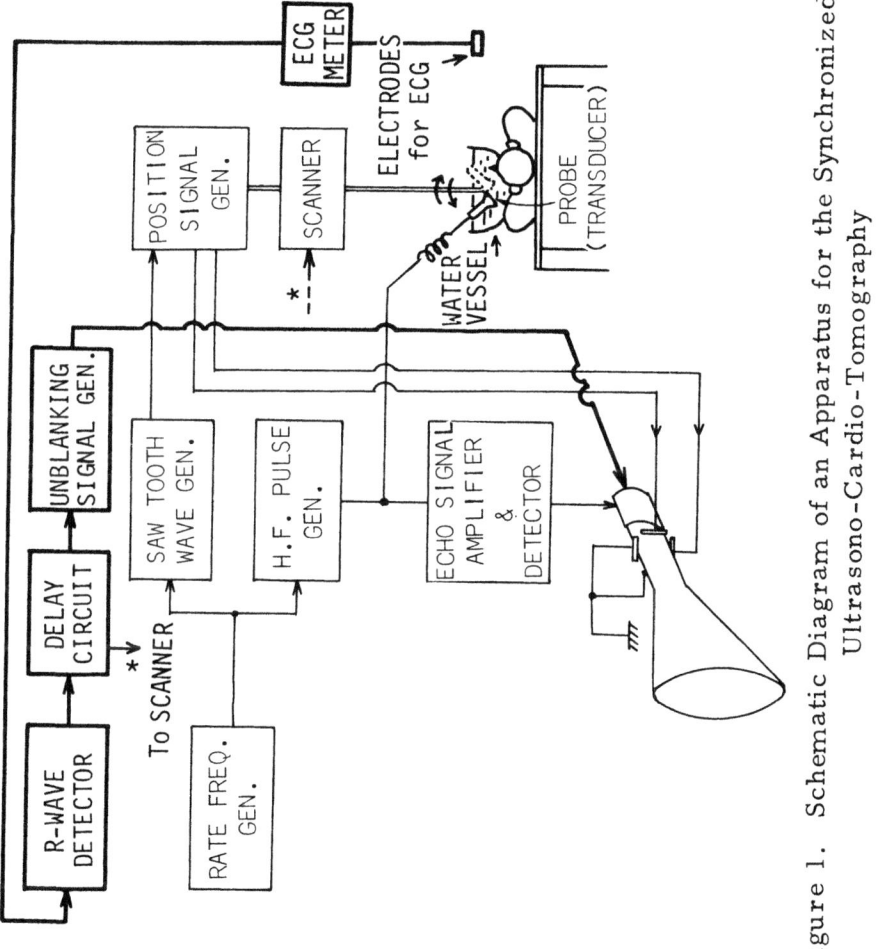

Figure 1. Schematic Diagram of an Apparatus for the Synchronized Ultrasono-Cardio-Tomography

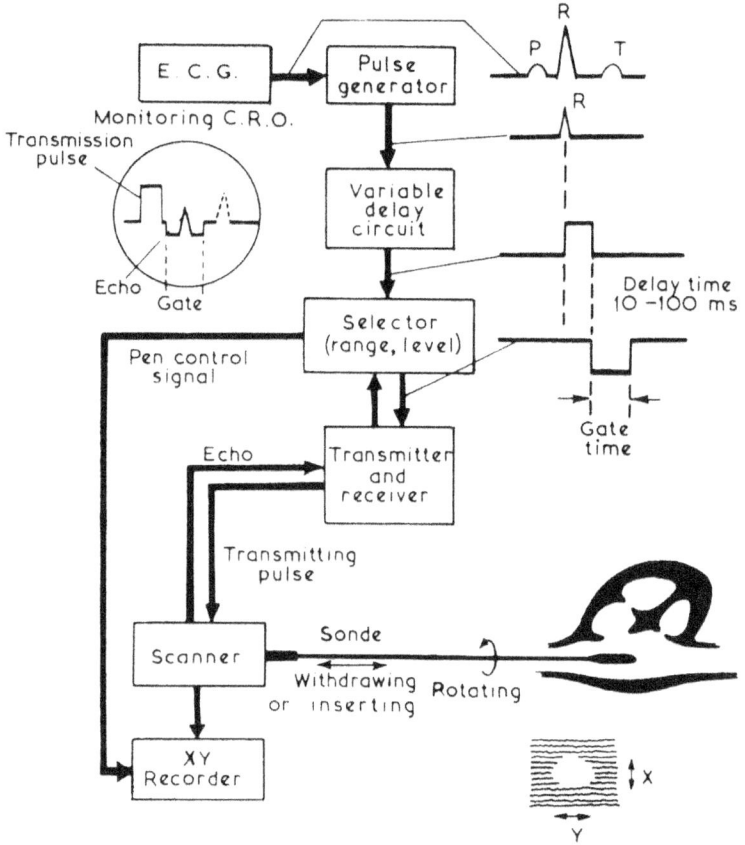

Figure 2. C-Scan System for the Examination of Atrial Septum Defect by "Cardiac Sonde"

© Kimoto, S., Omoto, R., Tsunemoto, M., Muroi, T., Atsumi, K., and Uchida, R., <u>Ultrasonic Tomography of the Liver and Detection of Heart Atrial Septal Defect with the Aid of Ultrasonic Intravenous Probes,</u> Ultrasonics, Vol. 2, P. 85, 1964, Figure 7.

* Dr Seiji Kimoto, Mitsui Memorial Hospital, Kanda Izumi-cho, Chiyodaku, Tokyo, Japan 101

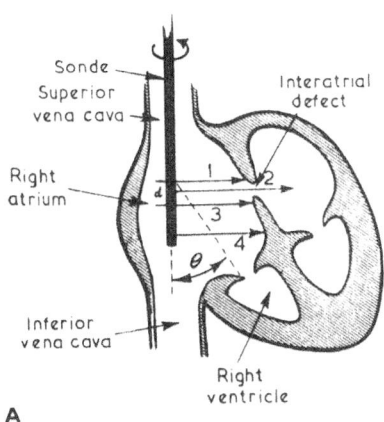

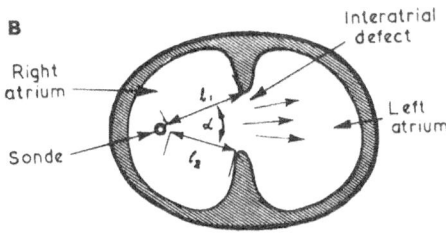

Figure 3. Illustration for the Scanning of Interatrial Septum by the Cardiac Sonde

© Kimoto, S., Omoto, R., Tsunemoto, M., Muroi, T., Atsumi, K., and Uchida, R., Ultrasonic Tomography of the Liver and Detection of Heart Atrial Septal Defect with the Aid of Ultrasonic Intravenous Probes, Ultrasonics, Vol. 2, P. 85, 1964, Figure 6.

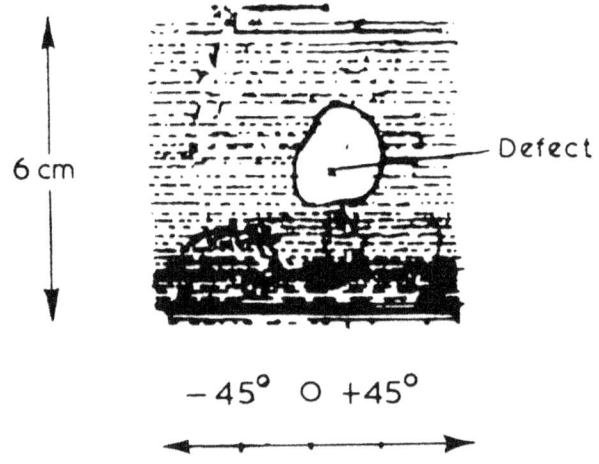

Figure 4. An Example of the C-Scan Image of Atrial Septum Defect

© Kimoto, S., Omoto, R., Tsunemoto, M., Muroi, T., Atsumi, K. and Uchida, R., Ultrasonic Tomography of the Liver and Detection of Heart Atrial Septal Defect with the Aid of Ultrasonic Intravenous Probes, Ultrasonics, Vol. 2, P. 86, 1964, Figure 9.

The C-scan indicator, however, is not suitable for an intuitive observation, because the abscissa of the pattern is the azimuth angle, and the ordinate is the distance of the cardiac sonde movement in the axial direction. Then they began to use a PPI presentation by the same procedure.[4]

b. Intra-Esophagal Method[5] - At about the same time, an intra-esophagal method was presented by Kikuchi, Ebina, Tanaka and others to obtain a PPI presentation of synchronized ultrasono-cardio-tomograms. This method is completely non-operative. They succeeded in obtaining a stationary tomogram of the heart at any cardiac phase from the inside of the esophagus by controlling the rotation of the ultrasonic scanner in synchronization with an electrocardiographic current. As will be described later in further detail, the method consists in the shifting of the ultrasonic beam direction so as to complete the scanning of the entire heart in a specified short time at a given cardiac phase. Figure 5 shows an example of the result when this method was applied to an anesthetized dog. This is an instantaneous ultrasono-tomogram of the mediastinum part, including the heart obtained by the radial scanning of the part. In the experiment, a concave transducer, 10 mm in diameter, and 60 mm in radius of curvature, was used at five Megahertz inside the esophagus. As is evident from this example, the pattern is rough because of the limited number of ultrasonic pulses during the ultrasonic beam movement.

c. Intra-Tracheal Method[6] - Similarly with the preceding method, an intra-tracheal PPI transducer has been developed so that the mediastinum organs can be ultrasonically scanned from the inside of the trachea. Experiments were reported as being successful with a dog.

d. Intracardiac Catheter for an Electronic High Speed Scanning[7] - Recently, Bom has shown a new type of intracardiac ultrasonic catheter. The catheter consists of a multi-element cylindrical acoustic transducer whose elements are to be switched electronically. When electronic scanning is fast, with a corresponding display, an instantaneous image of the echo structure is obtained.

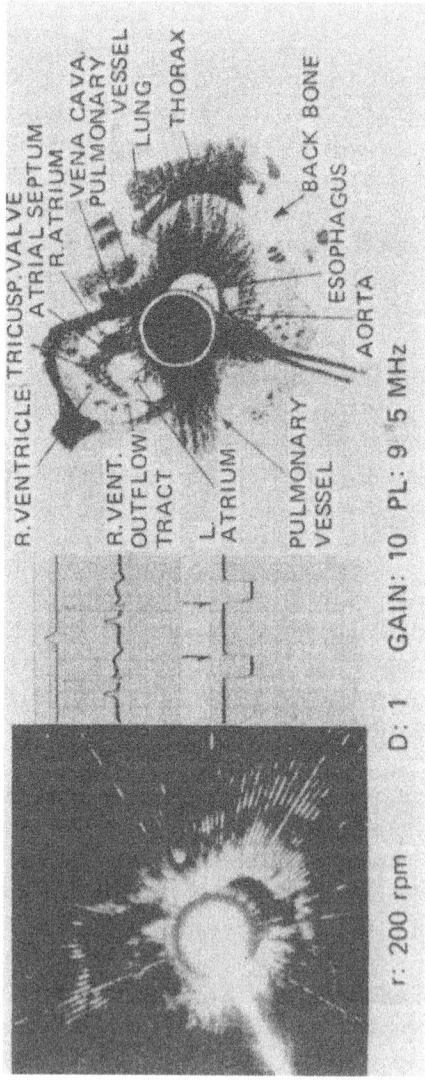

Figure 5. Ultrasono-Cardio-Tomogram of the Mediastinum of an Anesthetized Dog Obtained by the Intraesophagal Method

© Kikuchi, Y., Okuyama, D., Ultrasono-Cardio-Tomography, Japan Electronic Engineering, No. 47, October, 1970, P. 54, Figure 2

The picture of the catheter is shown in Figure 6. The echo pattern of a cylindrical mesh reflector is as shown in Figure 7. No data for clinical use, however, have been reported.

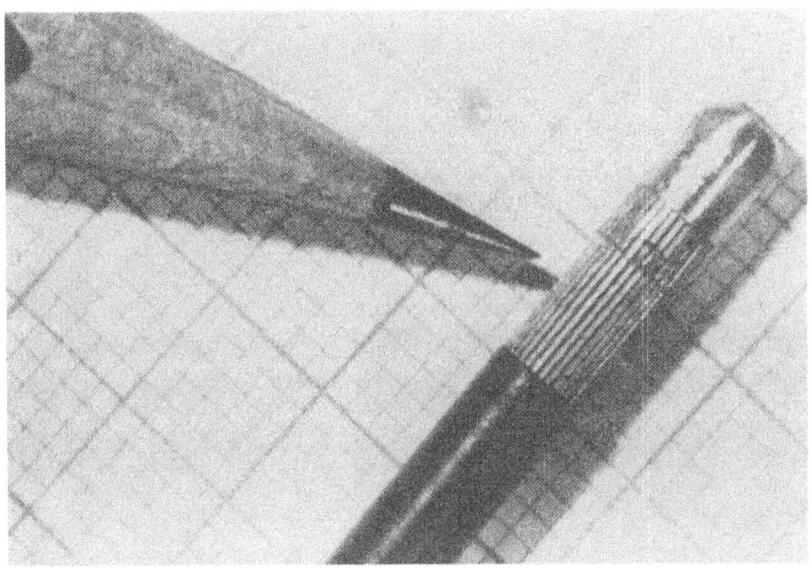

Figure 6. Photograph of a 32 Element Catheter

© Bom, Ir. N.*, <u>New Concepts in Electrocardiography</u>, H. E. Stenfert Kroese N. V., Leiden, 1972, P.54, Figures 5-11.

* Dr Ir. N. Bom, Department of Cardiology, University of Rotterdam, The Netherlands.

2. <u>Transthoracic Method</u>[1],[2]- Some transthoracic methods of synchronized ultrasono-cardio-tomography were proposed around 1963 by Kikuchi, Ebina, Tanaka and others, and developed accordingly. The principle is based

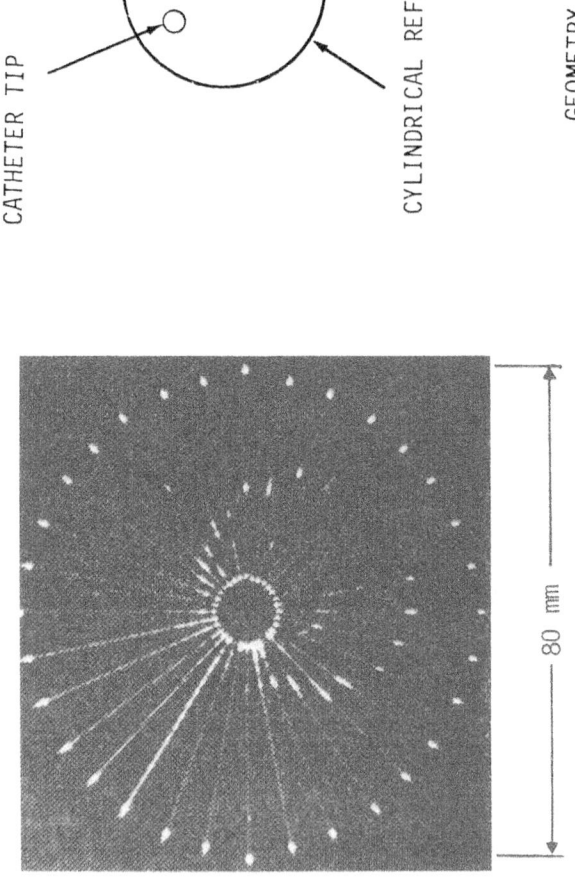

Figure 7. An Echo Pattern and the Configuration of a 32 Element Catheter
ⓒ Bom, Ir N., <u>New Concepts in Electrocardiography</u>, H. E. Stenfert Kroese, N.V., Leiden, 1972, P. 55, Figures 5-12.

on an idea that an ultrasonic apparatus for the usual ultrasonotomography is so modified that it is operated only within a short interval at a selected instant of a cardiac cycle.

Ultrasono-cardio-tomography by cardiac sychronization is the method for describing ultrasono-tomograms synchronously at any phase of the cardiac cycle by operating the apparatus for a short period of time in which the heart is deemed stationary. When the description of a wider area is required, a method similar to a panorama photographic method is combined with the former.

The principal points in these systems consist in the detection of the cardiac phase and instantaneous operations. Three types of apparatus operations have been proposed by the authors. These are: (1), synchronized single scanning; (2) synchronized repetition of the single scanning; and (3), the repeated synchronized unblanking of the echo display. These have been named Operations I, II, and III, respectively.

Operation I has been mentioned in the paragraphs on intra-esophagal and intratracheal methods. As was described there, the patterns obtained are usually rough because of insufficient density of the cathode ray sweeps.

The aim of Operation II is to make a smoother pattern by repeating the synchronized single scanning several times at the same cardiac phase of successive pulsations so as to fill the sweep gaps left during the preceding scannings.

In Operation III, though the short, unblanking of the display is repeated at a given cardiac phase, the transducer position, consequently the ultrasonic beam position, is free running; so one complete scanning of a given area in a human organ is made by patching a certain number of partial tomograms of the area. The clinical operator (physician) can adjust the free running phase and speed so that the required time may be minimized with a compromise to the cathode-ray sweep density for obtaining a smooth pattern. In this operation, however, the identical succession of the cardiac cycle must be assumed.

In the case of the transthoracic method, the ultrasonic beam travels through the intercostal space from an ultrasonic transducer that is located in very close vicinity to the chest wall, as shown in Figure 8. This method is named "proximity-immersed method". In this case, Operation III is usually employed, and the ultrasonic transducer makes sector swings continuously without being in synchronization with the heart movement; but the unblanking of the cathode ray tube is repeated for a short period of time at the selected phase of every cardiac cycle determined by the output signal of the delay circuit as already shown in Figure 1. As generally described in the foregoing, only a partial tomogram of the heart is obtained for a single cardiac cycle; so the repetition over 20 to 30 times covers the whole heart, and the pattern thus obtained is a positionally patched pattern of several numbers of partial tomograms for a given cardiac phase. Figure 9 represents an example of the application of this method to a normal heart. Figure 9A shows the tomograms for the systolic phase, and 9B is for the diastolic phase. Figures 9C and 9D are roentgenograms showing the plane of scanning. The arrow indicates the plane scanned by the ultrasonic beam in the thoracic cavity.

Figure 10 shows a set of the stationary tomograms for one cardiac cycle when the cycle is divided into nine intervals. The phase relation of each tomogram to ECG is given on the lower left, each with a dot. In this figure, (1) is for early systole, (6) is for late systole, (7) is for early diastole, and (9) is for late diastole. When this set of tomograms is compared with one after another in the numbered order, one can observe the movement of various parts of the heart.

Numerous interesting findings were observed on a set of the stationary tomograms.[8] The movement of each part of the heart in relation to the movement of the whole heart can be clarified; any movement, not only the movement in the ultrasonic beam direction, but in the other directions, is observable. This is not possible by a single UCG. When attention is focused on the mechanical movement of the left ventricle, both in the case of normal and of atrial septal defect, it is observed that the displacement amplitudes of the left ventricular wall and septum are almost equal in magnitude, and that the function of the right ventricle is largely affected by that of the left ventricle. The motion of the left ventricle is almost concentric in both the systole and the diastole in normal function, as seen in Figure 11A, whereas the motion is eccentric in the case of atrial septal defect as seen in Figure 11B.

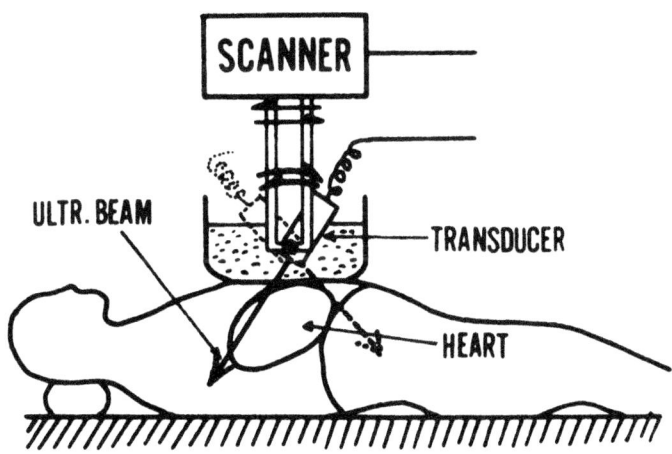

Figure 8. Schematic View of the Configuration of the Transducer for "Proximity-Immersed Method"

© Kikuchi, Y., Okuyama, D., Tanaka, M., Ebina, T. and Oka, S., <u>Ultrasono-Cardio-Tomography and Its Application to Mophological Measurement of the Heart,</u> Ultrasono Graphia Medica, Vol. III, Verlag Der Wiener Medizinischen Akademie, 1971, P. 426, Figure 2.

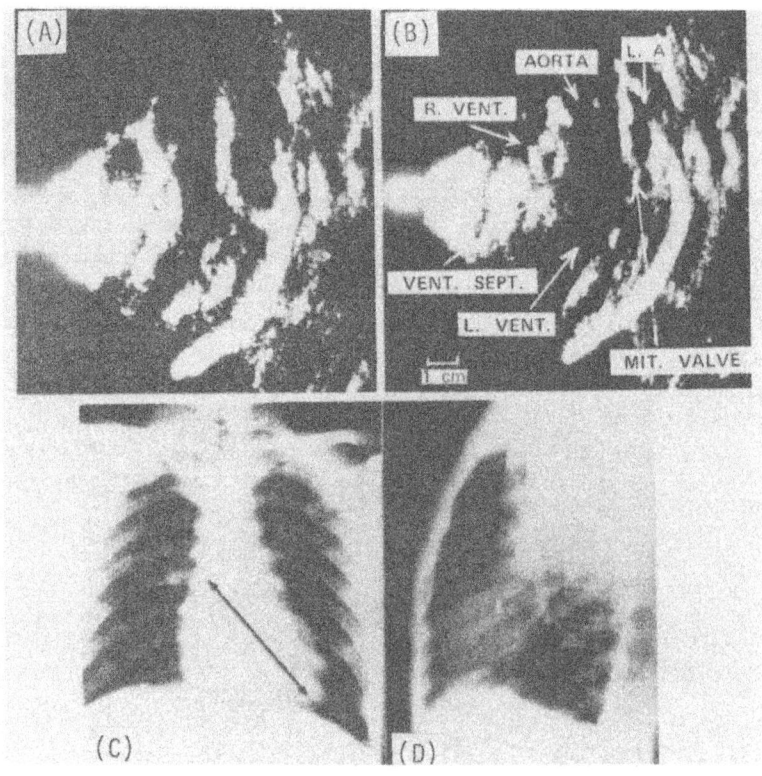

Figure 9. Ultrasono-Cardio-Tomograms and Roentgenograms in a Normal Heart. Arrow shows the direction of ultrasonic scanning.
ⓒ Kikuchi, y., Okuyama, D., Ultrasono-Cardio-Tomography, Japan Electronic Engineering, No. 47, October, 1970, P. 55, Photograph No. 1.

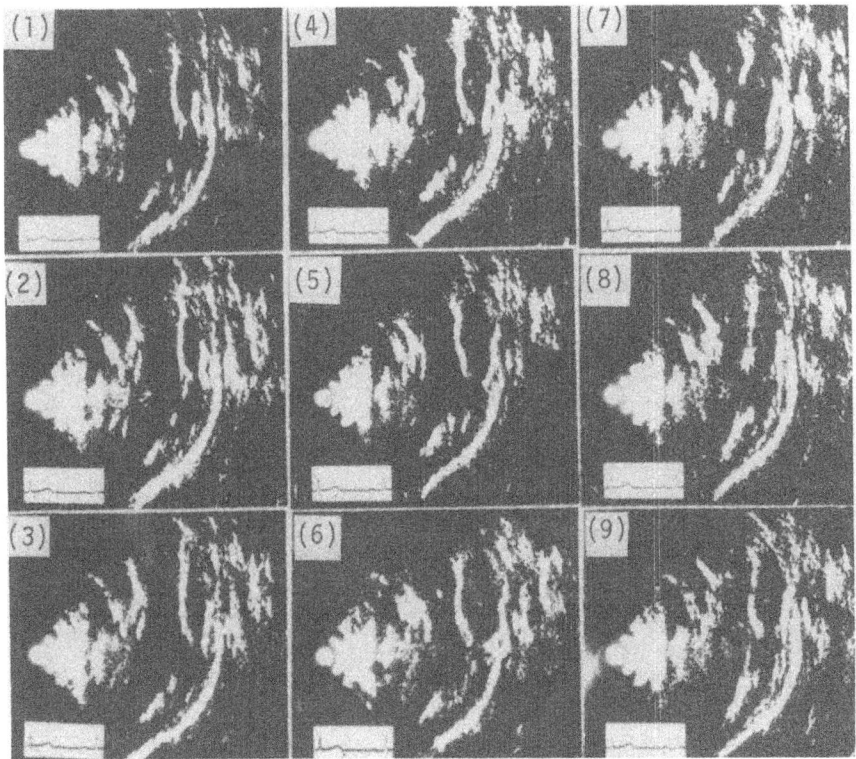

Figure 10. Ultrasono-Cardio-Tomograms Obtained at Nine Different Cardiac Phases of One Cardiac Cycle

© Kikuchi, Y.*, Okuyama, D., Kasai, C., Ebina, T., Tanaka, M., Terasawa, Y. and Uchida, R., <u>Multi-Information Recording and Reproduction in the Ultrasono-Cardio-Tomography,</u> Acoustical Holography, Vol. 4 (edited by G. Wade), Plenum Press, New York-London, P. 118, 1972, Figure 4.

* Dr Yoshimitsu Kikuchi, Res. Inst. of Electrical Communication, Tohoku University, Senda, Japan 980.

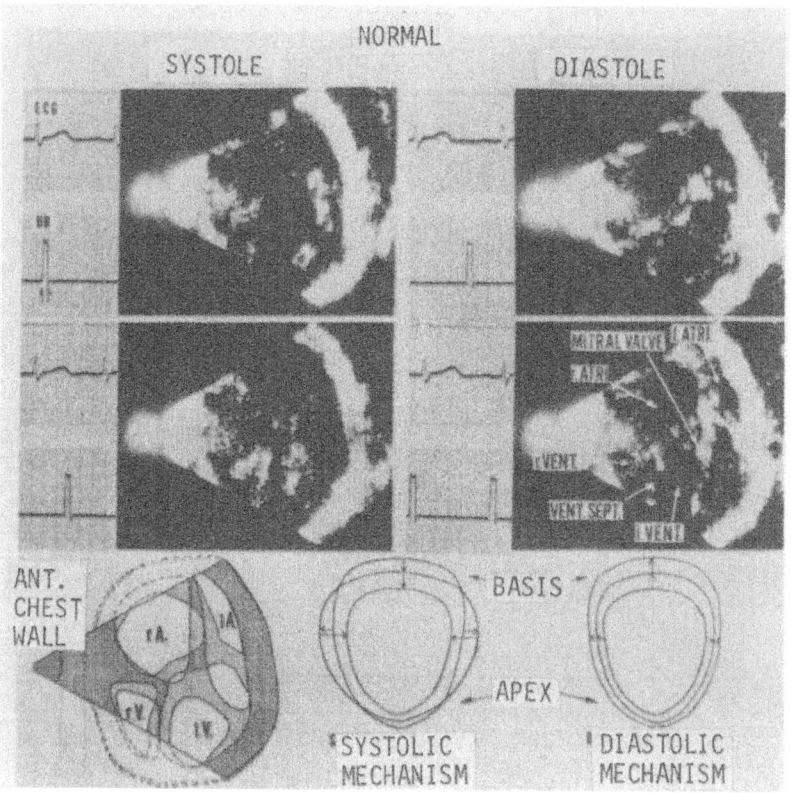

Figure 11A. Analysis of the Left Ventricular Movement, Motion is concentric in a normal heart

ⓒ Kikuchi, Y., Okuyama, D., Tanaka, M., Ebina, T., and Oka, S., Kineto-Ultrasono-Tomography of the Heart, Ultrasono Graphia Medica, Vol. III, Verlag Der Wiener Medizinischen Akademie, 1971, P. 478, Figure 3.

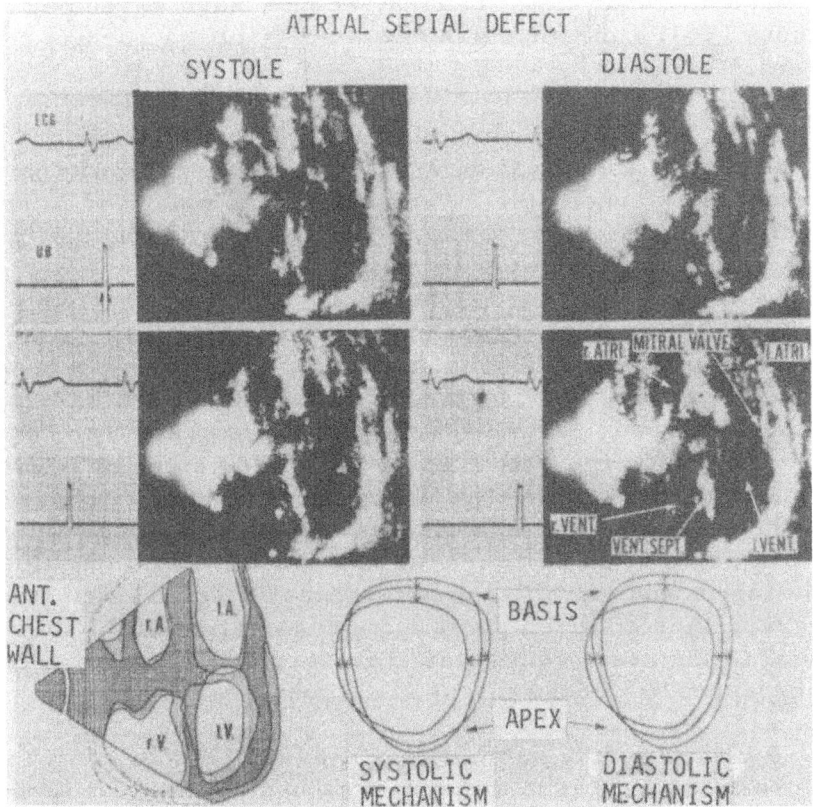

Figure 11B. Analysis of the Left Ventricular Movement, Motion is eccentric in a case of atrial septal defect.

© Kikuchi, Y., Okuyama, D., Tanaka, M., Ebina, T., and Oka, S., Kineto-Ultrasono-Tomography of the Heart, Ultrasono Graphis Medica, Vol. III, Verlag Der Wiener Medizinischen Akademie, 1971, P. 479, Figure 4.

III. KINETO-ULTRASONO-TOMOGRAPHY[9,10]

Kikuchi, Okuyama, Tanaka, et al., have developed (about 1967) a cinematographic display of the cardiac tomogram.

If the set of synchronized ultrasono-cardio-tomograms is made into an animation as shown in Figure 12 and projected on a screen, the movement can be observed intuitively. In this chapter, the technique for making the cinematograph is illustrated briefly.

In the first step, ultrasono-cardio-tomograms of the heart in action are taken at a number of different phases for one cardiac cycle by means of the synchronized tomography already described. Each tomogram is then taken by a cine camera, one by one; thus, one series of animations is made for one cardiac cycle. In order to observe the heart movement on a cinematograph, this one series is repeated as often as necessary for observation. Endless filming is also advantageous for repetition. The cinematograph thus obtained gives a representation of the movement of the heart section as if it were presented by a real time operation of the ultrasonic apparatus.

In practice, the number of individual tomograms can be decreased when each tomogram is used for two or three frames of successive animation. In normal speed observation, 10 to 20 individual tomograms are suitable for one cardiac cycle when dual or triple use is employed. Figure 12 is an example of this triple use. If both the numbers of the tomograms and multiple use are increased, a slow-motion picture is obtained, although the motion appears somewhat unnatural. (At the meeting at which this report was presented, the result was shown by a roll of 16 mm film on which the cases of normal, atrial septal defect, the tetralogy of Fallot, and patent ductus arteriosus had been edited.)

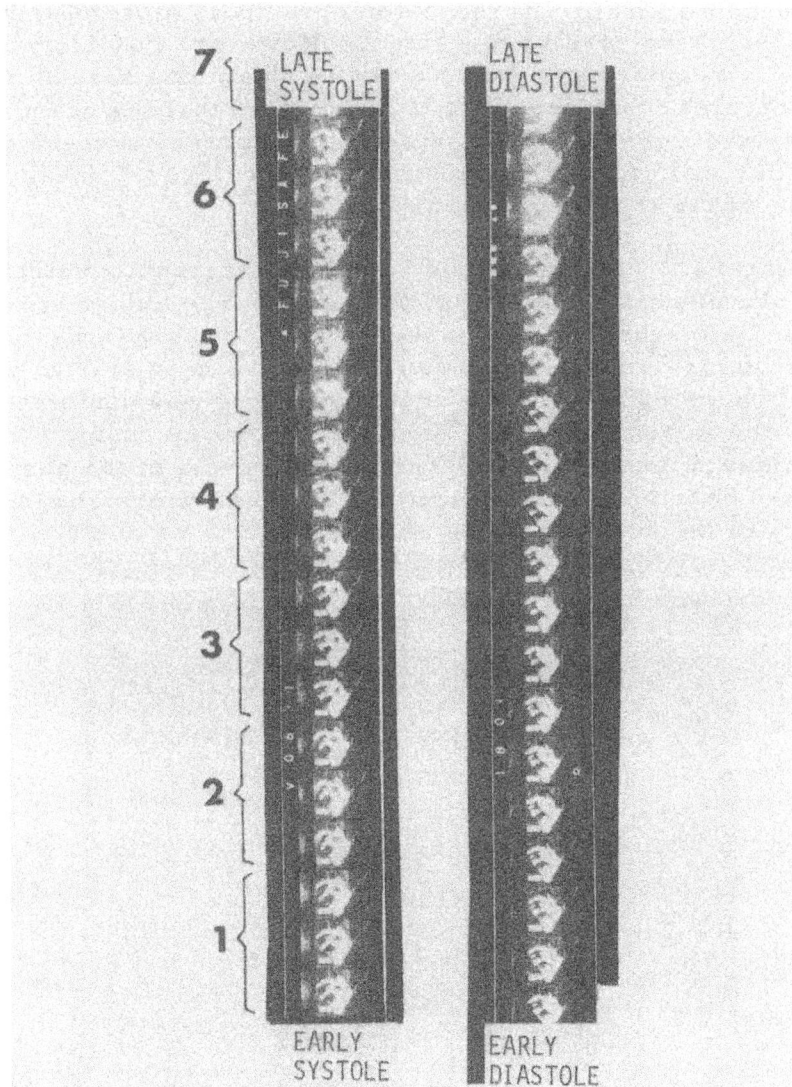

Figure 12. A Part of Kineto-Ultrasono-Tomographic Film to be Presented Cinematographically

Cardiac Kineto-ultrasono-tomography is a useful method for observing cardiac movement. However, considerable time is required at the clinic bed to obtain one series of tomograms. Moreover, it is anticipated that the heart movements may differ to some extent between successive cardiac cycles. Improvements are therefore necessary to decrease the time needed.

Asberg[11] presented a sort of cinematographic method for ultrasono-tomography of the heart. The author employed a mechanical high-speed scanner, using Olofsson's type ultrasonic mirror system.[12] The mirror system was set in a water tank, and a sector scanning was performed with the mirror system. Figure 13 shows an example of a series of tomograms for one cardiac cycle of the normal human heart. The movements of the interventricular septum can be observed in the tomogram.

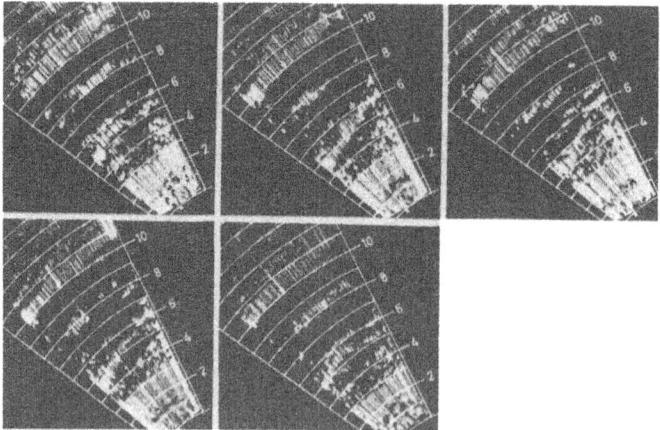

Figure 13. An Example of a Series of the Ultrasono-Cardio-Tomograms during one Cardiac Cycle Obtained by the Ultrasonic Mirror System (11)

© Asberg, A., Ultrasonic Cinematography of the Living Heart, Ultrasonics, Vol. 5, P. 116, 1967, Figure 6.

* Dr Asberg, Department of Electrical Measurements Lund Institute of Technology, Lund, Sweden.

Asberg, however, pointed out some disadvantages in this method. One is that the tomogram has some distortion because the scanning speed is not fast enough; the other is that the azimuthal resolving power of the mirror system varies along the mirror axis too much, to cover an entire heart.

IV. ULTRASONO-CARDIO-TOMOGRAPHY BY MEANS OF MULTI-INFORMATION RECORDING SYSTEM[13]

As is evident from the preceding paragraph, in syncronized ultrasono-cardio-tomography the partial tomograms for 20 to 30 heart beats, each of which is taken in a specified short time at a required phase, must be patched together panoramically in order to obtain a complete tomogram. For this reason, 0.5 to 1.0 minute is required for obtaining one stationary tomogram which covers the entire heart. Depending upon heart diseases, necessary numbers of tomograms become a few dozen in some cases;[13] thus, it takes at least two to three hours for one patient. The cause for such a long examination is based on the intermittent use of echo information, despite the fact that some echo information of the ultrasonic pulse is continuously coming from the inside of the beating heart.

As a means of solving the problem of information loss, a system of multi-information recording and reproduction has been introduced into synchronized ultrasono-cardiotomography. In this method, the entire ultrasonic echo information is recorded, together with the ECG at the clinic. The recorded information is repeatedly reproduced later to obtain any number of stationary cardio-tomograms for every cardiac phase, by employing the synchronization method in which the recorded ECG is, of course, used for the reproduction signal for synchronization. Even when a number of tomograms for various phases are required for a certain cross-section of the heart, the time required at the clinic would be only that required for obtaining a single tomogram, no matter how many phases might be required.

The block diagram of the system is shown in Figure 14. In the figure, the diagram below the broken line is the usual apparatus for the synchronized ultrasono-cardio-tomography, while the unit above the broken line consists of the information recording and reproduction device, and the ECG signal processing device. When the changeover switch at the center of the figure is shifted to RECORDING, the recording of information takes place, while at the same time, the display unit serves as the monitor.

Figure 15 shows both examples of the original tomogram obtained with the usual apparatus and the reproduced tomogram thereof. It may be considered as being satisfactory.

V. ULTRASONO-TOMO-KYMOGRAPHY[14]

In addition to the method of ultrasono-cardio-tomography, kineto-ultrasono-tomography, and the combination of tomography with time-position-indication, a method in which the displacement amplitudes of cardiac parts can be presented in a superimposing way on an ultrasonic tomogram has been developed, also by Kikuchi, Okuyama, Tanaka, et al., and named, "Ultrasono-Tomo-Kymography". The cardiac mobility, the maximum displacement of various parts of the heart, the variation in radii of the cardiac chambers, the curvature change of the ventricular wall, etc., can be represented in a single sheet of the tomogram; which is, therefor, useful for the analysis of the mechanical movement of the heart. This method is simpler in a clinical sense than the ultrasono-cardio-tomographical analysis, and it still gives more information[14] than the usual UCG.

By using the usual ultrasonic apparatus for tomography, the heart is now scanned by the ultrasonic beam in a very slow, angular velocity (in the case of sector scanning) as slow as three to five degrees per one cardiac cycle; or in a very slow linear velocity, in the case of linear scanning. On the cathode ray screen, every part of the heart is represented by a wave pattern indicating its maximum and minimum positions for one cardiac cycle. The entire pattern, however, still shows the tomographical section of the heart.

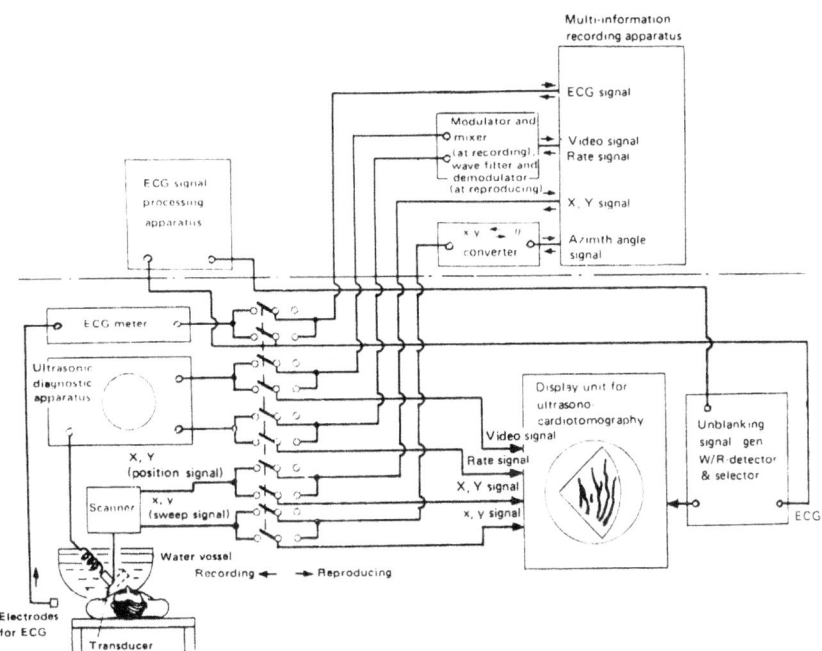

Figure 14. Schematic Diagram of Equipment for Ultrasono-Cardio Tomography when "Multi-Information Recording System is Employed

© Kikuchi, Y., Okuyama, D., Kasai, C., Ebina, T., Tanaka, M., Terasawa, Y., and Uchida, R., <u>Multi-information Recording and Reproduction in the Ultrasono-Cardio-Tomography,</u> Acoustical Holography, Vol. 4, (edited by G. Wade) Plenum Press, New York-London, P. 119, 1972, Figure 5.

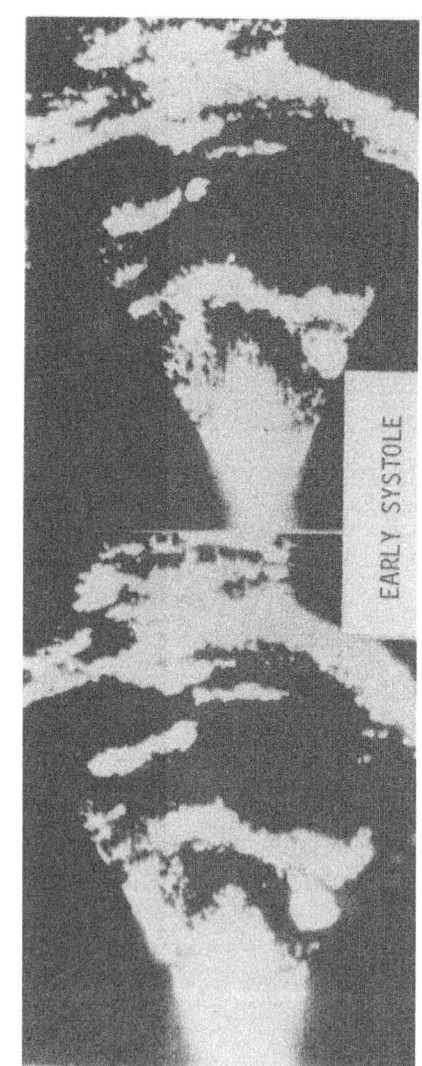

Figure 15. Ultrasono-Cardio-Tomograms at an Early Systole, Showing the Comparison with the Reproduced Tomogram by Means of the "Multi-Information Recording System."

© Kikuchi, Y., Okuyama, D., Kasai, C., Ebina, T., Tanaka, M., Terasawa, Y., and Uchida, R., Multi-information Recording and Reproduction in the Ultrasono-Cardio-Tomography, Acoustical Holography, Vol. 4 (edited by G. Wade), Plenum Press, New York-London, P. 122, 1972, Figure 7.

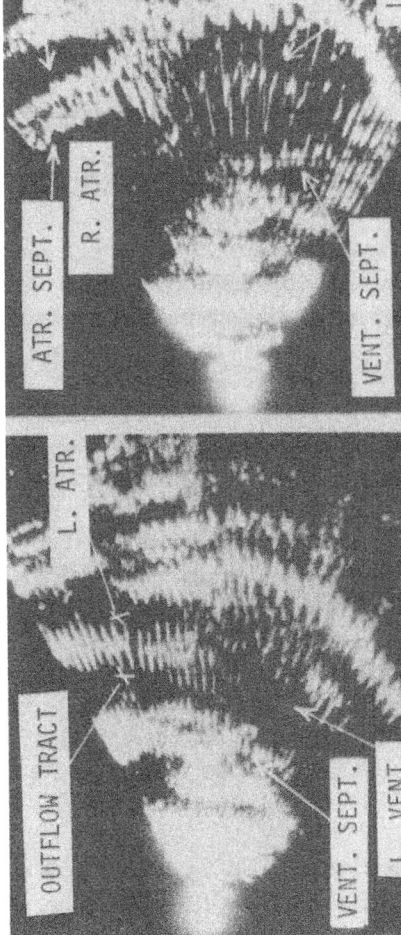

Figure 16. Ultrasono-Tomo-Kymograms of the Beat in a Normal Heart

© Kikuchi, Y., Okuyama, D., Tanaka, M., Ebina, T., and Oka, S., Ultrasono-Tomo-Kymography of the Heart, Ultrasono Graphis Medica, Vol. III, Verlag Der Wiener, Medizinischen Akademie, 1971, P. 483, Figure 2.

Figure 16 shows the wave tomograms for a normal heart obtained by the proposed method. As the method is based on the accumulated recording of the positions at which the echoes of the ultrasonic pulses occur, and changing with cardiac motion, the perpendicular distance between the envelopes touching the wave pattern at its maximum and minimum positions shows the maximum displacement of the corresponding cardiac part, notwithstanding the direction of the ultrasonic beam.

VI. INSTANTANEOUS OBSERVATION OF THE HEART BY MEANS OF HIGH-SPEED LINEAR SCANNING OF ULTRASONIC BEAM[15]

Recently, Bom has proposed to use high-speed scanning of an ultrasonic beam for obtaining a cardio-tomogram. A multi-element linear array transducer to be controlled by the electronic switch is employed. The repetition rate of electronic scanning is 190 frames per second.

Some interesting data have already been obtained in vivo on a small number of patients. With the transducer centered over the left fourth intercostal area, the moving left ventricular structures such as posterior and anterior walls were recognized in most cases, and a moving tomographical pattern was displayed continuously.

Figure 17 shows one systolic frame of the tomograms video-recorded. In the figure is also shown the corresponding sagittal section of the thorax and the heart. In this case, a fifteen-element transducer was used at 4.5 Megahertz.

Bom pointed out that the limitation of this system is caused by poor resolution and specular reflection, and that only the gross movement of larger structures can be viewed with the present apparatus. However, he also mentioned that it might be stated that there is much to be learned in the interpretation of the obtained data.

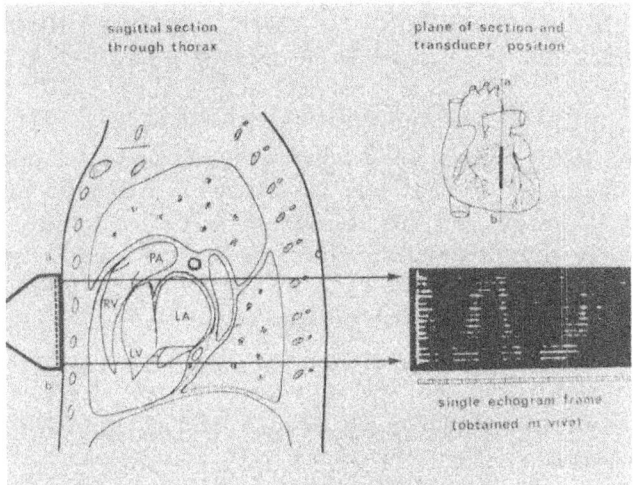

Figure 17. An Example of a Resulting Single Frame Tomogram and the Corresponding Sagittal Section and the Heart

ⓒ Bom, Ir. N., <u>New Concepts in Electrocardiography,</u> H. E. Stenfert Kroese N.V., Leiden, 1972, P. 41, Figure 4-8.

ray tube screen. As this method is based on the multiplex use of echo information, the time required for obtaining one series of tomograms for one cardiac cycle is equal to the time required for any one tomogram, regardless of the number of parts in a series. As is easily understood from the preceding articles, the required time is around 0.5 to one minute in the case of the transthoracic method described in Operation III. The details will soon be published in a separate paper.

VII. MULTIPLEX PRESENTATION OF ULTRASONO-CARDIO-TOMOGRAMS FOR DIFFERENT CARDIAC PHASES

Recently, the present authors have proposed a method named multiplex presentation. A suitable number of ultrasono-cardio-tomograms for different cardiac phases are simultaneously presented on separate areas of one cathode-

VIII. SUMMARY

Synchronized ultrasono-cardio-tomography is based on the cardiac synchronization principle and presents the instantaneous cross-section of the pulsating heart. The intracavitary method and the transthoracic method have been proposed and developed for clinical examination by this principle.

Kineto-ultrasono-tomography is a useful method for intuitive observation of the heart movement.

Ultrasono-cardio-tomography by means of a multi-information recording system eliminates the problem of the comparatively prolonged clinical examination.

Ultrasono-tomo-kymography in an asynchronized cardiac method is simpler than the synchronized ultrasono-cardio-tomography. It can also be used for clinical examination.

High speed scanning methods are tried for use in ultrasono-cardio-tomography.

Multiplex presentation of cardio-tomograms is proposed to overcome the prolonged clinical examination.

REFERENCES

1. Kikuchi, Y., and Tanaka, M., Some Improvements in Ultrasonotomograph for the Heart and Great Vessels, Part II, Tomography Synchronized with any Cardiac Phase, 1966 IEEE Symposium on Ultrasonics, J-3, Cleveland, Ohio, October 12-14, 1966.

2. Kikuchi, Y., Okuyama, D., Tanaka, M., Ebina, T., and Oka, S., Ultrasono-Cardio-Tomography and Its Application to Mophological Measurement of the Heart, Ultrasono Graphia Medica, Volume III, Verlag Der Wiener Medizinischen Akademie, P. 423, 1971.

3. Omoto, R., Atsumi, K., et al., Ultrasonic Intravenous Sonde, Japan Journal of Medical Electronics and Biological Engineering, Volume 1, P. 90, 1963; Kimoto, S., Omoto, R., Tsunemoto, M., Muroi, T., Atsumi, K., and Uchida, R., Ultrasonic Tomography of the Liver and Detection of Heart Atrial Septal Defect with the Aid of Ultrasonic Intravenous Probes, Ultrasonics, Volume 2, P. 82, 1964.

4. Omoto, R., Ultrasonic Tomography of the Heart-An Intracardiac Scan Method, Ultrasonics, Volume 5, P. 80, 1967.

5. Ebina, T., Oka, S., Tanaka, M., Kosaka, S., Kikuchi, Y., Uchida, R., and Hagiwara, Y., The Diagnostic Application of Ultrasound to the Disease in Mediastinal Organs, Science Rep. Research Institute, Tohoku University, Series C., Volume 12, P. 199, 1965.

6. Haneda, Y., Tanaka, M., et al., Ultrasonic Visualization of Mediastinal Tumor (Especially Regarding the Transtracheal Method), Medical Ultrasonics, Volume 6, No. 1, P. 54, 1968.

7. Bom, Ir. N., New Concepts in Electrocardiography, H. E. Stenfort Kroese, N.V., Leiden, P. 44, 1972.

8. Kikuchi, Y., Okuyama, D., Tanaka, M., Ebina, T., and Oka, S., Kineto-Ultrasono-Tomography of the Heart, Ultrasono Graphia Medica, Volume III, Verlag Der Wiener, Medizinischen Akademie, P. 475, 1971.

9. Tanaka, M., Oka, S., Kikuchi, Y., Okuyama, D., Ebina, T., et al., Cardiac Kineto-Ultrasonotomography (in Japanese), Reports of the 13th Meeting of the Japan Society of Ultrasonics in Medicine, P. 7, 1968.

10. Kikuchi, Y., Okuyama, D., Ebina, T., Oka, S., and Tanaka, M., Cardiac Kineto-Ultrasono-Tomography, Reports of the Sixth International Congress on Acoustics, Tokyo, M-1-8, 1968, and Reference 8.

11. Asberg, A., Ultrasonic Cinematography of the Living Heart, Ultrasonics, Volume 5, P. 113, 1967.

12. Olofsson, S., An Ultrasonic Optical Mirror System, Acoustica, Volume 13, P. 361, 1963.

13. Kikuchi, Y., Okuyama, D., Kasai, C., Ebina, T., Oka, S., Tanaka, M., Terasawa, Y., and Uchida, R., Ultrasono-cardio-tomography for the Heart and Great Vessels by Means of "Multi-information Recording System" (in Japanese), Reports of the 17th Meeting of the Japan Society of Ultrasonics in Medicine, P. 23, 1970;

    Kikuchi, Y., Okuyama, D., Kasai, C., Ebina, T., Tanaka, M., Terasawa, Y., and Uchida, R., Multi-information Recording and Reproduction in the Ultrasono-cardio-tomography, Acoustical Holography, Volume 4 (edited by G. Wade), Plenum Press, New York-London, P. 113, 1972.

14. Kikuchi, Y., Okuyama, D., Tanaka, M., Ebina, T., and Oka, S., Ultrasono-Tomo-Kymography of the Heart, Ultrasono Graphis Medica, Vol. III, Verlag Der Wiener Medizinischen Akademie, P. 481, 1971.

15. Ibid (Reference 7), P. 34.

COMPUTER PROCESSING OF ULTRASONIC IMAGES

Morio Onoe

Institute of Industrial Science

University of Tokyo, Roppongi, Tokyo, Japan

I. INTRODUCTION

No ultrasonic image system gives an image of perfect quality. There is much room for improvement of its quality by either optical or digital image processing. Advantages of digital processing are accuracy, repeatability and flexibility, whereas, a large memory capacity required to store whole picture elements, and a great length of time required for serial processing are often mentioned as disadvantages. Professors Stroke and Lohmann have demonstrated well the superiority of optical processing over digital processing in speed and parallel processing capabilities.[1] Indeed, digital processing of images has been both expensive and time consuming, but the situation has been changing rapidly in recent years.

First, the cost of digital computation is sharply decreasing because of the advent of larger and faster computers supported by better software. In a table shown by Professor Stroke the acceleration of processing time from two hours to some 20 minutes, even for images consisting of 1,000,000 picture elements, was witnessed.

Second, the high cost of digital image processing is partly due to the fact that input and output equipment for images have been very expensive; but now, integrated circuits and minicomputers have become available as hardware components. They are inexpensive, and they offer a good chance to reduce the cost of input and output equipment. An example will be shown in Section 2 of this paper.

Even after taking these points into account, the digital processing of <u>optical</u> images is still a hard task. This is because picture elements required to represent most optical images occupy a large computer memory and the time required for serial processing of these elements is long. But the task is much easier in the case of <u>ultrasonic</u> images.

Ultrasonic imaging is characterized by long wave length in comparison with optical imaging. Since the information contained in an image is proportional to the ratio of the aperture size of the imaging system to the wavelength, memory capacity required for ultrasonic images is far less than that of optical images. Also, because of the lack of sensors with large apertures and high sensitivity, as well as high quality lenses for ultrasonic imaging, the use of mechanically scanning transducers with or without synthetic aperture techniques, is very common. The speed of scanning should be slow in order to be compatible with the speed of sound, and sometimes, to avoid causing turbulance within a water tank; hence, the scanning time is usually much longer than the processing time.

Furthermore, ultrasonic sensors that respond to both amplitude and phase of field instead of intensity only are available; so ultrasonic fields can be recorded and reconstructed without forming spurious images. Since a spatial carrier used in optical off-axis hologram is not necessary in the case of digital reconstruction of ultrasonic holograms, sampling density, and hence, computer memory and processing time can be reduced accordingly.

These less demanding requirements for memory capacity and processing time make easier than optical imaging the introduction of computer processing into ultrasonic imaging. Once a computer is used, full advantages of various techniques of digital image processing; amplitude manipulation, geometrical manipulation, spatial and transform domain filtering, multi-image operation, etc.; can be taken without additional equipment. Excellent reviews of digital image processing with an extensive list of references are available.[2,3]

Amplitude manipulation is a simple but powerful tool for contrast enhancement of ultrasonic images. Figure 1 shows a few examples. A choice of characteristics is available, even in non-linear manipulations; this is a contrast to optical computation which essentially is a linear process and photographic techniques in which nonlinearity is hard to control. This manipulation is called, "Point Operation," because one, and only one picture element is involved. For filtering, manipulation is extended to neighboring points. For example, a spatial, low pass filter is obtained by taking an average of eight surrounding points and substituting it into the center. With appropriate weighting, various linear or non-linear filters are realized. When the neighborhood is small, the operation is performed by direct convolution; whereas, when the neighborhood becomes large, the filtering is done in the frequency domain after Fourier transform, and the result is again inversely transformed into the spatial domain. Thanks to fast Fourier transform, this indirect method is much faster than the direct convolution if the neighborhood involved becomes, say, more than 32 x 32 picture elements.

This paper presents three examples of computer processing of ultrasonic images. In the first example, the output of a pulsed exho ultrasonic flaw detector is digitized and stored in a computer memory, together with positional information of a scanning transducer. Then the computer processes the data and presents the results on a graphic display. Any cross-sectional view, as well as perspective or sterographic view can be displayed. In the angle beam method for testing welds, true location of defects in a structure can be displayed. Such displays facilitate a thorough evaluation of defects.

Wavefront reconstruction by computer is applied to scanned continuous wave holograms in the second example, and to synthetic aperture side-looking pulsed sonar signals in the third example. Computer reconstruction makes it possible to eliminate such intermediate steps as photo-reduction, and also to incorporate image enhancement techniques. Although the present work is still in progress, preliminary results obtained show promising improvement.

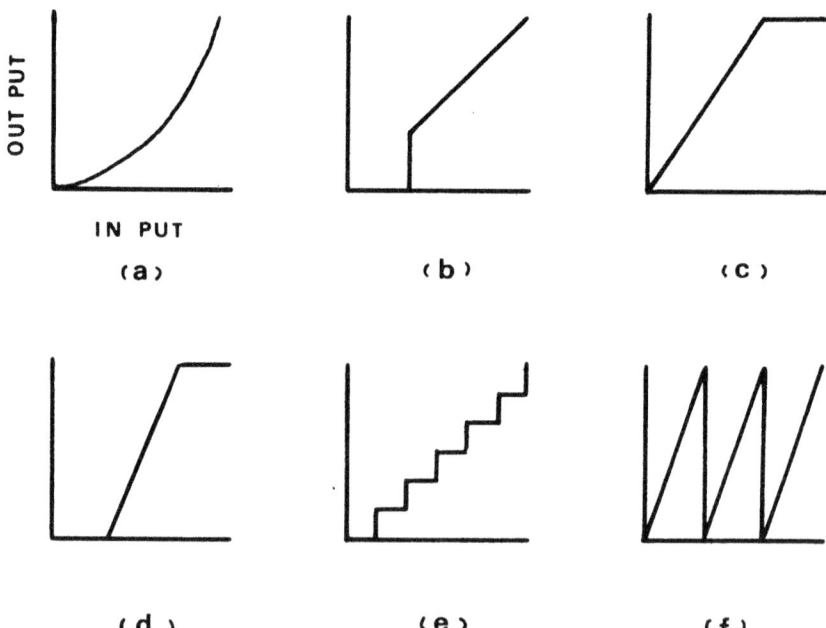

Figure 1. Examples of Amplitude Manipulation
Explanation of curves:
(a) Gamma Correction
(b) Extraction of High Light Area
(c) Enhancement of Low Light Area
(d) Stretch
(e) Quantization
(f) Folding

II. IMAGE PROCESSING SYSTEM

Before getting into details of examples, a brief description of input-output equipment to digitize and display images will be in order. Figure 2 shows a schematic diagram of an image processing laboratory which Dr N. Takagi and the author are now building. In the right column are various input-output equipment, some of them not directly related to ultrasonic images, but included here because they are used in biomedical image processing such as karyotyping of human chromosomes, elsewhere referred to by Professor Atsumi.[4] All equipment is under the control of a minicomputer, shown at the left. The basic philosophy here is to perform almost everything in software and to keep hardware to a bare minimum. This is effective for reduction of cost.

At the top in the righthand column is a mechanical scanner, modified from a facsimile of a rotating drum type, for digitizing a photograph. A rotary encoder is added to ensure the circumferential resolution of more than 5,000 picture elements. Axial translation of the optical head is disconnected from the rotation and stepped by a pulse motor under the control of the computer. This provides more flexibility in scanning formats in a comparison with commercial digitizers. Since the volume of data is very large in photographs, a high speed A/D converter is provided. This is also used for data acquisition in ultrasonic holography.

The second is a modified cathode ray tube capable of half-tone display of 32 gray levels. Since the adjustment is rather critical and direct viewing is not possible, it is rarely used for ultrasonic images requiring fewer levels of gray.

An XY recorder and a digital plotter are conventional and are used mostly for line drawing. A modified facsimile uses wet-type recording paper and yields hard copy without development. Only eight gray levels are available, but are adequate for display of nondestructive testing type data.

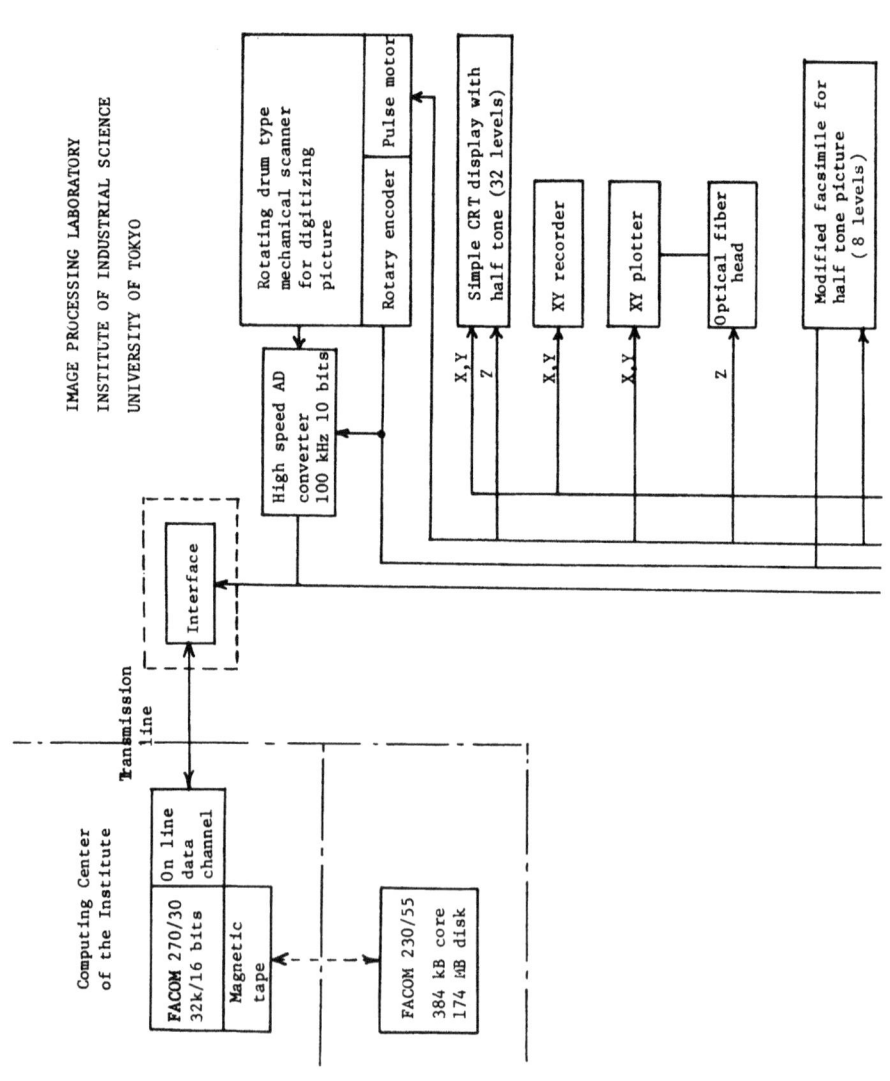

COMPUTER PROCESSING OF ULTRASONIC IMAGES

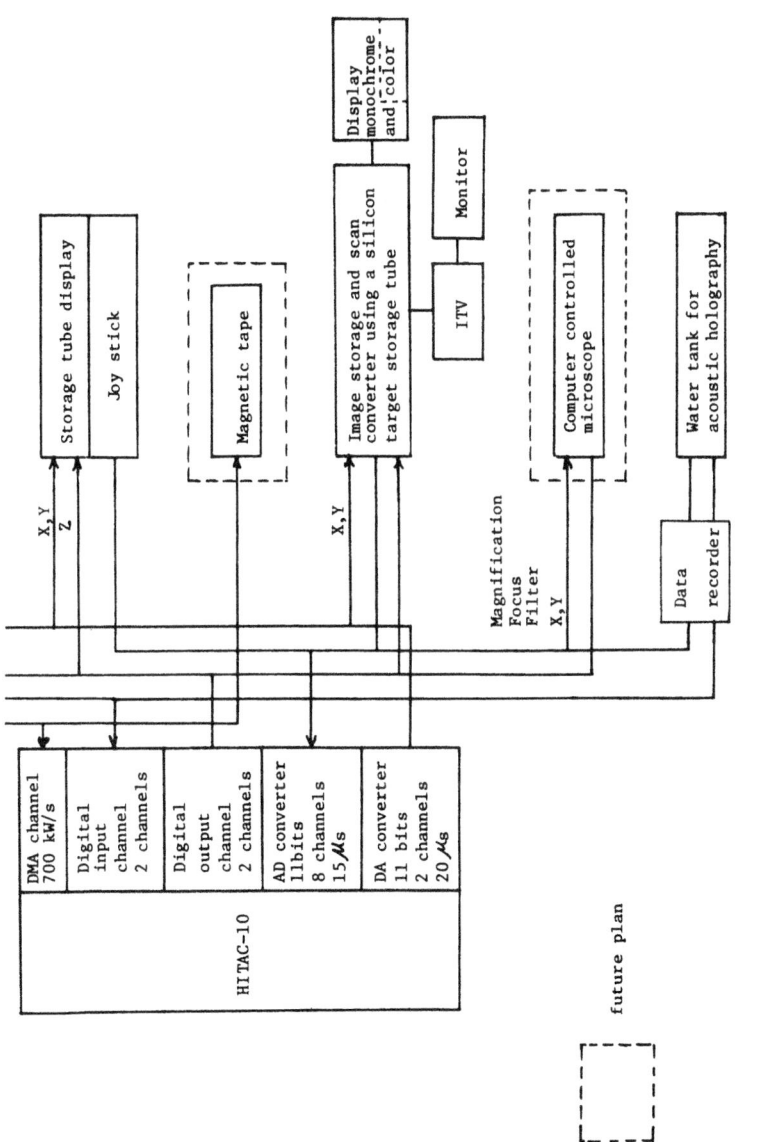

Figure 2

A storage tube display yields a resolution of more than 1,000 x 1,000 picture elements, but only binary gray levels. Since ultrasonic images require less resolution we can trade the resolution into gray levels, which are expressed by the density of dots within 4 x 4, or 8 x 8 matrixes, as shown in Figure 3. In the latter case at least 16 gray levels, as shown in Figure 4, are realized. Figure 5 shows an example of display, admittedly crude as an optical image, but adequate as an ultrasonic image. With a little imagination, a face of a mermaid can be seen. A joystick is provided for man-machine interactive communication through this display.

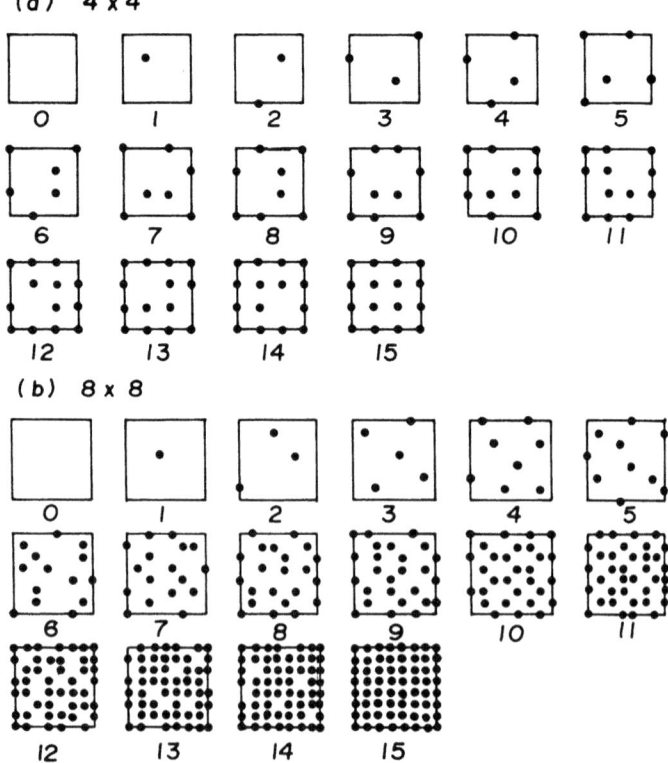

Figure 3. Representation of Gray Levels by Binary Dot Density

A silicon target tube can store one frame of television video image and is used in the following manner:

1. Scan conversion of television video signal

2. Refreshment of memory for displaying an output from the computer on conventional television monitors

3. Temporary analog storage of images

4. Simple analog image processing such as the subtraction and the edge enhancement by defocusing mask technique

The last one is a water tank for acoustical holography and will be discussed in Sections 4 and 5.

Figure 4. 16 Gray Levels Realized by Dot Density

Figure 5. Example of Gray Level Display

III. GRAPHIC DISPLAY FOR PULSED ULTRASONIC TEST

The capability of nondestructive testing techniques depends heavily on the manner of collecting, analyzing and presenting data for defect evaluation. More dependable evaluation calls for more thorough information. It is often desirable to present data in graphical form so the location and the shape of defects within the sample under test can be seen at a glance. Figure 6 shows such a system of pulsed ultrasonic testing which can present a complete image of defects on a graphic display controlled by a computer. Data are collected in the field and sent to a center for processing.

The signal bandwidth of the original A-scan is several hundred Kilohertz, yielding the distance resolution of a few millimeters in steel. Fortunately, the scanning speed of an ultrasonic probe is much slower than the repetition rate of the signal; hence, the signal waveform is essentially repetitive and can be read out at a slow speed after the manner of a sampling oscilloscope. The bandwidth of the output is less than one hundred Hz when the readout time is one second. The output, together with the positional information of the probe, are sent to the center via telephone line or by paper tape or magnetic cassette tape.

At the center, the data are first stored in a memory of a computer. Since the depth information or defects is obtained from A-scan signal, only the two-dimensional scanning of the probe yields the whole three-dimensional information. The computer rearranges the data and presents them on a graphic display in a specified format.

Since details of the system were already reported elsewhere,[5] only a few examples of display will be shown. Figure 7 shows dimensions of a steel specimen, which is the Japanese standard block V 15-4. There is a flat bottom hole of four millimeters in diameter at 15 mm under the surface. A vertical probe of five MHz scans the upper surface and data are taken at every two mm in both X and Y directions. Figure 8 presents perspective views from two different angles.

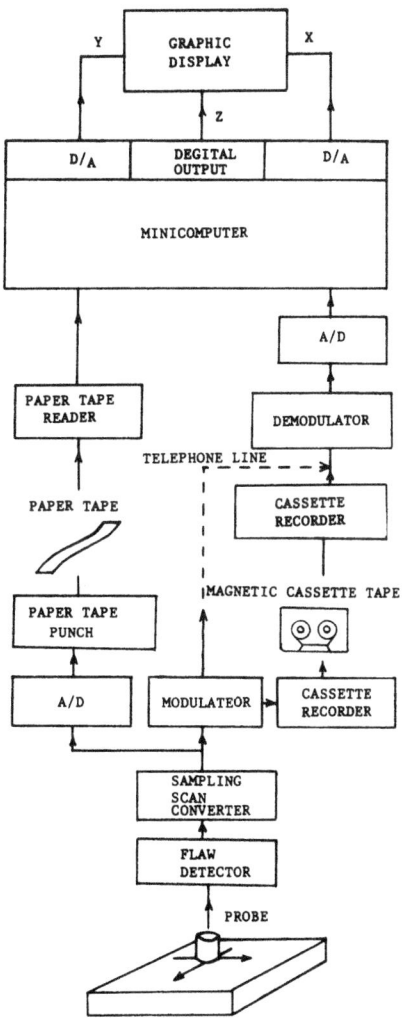

Figure 6. Schematic Diagram of Pulsed Ultrasonic Testing

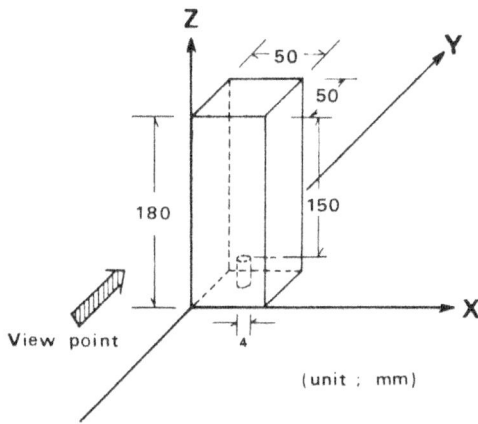

Figure 7. Dimensions of Steel Specimen, STB V 15-4

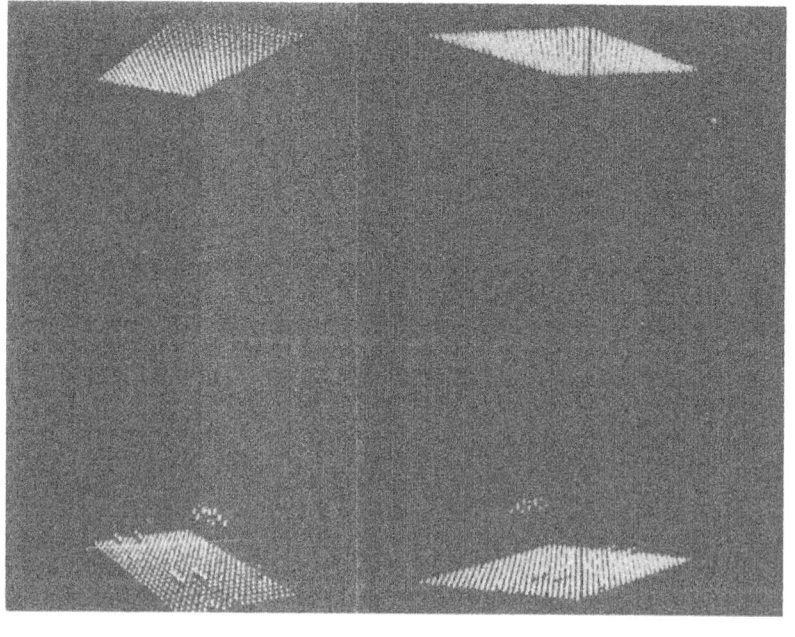

(a) (b)

Figure 8. Perspective Views of a Flaw in the Specimen Shown in Figure 7

A natural flaw called "lamination," in a large steel plate of 30 mm in thickness is shown by stereographic display in Figure 9. A probe of five MHz in frequency and 30 mm in diameter scans the upper surface and data are taken at every 10 mm in both X and Y directions. Figure 10 shows a C-scan display of the same flaw.

Figure 11 shows steel plates, but welded together at the center. The angle beam method is commonly used for detecting defects within the weld. There is a mode conversion from longitudinal wave to shear wave at the interface between the probe and the specimen. Furthermore, the beam reflects at the bottom when the distance between the weld and the probe increases. The computer performs the velocity correction and the geometrical correction and presents a true location of defects in the weld as shown in Figure 12. A probe of five MHz is used.

These examples show that the pulsed ultrasonic method, combined with graphic display controlled by a computer can present a complete image of three-dimensional ultrasonic fields. In the present state of the art, this system has the following advantages over the acoustical holography, using a scanning transducer and digital data processing:

a. A transducer with higher directivity can be used, yielding higher sensitivity and higher signal-to-noise.

b. There are both extensional and shear waves in solid. Reflection and transmission at a boundary are usually accompanied by mode conversion between two types of waves; hence, the ultrasonic field to be recorded by acoustical holography may become complicated when there are boundaries in the specimen. A testing weld with an angle beam probe is an example, whereas, the pulsed ultrasonic method can avoid this trouble, thanks to its high temporal resolution.

Figure 9. Stereographic View of Lamination in a Steel Plate

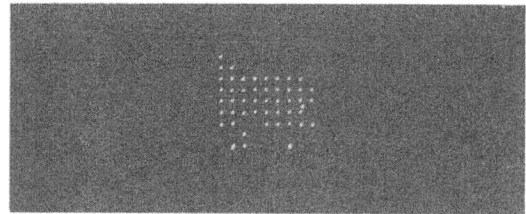

Figure 10. C-scan Display of Lamination Shown in Figure 9

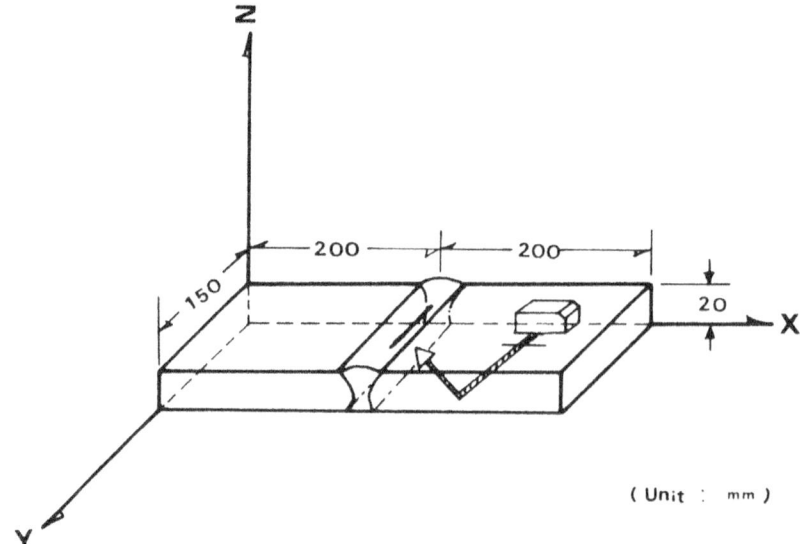

Figure 11. Inspection of Butt Weld by Angle Beam Method

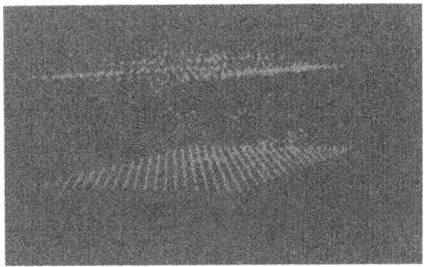

Figure 12. Perspective View of Defects in Weld

c. The data processing in the present system is mostly simple transformation of coordinates and is much faster than the Fourier transform required in the reconstruction of images from a hologram.

d. Pulsed flow detectors with the capability of C-scan have been widely used in the field. They can be expanded into the present system without much modification.

Whereas, at a low frequency range where transducers with good directivity are not available, holographic and synthetic aperture techniques will find much usefulness. Hence, the following two experiments were conducted. Because of the availability of a water tank and electronic equipment, however, experiments were carried out at a higher frequency than were potentially most useful.

IV. RECONSTRUCTION AND IMAGE ENHANCEMENT OF ACOUSTICAL HOLOGRAMS

The experimental system for acoustical holography is similar to the one reported in the literature.[6] Figure 13 shows a schematic diagram of the system. The output of a continuous wave oscillator is divided into two, one of which is fed through a power amplifier to the transmitting transducer in a water tank, as shown in Figure 14. The signal picked up by the receiving transducer is first amplified and then compared in a vector voltmeter with the reference signal derived from the oscillator. Both the amplitude and the phase of the received signal are obtained. They are first recorded in a data recorder, together with the positional information of the receiving transducers, and then played back, ten times faster, to be read into the computer for processing through a high speed A/D converter. This off-line data acquisition scheme is preferred to a direct on-line connection between the computer and the water tank system because of a long scanning time ranging from a fraction of an hour to a few hours, depending upon the area of scan. In the following examples, the transmitter and the object are fixed and the receiver scans two-dimensionally. Other configurations as well as pulsed operations are possible, but are omitted here for the sake of brevity.

The scattered wavefront at the object is reconstructed by the inverse Fresnel transform from a two-dimensional distribution pattern of the amplitude and the phase at the scanning plane. Formulas are available in the literature.[7] Rigorously speaking, this two-dimensional pattern is not a hologram, which is essentially an intensity pattern preserving the phase information in the form of spatial interference. Hence, a new name, "salogram" is proposed after the Greek word σάλοσ meaning vibration and wave. Each salogram consists of a pair of components which are the amplitude part and the phase part or the real (or inphase) part and the imaginary (or quadrature) part as described in Section V.

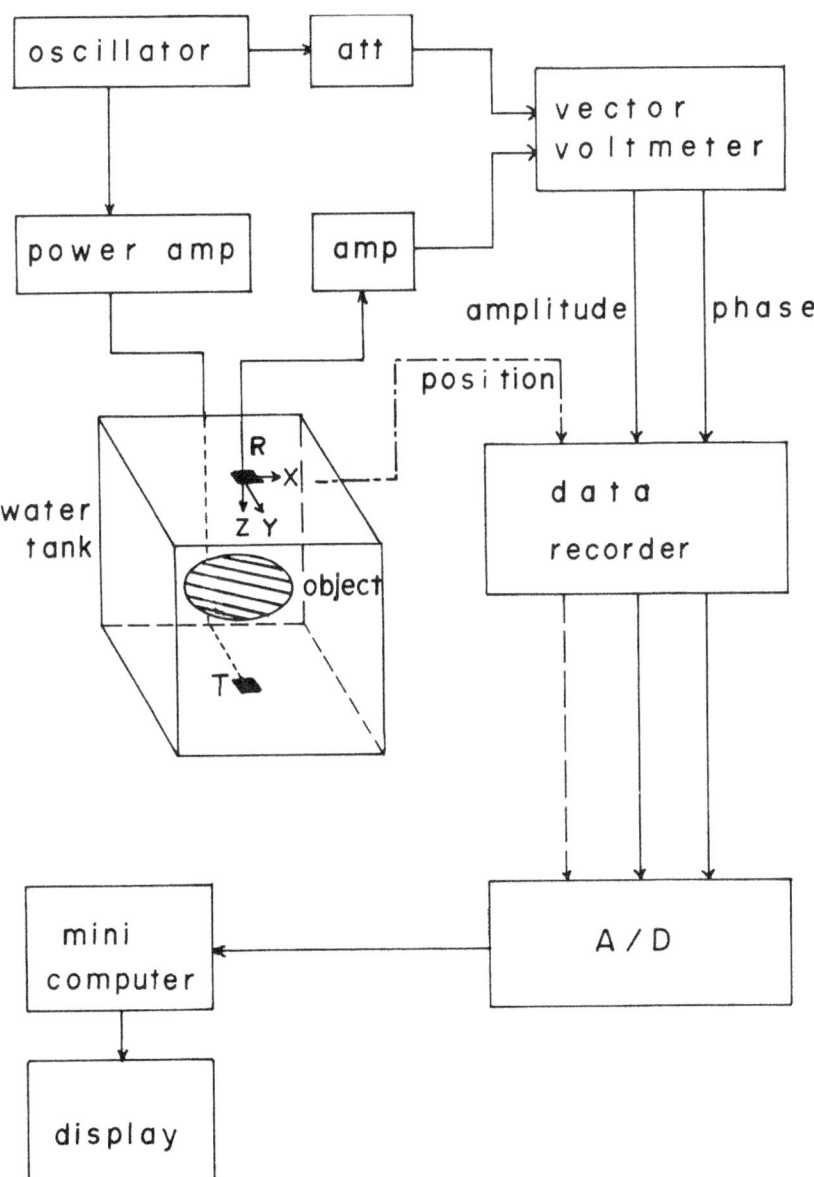

Figure 13. Schematic Diagram of the System for Acoustical Holography

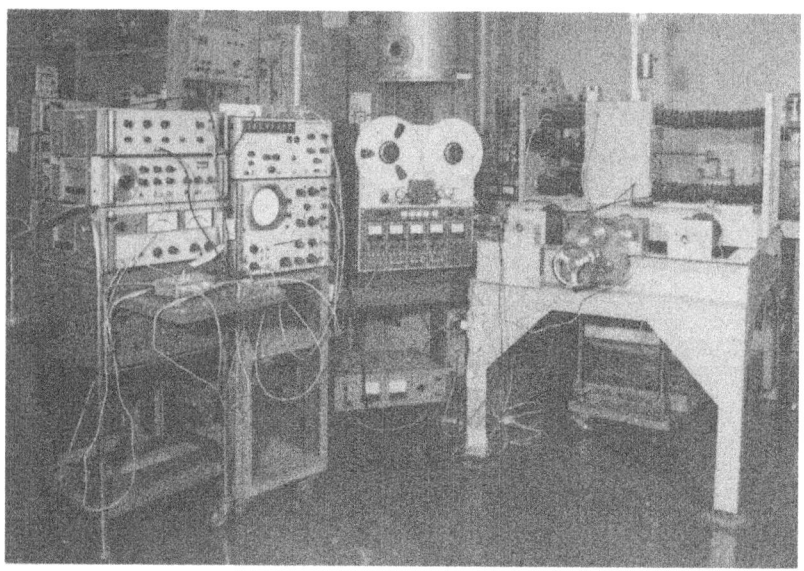

Figure 14. Water Tank and Associated Instruments For Acoustical Holography

Reconstructed images from an in-line hologram are degraded by the conjugate image and the central order, which does not exist in reconstructed images from salograms, while an off-axis hologram requires at least two times more sampling points than a salogram does, because of the use of a high spatial carrier; hence, salograms are more suitable than holograms for wavefront reconstruction when sensors responding to both amplitude and phase are available.

As a check of the system, a photograph of an opening cross window shown in Figure 15 is digitized by the mechanical scanner. The sound field is calculated by Fresnel transform, assuming this cross window is back illuminated by a point source 200 millimeters apart from the window. Figure 16 shows a salogram at a scanning plane 195 mm apart from the window. The number of sampling is 64x64.

Figure 15. Photograph of an Opening Cross Window

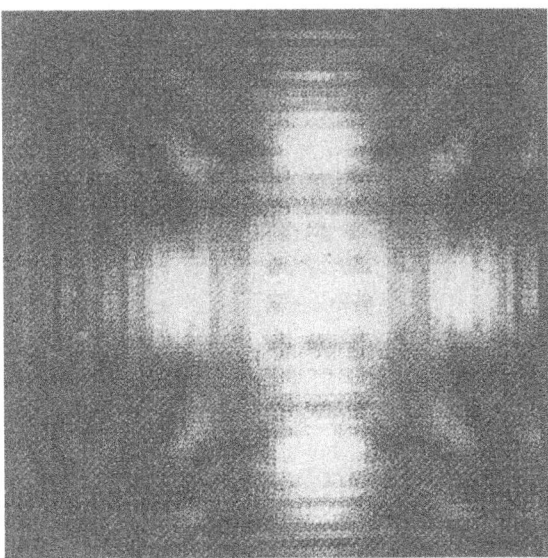

Figure 16a. Computer Generated Salogram; Amplitude

Figure 16b. Computer Generated Salogram; Phase

An image reconstructed from the salogram is shown in Figure 17. The reconstruction seems perfect, indicating that the sampling density and the quantization are adequate for the present purpose.

Various causes of degradation of images can be simulated in the computer. Figure 18 shows an out-of-focus reconstruction at a plane 10 centimeters in front of the object.

The real part of a salogram is equivalent to an in-line hologram. Figure 19 shows an image from only the real part, obtained by the conventional holographic reconstruction technique. Degradation due to conjugate image, and low contrast due to the central order are apparent.

COMPUTER PROCESSING OF ULTRASONIC IMAGES

Figure 17. Reconstructed Image from the Salogram Shown in Figure 16

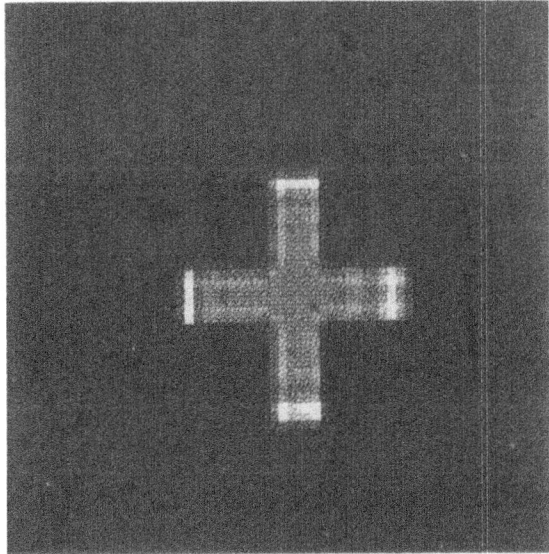

Figure 18. Degradation of Reconstructed Image Due to Defocusing by 10 Centimeters

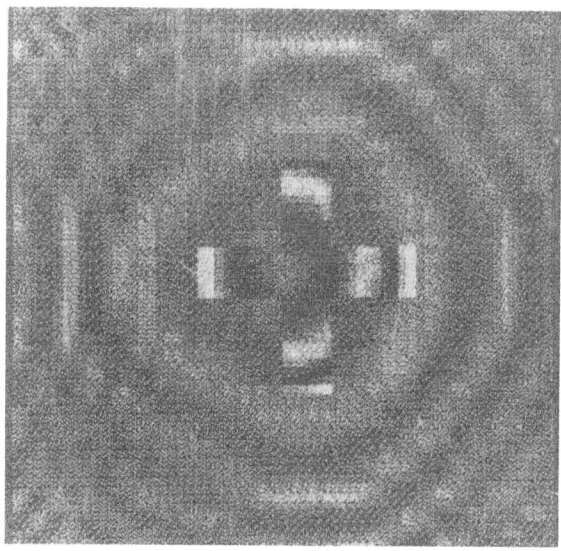

Figure 19. Degradation of Image Reconstructed from In-line Hologram Due to Conjugate Image and the Central Order

In the following, images taken in the water tank will be shown. The transmitter and the receiver are piezoelectric ceramic transducers of five millimeters in diameter. The frequency used is about one Megahertz.

A salogram without an object, at a plane 418 mm apart from the transmitter, is shown in Figure 20. Samples 64 x 64 in size are taken at every two mm interval. Weak illumination at the left side is due probably to a little tilt of the transmitter or its asymmetry in construction.

An aluminum sheet two mm in thickness, with a cross opening at the center as shown in Figure 21, is inserted in the half point between the transmitter and the scanning plane.

COMPUTER PROCESSING OF ULTRASONIC IMAGES

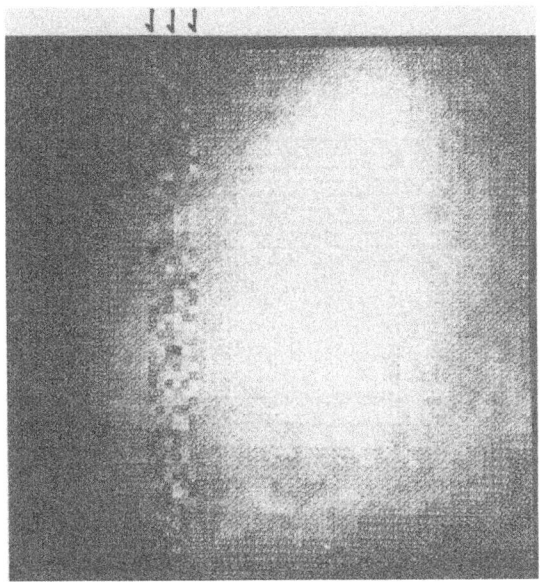

Figure 20a. Salogram for Field Without Object; Amplitude (Portion under arrows is to be disregarded because of malfunction of display)

Figure 20b. Salogram for Field Without Object; Phase

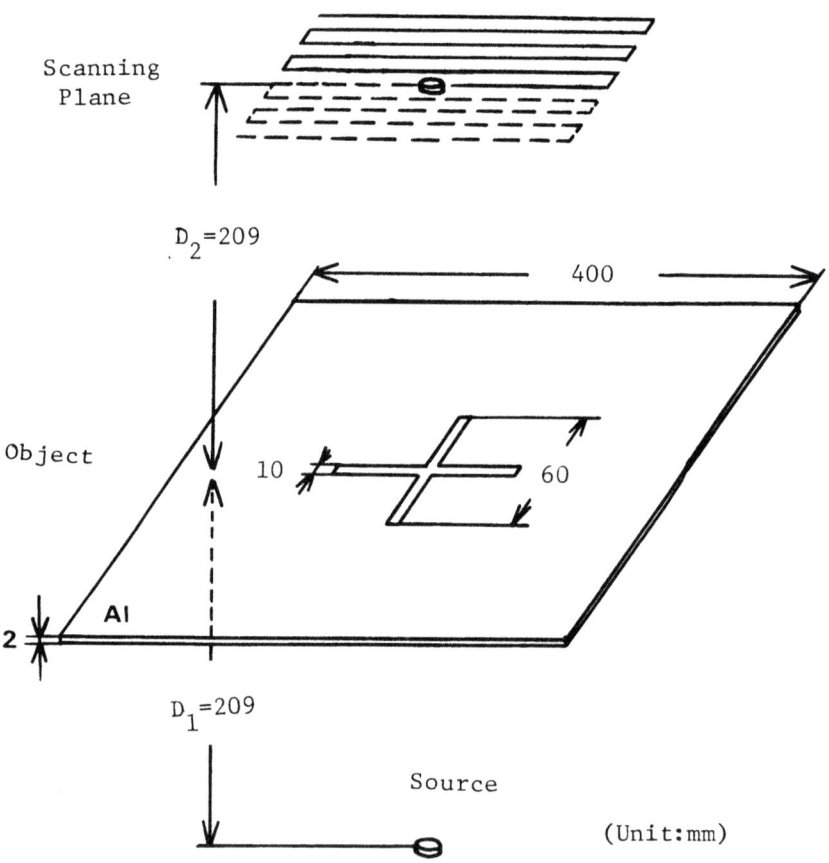

Figure 21. Arrangement of Aluminum Sheet with a Cross Opening in Water Tank

COMPUTER PROCESSING OF ULTRASONIC IMAGES

A salogram shown in Figure 22 is obtained which corresponds to the simulated salogram shown in Figure 16. Conventional holographic reconstruction from only the real part yields an image as shown in Figure 23, which corresponds to the simulated one shown in Figure 19; whereas, the reconstruction from both the amplitude and the phase parts yields an image as shown in Figure 24. The left side is scarecely seen because of the lack of illumination already mentioned. Reconstruction from the phase part only, with an assumption of uniform amplitude, yields the image shown in Figure 25. Now the left side is visible, but the overall quality of the image is not good. When the sound field correction based on the data in Figure 20 is applied to the amplitude part, a reconstructed image as shown in Figure 26 is obtained. A slight improvement over the image shown in Figure 24, at the left side, can be seen. Furthermore, the amplitude stretch manipulation, as shown in Figure 1d is applied to images as in Figure 24 and Figure 26, yielding Figures 27 and 28, respectively. Now the effect of sound field correction becomes more apparent, as in Figure 28.

Next, a T-shaped object made of aluminum sheet, one millimeter in thickness as shown in Figure 29, is placed as an object. The distances from the transmitter and the scanning plane are 213 mm and 189 mm, respectively. Figure 30 shows a salogram from which an image as in Figure 31 is reconstructed. Sound field correction yields Figure 32, and amplitude stretch manipulation yields Figure 33; whereas, the combination of both operations yields Figure 34 in which the ends of the T-bar are clearly seen.

As a last example, three letters, IIS, made of aluminum sheet of one mm in thickness are placed at different distances from the transmitter as shown in Figure 35. A salogram shown in Figure 36 is obtained. Figures 37, 38, and 39 show reconstructed images at planes which are 160, 200 and 240mm, respectively, apart from the scanning plane. Figures 40, 41 and 42 are similar images after amplitude stretch. Now it is easier to see that the focus is on each letter.

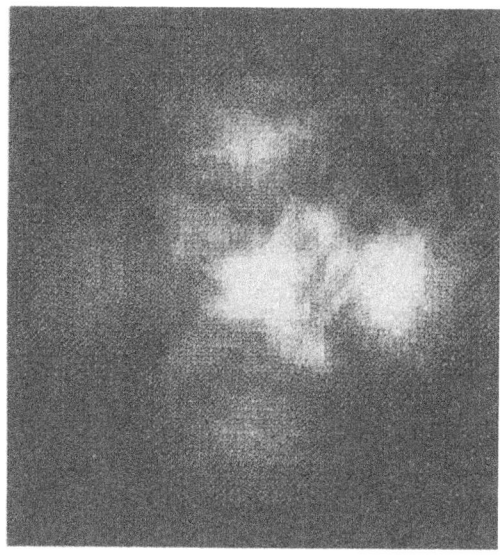

Figure 22a. Salogram of Sheet with Cross Opening; Amplitude

Figure 22b. Salogram of Sheet with Cross Opening; Phase

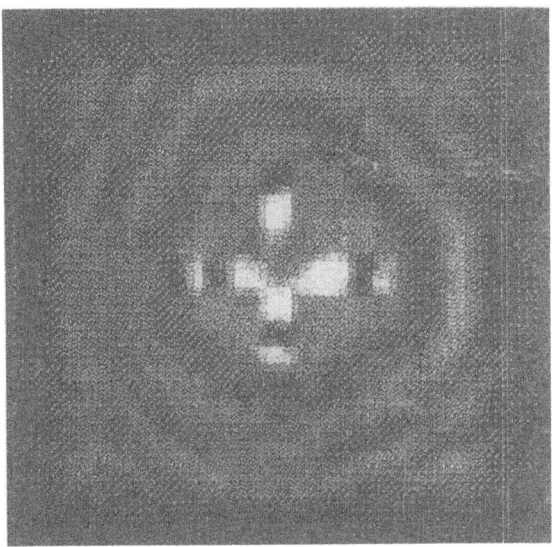

Figure 23. Degraded Image Reconstructed from only Real Part of Salogram, shown in Figure 22

Figure 24. Reconstructed Image from Salogram Shown in Figure 22

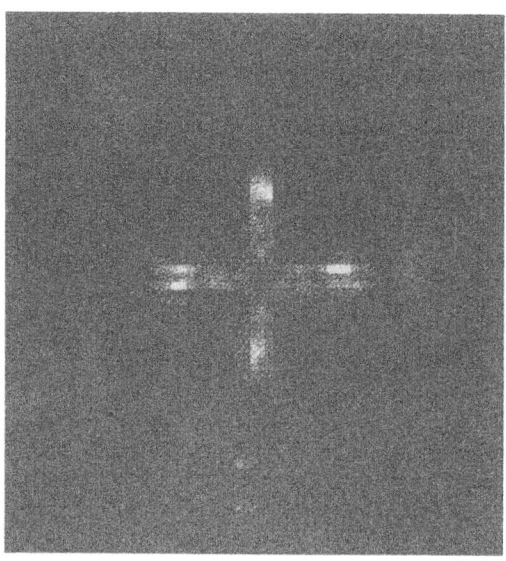

Figure 25. Image Reconstructed from only Phase Part of Salogram in Figure 22b

Figure 26. Enhanced Image of Figure 24 After Sound Field Correction

COMPUTER PROCESSING OF ULTRASONIC IMAGES 483

Figure 27. Enhanced Image of Figure 24 After Amplitude Stretch Manipulation

Figure 28. Enhanced Image of Figure 24 after Both Sound Field Correction and Amplitude Stretch Manipulation

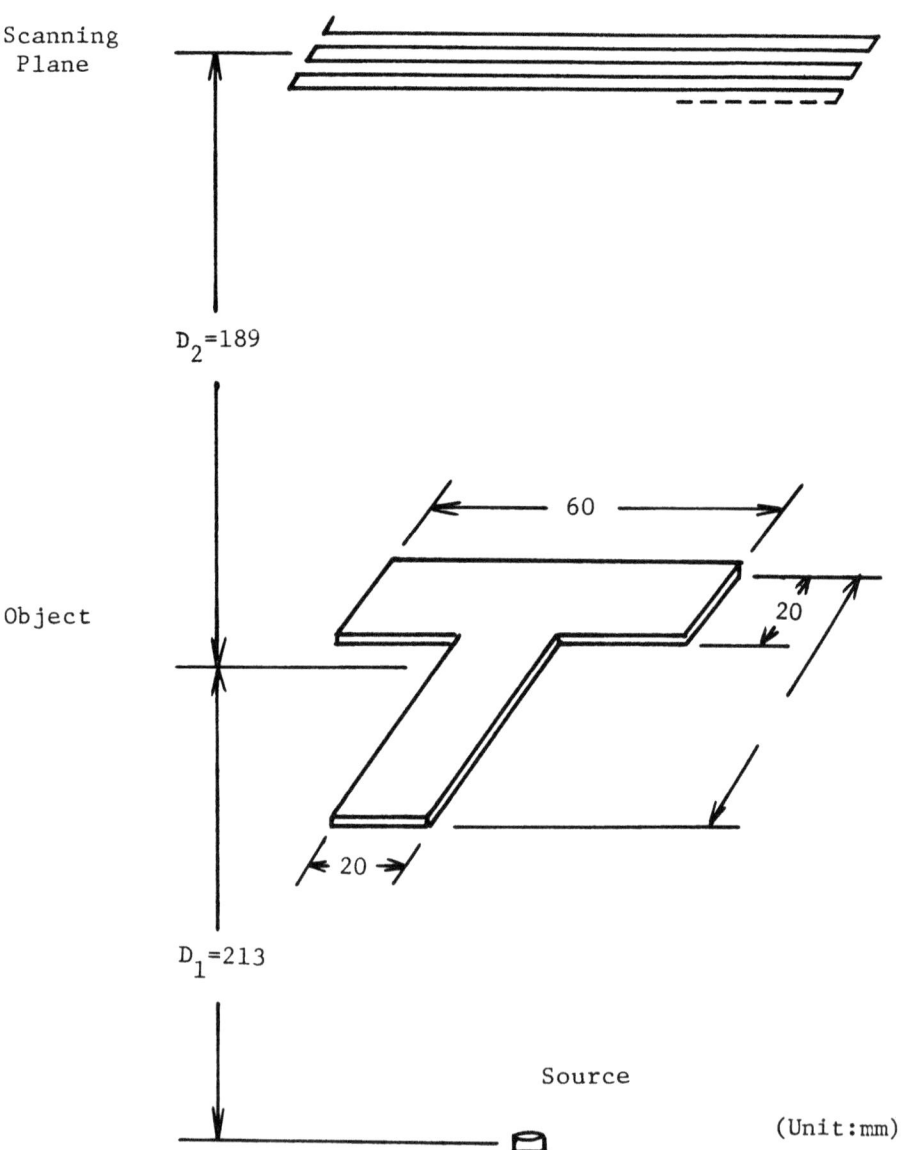

Figure 29. Arrangement of Letter "T" in Water Tank

COMPUTER PROCESSING OF ULTRASONIC IMAGES

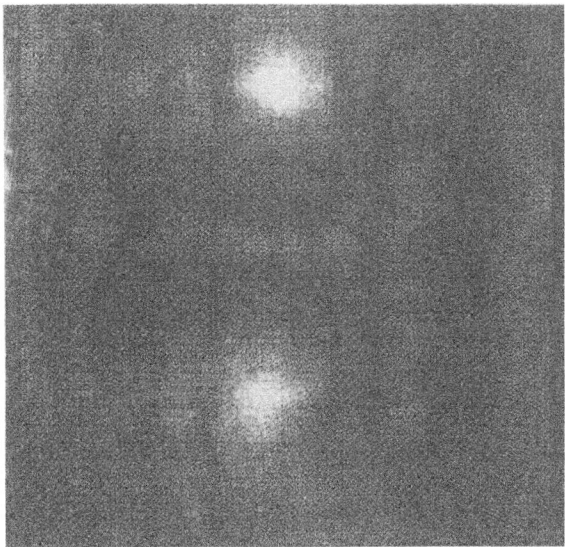

Figure 30a. Salogram of Letter "T"; Amplitude

Figure 30b. Salogram of Letter "T"; Phase

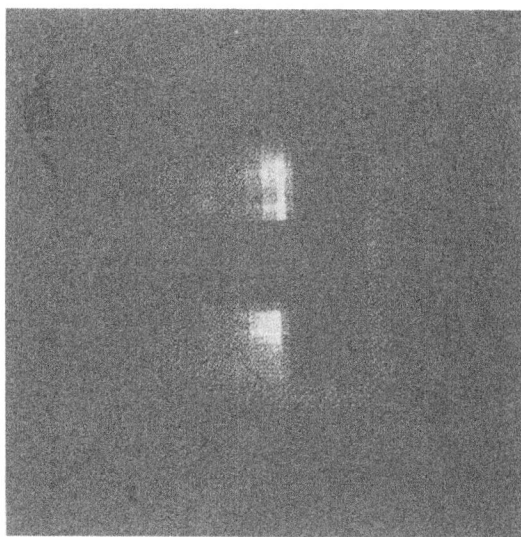

Figure 31. Reconstructed Image from Salogram, Figure 30

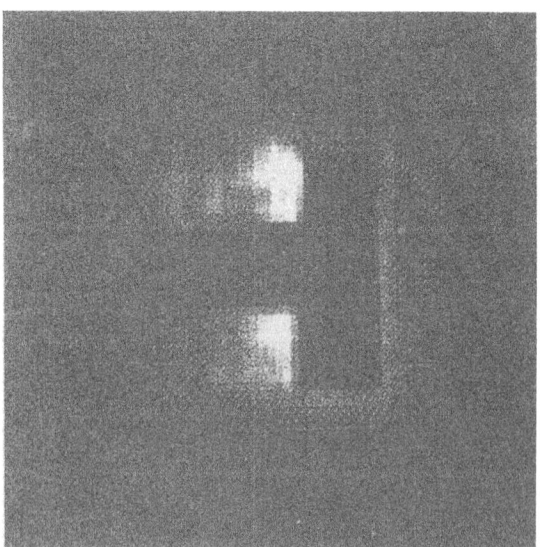

Figure 32. Enhanced Image of Figure 31 After Sound Field Correction

COMPUTER PROCESSING OF ULTRASONIC IMAGES

Figure 33. Enhanced Image of Figure 31 after Amplitude Stretch Manipulation

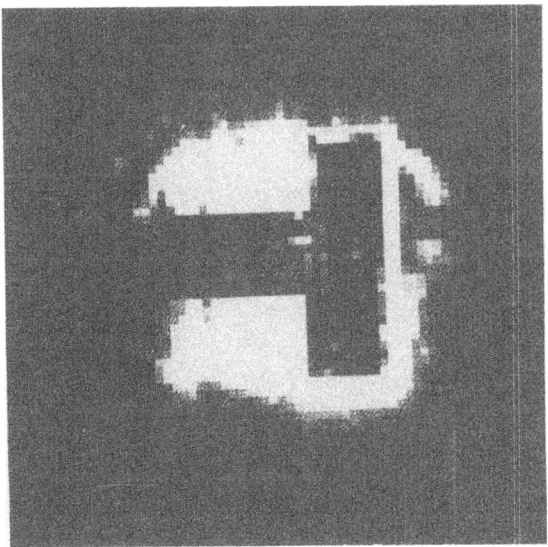

Figure 34. Enhanced Image of Figure 31 after both Sound Field Corrention and Amplitude Stretch Manipulation

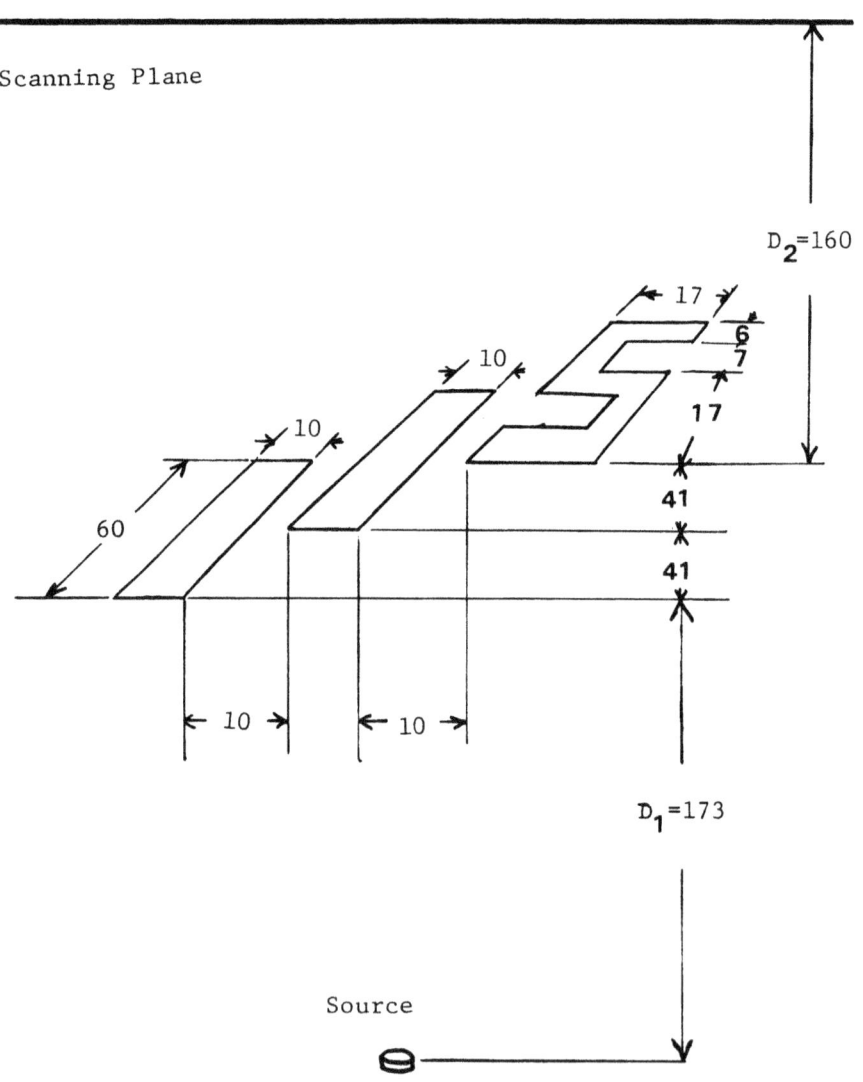

Figure 35. Arrangement of Letters "IIS" in Water Tank

COMPUTER PROCESSING OF ULTRASONIC IMAGES

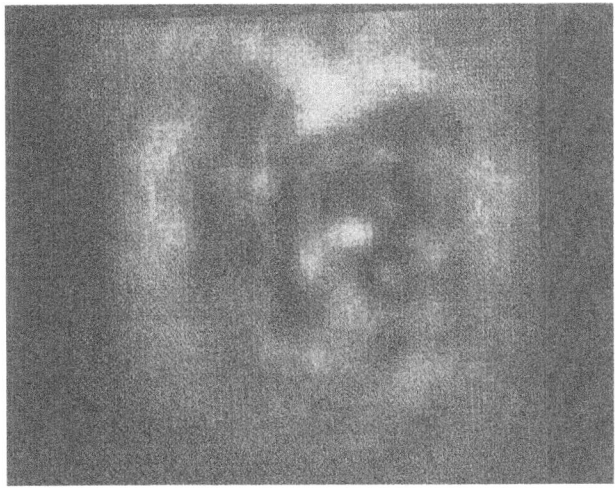

Figure 36a. Salogram of Letters "IIS"; Amplitude

Figure 36b. Salogram of Letters "IIS"; Phase

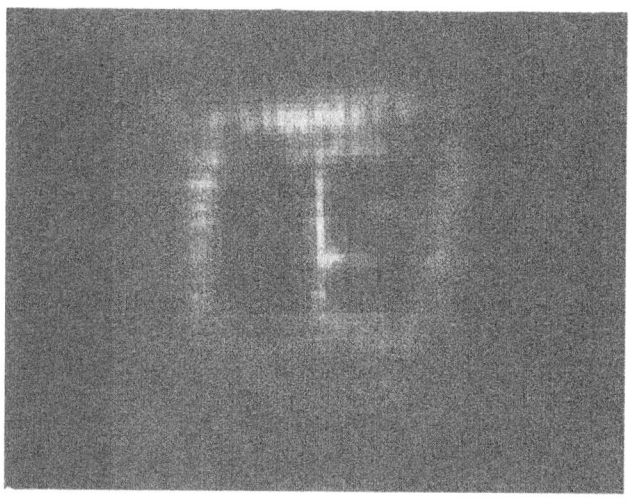

Figure 37. Reconstructed Image at 160 mm from Scanning Plane

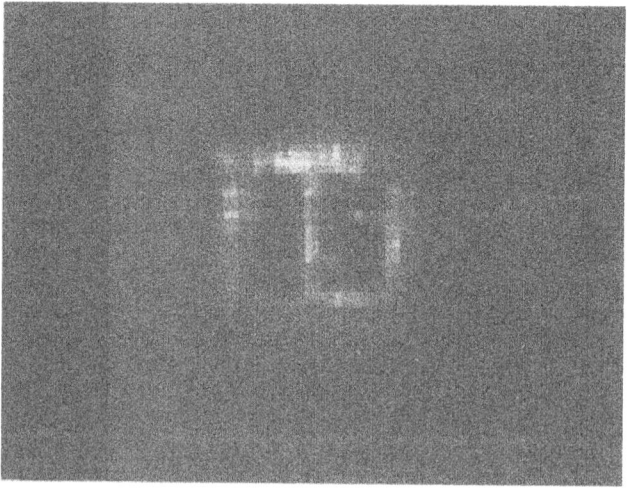

Figure 38. Reconstructed Image at 200 mm from Scanning Plane

COMPUTER PROCESSING OF ULTRASONIC IMAGES 491

Figure 39. Reconstructed Image at 240 mm from Scanning Plane

Figure 40. Enhanced Image of Figure 37 after Amplitude Stretch Manipulation

Figure 41. Enhanced Image of Figure 38 after Amplitude Stretch Manipulation

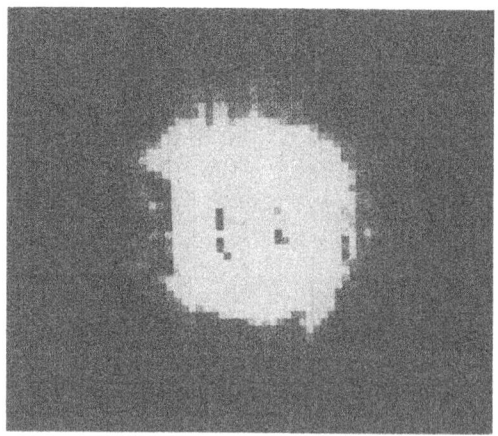

Figure 42. Enhanced Image of Figure 39. after Amplitude Stretch Manipulation

These examples show that digital reconstruction of ultrasonic salograms, combined with image enhancement techniques, is quite promising. Applications of more sophisticated filtering techniques are planned so better images can be obtained.

V. RECONSTRUCTION OF SYNTHETIC APERTURE SIDE-LOOKING PULSED SONAR SIGNALS

The success of microwave side-looking radars have prompted us to apply the same synthetic aperture technique to sonars. In radars, fine resolution images are reconstructed from stored video data by a coherent optical processor.[8] Because of a large volume of data, no other processing technique would match the optical technique in speed and resolution. One difficulty in the optical processing is the need for a very special conical lens which is difficult to manufacture. In the case of sonars, however, there is much room for the introduction of digital processing because of smaller volume of data and longer acquisition time as mentioned in the introduction; hence, the following experiment was conducted to see the feasibility of digital processing.

A schematic diagram of the experimental system is shown in Figure 43, and is similar to the system for acoustical holography shown in Figure 13. The present system, however, operates in pulsed mode and uses a pair of synchronous detectors and a scan converter instead of the vector voltmeter.

The output of a continuous wave oscillator was divided into three parts, one of which was gated by a pulse generator to form repetitive modulated RF pulses, then fed into the transmitting transducer in a water tank after power amplification. The other two were fed to synchronous detectors as references; one directly and another after the phase shift of 90 degrees. A receiving transducer next to the transmitter picked up echo returns from objects. Its output was amplified and fed to synchronous detectors which yielded the inphase (real) part and the quadrature (imaginary) part. The transmitter and the receiver moved together, along a straight path; i.e., X-axis. The speed of motion was constant and slow; hence, the outputs of synchronous detectors are nearly repetitive and can be scan converted after the manner of oscilloscope sampling. The outputs of the scan

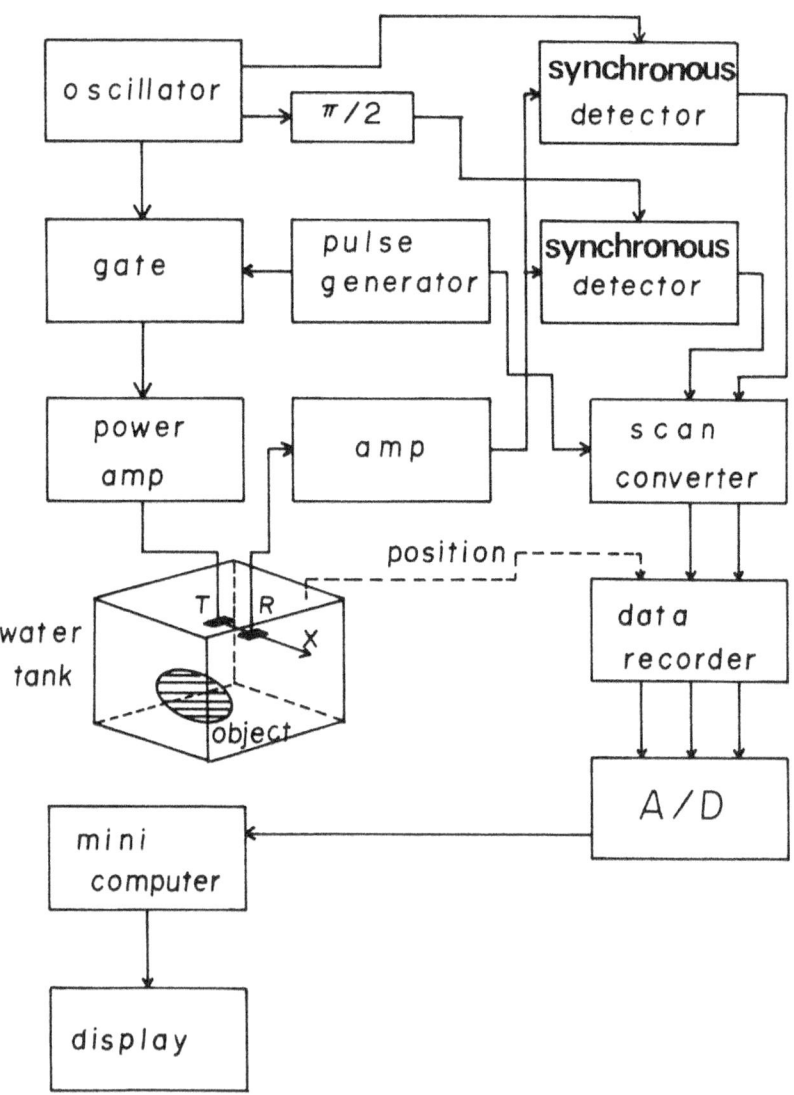

Figure 43. Schematic Diagram of the Experimental System for Side-Looking Sonar

converter were first recorded by a data recorder, together with positional information of transducers, and then played back, to be read into the computer.

In the case of radar, only one synchronous detector was used and its output modulated the intensity of the spot of a cathode ray tube for photographic recording. The dimension across the film represents range and the along the film corresponds to the position of the transmitter and the receiver. The range information was derived through pulsing as in the case of conventional radars, whereas, the azimuth information was obtained by processing data along a line of each range interval, which constituted one dimensional hologram.[9] The reference signal to the synchronous detector was frequency shifted to leave a carrier frequency in the output, making the recording an off-axis hologram, so that the degradation due to conjugate image is avoided at the reconstruction.

In the case of the present sonar, the recording is a salogram with the real and imaginary parts, and can be reconstructed without conjugate image. No frequency shifting of the reference is needed.

A letter "K" made of aluminum, one millimeter thick, was supported by two wires of one millimeter in diameter in the water tank as shown in Figure 44. Shallow indentations were made by a drill all over the surface in order to avoid a specular reflection of obliquely incident ultrasonic beam. Figure 45 is its photograph. Thirty-two samples at every four microseconds along the range, and 64 samples at every two millimeters along the track were taken. Figure 46 shows a salogram obtained. The upper is the real part and the lower is the imaginary part.

Reconstructed images are shown in Figure 47. The upper is the amplitude display and the lower is the intensity display. Supporting wires are just barely seen in the upper, but are weakened by a squaring operation in the lower. Figure 48 shows the enhanced images after the

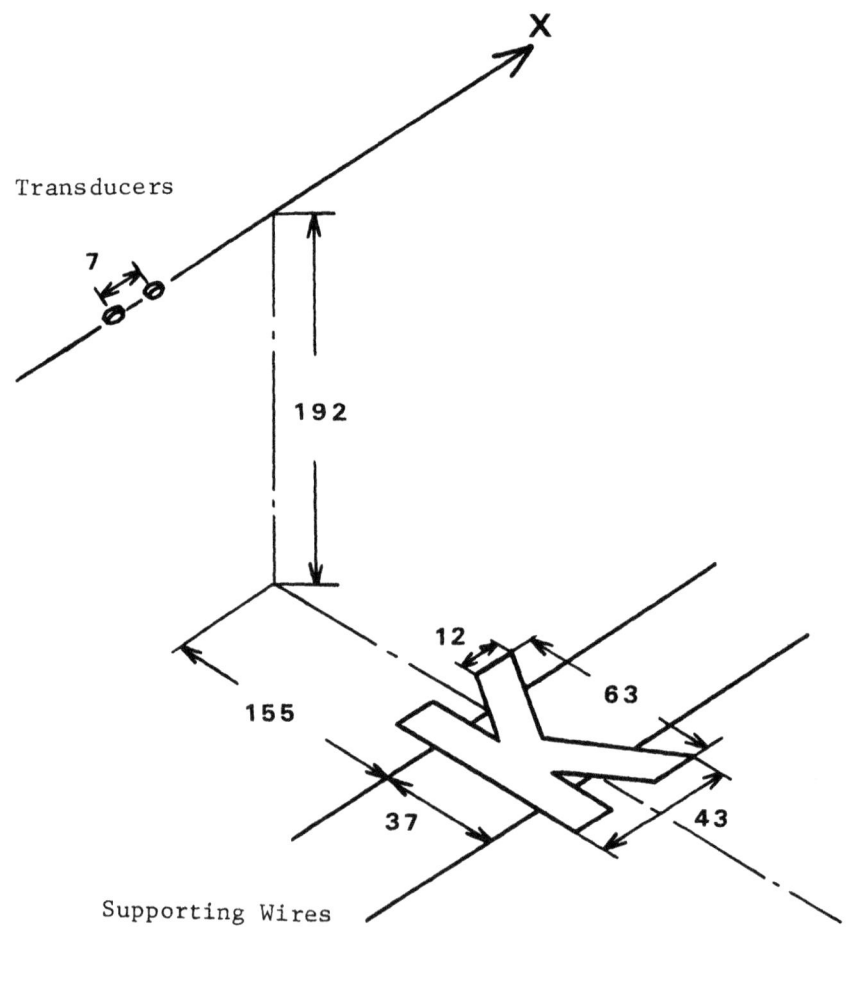

Figure 44. Arrangement for Making Side-Looking Salogram in Water Tank

COMPUTER PROCESSING OF ULTRASONIC IMAGES

Figure 45. Photograph of the Letter "K" with Indented Surface

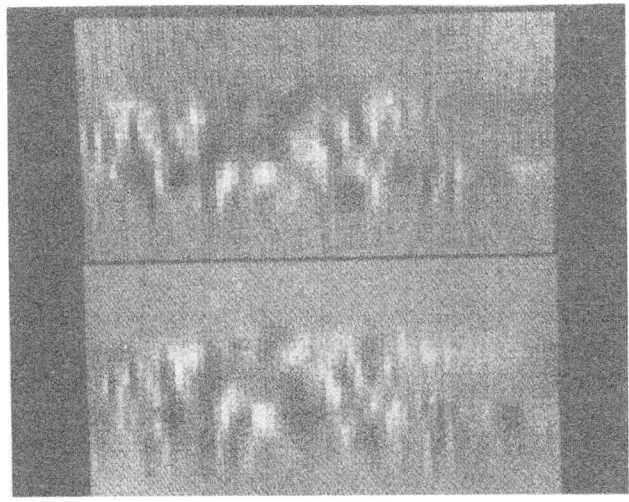

Figure 46. Side-Looking Salogram of Letter "K"
Upper: Real Part
Lower: Imaginary Part

amplitude stretch manipulation shown in Figure 1d. The upper and lower correspond to the same in the previous figure. In the upper, both the letter and the wires are clearly seen. Since the vertical scale is proportional to the range, these images are really projected ones on a tilted plane. The geometrical correction yields the true image on the ground plane shown in Figure 49. Figure 50 shows another image which is obtained after logarithmical enhancement of the contrast, amplitude stretch manipulation, geometrical correction, and interpolation between picture elements. Some improvement at the upper and lower right corner is noticed.

Fringes are visible on reconstructed images of supporting wires. They are artifacts related to discrete transform. Since wires are parallel to the track of transducers, the return signal is uniform over the scanning aperture. Sampling density becomes insufficient for the reconstruction of such an extended salogram by inverse Fresnel transform. Details will be reported elsewhere.

These experiments show that the digital reconstruction and the image enhancement are useful for processing synthetic aperture side-looking sonar signals.

VI. CONCLUSIONS

It has been shown that the computer image processing is especially suitable for ultrasonic images, which require less computer memory and tolerate longer processing time in comparison with optical images. As an example, a graphic display is applied to a pulsed echo ultrasonic flaw detector to realize flexible presentation of defects for thorough evaluation.

The wavefront reconstruction by computer from salograms, which are two-dimensional distribution of wave vector components at a scanning plane, is not degraded by

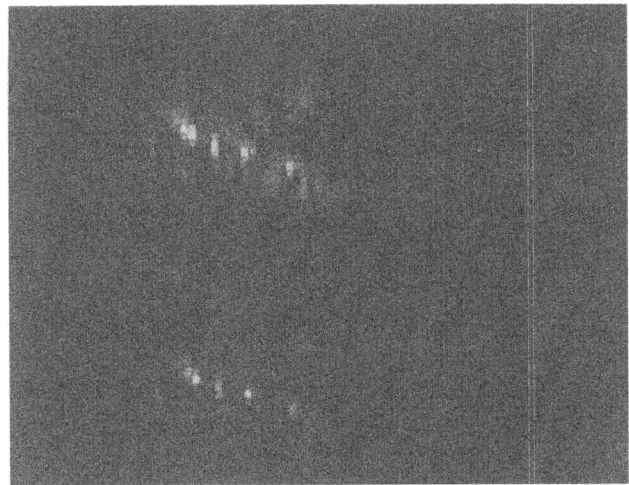

Figure 47. Reconstructed Images
Upper: Amplitude Representation
Lower: Intensity Representation

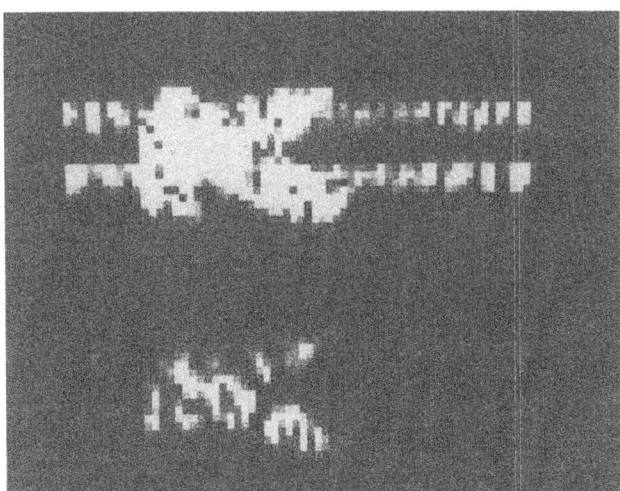

Figure 48. Enhanced Images after Amplitude Stretch
Manipulation
Upper: Amplitude; Lower: Intensity

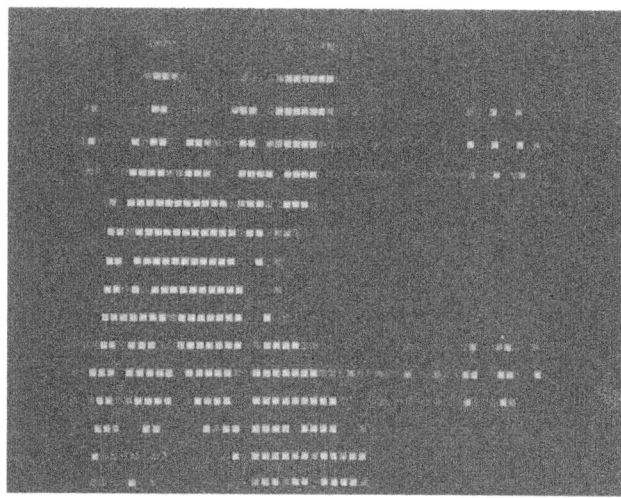

Figure 49. Geometrically Corrected Image of Figure 48

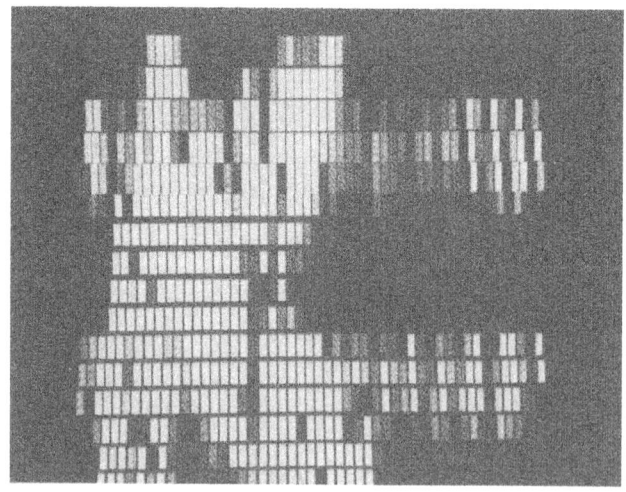

Figure 50. Image after Logarithmical Contrast Enhancement, Amplitude Stretch Manipulation, Geometrical Correction and Interpolation between Picture Elements

conjugate images and requires no spatial carrier. Various image enhancement techniques are easily incorporated with the reconstruction process. As examples, scanned continuous wave salograms and synthetic aperture side-looking sonar signals are reconstructed and enhanced. Results obtained have been very encouraging.

ACKNOWLEDGEMENTS

Many persons have contributed to the work presented here. Special thanks are due to Dr M. Takagi for his cooperation in building the image processing laboratory; to Mr N. Hamano, for the operation of various equipment; and to Mr T. Ishikawa, for the reconstruction of salograms.

REFERENCES

1. Stroke, G. W., Optical Computing, IEEE Spectrum, 24-41, Dec., 1972; Stroke, G. W., Optical Image Improvement in Biomedical Electron Microscopy and Ultrasonics; and Lohmann, A. W., The Aspects of Optical Holography That Might Be of Interest for Acoustical Imaging, U.S.-Japan Seminar on Ultrasonic Imaging, Hawaii, January, 1973, Plenum Press.

2. Huang, T. S., Schreiber, W. F. and Tretiak, O. J., Image Processing, Proc. IEEE, Vol. 59, PP 1586-1609, Nov. 1971.

3. Andrews, H. C., Tescher, A. G. and Kruger, R. P., Image Processing by Digital Computer, IEEE Spectrum, PP 20-32, July, 1972.

4. Atsumi, K., Image Processing in Bio-Medical Engineering, U.S.-Japan Seminar on Ultrasonic Imaging, Hawaii, January, 1973, Plenum Press.

5. Onoe, M., Takagi, M., Matsumoto, T., and Hamano, N., Graphic Display for Ultrasonic Nondestructive Testing, Acoustical Holography, Vol. 4, PP 299-315, Plenum Press, New York, 1971.

6. Boyer, A. L., Jordan, J. A., Van Rooy, D. L., Hirsch, P. M., and Lesem, L. B., Computer Reconstruction of Images from Ultrasonic Holograms, Acoustical Holography, Vol. 2, PP 211-223, Plenum Press, N. Y., 1970.

7. Ibid, P. 215.

8. Cutrona, L. J., Leith, E. N., Porcello, L. J. and Vivian, W. E., On the Application of Coherent Optical Processing Techniques to Synthetic-Aperture Radar, Proc. IEEE, Vol. 54, PP 1026-1032, August, 1966.

9. Kock, W. E., Side-Looking Radar, Holography and Doppler Free Coherent Radar, Proc. IEEE, Vol. 56, PP 238-239, February, 1968.

OPTICAL IMAGE IMPROVEMENT
IN BIOMEDICAL ELECTRON MICROSCOPY
AND ULTRASONICS*

George W. Stroke and Maurice Halioua

State University of New York

Stony Brook, New York

ABSTRACT

The imaging improvement method recently proposed by the authors for transmission electron microscopy in the general case has now been successfully demonstrated with electron micrographs of a bacteriophage virus and of carbon-foil test specimens originally photographed with heretofore unsurmountable blurring. The original electron micrographs were recorded on two commercial high resolution electron microscopes. The theoretical limit of diffraction attainable with the microscopes used has been obtained by removal of artifacts, of contrast inversions, and of amplitude attenuations in the transfer function resulting from the unavoidable aberrations and the use of the needed defocusing phase contrast.

---

* A major advance in electron microscopy has been achieved since the time of the January 1973 meeting in Hawaii. Because of its particular relevance to ultrasonic imaging, this work, the fruit of a long and continuing collaborative effort of the authors, is presented for the first time here in the state of perfection and practicality attained by the end of July, 1973.

The improved electron micrographs are characterized by a dramatically visible improvement in the overall resolution, in addition to the objectively verified improvement using the Thon diffractograms and computer calculations.

The method may be extended to applications in a number of comparable image improvement problems ranging from engineering, astronomy and materials science among others, to biology and medicine including X-ray and gamma-ray radiography, thermography and ultrasonic imaging.

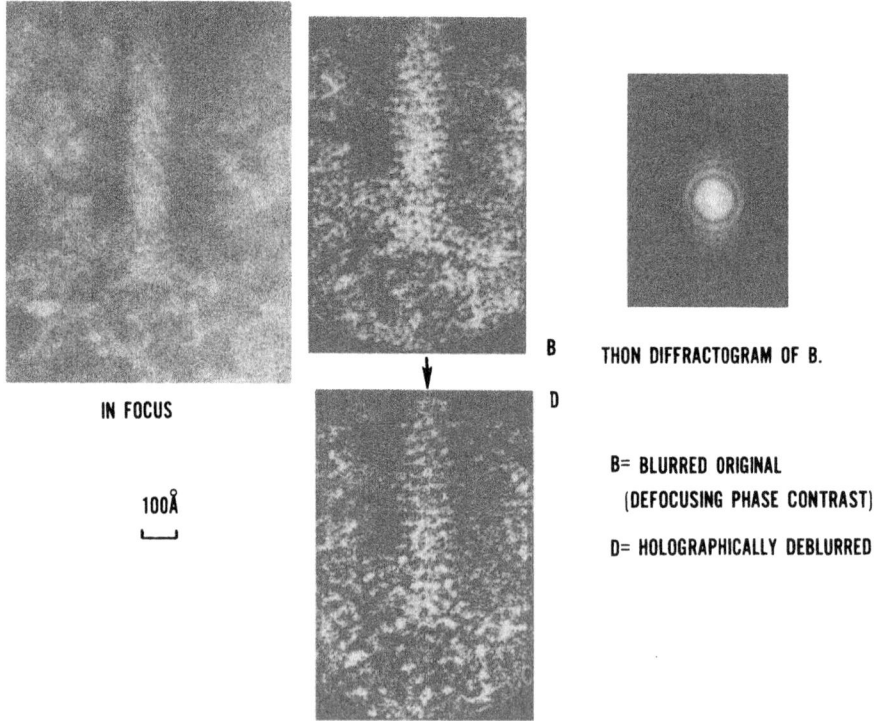

Figure 1. Improvement of Electron Micrograph of T2 Bacteriophage Virus (Recorded in High-Resolution "Coherent Mode" Electron Microscope (se text). Note the clearly evident 34 Angstrom period helical structure of the viral DNA in the sharpened image.

I. IMPROVEMENT OF ELECTRON MICROGRAPH OF T2 BACTERIOPHAGE VIRUS

"A posteriori" holographic image deblurring of electron micrographs can provide a general solution to the improvement of electron microscopes in "real-world" cases of practical importance.[1,2,3] The imaging improvement demonstrated by the holographic method appears to be presently unattainable by any other means; e.g., as reviewed by Hanszen[4] and Thon.[5]

Our most recent results are shown in Figures 1 to 6. Theoretical details are given in References 1, 2, 3, 8, 9, 12, 13 and 18.

Figure 1 shows the holographic improvement of an electron micrograph (B) of a T2 bacteriophage virus recorded in defocusing phase contrast by Professor G. B. Haydon, of Stanford University Medical School, on a Philips EM200 electron microscope and deblurred by us in Stony Brook. The holographically deblurred electron micrograph (D) shown here for the first time is the result of the final refinement of our method of which an early example was given in Reference 6. Figure 1 also shows the poorly contrasted "in focus" photograph, in illustration of the fact that the specimen appears essentially as a "phase object", notably in its high resolution details; this photograph also illustrates the necessity of defocusing the microscope in order to reveal the otherwise invisible object information by the method commonly known as "defocusing phase contrast" (Figure 1B). It should be clear that visible images are obtained in electron microscopy under such conditions at the price of an unavoidable "blurring."

Strictly speaking, the images produced by the electron microscope under such conditions are not immediately interpretable, especially because the image is degraded throughout its entire spatial frequency domain and not just at its extreme limit, by the combination of the well-known spherical (and other) aberrations, with the aberration resulting from the needed defocusing. One could easily think that such blurred images are unimprovable. In fact, the high resolution image (D) is encoded in the blurred photograph (B) in a decodable convolution-integral form.[1-3] The code, needed for the deblurring of the electron micrograph, is, in fact, also contained in the original blurred electron micrograph in a readily extractable form, using the method originated by F. Thon[7] as we proposed in Reference 3. The code (impulse response function in B) is extracted in the form of the transfer function (spatial Fourier transform of the impulse response function[8,9] by very easy experimental means. The modulus of the transfer function is made to appear by simple Fourier transformation of the micrograph, using a laser beam to illuminate the micrograph and recording the resulting image in the focal plane of a lens (the "Thon diffractogram" as shown in Figure 1). The sign of the all-real transfer function (+ and - regions, corresponding to the contrast inversions) is determined with the aid of a library of computed transfer functions and simple measurement of the zero-crossings in the diffractogram, taking into account also (in a small number of special cases) approximate recording conditions in the electron microscope (wavelength, spherical aberration constant and approximate amount of defocusing).

The deblurring is carried out by means of an optical (holographic) computer shown schematically in Figure 2, and in its experimental implementation as set up in our laboratory, in Figure 3.

Historically, it may be of interest to recall at this point that the spatial filtering concepts as used may be traced at least as far back as the work of Abbe, around 1873, and that of R. W. Wood, around the turn of the century. In its modern version, image deblurring for incoherent-mode images was first proposed and demonstrated by A. Maréchal in 1953[10] and for coherent-mode and partially coherent-mode images by Stroke and Halioua in 1972 and 1973,[2,3] notably as needed for the long-sought solution of the electron microscope improvement as first defined by Gabor in

1948.[11] The use of holographic filters for image deblurring was first proposed and demonstrated by Stroke in 1966 (for example, see Reference 12).

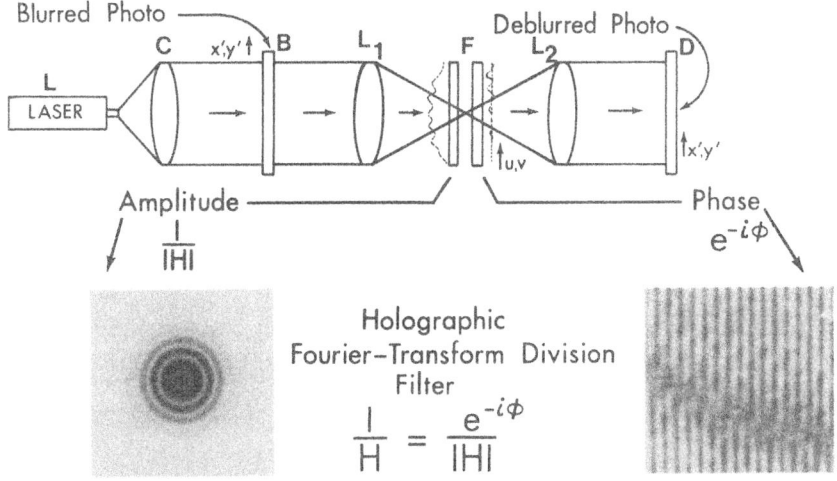

Figure 2.   Optical (Holographic) Image-Deblurring Computer (Schematic Diagram)

C=collimator lens; B=blurred photograph (transparency); L_1=Fourier Transforming Lens; F=holographic image-deblurring filter; L_2=Fourier Transforming Lens; D=deblurred photograph

Note: The photographically illustrated "amplitude component" of the filter is a greatly enlarged photograph of an actual filter component used for the improvement of a test electron micrograph. The holographic phase component is illustrated with a greatly enlarged portion of an actual holographic filter near the zero crossing between two rings, clearly indicating the 180° phase shift, as needed for the removal of the corresponding contrast inversions in the blurred photograph; this photograph is from the filter used for the deblurring of the fd bacteriophage virus of Figure 5.

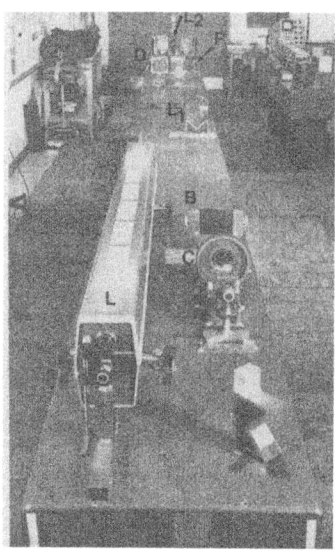

Figure 3.   Optical (Holographic) Image-Deblurring Computer (as assembled according to Figure 2, in Stony Brook)

It may be shown[6] that the best modern electron microscopes such as that used to obtain the micrograph of the bacteriophage virus of Figure 1 operate in completely coherent-mode illumination practice, notably because of the use of high-intensity, point-like sources and of small condenser apertures. The lack of in-focus contrast, the need for defocusing phase contrast and the possibility of deblurring the electron micrographs according to the linear approximation theory, which is applicable as defined in Reference 3, all result from the use of ultra-thin specimens (say, under 100 Angstroms) and the use of high voltages (say in the 100 kilovolt range and over). For details, see, for example, Reference 3 and its extensive bibliography.

The very considerable difference between the decoded (deblurred) micrograph (D) and the original (B) of Figure 1 becomes more believable perhaps by reference to the deblurred electron micrographs, improved with less complete degree such as the partially improved electron micrographs

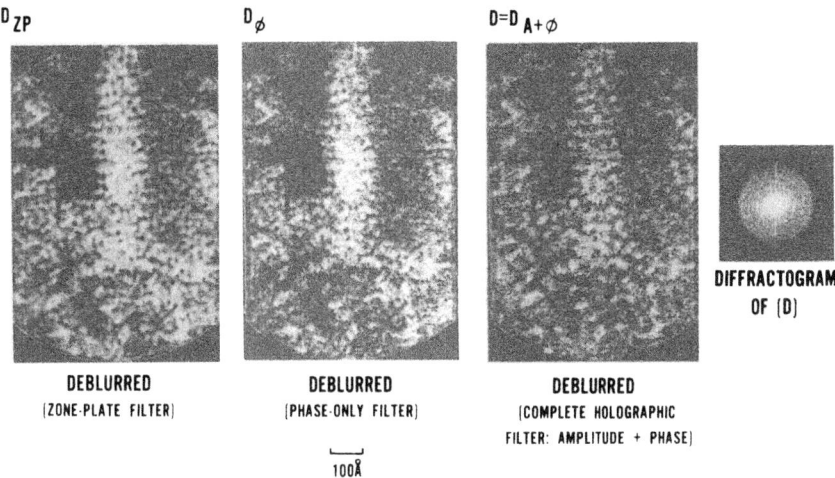

Figure 4. Improvement of Electron Micrograph of T2 Bacteriophage Virus; showing considerably greater improvement by use of the complete holographic (amplitude and phase) filter (Image (D) with corresponding diffractogram) in comparison to partially improved Images D_{ZP} and $D_\emptyset$, obtained by use of zone plate, phase-only filters (see text).

shown in Figure 4. In this figure we also show, together with the deblurred image of the virus (D), its diffractogram which demonstrates clearly the considerable improvement in the spatial frequency transfer function (as compared with the transfer function of the blurred virus electron micrograph of Figure 1); notably, the equalization of amplitudes and the long-sought important reduction of "gaps" (zeros of the original transfer function) essentially to their minimum limit as set by noise threshold. The deblurred image, $D_\emptyset$, of Figure 4 was extracted from the original (B) of Figure 1 by using only the phase component of the holographic filter (see Figure 2) which removed the contrast inversions but left unequalized the amplitude in the transfer function according to the partial method independently investigated by Hahn[14]; albeit using an evaporated Tsujiuchi[15] rather than the holographic filter.

The other deblurred electron micrograph in Figure 4, described by D_{ZP}, was obtained by using a Hoppe zone-plate binary (black and white) masking filter according to Thon and Siegel[16] which removes the negative (-) regions in the transfer function. The two partial methods of Figure 4 clearly illustrate that the complete electron micrograph deblurring method of Stroke and Halioua[3], using the complete (amplitude and phase) filter, produces by far the best improvements in practice, and this without any removal of any regions in the transfer function as is the case with the partial filters; i.e., zone-plate filter (removal of negative frequency regions from spectrum) or phase-only filter (loss of frequency bands around zero gaps by leaving uncompensated the amplitude in the filtering). For example, the diffractogram of phase-only filtered image, $D_{\emptyset}$ is identical to the diffractogram of the blurred electron micrograph (B) of Figure 1, and the diffractogram of the zone-plate filtered image, D_{ZP} has removed from it the entire first bright (negative frequency) ring of the diffractogram (B).

II. IMPROVEMENT OF ELECTRON MICROGRAPHS OF CARBON-FOIL TEST SPECIMENS*

The result of a collaborative project of the authors, with F. Thon and D. Willasch, and using the new, high-resolution SIEMENS electron microscope ELMISKOP 102 is being prepared in its final form for its joint, full scientific publication, with complete theoretical and experimental details[13]. The brief reference to this work here is given only in illustration of work under joint completion and is similar to the report given on behalf of the authors (Stroke and Halioua, Stony Brook; and Willasch and Thon, Siemens, Berlin) by Professor Dennis Gabor, Nobel Laureate, in his public lecture at the Carl Friedrich Siemens Stiftung in Munich, 15 June 1973. On this occasion, Professor Gabor showed that the authors had succeeded in dramatically visible improvement in the image of a thin, carbon-foil test specimen from the original electron micrograph recorded in defocusing phase contrast, and that the corresponding Thon diffractograms demonstrated the corresponding considerable improvement in the spatial frequency transfer function.*

III. IMPROVEMENT OF ELECTRON MICROGRAPHS RECORDED IN SCANNING TRANSMISSION ELECTRON MICROSCOPES

In Figure 5 we reproduce for completeness, the improvement of an electron micrograph of an fd filamentous bacteriophage virus that resulted from a collaborative effort of the authors with A. V. Crewe and J. Wall[1], together with the corresponding optical test photographs, blurred (optically) and deblurred to the same degree. The original electron micrograph was recorded on the five Angstrom, high resolution scanning transmission electron microscope of A. V. Crewe at the University of Chicago, and the deblurring was carried out in Stony Brook in the same apparatus (Figure 3), according to the same arrangement (Figure 2) as that used for the deblurring of the electron micrographs from the transmission electron microscopes described above, albeit with a different filter of course.[1] Two remarks are in order. First, it may be shown[1] that electron micrographs recorded in the scanning electron microscopes, under conditions such as those of Figure 5 may be considered as essentially complete, incoherent-mode photographs. Whereas, such imaging is linear,[1,8,12] and therefore permits deblurring deconvolution in the general case (without restrictions on object contrast), there exist as yet no ready implementations in practice for extracting the needed impulse response function from the photographs (notably, electron micrographs) comparable to the method we described above for the case of coherent-mode electron micrographs. Next, it follows that the impulse-response function must be derived by a number of other practical means such as those which we have developed[1,9] (see also the next section, with reference to Figure 6). In the case

---

* Private report of Professor Gabor's lecture by Dr D. Willasch, Siemens, to the authors in Stony Brook, on 6 August 1973.

of Figure 5, the impulse response function was simulated optically, based on detailed electron microscope operating data provided us by A. V. Crewe.

IV. IMPROVEMENT OF ACCIDENTALLY BLURRED ARCHEOLOGICAL PHOTOGRAPHS OF "SHEKEL" COINS[*]

Figure 6 shows the considerable degree of improvement in the readability of the characters in the images of the approximately 23 mm diameter coins from a private Vatican and Jordan collection, as seen in the "holographically deblurred" photographs. The originals, recorded by C. Colbert, were accidentally blurred by mechanical imperfection in the setting of the focus of the 35 mm Leica camera used in the Middle East, and apparently there was no further possibility of returning for a repetition of the original photography. The impulse-response function, needed for the deblurring shown in Figure 6, was obtained in this case by very careful experimentation, using the original Leica camera and some plaster replicas of comparable coins to produce a comparably blurred photograph; and then, with the same degree of defocusing, an impulse-response function with the aid of a point (white light) source placed into the plane of the plaster-replica object. Finally, the holographic image-deblurring filter (for use in the arrangement of Figure 2) was produced according to our method of Reference 1, of which clarifying details (in the process of preparation for our complete publication[18] were given in Reference 9.

V. CONCLUSION

We may add in conclusion that the practical realization of the crucial holographic image-deblurring filters from

---

[*] This work forms part of a collaborative effort of the authors and their collaborator, Morimasa Shinoda of NIPPON KOGAKU K. K., with Professor Charles Colbert of Wright State University, Radiological Research Laboratory, Dayton, Ohio. It is being prepared for joint publication.[17]

OPTICAL IMAGE IMPROVEMENT 513

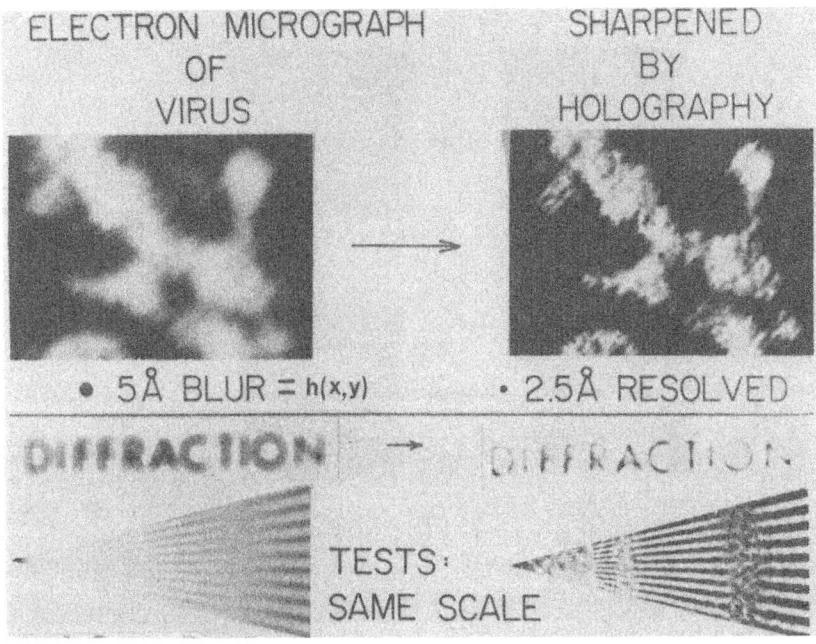

Figure 5.   Improvement of Electron Micrograph of fd
Filamentous Bacteriophage Test Specimen
(recorded in high-resolution "incoherent
mode" five Angstrom scanning transmission
electron microscope) Top, Together with
Corresponding Optical-Test Photograph
Deblurring; to indicate considerable degree
of improvement. Note, for example, the
remarkable removal of the image-degrading
contrast inversions responsible for well-
known artifacts, among other causes; from
References 1 and 9 (see text)

Note: This figure was first prepared for presentation by
Professor Dennis Gabor in his Nobel Lecture on
the occasion of the award to him of the Nobel
Prize in Physics, 10 December 1971, in Stockholm.
It is included there as Figure 33, in the published
version of his lecture under the title, Holography
1948-1971, PP 169-201, in Les Prix Nobel en
1971, Imprimerie Royale P. A. Norstedt & Söner,
Stockholm, 1972.

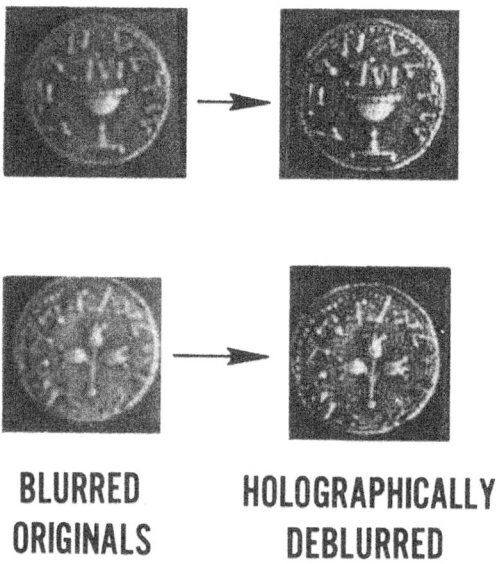

Figure 6. Holographic Deblurring of Archeological Shekel-Coin Specimens, Accidentally Blurred in Ordinary Incoherent-Mode, 35 mm Camera Photography (approximate diameter of coins = 23 mm. From Reference 17 (see text)

the impulse-response (respectively, transfer function) has now been shown to be possible in all cases in a matter of a day's work (using the great library of sensitometric data developed over the course of years in our laboratory). Moreover, the use of the optical (holographic) image-deblurring computer may be learned very easily, even by non-specialized users, literally in a matter of minutes without prior experience. Theoretical and experimental details are fully given in the references 1, 2, 3, 8, 9, 12 and 13.

Specifically, for the case of incoherent imaging, the filters are realized directly from the intensity impulse-response function by our previously described photo-holographic method.[1,9,19]

For the case of coherent and partially coherent imaging, the filters are made directly from the transfer functions; i.e., from the Thon diffractograms with the aid of the additional pre-computed data as described above, according to the detailed method given in Reference 3; this by a combination of the photo-mechanical arrangement and computer calculations, for the amplitude component of the filter, and interferometric synthesis of the holographic phase component described in Reference 3.

ACKNOWLEDGEMENTS

In addition to the acknowledgements of the collaborative efforts of Doctors F. Thon and Willasch, SIEMENS, AG, Berlin; Morimasa Shinoda, NIPPON KOGAKU, K.K.; and Professor C. Colbert, for parts of the work mentioned in the text, as well as of Professor G. B. Haydon and Dr A. J. Saffir, we wish expecially also to thank again, Professor Dennis Gabor, Nobel Laureate, for his continued kind interest in this work and for his many fruitful suggestions. We wish also to thank the National Science Foundation (Grant K-033-820-X01), and the National Aeronautics and Space Administration (Grant NGR-33-230-002) for their essential and most generous support without which none of this work would have been possible. A special acknowledgement is also due to the outstanding participation of our graduate student, Srinivasan.

REFERENCES

1. Stroke, G.W., and Halioua, M., Optik, 35, 1972, 50-65.

2. Ibid, PP 489-508.

3. Stroke, G.W., and Halioua, M., Optik, 37, 1973, PP 192-203 and PP 249-264.

4. Hanszen, K.J., Advances in Optical and Electron Microscopy, Editors, R. Barer and V. E. Cosslet, Academic Press, 1971, Vol. 4, 1-84.

5. Thon, F., Electron Microscopy in Material Science, Editor, U. Valdre, Academic Press, 1971, 570-625.

6. Stroke, G. W., Halioua, M., and Saffir, A. J., Phys. Letters, 44A, 1973, 115-117.

7. Thon, F., Naturforschung, Z, 21A, 1966, 476-

8. Stroke, G. W., An Introduction to Coherent Optics and Holography, Second Edition, Academic Press, 1969.

9. Stroke, G. W., IEEE Spectrum, 9, 1972, 24-41.

10. Maréchal, A., et Croce, P., C. r. Ac. Sc., Paris, 237, 1953, 607-

11. Gabor, D., Nature, 161, 1948, 777-778.

12. Stroke, G. W., Optica Acta, 16, 1969, 401-422.

13. Thon, F., Willasch, D., Stroke, G. W., and Halioua, M., 1973 (to be completed).

14. Hahn, M. H., Optik, 35, 1972, 326-

15. Tsujiuchi, J., Progress in Optics, Editor, E. Wolf, North Holland Publishing Co., 1963, Vol. 2, 131-180.

16. Thon, F., and Siegel, B. M., Berichte der Bunsen Gesellschaft f. physikalishche Chemie, 74, 1970, 1116-1120.

17. Stroke, G. W., Halioua, M., Shinoda, M., and Colbert, C., 1973 (to be completed).

18. Stroke, G. W., and Halioua, M. (in preparation).

19. Stroke, G. W., and Halioua, M. Physics Letters, 39A, 1972, 269-270.

ULTRASONIC HOLOGRAPHY BY TRANSDUCER ARRAY AND LIQUID-CRYSTAL DEVICE

M. Suzuki, T. Iwasaki, S. Fujiki and A. Hakoyama

Faculty of Engineering, Hokkaido University
Sapporo, Japan

ABSTRACT

Recently, a real-time holographic imaging system including a transducer array and a real-time reconstructor tube were reported. The reconstructor tube consists of a thick (DKDP) crystal, an off-axis scanning electron gun, and associated optics.

This paper proposes the use of a matrix-type (nematic) liquid crystal display plate[1,2] as a component, permitting the reconstruction of acoustic or microwave targets.

A preliminary experiment is also described.

I. PROPOSED ARRANGEMENT FOR REAL-TIME RECONSTRUCTION AND ITS MATHEMATICAL ANALYSIS

A matrix-type liquid crystal plate, as shown in Figure 1, is experimentally used as the component of an arrangement for displaying the alphabetical characters and numerals.

We propose here the use of such a liquid crystal plate as an image reconstruction device. The proposed arrangement is shown in Figure 2.

Before showing the results of our experimental work, an analysis of the reconstruction of images by the matrix-type liquid crystal plate is presented. The conjugate image component of the light wave, after passing through the display plate as shown in Figure 3 is:

$$U(x_3,y_3) = D \int_{-a/2}^{a/2} \int_{-b/2}^{b/2} \exp[-j\alpha(\rho_2,\rho_3;z)]dx_2dy_2 \times$$

$$\iint O(x_1y_1)\exp[j\beta(\rho_1)]S(\rho_1,\rho_2,\rho_3;z)dx_1dy_1 \qquad (1)$$

where

$$\alpha(\rho_2,\rho_3;z) = k_2 \frac{(x_2-x_3)^2 + (y_2-y_3)^2}{2z} \qquad k_2 = \frac{2\pi}{\lambda}$$

$$\beta(\rho_1) = \frac{k_1 z_1}{2z_0(z_1-z_0)}(x_1^2+y_1^2) \qquad k_1 = \frac{2\pi}{\Lambda}$$

$$S(\rho_1,\rho_2,\rho_3;z) = \sum_{m=-M}^{M} \sum_{n=-N}^{N} \exp[-jmX\{\frac{k_2(x_3-x_2)}{z} + \frac{k_1 M_0 y_1}{z_0}\}$$

$$-jnY\{\frac{k_2(y_3-y_2)}{z} + k_1 \frac{M_0 y_1}{z_0}\}]$$

under the perfect reconstruction condition of the conjugate image, which is given by

$$\frac{1}{z} = \frac{\lambda}{\Lambda} \frac{M_0^2(z_1-z_0)}{z_1 z_0} \qquad (2)$$

where Λ and λ are the wavelengths of acoustic wave and light wave, respectively. X and Y are the spacing of the transparent conductive strips of the display. M_0 is the reduction rate of the hologram.

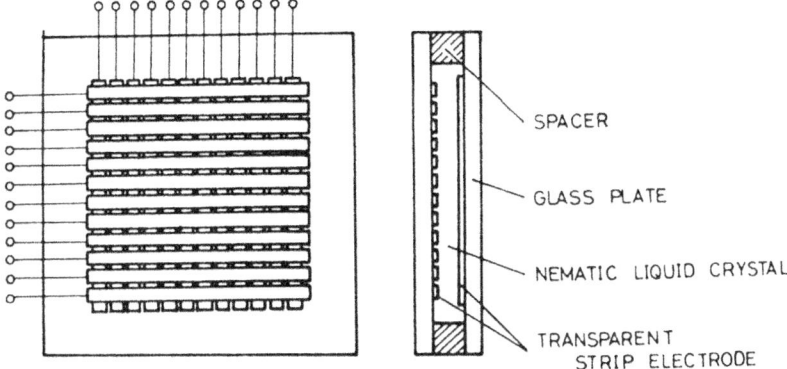

Figure 1. Matrix-Type Nematic Liquid-Crystal Display Plate

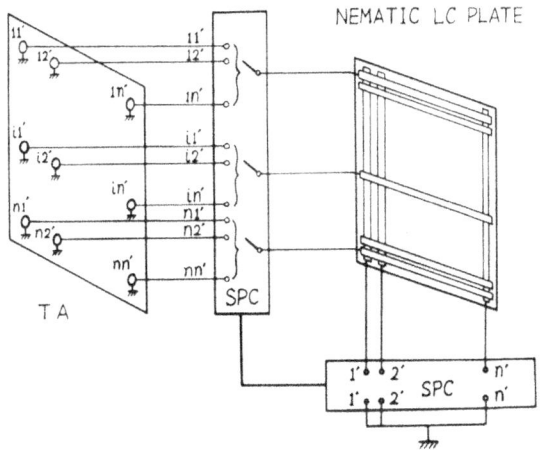

Figure 2. Proposed Arrangement Using a Matrix-Type Nematic Liquid Crystal Display Plate. TA is the Transducer with Amplifier. SPC is a Signal Processing Circuit.

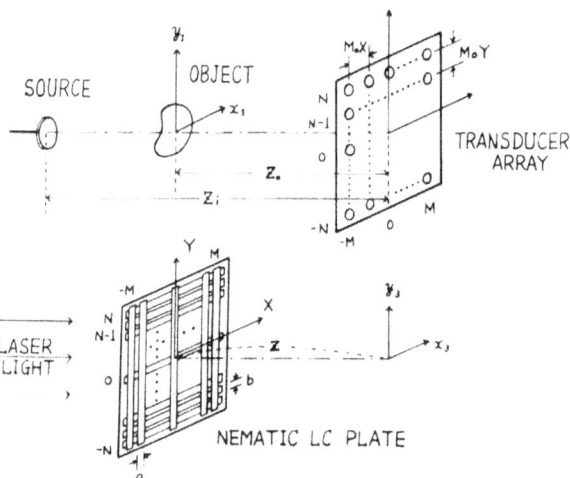

Figure 3. Diagram of Holography System using a Transducer Array and a Liquid Crystal Display Plate

When M and N are infinite, Equation (1) becomes:

$$U(x_3,y_3) = D \frac{(2\pi z_0)^2}{(k_1 M_0)^2 XY} \sum_{\ell=-\infty}^{\infty} \sum_{\ell'=-\infty}^{\infty} \exp[-j\gamma_{\ell\ell'}(\rho_3,z)] \times$$

$$\int_{-a/2}^{a/2} \int_{-b/2}^{b/2} O\{\frac{z_0}{k_1 M_0 X}(2\ell\pi + k_2 X\frac{x_3}{z}) - M_0\frac{z_i-z_0}{z_i}x_2, \frac{z_0}{k_1 M_0 X} \times$$

$$(2\ell'\pi + k_2 Y\frac{y_3}{z}) - M_0\frac{z_i-z_0}{z}y_2\} \exp[-j(\frac{2\ell\pi x_2}{X} + \frac{2\ell'\pi y_2}{Y})]dx_2 dy_2$$

(3)

by considering the following formula:

$$\sum_{m=-\infty}^{\infty} e^{jmx} = 2\pi \sum_{\ell=-\infty}^{\infty} \delta(x - 2\ell\pi) \qquad (4)$$

where

$$\gamma_{\ell\ell}{}'(\rho_3,z) = \frac{k_2(x_3{}^2+y_3{}^2)}{2z} - \frac{z_o z_i}{2k_1(M_o X)^2 (z_i-z_o)} \times$$

$$[(2\ell\pi+k_2 X \frac{x_3}{z})^2 + (2\ell'\pi+k_2 Y \frac{y_3}{z})^2]$$

and $\hat{o}(x)$ is the Kronecker notation.

In case of a point object located at the origin, we obtain the following result:

$$U(x_3,y_3) = D \frac{(2\pi z_o)^2}{(k_1 M_o)^2 XY} \{\frac{z_i}{M_o(z_i-z_o)}\}^2 \times$$

$$\sum_{\ell=-\infty}^{\infty} \sum_{\ell'=-\infty}^{\infty} \exp[-j\phi_{\ell\ell}{}'(\rho_3,z)].$$

for

$$-\frac{z}{k_2 X}(2\ell\pi+k_1 M_o{}^2 X\frac{z_i-z_o}{z_i z_o}\frac{a}{2}) \leq x_3 \leq \frac{z}{k_2 X}(k_1 M_o{}^2 X\frac{z_i-z_o}{z_o z_i}\frac{a}{2}-2\ell\pi)$$

$$-\frac{z}{k_2 Y}(2\ell'\pi+k_1 M_o{}^2 Y\frac{z_i-z_o}{z_i z_o}\frac{b}{2}) \leq y_3 \leq \frac{z}{k_2 Y}(k_1 Mo^2 Y\frac{z_i-z_o}{z_o z_i}\frac{b}{2}-2\ell'\pi)$$

$$U(x_3,y_3) = 0 \qquad (5)$$

for the regions excluding the above. Otherwise, where $\psi_{\ell\ell}{}'(\rho_3, z)$ is given by

$$\psi_{\ell\ell}{}'(\rho_3,z) = \gamma_{\ell\ell}{}'(\rho_3,z) + \frac{z_o z_i}{k_1 M_o{}^2(z_i-z_o)}[\frac{2\pi\ell}{X}(2\ell\pi+k_2 X \frac{x_3}{z}) + \frac{2\pi\ell'}{Y}(2\ell'\pi+k_2 Y\frac{y_3}{z})].$$

The above result shows that the point object is reconstructed as an image which has a finite size of $L_x \times L_y$, where

$$L_x = M_o \frac{2k_1}{k_2} \frac{z(z_i-z_o)}{z_i z_o} a, \quad L_y = M_o \frac{2k_1}{k_2} \frac{z(z_i-z_o)}{z_i z_o} b \qquad (6)$$

in case of $l = 0$ and $l' = 0$, even if the hologram aperture is infinitely large.

This means that the reconstructed image of a target becomes blurred because of the matrix-type structure of the display plate. The same situation arises in the case of a mechanical scanning type hologram. The blurring effect described in the foregoing may be reduced by making the strip spacings X and Y smaller.

II. PRELIMINARY EXPERIMENT

The one-dimensional transducer array system consisting of 30 elements and the matrix-type display device used in our experiment are shown in Figures 4 and 5, respectively. The transmitting and receiving transducers with five millimeter diameter PZT disks were used without a converging lens system.

The matrix-type liquid crystal plate used in our experiment was constructed of glass substrate, 25 micron nematic liquid crystal film, and 36 x 36 matrix-type transparent conductive electrodes. The transparency versus applied DC voltage characteristics of this plate is shown in Figure 6. The applied voltages were limited to between 0 volt and 30 volts.

In our experiment, the output voltage of the holographic signal processing circuit was not directly applied to the reconstructor plate but applied to the cathode ray tube; then the one-dimensional hologram was produced as a photograph. The hologram made by the above process was quantized and sampled appropriately, and the DC voltages proportional to the sampled values were applied to the corresponding terminals of the liquid crystal device. The object image was reconstructed by illuminating the display plate with a He-Ne laser.

Figures 7a and 7b show a one-dimensional holographic signal output and reconstruction taken with the above arrangement. The acoustic wavelength in water was 1.5 mm, and the object was a long sponge strip, five millimeters wide. Figures 8a and 8b show the computed hologram of the above object and its reconstructed image, respectively. Figures 9a and 9b show the two-dimensional hologram of the same object produced by the usual method and the reconstruction image, respectively.

Figures 10 and 11 show the images reconstructed from the computed holograms in the case in which, the illuminating wavelength being three centimeters, the first object was a very long metal strip, 40 centimeters wide; the second object was composed of two very long strips, each 40 centimeters wide, with 40 centimeters spacing.

Figure 12 shows the reconstructed image of a narrow slit with Λ = three centimeters.

Figure 4. One-dimensional, 30 Element Transducer Array

Figure 5. Matrix-type Nematic Liquid-Crystal Display Device

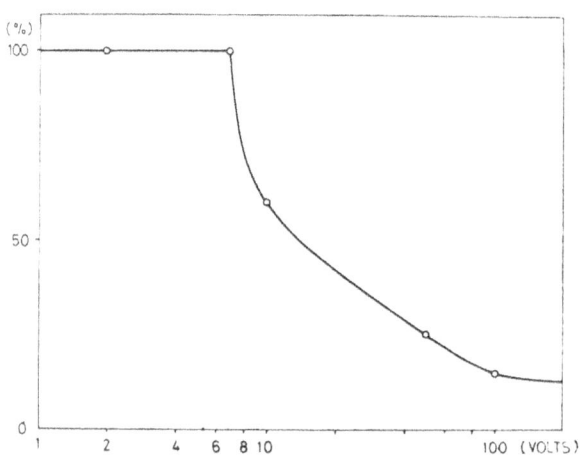

Figure 6. Transmission Coefficient of the Display Plate Versus Applied DC Voltage Characteristic

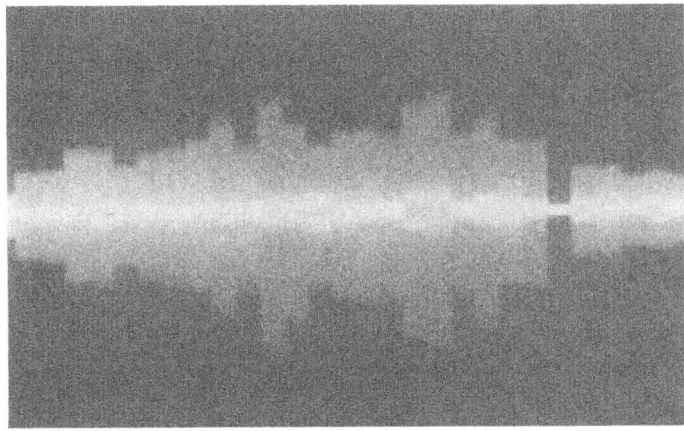

Figure 7a. One-dimensional Holographic Signal Output When $\Lambda = 1.5$ mm with a long sponge strip, 5 mm wide

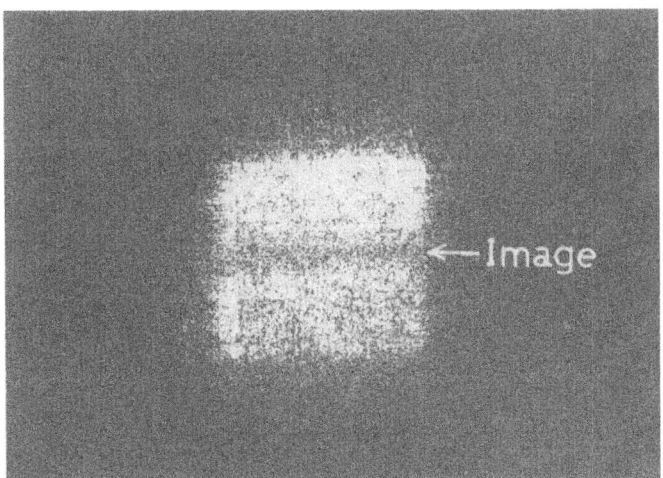

Figure 7b. Reconstructed Image from Previous Data

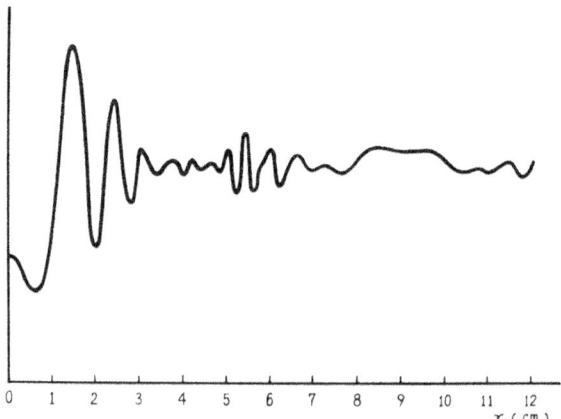

Figure 8a. Computed Hologram of the Previous Object

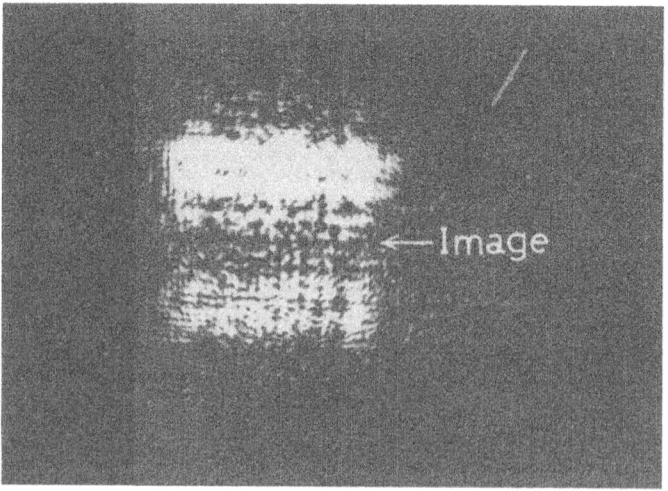

Figure 8b. The Reconstructed Image from the Hologram in Figure 8a

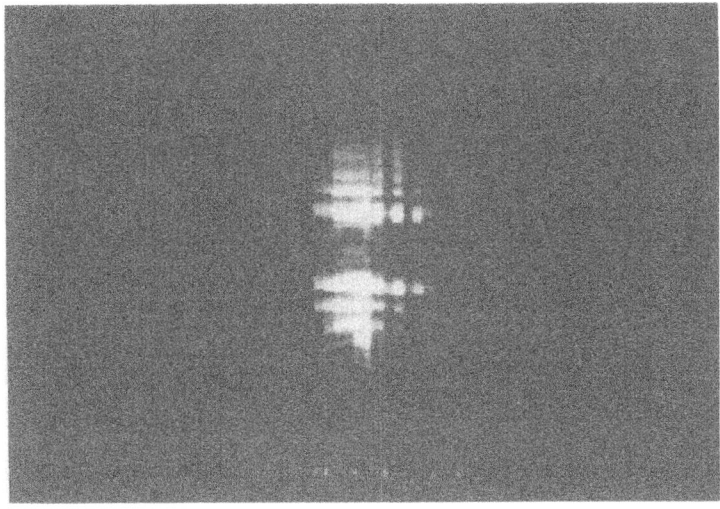

Figure 9a. Usual Hologram of the Previous Object

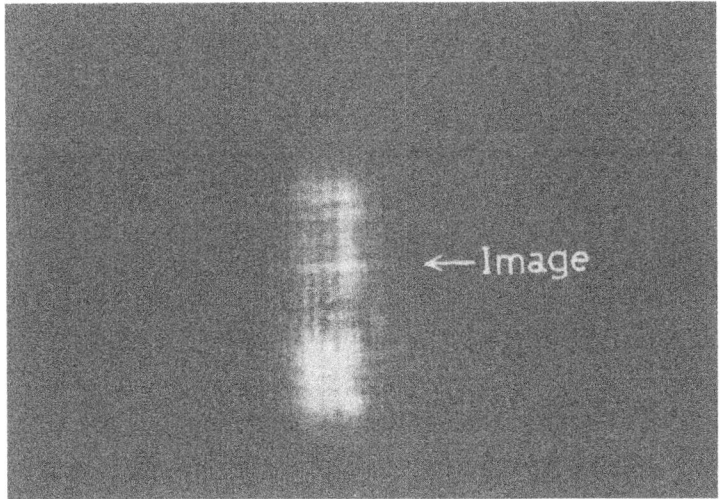

Figure 9b. Reconstructed Image from Above Data

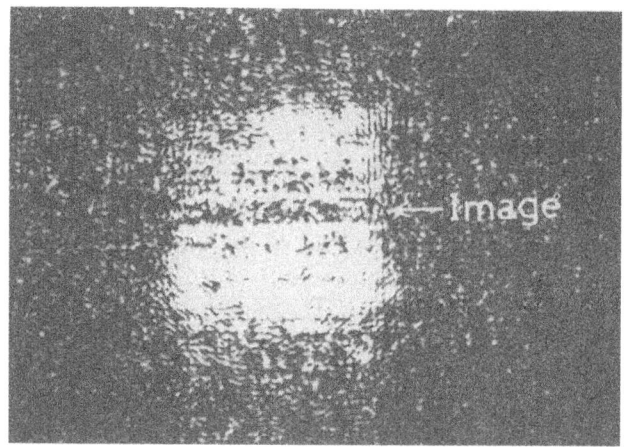

Figure 10. Image Reconstructed from Computed Hologram
Λ = 3 cm, Long Metal Strip, 40 cm Wide

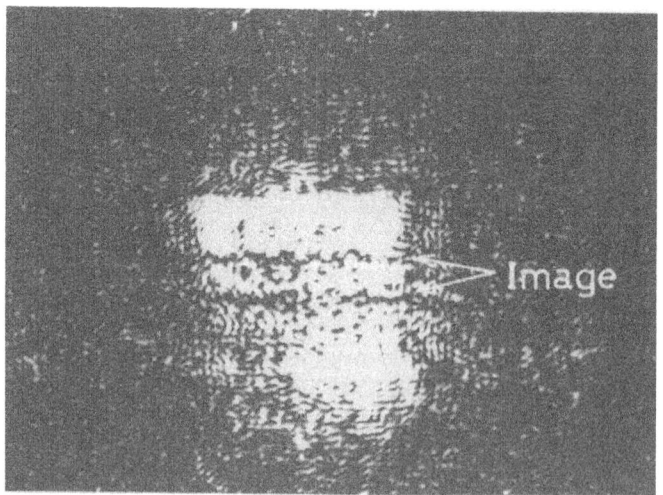

Figure 11. Image Reconstructed from Computed Hologram
From Two Long Metal Strips, Each 40 cm Wide
40 cm Spacing

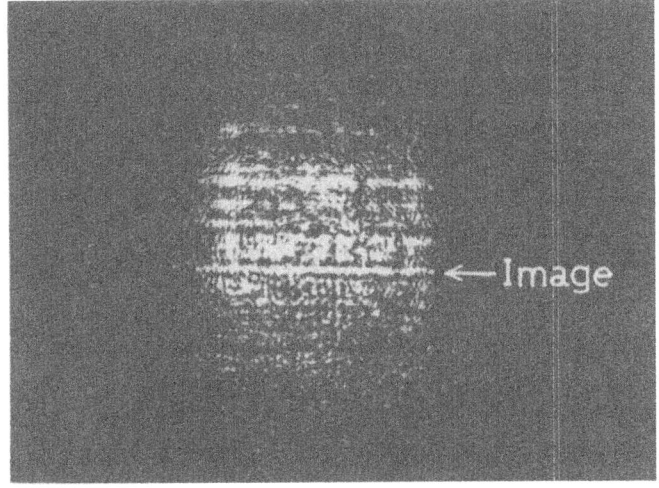

Figure 12. Reconstructed Image of a Narrow Slit with $\Lambda = 3$ cm

III. CONCLUSION

One proposal of real-time holography which consists of a transducer array and a matrix-type liquid crystal device has been presented. The experimental work in the case of one-dimensional holography shows the possibilities of such a device.

REFERENCES

1. Fergason, J.L., <u>Liquid-Crystal Detectors</u>, Acoustical Holography, Vol. 2, Plenum Press, 1970, Page 53.

2. Lechner, B.J., et al., 1969 IEEE International Solid State Circuit Conference.

HOLOGRAPHIC SYNTHETIC APERTURE SONAR SYSTEM

J. Tsujiuchi, S. Ueha and K. Ueno

Imaging Science and Engineering Laboratory
Tokyo Institute of Technology
O-Okayama, Meguro-ku, Tokyo, Japan

I. INTRODUCTION

As one of the ultrasonic imaging techniques, a sonar system is widely used in oceanography to measure the depth of bottom and to locate schools of fish or other submerged objects. In the conventional sonar system which uses the ultrasonic pulse-echo method, the resolution limit of the object in range, the longitudinal resolution, is determined by $VT/2$; where V is the sound velocity in the water, and T is the duration of ultrasonic pulse, and the improvement of longitudinal resolution can be brought about by shortening the duration of pulse. But the resolution limit in angle, the lateral resolution, is λ/D; where λ is the ultrasonic wavelength and D is the dimension of the receiver; so the improvement is to be made by the increase of the dimension of the receiver by use of a special shape of the receiver, or by the choice of a shorter wavelength. This seems very difficult to realize.

In ultrasonic holography, on the other hand, the lateral resolution can be made much higher than that of the abovementioned pulse-echo method, but the longitudinal resolution may be lower because the relative aperture of the hologram has a tendency to decrease along the longitudinal position of the reconstructed image.

This decrease cannot be determined because of the change of wavelength from ultrasonic wave to laser light. These difficulties in ultrasonic imaging can be eliminated by the use of holographic synthetic aperture techniques which are realized by the combination of the range determination by the pulse-echo method and the image formation by ultrasonic holography, which have been successfully used in the processing of side-looking radar systems.[1,2]

The use of this holographic synthetic aperture sonar system, however, is limited to the case in which the range of the object is not so large because of the smallness of the velocity of an acoustic wave in comparison with that of an electromagnetic wave. This difficulty can be avoided by using an additional receive-only transducer placed ahead of the ordinary transmitter-receiver transducer.[3] Pekau et al.[4] have reported the proposition of this type of receiving transducer array and the results of experiments using a single receiver system.

This paper deals with some experimental considerations concerning the holographic synthetic aperture sonar system with a single receiver applied to the object, whose range is not so large. Some experimental results are also shown.

II. HOLOGRAPHIC SYNTHETIC APERTURE SONAR SYSTEM WITH COMPUTER RECONSTRUCTION

Sato et al.[5] have investigated a simplified two-dimensional holographic synthetic aperture sonar system in which a computer was used for the reconstruction. We will first introduce this work for the reference.

1. <u>Principles</u> - As shown in Figure 1, an ultrasonic transmitter, A, placed on the x axis, emits an ultrasonic pulse

where
$$P_T(t) \cos 2\pi ft,$$

$$P_T(t) = 1 \quad 0 \leq t \leq T$$
$$ = 0 \quad \text{otherwise,} \tag{1}$$

HOLOGRAPHIC SYNTHETIC APERTURE SONAR SYSTEM

and the reflected wave from an object is detected by a receiver, B, which is placed at a distance of 2a from A in x direction. If the object consists of many points and g_n is the reflectivity of the nth point, object 0_n, which is located at (x_n, z_n), the returned signal becomes

$$S(x,t) = \sum_n g_n P_T \left(t - \frac{r_{1n} + r_{2n}}{V}\right) \cos 2\pi f \left(t - \frac{r_{1n} + r_{2n}}{V}\right) \quad (2)$$

where

r_{1n} and r_{2n} are the distances from 0_n to A and B respectively;

$$r_{1n} = [(x - a - x_n)^2 + z_n^2]^{1/2}$$
$$r_{2n} = [(x + a - x_n)^2 + z_n^2]^{1/2} \quad (3)$$

If the relations

$$z_n \gg (x + a - x_n)$$
$$T \gg [(x - x_n)^2 + a^2]/z_n V \quad (4)$$

hold, we have from (2), (3) and (4)

$$S(x,t) = \sum_n g_n P_T \left(t - \frac{2z_n}{V}\right) \cos 2\pi f \left[t - \frac{2z_n}{V} - \frac{(x-x_n)^2 + a^2}{z_n V}\right] \quad (5)$$

This returned signal is multiplied by the reference signal, $\cos 2\pi ft$, and is passed through a low pass filter. We have then

$$S_H(x,t) = \sum_n g_n P_T \left(t - \frac{2z_n}{V}\right) \cos 2\pi f \left[\frac{2z_n}{V} + \frac{(x-x_n)^2 + a^2}{z_n V}\right] \quad (6)$$

and this output signal is recorded by a data recorder while the transmitter-receiver array, A-B moves to scan along x axis.

To reconstruct the image of the object which is located at $z = z_n$, only the values $S_H(x, 2z_n/V)$ in the record are used. That is

$$S_H\left(x, \frac{2z_n}{V}\right) = g_n \cos 2\pi f \frac{(x-x_n)^2}{z_n V} \tag{7}$$

where the constant phase term is excepted, and it is regarded as a one-dimensional in-line Fresnel hologram of the object.

The reconstruction is represented by the following integral:

$$g'_n(x', z_n) = \int_{-l/2}^{l/2} S_H\left(x, \frac{2z_n}{V}\right) \exp\left[-2\pi i f(x-x')^2/z_n V\right] dx \tag{8}$$

where x' is the coordinate taken in the reconstructed image plane and the function g'(x', z_n) shows the amplitude distribution of the reconstructed image at $z = z_n$.

2. **Practical System and Experimental Results** - Figure 2 shows a schematic diagram of the experimental setup. In this technique, the time variable signal, $S_H(x, t)$ must be recorded at each sampling point on the hologram. But, as the frequency spectrum of $S_H(x, t)$ is much higher than that of a conventional data recorder, a technique of time-stretcher which is shown in Figure 3 is used. A series of ultrasonic pulses is emitted from the transmitter with the interval T_p at each sampling point, and the same signal $S_H(x, t)$ is obtained for each pulse. If a series of the signal $S_H(x, t)$ is sampled by a sample holder whose sampling command pulses are shifted by a time of T_s for each pulse, the time-stretched signal $S_H(x, (T_s/T_p)t)$ is obtained as the output of the sample holder and the signal can be easily recorded by a conventional data recorder as a result of a suitable choice of T_s/T_p value.

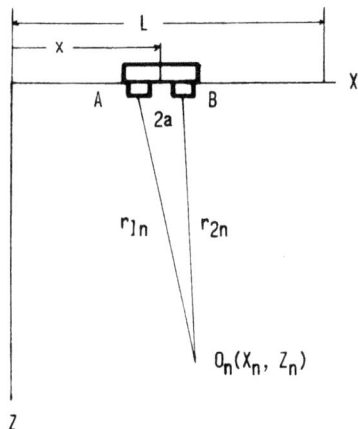

Figure 1. Geometrical Arrangement of Sonar System

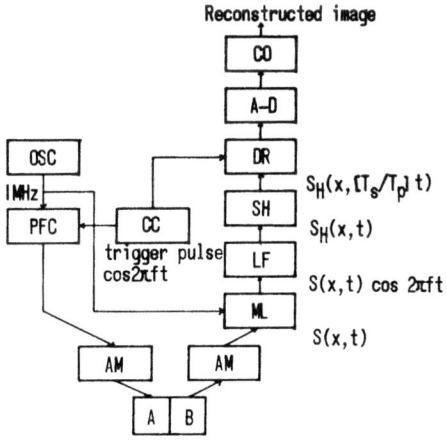

Figure 2. Schematic Diagram of System:
- OSC: Oscillator
- PFC: Pulse Forming Circuit
- CC: Control Circuit
- AM: Amplifier
- ML: Multiplier
- LF: Low Pass Filter
- SH: Sample Holder
- DR: Data Recorder
- A-D: A-D Converter
- CO: Computer
- A: Transmitter
- B: Receiver

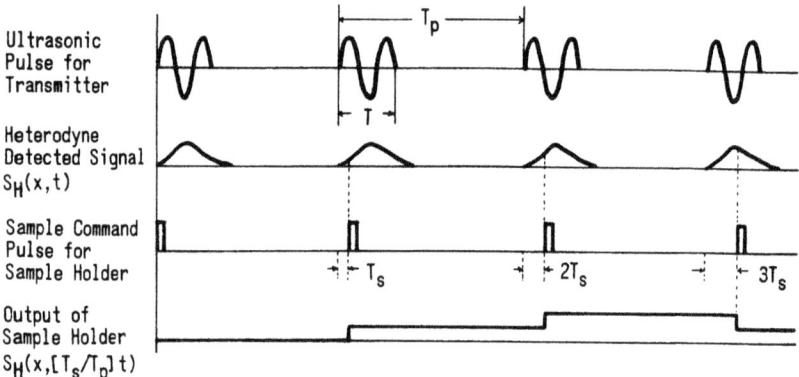

Figure 3. Principles of Time Stretcher

As the transmitter and receiver, two circular PZT transducers, nine mm in diameter, are used and separated by 2a = 15 mm. Experiments are made at an ultrasonic frequency of one MHz, using four mm diameter steel wire as an object. The wire is placed at a distance of 158 mm below the transmitter-receiver array which moves to scan over the hologram perture (L = 50 mm) across the wire. The reflected ultrasonic signal is sampled every one millimeter movement of the array. The recorded signals $S_H(n\Delta x, t)$ (n = 1, 2, ... 50), where Δx(1mm) is the sampling interval, are digitized by an A-D converter and the reconstruction of the image is carried out by using a computer. The computation is made by the equation

$$|g'(x', z_n)|^2 = \left| \sum_{n=1}^{50} S_H(n\Delta x, \frac{2z_n}{V}) \exp[-2\pi i f \frac{(n\Delta x - x')^2}{z_n V}] \right|^2 \quad (9)$$

which is modified from the equation (8).

Figure 4 shows the result of the experiment. This figure contains two loci of 70% and 50% intensities of the peak value from the reconstructed images for various z_n positions, and this corresponds to the cross section of the wire. The dotted line contour is the geometrical image of the wire.

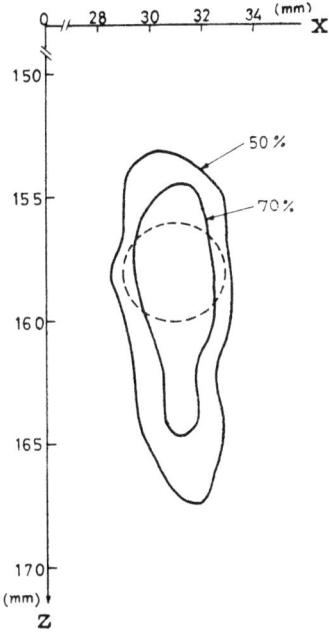

Figure 4. Reconstructed Image of a Wire. The dotted line is the geometrical image of the object.

© Bulletin of the Tokyo Institute of Technology, No. 107, 1971, 39, Takuso, Professor Sato, Research Laboratory of Precision Machinery and Electronics, Tokyo Institute of Technology, O-Okayama, Meguro-ku, Tokyo, Japan. Published by Director of Library, Tokyo Institute of Technology.

It is seen that the lateral resolution is fairly good but the duration of pulse is limited more than 12 microseconds, owing to the frequency characteristics of the transducer. This method is to be extended to the three-dimensional objects by using the two-dimensional scanning and the cross-sections of the object. These are parallel to the scanning plane but can be obtained by using each $g'(x, y, z_n)$ values; but the two-dimensional Fresnel transformation by a computer is not so easy to use in practical applications.

III. HOLOGRAPHIC SIDE-LOOKING SONAR WITH OPTICAL RECONSTRUCTION

Another feature of the holographic synthetic aperture sonar system is the holographic side-looking sonar, the principles of which are the same as that of the holographic side-looking radar[1,2] in which the image of a plane object can be obtained by only one linear scanning and the coherent optical processing. Some experimental considerations concerning the applicability of this technique to the sonar system have been discussed.

1. <u>Principles</u> - Figure 5 shows the geometrical arrangement of the system. An ultrasonic transmitter, A, placed on the x axis, emits an ultrasonic pulse, $P_T(t) \cos 2\pi f t$, and the reflected waves from the object points O_n, which are distributed in the (u, v) plane return to the receiver, B, which is placed adjacent to the transmitter, A. The return signal can be expressed similarly as the previous case (2)

$$S(t) = \sum_n g_n P_T\left(t - \frac{2r_n}{V}\right) \cos 2\pi f \left(t - \frac{2r_n}{V}\right). \qquad (10)$$

If we consider that the transmitter-receiver array A-B moves to scan at a constant velocity of V_s along the x axis, and O_n is at a distance of r_{on} from the x axis, we have

$$r_n = r_{on} + (x - x_{on})^2 / 2r_{on}, \qquad (11)$$

where x an x_{on} are the x coordinates of A-B and O_n, respectively. The return signal from the object points $O_n(x_{on}, r_{ol})$ which are located at a distance of r_{ol} from the x axis becomes

$$S(t, r_{ol}) = \sum_n g(x_{on}, r_{ol}) P_T\left(t - \frac{2r_{ol}}{V}\right) \cos 2\pi \left[ft - \frac{2r_{ol}}{\lambda_0} - \frac{(V_s t - x_{on})^2}{\lambda_0 r_{ol}}\right]. \qquad (12)$$

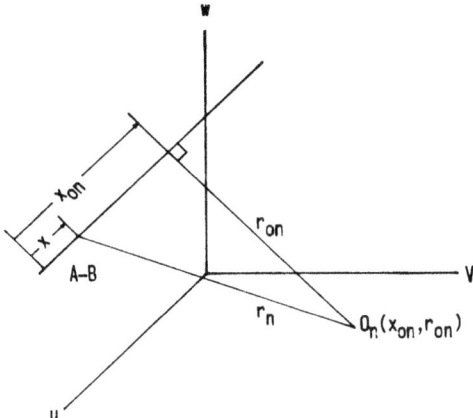

Figure 5. Geometrical Arrangement of Side-Looking Sonar System

This return signal is multiplied by a reference signal $\cos 2\pi f_r t$, and is passed through a low pass filter. We have as a resultant signal:

$$S_H(t, r_{o1}) = \sum_n g(x_{on}, r_{o1}) P_T(t - \tfrac{2 r_{o1}}{V}) \cos 2\pi [f_b t - \tfrac{2 r_{o1}}{\lambda_o} - \tfrac{(V_s t - x_{on})^2}{\lambda_c r_{o1}}] \quad (13)$$

where $f_b = f - f_r$. The beat frequency produced by the interference between the signal and the reference waves may be negative as well as positive.

This output signal is displayed, adding a constant term, C, to make the signal non-negative, in an intensity-modulated cathode ray tube which is photographed by a film drawn vertically to the direction of sweep by a constant velocity of V_f. Taking (p, q) coordinates in the film thus obtained, every pulse in the q direction is in rows, at a constant interval of $V_f T_p$, as shown in figure 6; and (p, q) coordinates are expressed by

$$\left. \begin{array}{l} p = V_f t \\ q = 2 V_c m r_o / V \end{array} \right\} \quad (14)$$

where V_c is the sweep velocity of the electron beam in the CRT screen, and m is the lateral magnification of the image-forming device from CRT screen to film. The amplitude transmission of the film in p direction at a fixed value of r_{o1} becomes

$$T_H(p, q_o) = |C + S_H(t, r_{o1})|^{-1/2} \tag{15}$$
$$= C - \frac{\gamma}{2} \sum_n g(z_{on}, r_{o1}) \cos 2\pi [\frac{f_b}{V_f} p - \frac{1}{\lambda_o r_{o1}} (\frac{V_s}{V_f})^2 (p - \frac{V_f}{V_s} z_{on})^2].$$

This amplitude transmission shows that the film thus obtained is a one-dimensional Fresnel hologram with a spatial carrier frequency of f_b/V_f; and if a parallel laser light with a wavelength of λ is incident vertically upon this hologram H, as shown in Figure 7, the image is reconstructed at a distance of

$$r' = \pm \frac{1}{2} \frac{\lambda_o}{\lambda} (\frac{V_f}{V_s})^2 r_{o1} \tag{16}$$

from the hologram in a direction of

$$\theta' = \pm \frac{f_b}{V_f} \lambda. \tag{17}$$

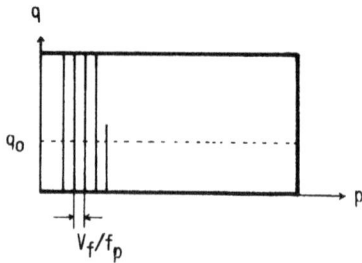

Figure 6. Recording of Holographic Signal

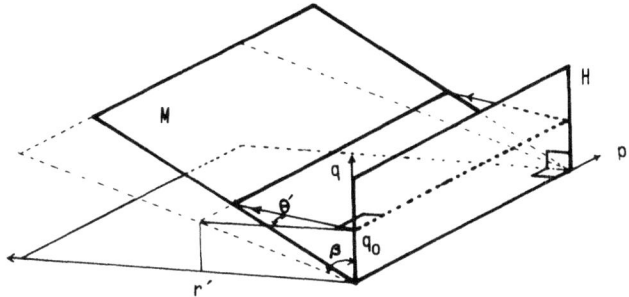

Figure 7. Reconstruction of Hologram

2. Resolution and Ambiguity of Reconstructed Image

As the object is illuminated by a pulse train with an interval of T_p, the range ambiguity will occur when the reflected signal of a pulse from the nearest end of the object is arrived at before the signal of the previous pulse is reflected from the farthest end of the object. This difficulty can be avoided if the following relation is satisfied:

$$T_p \geq 2 \Delta r_o / V \tag{18}$$

where Δr_o is the range interval of the object.

The lateral resolution of the system for a perfectly diffused object can be evaluated by

$$\rho = \lambda_0 / 2 \sin \alpha, \tag{19}$$

where 2α is the maximum angle at which the scanning width L is seen from the object 0 (Figure 8). If the transmitter and the receiver, both with the dimension D, are diffraction-limited, and the angle α can be evaluated effectively by the half-power angle of the transmitter and receiver beams, we have

$$\sin \alpha \doteqdot \lambda_0 / D \tag{20}$$

Finally, $\rho = D$. (21)

Namely, the resolution of the object in p direction is determined by the dimension of the transmitter and receiver if the scanning is made over the whole half-power angle extent, L, of a diffraction-limited transducer.

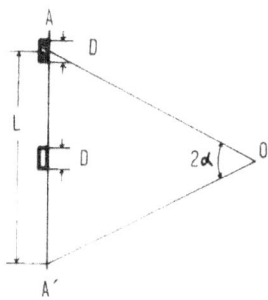

Figure 8. Lateral Resolution of System

Suppose a transmitter-receiver array, A, which is located at the end of the above-mentioned scanning aperture receives the reflected signal from a point, O, the holographic signal S_H can be expressed similarly as in (13), taking another end, A', as the origin of x axis

$$S_H(t, r_{o1}) = P_T(t - \tfrac{2 r_{o1}'}{v}) \cos 2\pi [f_b t - \tfrac{2 r_{o1}'}{\lambda_o} - \tfrac{(v_s t - l/2)^2}{\lambda_o r_{o1}}]$$ (22)

and the instantaneous frequency at A becomes

$$f_i = f_b - L V_s / \lambda_o r_{o1} = f_b - V_s / D,$$ (23)

where the sign of the second term is changed by the choice of the scanning direction. So in order to obtain the above-mentioned resolution, it is preferable from the sampling theorem, to choose T_p as

$$T_p \leq 1/2(|f_b| + |V_s/D|)$$ (24)

for the sake of safety.

HOLOGRAPHIC SYNTHETIC APERTURE SONAR SYSTEM

3. <u>Choice of Beat Frequency</u> - The beat frequency, f_b, is obtained by the difference of signal and reference frequencies if the Doppler shift is neglected, and this frequency contributes to separation of the reconstructed image from the background light and another unnecessary image. For that purpose, the Fourier spectrum of the reconstructed image must be confined to the carrier frequency of the hologram. We have then from (23)

$$|f_b| \geq |V_S/D|. \qquad (25)$$

Although the relation between f_b and $f_p = 1/T_p$ is already given by the equation (24), where $f_p \geq f_b$, a beat frequency which is higher than f_p is also usable. Figure 9 is a schematic diagram which shows the sampling of a holographic carrier wave with a period of V_f/f_b, by using an ultrasonic pulse with an interval of V_f/f_p. As the sampling is made by more than two pulses, P_1 and P_2 for example, spatial phase values of the carrier wave at each sampling point becomes

$$\left. \begin{array}{l} \phi_1 = 2\pi(M_1 V_f/f_p - M_2 V_f/f_b)/(V_f/f_b) = 2\pi M_1 f_b/f_p \\ \phi_2 = \phi_1 + 2\pi(V_f/f_p)/(V_f/f_b) = 2\pi(M_1 + 1)f_b/f_p, \end{array} \right\} \qquad (26)$$

where M_1 and M_2 are positive integers and the integral multiples of 2π are excepted. Then, if we take

$$f_b = f_b' + N f_p, \qquad (27)$$

where N is a positive or negative integer. Equation (26) then becomes

$$\left. \begin{array}{l} \phi_1' = 2\pi M_1 f_b'/f_p + 2\pi M_1 N = 2\pi M_1 f_b'/f_p \\ \phi_2' = 2\pi(M_1+1)f_b'/f_p + 2\pi(M_1+1)N = 2\pi(M_1+1)f_b'/f_p \end{array} \right\} \qquad (28)$$

Accordingly, this hologram has the same properties as a hologram which has a beat frequency of f_b', and f_b higher than f_p can be used. This effective beat frequency, f_b', of course, must also satisfy the conditions of (24) and (25).

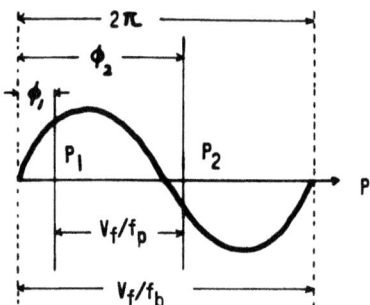

Figure 9. Sampling of Hologram

In experimental fact, the reconstruction has been made for even N, but not for odd N. This reason has not been well understood.

As the intensity distribution of the CRT beam in the p direction can be considered as a gaussian form exponent $(-p^2/2\sigma^2)$, the amplitude distribution of the sampled hologram can be expressed by

$$T_H' = [\text{Comb}(p/(V_f/f_p)) * \exp(-p^2/2\sigma^2)] T_H(p), \quad (29)$$

If we make an approximation that the image is reconstructed by the Fourier transformation of (29), the amplitude distribution of the reconstructed image becomes

$$g'(x') = \text{const.} \text{Comb}\left(\frac{V_f x'^2}{f_p \lambda r'}\right) \exp(-2\pi^2\sigma^2 x'^2/\lambda^2 r'^2)$$
$$* [\delta(\tfrac{x'}{\lambda r'}) + ag(\tfrac{x'}{\lambda r'} - \tfrac{f_b}{V_f}) + ag^*(\tfrac{x'}{\lambda r'} + \tfrac{f_b}{V_f})], \quad (30)$$

where a is a constant due to the modulation degree of CRT beam intensity and to the photographic gamma of the film.

Figure 10 shows a schematic diagram of the reconstructed image, and it is understood that the efficiency of the reconstructed image is influenced by not only the modulation degree, but also by the shape of the beam of the cathode ray tube. In this experiment, as the effective width of the CRT beam, $2\sqrt{2}\sigma$ is 150 μm, reconstruction is obtained for f_b (or f_b')≦15 Hz.

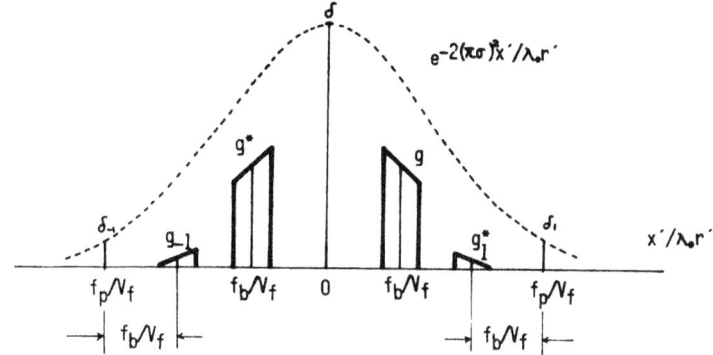

Figure 10. Structure of Reconstructed Image

Normal image (δ: Background light, g: Direct image, g^*: Conjugate image)

First order diffracted image by sampling structure (δ_1: Background light, g_1^*: Conjugate image)

-First image (δ_{-1}: Background light, g_{-1}: Direct image)

4. <u>Reconstruction Process</u> - In practical cases, the object is distributed in (u, v) plane and the hologram contains this two-dimensional object information. As q is proportional to the range, r_o, of the object, if the reconstruction is made using the whole plane of the hologram the distance of reconstructed image, r', changes linearly in accordance with the position q; namely, the one-dimensionally reconstructed image in p direction is contained in a plane, M, which is inclined by β to the hologram plane (Figure 7), where

$$\tan \beta = \frac{dr'}{dq} = \frac{dr'}{dr_o}\frac{dr_o}{dq} = \frac{1}{4m}\frac{\lambda_o}{\lambda}\left(\frac{V_f}{V_s}\right)^2 \frac{V}{V_c}. \qquad (31)$$

However, as the image in q direction is recorded on the hologram plane itself, the reconstruction of a perfect two-dimensional image is not so easy.

Several methods have been reported for the reconstruction of such a hologram. Cutrona, et al.[1] have used a conical lens placed adjacent to the hologram, Castella[6] has suggested the possibility of using a combination of two tilted cylindrical lenses, and Kozma, et al.[7] have developed a tilted plane optical processor by using multi-lens anamorphic telescopes. Other possibilities have been investigated and a system which is shown in Figure 11 has been proposed for the purpose.

If the hologram, H, is illuminated by a collimator lens L_1, and a line source, LS, is placed in the conjugate plane of M in regard to L_1 and has the property of emitting a light beam divergent to p direction but parallel to the optical axis of L_1 in q direction, the reconstructed image in p direction is produced at infinity and the complete two-dimensional image is obtained by a cylindrical lens, L_2, and an ordinary lens, L_3, the focal lengths of which are chosen to obtain the proper aspect ratio of the final image. This type of line source can be approximately realized by setting a narrow slit in a parallel intense laser beam which is obtained by another collimator lens, L_o, and the center of LS is placed at a distance of

$$l = F_1^2/r_m' \tag{32}$$

inside or outside the focus of L_1, and the inclination of LS is

$$\tan \beta' = (F_1/r_m') \tan \beta = F_1 V/(2mV_c r_{om}), \tag{33}$$

where F_1 is the focal length of L_1, or r_{om} is the range of the object which is recorded at the center of q axis of the hologram, and r_m' is the position of the reconstructed image calculated from r_{om} by using Equation (16).

5. <u>Experiments and Results</u> - Figure 12 shows a schematic diagram of the experimental setup. A continuous wave

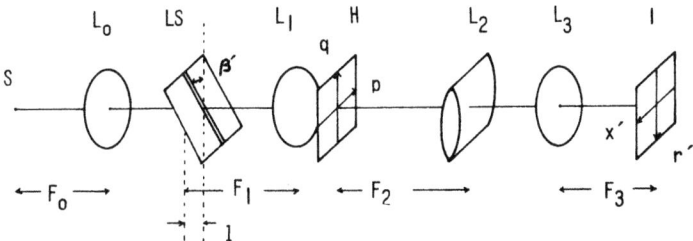

Figure 11. Optical System for Two-Dimensional Reconstruction

signal (f = 455 kHz) from an oscillator, OSC 1, is changed into a pulse wave (f_p = 130 Hz, T = 45 µs) by passing through a chopper, and is supplied to a transmitter A (PZT, D=3mm) which moves in the water with a constant velocity (V_s=45mm/s) and emits the ultrasonic wave (V = 1483 m/s in 20°C water, λ. = 3.3 mm). The return signal from an object is detected by a receiver, B (PZT, D = 3 mm), which moves with A and is multiplied by the reference wave (f_r = 455 kHz + f_b) generated by a local oscillator, OSC 2. The resultant signal is passed through a low pass filter and is displayed in an intensity modulated CRT (V_c = 100 m/s) which is photographed (m = 1/4) onto a film drawn at a constant velocity (V_f = 2.3 mm/s). An object that is made of aluminum in a shape of the character "H" and is covered by a thin plastic film, with an air gap, is placed in the water at a distance of r_o = 350 - 500 mm from the scanning line, the maximum stroke of which is about 360 mm.

Holograms thus obtained have a dimension of about 8 x 16 mm, and the reconstruction is made by the normal incidence of a parallel He-Ne laser light. The reconstructed image in p direction is to be formed at r' = 2.3 - 3.3 m, but the position of the reconstructed image is not distinct because of the smallness of the hologram aperture.

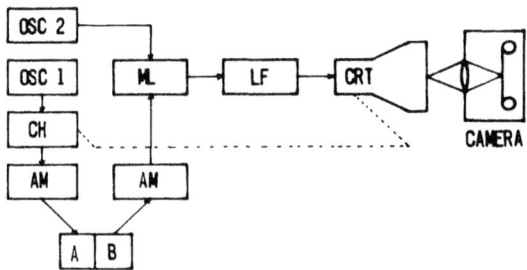

Figure 12. Schematic Diagram of System

OSC 1: Signal Oscillator, CH: Chopper,
AM: Amplifier; OSC 2: Local Oscillator,
ML: Multiplier, LF: Low Pass Filter,
CRT: Cathode Ray Tube.

Owing to this vagueness of the image position and the smallness of the object depth, the two-dimensional reconstruction can be made by using a combination of an ordinary lens and a cylindrical lens within its depth of focus. The result is shown in Figure 13. Figure 13a is a hologram obtained by the beat frequency, f_b = 10 Hz; Figure 13b is the reconstructed image from the hologram, a, without spatial filtering, where the focused direct image, defocused background light and defocused conjugate image can be observed side by side; and Figure 13c is the final image through spatial filtering. The lateral resolution in this case is about 10 mm, which is not so high because of sharp reflection directivity of the object and low signal-to-noise ratio of the system.

In the case of deeper objects, reconstruction cannot be made by this simple method, and the proper method such as the method of Cutrona[1], Castella[8], Kozma[7], or the proposed method in this paper is to be used.

IV. CONCLUSION

The holographic synthetic aperture sonar system which is realized as a combination of the pulse-echo method and

HOLOGRAPHIC SYNTHETIC APERTURE SONAR SYSTEM

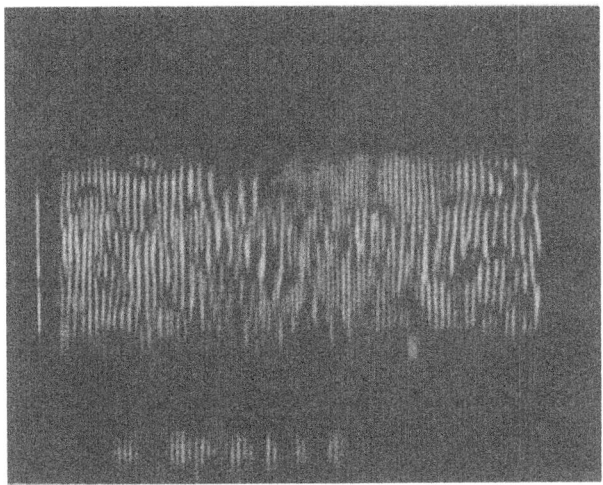

Figure 13. Results of Experiment
(a) Hologram with $f_b = 10$ Hz

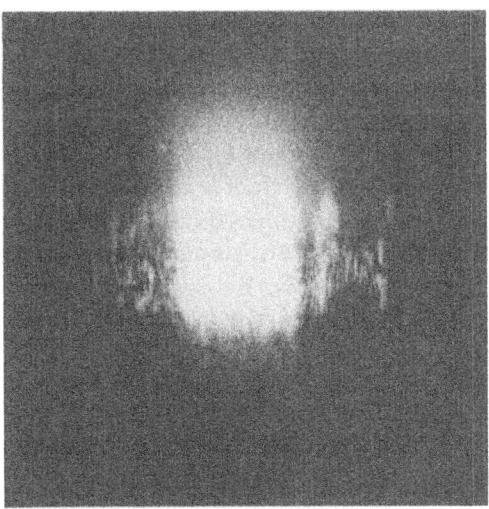

Figure 13b. Reconstructed Image without Spatial Filtering

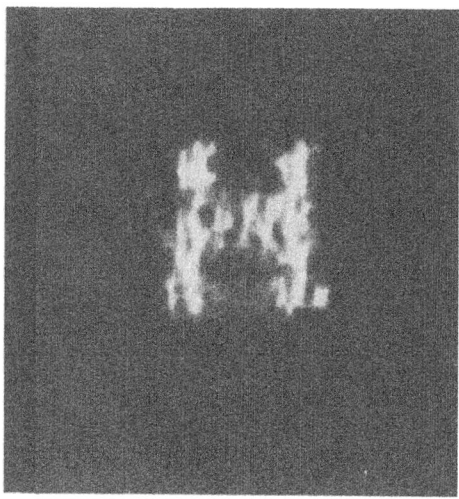

Figure 13c. Reconstructed Image through Spatial Filtering

ultrasonic holography is an ultrasonic imaging method with high performance. It is to be used in practical applications such as medical diagnosis, non-destructive testing and sea bottom exploration, but is prevented from practical use by difficulties of instrumentation.

One of the difficulties of this system is its troublesome signal processing, and experimental considerations concerning the optical signal processing have been given in this paper as compared with the computer processing. The computer processing will be the best way from the point of view of accuracy, but the practical use will be limited in special cases because the two-dimensional reconstruction by computer is not so easy. The side-looking holographic sonar system with optical processing, on the other hand, will be easier than the previous case because its merit is in needing only one-dimensional scanning to make the side-looking image of an object, and its signal processing is more simple and rapid.

The synthetic aperture holographic sonar system has another merit; to be able to use the additional signal processings such as noise elimination and image enhancement. Noise in this case is produced mainly by the turbulence of the water in which the ultrasonic signal propagates, and this type of noise can be eliminated by using a stochastic processing[8] which is suitably applicable in this case. Image enhancement is to be accomplished by the spatial high pass filtering which can be made by optical processing[9], as well as by computer processing.[10]

ACKNOWLEDGEMENTS

The authors wish to express their sincere thanks to Professors T. Sato and M. Ueda of the Tokyo Institute of Technology for their kindly permission to quote their work in this paper.

REFERENCES

1. Cutrona, L. J., Leith, E. N., Porcello, L. J., and Vivan, W. E., Proceedings IEEE, 54, 1966, P. 1026.

2. Kock, W. E., Applications of Holography, Edited by E. S. Barrekette, et al., Plenum Press, 1971, 323.

3. Kock, W. E., Proceedings IEEE, 60, 1972, 1459.

4. Pekau, D. F., and Diehl, R., Applications de l'Holographie, edited by J.-Ch. Vienot, et al., University Besancon, 1970, 15-4.

5. Sato, T., Ueda, M., and Fukuda, S., Bulletin Tokyo Institute Technology, No. 107, 1971, 39.

6. Castella, F. R., Acoustical Holography, Vol. 3, edited by A. F. Metherell, Plenum Press, 1971, 247.

7. Kozma, A., Leith, E. N., and Massey, N. G., Applied Optics, 11, 1972, 1766.

8. Ueda, M., Ikeda, O., and Sato, T., Bulletin Tokyo Institute Technology, No. 112, 1972, 27; No. 113, 1972, 41.

9. Tsujiuchi, J., and Stroke, G.W., Applications of Holography, edited by Barrekette, E.S., et al., Plenum Press, 1971, 259.

10. Billingsley, F.C., Proceedings of Computerized Imaging Techniques Seminar, SPIE, 1967.

PRESENT STATE OF THE CLINICAL APPLICATION OF ULTRASONOTOMOGRAPHY

Toshio Wagai

Medical Ultrasonics Research Center
Juntendo University School of Medicine
Hongo, Bunkyo-Ku, Tokyo, Japan

I. INTRODUCTION

The ultrasonic tissue visualization technique by means of the reflection method (Ultrasonotomography, Echography) has been utilized effectively for clinical diagnoses throughout the entire world. It has various kinds of advantages, and is also highly valued among many techniques in the application of diagnostic ultrasound. According to this trend, ultrasonotomography has been accepted as a new, routine, diagnostic tool, particularly in the fields of internal medicine, neurology, obstetrics and others. At the same time, this method has been discussed in its practical evaluation and limitation in comparison with other usual tissue visualization techniques such as X-ray, Scintigraphy and thermography. In this paper, the outline of the present state of diagnostic application of ultrasonotomography which has been carried out at the Research Center, as well as future prognosis, will be described from a clinical point of view.

II. ULTRASONOTOMOGRAMS

The ultrasonotomograms (echograms) discussed in this paper mean the cross-sectional tomograms of a living human body, or tissue displayed by ultrasonic scanning. As far as the pulse echo method is concerned, echograms theoretically mean the distribution and gradient of acoustic impedance of various tissues which construct the human body. The practical echograms obtained by using usual equipment, however, may be affected by various performances of the equipment such as resolving power, sensitivity, time control, characteristics of the cathode ray tube, signal processing techniques, and others.

On the other hand, there are factors in living tissues that affect the echograms. Various acoustic properties such as acoustic impedance, propagation velocity and ultrasonic attenuation of different tissues are not exactly clear, yet they may delicately affect the echograms. As a lot of echoes can be detected from inside the human body, care must be taken while interpretation of echograms is being performed. Under these conditions, though the investigation of the exact correspondence between echograms and biological tissue structure in connection with the performance of equipment is being carried out from all approaches, it can be recognized that echograms obtained by using recent improved equipment have been satisfactorily utilized in various clinical diagnostics as a sort of tissue visualization showing cross-sectional tomograms of a living body, with such resolution as a macroscopic anatomical level (Figures 1 and 2). Moreover, the ultrasonic tissue visualization shows the excellent ability that enables us to analyze the soft tissue structure more than with X-ray, and to detect pathological tissues; e.g., cancer, with direct information. Such characteristics of diagnostic ultrasound have been remarkably evaluated from a clinical point of view. Furthermore, the fact that pulsed ultrasonic wave gives no hazard nor pain to the patients is one of the big advantages in clinical medicine.

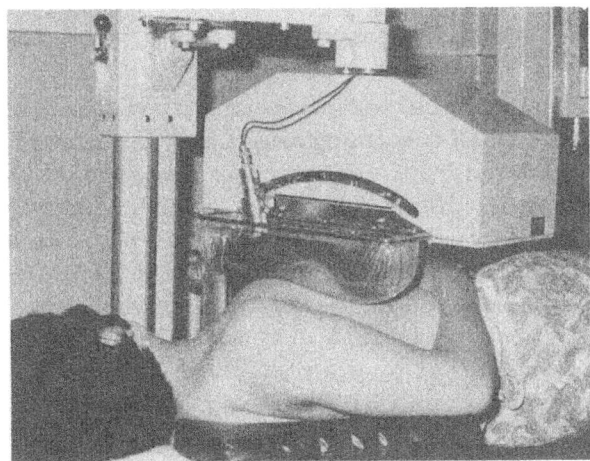

Figure 1. Scanner of the Ultrasonotomographic Equipment for the Examination of the Breast, with Arc Scan

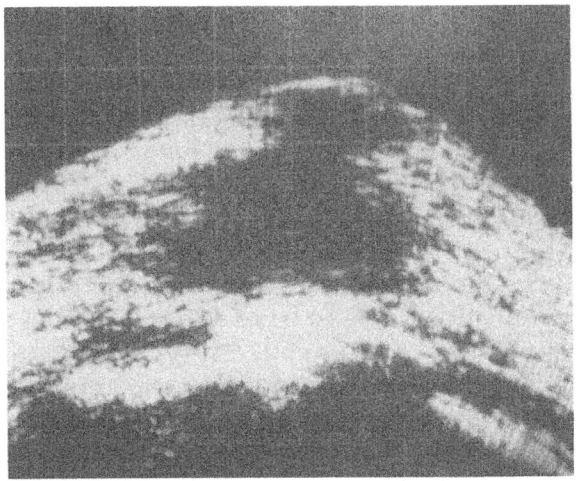

Figure 2. Echogram of Chronic Breast Abscess
5 MHz PZT 10ϕ 75R, Arc Scan

III. CLINICAL APPLICATION AND EVALUATION OF ULTRASONOTOMOGRAPHY

The various characteristics of ultrasonic tissue visualization described in the foregoing have been applied in clinics as a new diagnostic tool. The main purpose of utilizing echography can be divided into two such categories as, the measurement of distance or volume of organs and tissues, and the detection of pathological condition of tissues. Concerning the first category of measuring distance, which is very important in clinical diagnoses, although X-ray has been mainly applied at present, echography is being utilized with simple employment and better accuracy. One example is shown in Figure 3. These transverse echograms were obtained serially by using the contact compound scanning method (Figure 4) in the case of a left kidney tumor (hydronephrosis) with the same intervals. By using these tomograms, not only the analysis of internal structure of the tumor, whether it is cystic or solid, but also the measurement of tumor volume can be easily performed by means of a reconstruction method. The measurements of tissue distance and volume by using echography have been utilized effectively with better accuracy in the case of thyroid gland, liver, spleen, kidney, placenta and various tumors, according to the improvement of resolution on the echograms. The fetal cephalometry by echography is becoming very popular in the world, with error of one percent.

Concerning the second category of detection, and diagnosis of abnormal or pathological condition of tissues, echography has been of great use in clinical medicine. In order to analyze the echograms for such purpose, the distribution of echo intensity and also the shape of abnormal echo image should be considered. In practice, however, the exact and relative comparison of echo intensity on the echograms is not so easy without the establishment of a feedback system which compensates the ultrasonic attenuation in tissue along each different direction of the ultrasonic beam, in addition to the improvement of the usual STC characteristics.

CLINICAL APPLICATION OF ULTRASONOTOMOGRAPHY

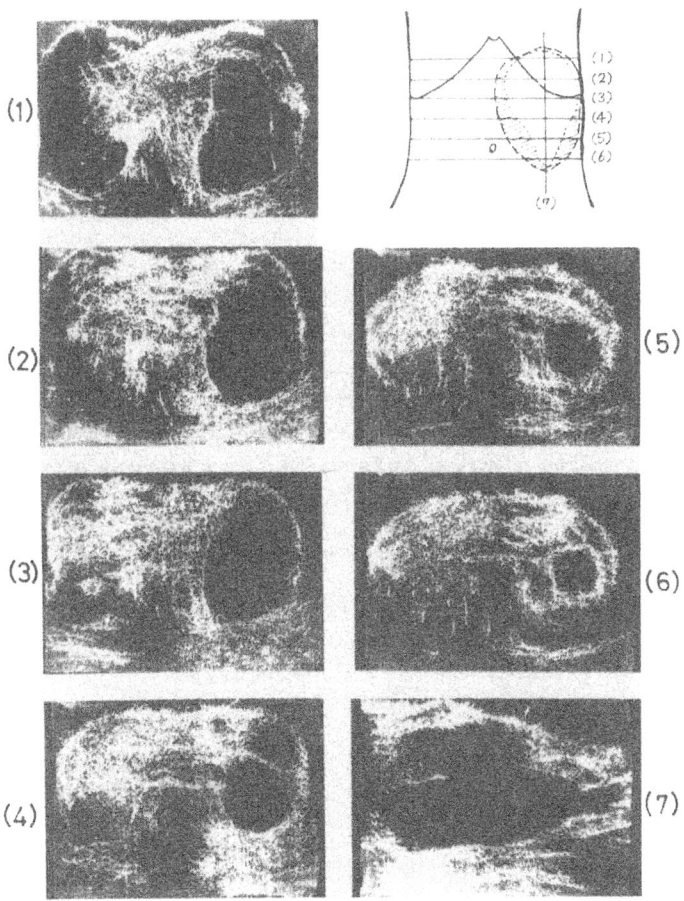

Figure 3. Echograms of Left Kidney Tumor (Hydronephrosis), 2 MHz PZT 20φ, Manual Compound Scan

Furthermore at present, the receiving systems, including CRT, are not entirely free from the saturation phenomenon. However, the method of sensitivity graded tomogram pairs[1] has been proved able to compare quantitatively the echo intensity which constructs the echograms as far as echo sources are located at the same depth from the body surface. This method has been utilized effectively in clinical diagnoses [2,3] Figure 5 shows an example of the sensitivity graded tomogram pairs in breast fibroadenoma (benign tumor). From these six echograms which are obtained by changing the receiving gain in a five dB step, it can be easily understood that the tumor tissue in this case shows rather less ultrasonic attenuation than the surrounding normal breast tissue. Figure 6 shows the echograms obtained by the same method in the case of breast cancer. The cancer tissue shows abnormal increase of ultrasonic attenuation compared with normal breast tissue. Cancer tissue is generally displayed as an abnormal image according to its acoustic characteristics. The acoustic impedance and ultrasonic attenuation are larger than in normal soft

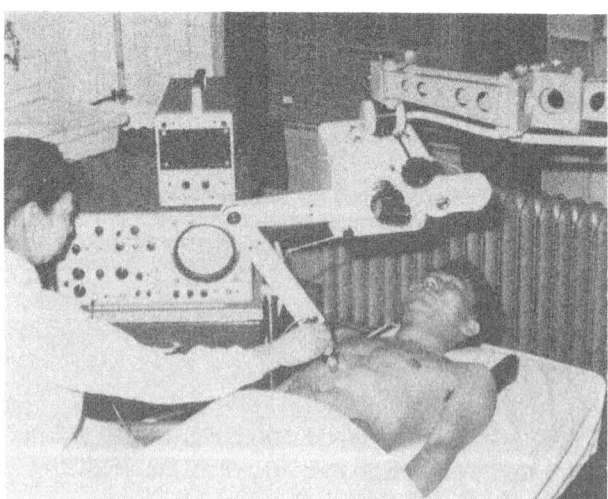

Figure 4. Whole View of Ultrasonotomographic Equipment With Manual Compound Scan and Examination of Abdomen

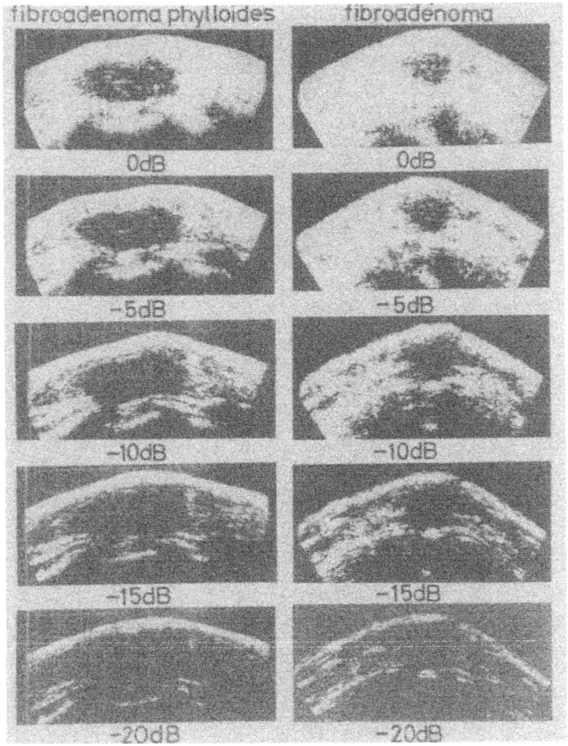

CLINICAL APPLICATION OF ULTRASONOTOMOGRAPHY 559

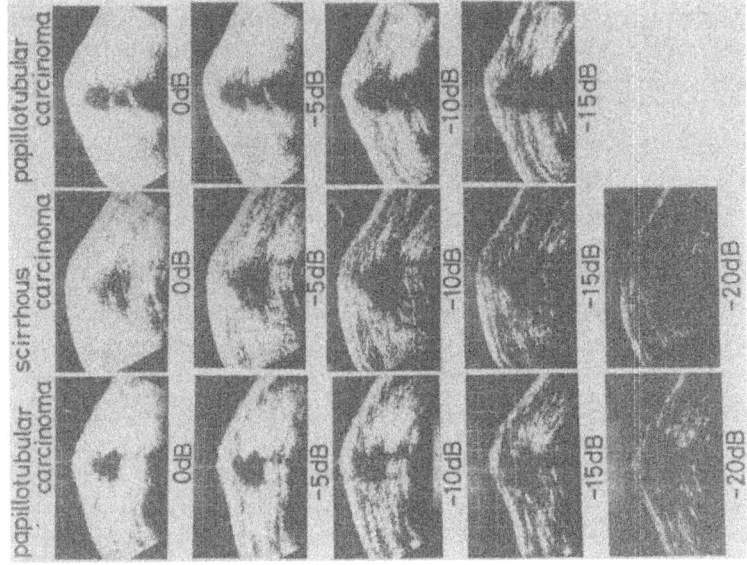

Figure 6. Echograms of Breast Cancer 5 MHz 10Φ75R, Arc Scan

Figure 5. Echograms of Breast Fibroadenoma 5 MHz PZT 10Φ75R Arc Scan

tissue. However, for instance in breast tumors, abnormal image of cancer tissue is much more influenced by its larger ultrasonic attenuation when higher frequency is used.

Color display technique is another sort of quantitative processing method for measuring the distribution of echo intensity which appears on the echograms.[3] It should be borne in mind that the different color corresponding to the different intensity makes it possible to recognize the distribution of echo intensity easily. Using the digital color display system, the grade of black and white which appears on the echograms can be easily displayed on the color monitor, which has different color steps. On the other hand, the video-tape recording of the echograms which is performed directly by electrical signals from ultrasonic equipment is being utilized. By using this technique, repeatable representation of echograms has become very easy, and at the same time, suitable image processing such as color display has become possible; however, the diagnostic evaluation of such image processing as color display technique will be left for future study, with essential development of echography.

On the other hand, the method of ultrasonic diagnosis based on analyzing the shape of abnormal image is being achieved clinically according to the improvement of resolution in the echograms. For improving resolution, transducer, transmitted pulse, beam width, various techniques for signal processing and others were technically discussed, and many available effects were obtained. The signal processing technique has been utilized also for the manual compound scanning scanning which is very popular, along with the use of the storage tube, particularly for the examination of abdominal diseases.

Figure 7 is an example of echograms showing gallstones. Gallstones are detected as abnormal images in the gallbladder. In recent years, the echography for the diagnosis of gallstones has been effectively used at The Research Center as first aid more than for intravenous cholecystography.[2]

CLINICAL APPLICATION OF ULTRASONOTOMOGRAPHY

Figure 8 is an example of echograms obtained in the case of metastatic liver cancer displayed on the storage tube with considerable signal processing. Echograms obtained from the equipment with such properties are generally constructed of rough spots and of linear drawing nature, and the gradient of echo intensity is poorly displayed according to the narrow gray scale of the storage tube. However, abnormal image of metastatic cancer can be diagnosed by attending only to its abnormal and irregular shape, particularly at its boundary. This is due to better resolution in recent equipment. In order to cover the disadvantages of using the storage tube mentioned above, echo signals were displayed simultaneously on the conventional tube. In the latter echograms, spots appeared sharper and the gradient of echo intensity was better, as shown in Figure 9. In this sense, it can be said that the storage tube is very useful in the clinical routine as a monitor of echograms.

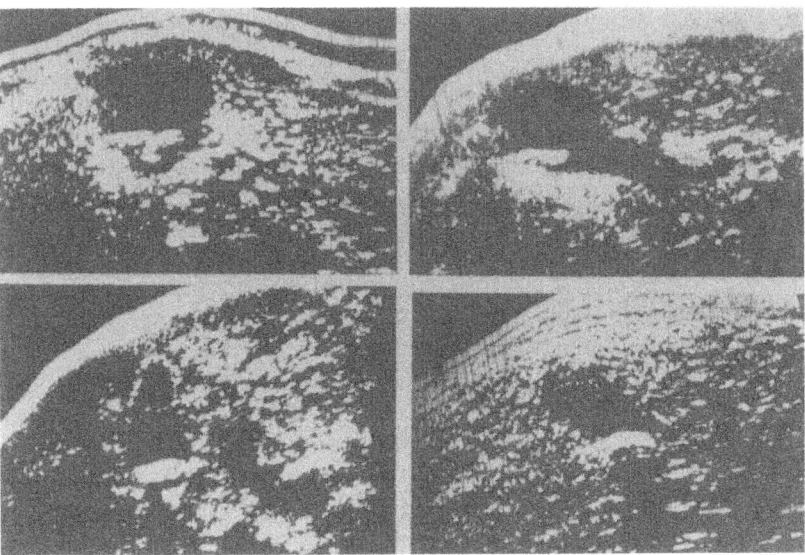

Figure 7. Echograms of Gallstones, 2 MHz PZT 10 Φ
Manual Compound Scan

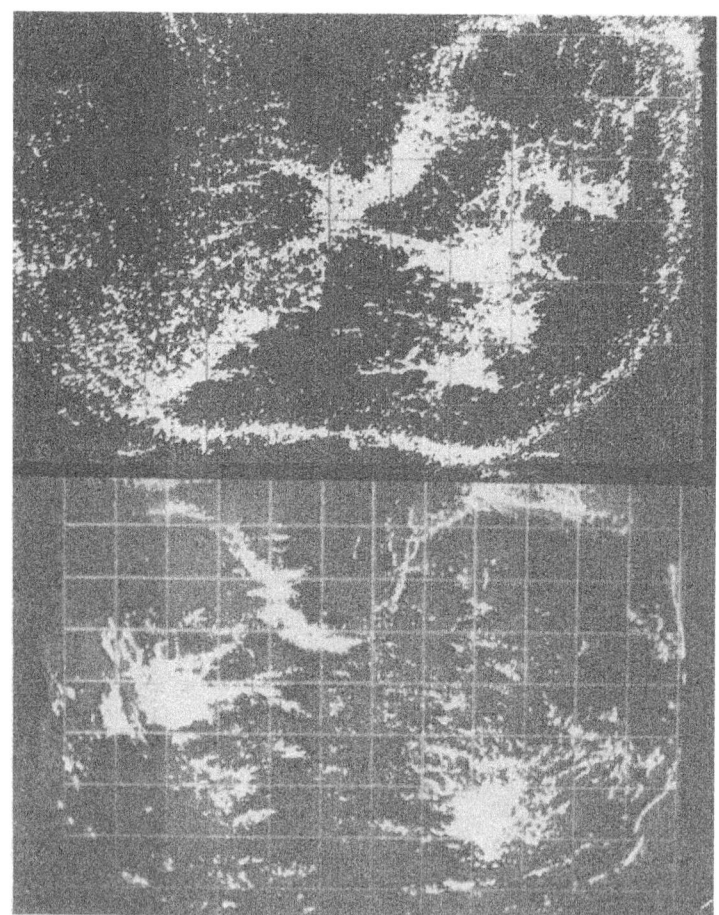

Figure 8. Echograms of Metastatic Liver Cancer
2 MHz PZT 20Φ, Manual Compound Scan

CLINICAL APPLICATION OF ULTRASONOTOMOGRAPHY

(a)
(b)

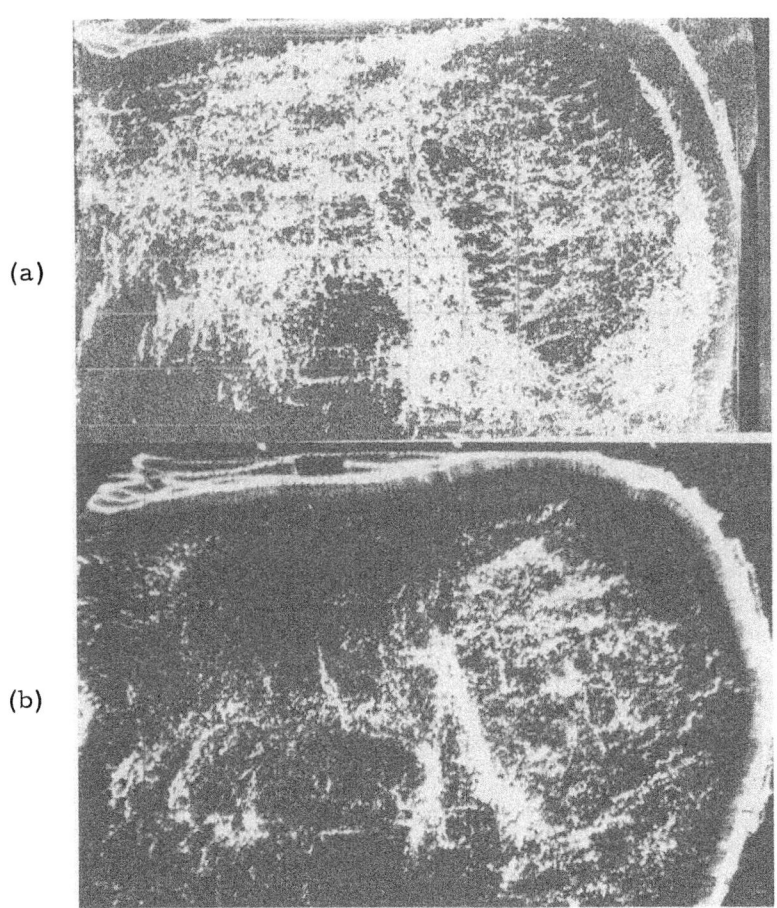

Figure 9. Echograms of Left Gravitz Tumor
2 MHz PZT 20Φ, Manual Compound Scan
(a) is displayed on the storage tube
(b) is displayed on a conventional tube

In order to display much information on one echogram simultaneously, the amplifier with wide dynamic range and the cathode ray tube with a wide gray scale are being used. Figure 10 shows an example of an echogram obtained in benign breast fibroadenoma. Echogram (a) is obtained by using the usual linear amplifier, and echogram (b) is obtained by using a logarithmic amplifier with improved gray scale. As one of the results of using the logarithmic amplifier, it can be easily recognized that the echogram (b) shows fine structure of tumor tissue and better gradient of echo intensity when compared with echogram (a).

Figure 11 shows an example of an echogram with better gray scale obtained in breast cancer. The internal irregular structure can be easily recognized in contrast to benign tumor.

As one of the diagnostic results in the use of echography, The Research Center has achieved a diagnostic accuracy of 90.2% in 210 cases of breast cancer. Even in the early stage of breast cancer, 75.5% in 53 cases was achieved.[3] Such accuracy may be superior to X-ray mammography for the detection of breast cancer. Moreover, echography has been effectively utilized for the detection of cancer in the thyroid gland, the liver, the pancreas, the ovaries, the uterus, and others.

IV. CONCLUSION

The present state of application and clinical evaluation of ultrasonotomography has been described from the standpoint of clinical medicine. The development and popularization of ultrasonotomography as a routine clinical diagnostic tool are very much indebted to the remarkable improvement of equipment, particularly of resolution in echograms. In the near future, improvement in technology, as well as an increase in medical knowledge in this field would make a promising prospect in clinical diagnostics. On the other hand, as the ultrasonotomography reveals only the

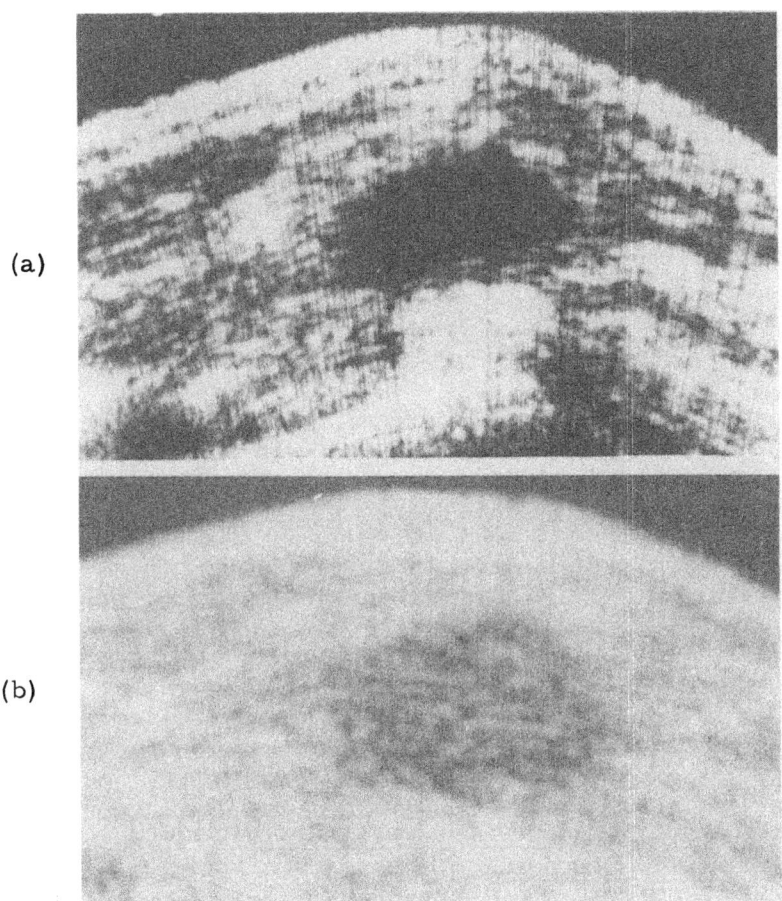

Figure 10. Echograms of Breast Fibroadenoma
5 MHz PZT 10Φ75R, Arc Scan
(a) is obtained by using the usual linear amplifier; (b) is obtained by using a logarithmic amplifier with better gray scale.

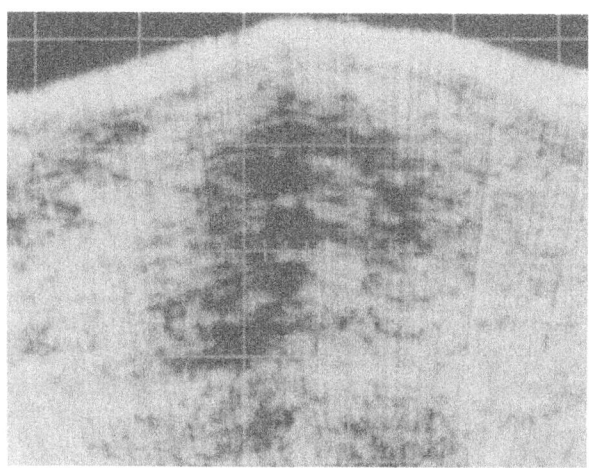

Figure 11. Echogram of Breast Cancer Obtained by Using a Logarithmic Amplifier, with better gray scale; 5 MHz PZT 10ϕ 75R, Arc Scan.

cross-sectional image, the inconvenience of not recognizing the three-dimensional image of the human body must be removed. If possible, more improvement in resolution will be expected in ultrasonic holography in the future.

REFERENCES

1. Kikuchi, Y., Way to Quantitative Examination in Ultrasonic Diagnosis, Medical Ultrasonics, P. 1-8, Vol. 6, No. 1, June, 1968.

2. Wagai, T., et al., Annual Report of the Medical Ultrasonics Research Center, Juntendo University School of Medicine, April, 1971.

3. Wagai, T., et al., Detection of the Breast Cancer at the Early Stage by Ultrasonic Scanning Method, Annual Report of the Medical Ultrasonics Research Center, Juntendo University School of Medicine, No. 1, June, 1972.

SOME PATHOBIOLOGICAL CONSIDERATIONS OF DETECTION OF BREAST CANCER BY ULTRASONIC HOLOGRAPHY

Leonard Weiss

Department of Experimental Pathology

Roswell Park Memorial Institute, Buffalo, NY

In this presentation some of the expected problems associated with the early diagnosis of carcinoma of the breast, using ultrasonic holography, will be considered. In the absence of much practical experience at a clinical level with ultrasonic holography, I think that some useful general lessons on procedure, and particularly the frightening logistic problems of mass screening, may be learned from the now considerable body of data available on the early diagnosis of mammary cancer by X-ray techniques. I will not deal with the engineering problems associated with ultrasonic holography, and others are much better qualified than I to comment on its feasibility for this purpose. However, I hope to show that from a consideration of some of the basic pathobiological aspects of the disease, a reasonably clear idea of the engineering requirements can be obtained.

THE MAGNITUDE OF THE PROBLEM

Some idea of the numerical problem of cancer may be obtained from " '73 Cancer Facts and Figures" published by the American Cancer Society.

About 53 million Americans now living will eventually have cancer. That is, one in four persons. In 1972, an estimated 344,000 Americans died of cancer; and in 1973, it is estimated that about 350,000 will die of the disease.

Cancer of the breast is the commonest cause of cancer and death among women, and is the main cause of all deaths among women between the ages of 40 and 44. It is estimated that 73,000 new cases of cancer of the breast in American women will occur in 1973, and that during this period, 32,400 women will die of the disease.

In their early stages, most types of cancer are localized disease processes which, in the breast, may originate in a duct. Such "in situ" cancers are not visible to the naked eye, and after removal for whatever reason, the patient is cured. The next step in the natural progress of the disease is when the growing cancer penetrates beyond the surface from which it arises, and invades the underlying tissue. For a time, the cancer cells remain localized. If diagnosed at this stage and treated, the 5-year survival rate is around 85%. If undiagnosed, the cancer cells spread to other parts of the body to form secondary growths which are not in contact with the primary cancer. This process is known as metastasis. Cancer cells which have become detached from the primary tumor may be trapped in one of the lymph nodes which drains the breast. Often these nodes are in the axilla. The trapping of tumor cells in this manner slows down the spread of the cancer, and is known as "regional involvement". The 5-year survival rate for a patient treated with carcinoma of the breast with regional involvement is approximately 53%. If the cancer is left untreated at this stage, cancer cells are distributed to other parts of the body where they may grow. This is known as "advanced cancer", which is virtually incurable, although the time taken for a patient to die is variable.

TUMOUR SIZE AND DIAGNOSIS

It is fairly obvious that patients with advanced cancer will not do as well under treatment as patients with early localized disease. A more subtle approach illustrating the effect of breast cancer size at diagnosis, on the expected course of the disease comes from the statistical studies of Slack et al (1). These workers have classified and analysed clinical data from large numbers of patients with breast cancer, and on the basis of the time taken for a tumour to double in size, and the presence or absence of axillary node (regional) and/or distant metastases, they have classified breast cancer into two clinical types as follows:

	DOUBLING TIME	RISK OF AXILLARY NODE METASTASIS	RISK OF DISTANT METASTASIS
TYPE A	SHORTER (c. 0.7 months)	HIGHER (7.7)	HIGHER (3.2)
TYPE B	LONGER (c. 1.5 months)	LOWER (1.0)	LOWER (1.0)

On clinical data, it can hopefully be determined whether a tumour falls into type A or B, and from its size at diagnosis, the probability of a distant metastasis may be calculated as a percentage, as follows (2):

DIAMETER OF PRIMARY LESION	PROBABILITIES OF DISTANT METASTASIS	
	TYPE A	TYPE B
1 mm	52%	22%
2 mm	62%	28%
3 mm	67%	32%
4 mm	70%	34%
5 mm	73%	36%
6 mm	75%	38%
7 mm	77%	39%
8 mm	78%	41%
9 mm	79%	42%
1 cm	80%	43%
1.5 cm	84%	47%
2 cm	86%	49%
3 cm	89%	53%
4 cm	90%	55%
5 cm	91%	57%

It can be readily appreciated from the tabulated increased probabilities of distant spread, with the associated, progressively worse prognosis, that it is very worthwhile to aim for methods of detection of breast cancers, while they are still barely visible to the naked eye. According to these computations, the difference in size at diagnosis between 1 mm and 1 cm diameter cancers, can be reflected in very considerable differences in the course of the disease.

Projections such as these serve to crystallize the doubts expressed by clinicians for many years on the adequacy of manual palpation alone, in the early diagnosis of breast cancer, whether this examination is performed by the patient herself, or by her clinician. The average thickness of two layers of skin and associated tissues covering a cancer within the breast is approximately 1 cm, which in itself is a major factor hindering thepalpation of tumours less than 1 cm in diameter. It is probably generally accepted that the smallest size lump - which may consist partly of cancer and partly of non-cancerous reactive tissues - which can be detected by manual palpation is not much less than 1 cm diameter in a "suitable" breast. The meaning of "suitable" will be discussed shortly. The relevance of these remarks is substantiated by consideration of the approximate 95% of patients with breast cancer, who have discovered the condition themselves through self-examination of their breasts. Unfortunately by this time, in 60% of them, the cancer has spread to the axillary lymph nodes and as stated, the 5-year survival rate has dropped to approximately 45% in comparison to the 85% survival rate in patients with localized cancer. It is recognized that this situation is desperate, and to promote the early detection of cancer of the breast while it is still localized and at such an early stage as to be 100% curable, the American Cancer Society formed a Task Force on Breast Cancer in 1972. Among existing diagnostic techniques to be fully utilized in addition to manual palpation, are standard X-ray mammography, Xeroradiography and thermography.

Following the pioneering studies of Egan (3), it has become generally accepted that X-ray mammography is a valuable adjunct in the early diagnosis of breast cancer. It would appear that Xeroradiographic techniques may extend the usefulness of this approach. As approximately 30% of malignant tumours of the breast show rather characteristic calcification patterns, the usefulness of X-ray mammography is enhanced (4).

Another current technique is thermography, in which local elevation of breast temperature is sought as an early diagnostic sign. In a review of the results obtained by a number of workers over the period 1964-1968, Haberman (5) notes that in a series of 16,409 patients in which 795 had proven breast cancer, thermography revealed 86.9% of them.

If ultrasonic holography is to take its place in this programme, it must show promise of adding to the armentarium of established diagnostic techniques.

BREAST-TYPE AND DIAGNOSIS

The earlier comment that lumps of approximately 1 cm diameter can be palpated in "suitable" breasts, implies that the characteristics of the non-cancerous tissues surrounding these lumps will play a significant role in determining those patients in whom diagnostic methds other than palpation will be of increasing importance. The parameters of "unsuitability" should therefore be carefully considered in this Symposium, since they represent the real challenge to physical diagnostic techniques. Various factors discussed by Thiessen (6) which contribute to "suitability" for manual palpation of tumours, are shown in the following table:

MACROSCOPIC PARAMETERS	"SUITABLE"	"UNSUITABLE"
BREAST SIZE	SMALL to MEDIUM	MEDIUM to LARGE
"SUPPORT"	PENDULOUS, POORLY SUPPORTED	FIRM, WELL SUPPORTED
"DENSITY"	SOFT or FATTY	FIBROUS or DENSE MATRIX
"NODULARITY"	SMOOTH	NODULAR

Thiessen considers that size is the most important parameter determining suitability for self examination, and "support" best summarized the other factors. The type of breast with which we should be particularly concerned, as material for physical diagnostic methods is therefore medium to large in size, well-supported, with a dense and nodular matrix.

MASS SCREENING

The well-recognized inadequacies of palpation of the breast for the early diagnosis of cancer has been discussed above. A case for mass screening by physical techniques, to reinforce palpation comes from studies made on the effectiveness of X-ray mammography, in this connexion. Strax (7) has reported the results of a study on 6604 women over the age of 35, in whom 55 cancer were found, corresponding to an incidence of 8.3 per 1,000. Of these 55 breast cancers, 19 were positive on clinical examination only, 10 were positive on X-ray mammography only and 26 were positive by 2 or more diagnostic methods. The results of a 4 year study of 62,000 women, aged 40 to 64, by New York's Health Insurance Plan (reported in Medical World News, November 17, 1972) show that X-ray mammography contributed substantially to detection. For example, in women under 50 with smaller, glandular breasts, 19% of the cancers were only detectable by mammography. However, the shortcomings of X-ray mammography are illustrated by the fact that in this latter group of patients, 61% of the breast cancers were detectable only by clinical examination. In women over 50, the two methods yielded approximately equal detection rates. A very important finding was that of the cancers detected by mammography, only approximately 20% had spread to the regional lymph-nodes, compared with approximately 55% in comparable groups not receiving regular X-ray screening. When this earlier effective diagnosis is translated into expected better therapeutic results, the indication is that by a combination of clinical examination and mammography, 30% of deaths due to cancer of the breast-say 10,000 women per year - could be avoided. It is of course possible that other diagnostic procedures, including acoustic holography could produce even further improvements.

Some idea of the time and resources required to implement such a mass screening programme can be obtained from experience and projections in X-ray mammography. Taking Strax's (8) estimates, 10 patients per hour can be examined radiographically in hospitals, and 20 per hour in a mobile unit. Dowdy et al (9) calculate that to annually screen the 38 million women in the U.S.A. who are 40 years of age and older, 1,900 centres screening 100 patients per day would be required. At present this would require 3,800 mammographers plus 26,000 technologists, secretaries and allied health personnel, at a projected annual cost exceeding $620,000,000! Even cutting the number of centres by speeding up the examinations would still require more trained radiologists than are available, although paramedicals can be trained to read mammograms. Consideration of the vast numbers of patients involved, clearly indicates that the apparatus used to reinforce manual palpation must be simple enough to operate so that patients should be examined at rates comparable to those achieved by X-ray mammography (i.e. 3 to 6 minutes).

A technique which currently shows great potential for mass screening is Xeroradiography. In essence, X-rays passing through the breast impinge on a charged selenium photoconductor, discharging it proportionally to the localized radiation. The image is rendered visible and transferred to paper. Compared to conventional mammograms, the xeroradiograms are said to be easier to produce and interpret. In a trial reported by Wolfe et al (10), 84.3% of true positive breast cancers were diagnosed by xerographic techniques, compared with 72.2% diagnosed by conventional X-ray mammography. In common with mammograms, xeroradiograms currently involve the human element in interpretation, and this brings about a major bottleneck. An interesting development therefore, is the attempt by Ackerman and Gose (11) to classify lesions seen on xero-radiograms by computer. In essence, a number of proven lesions were digitized by computer and recorded on film, and a number of "malignant" characteristics were mathematically extracted from the digitized lesions. Using 3 different methods of classification, these were identified by the computer as malignant and benign. In a trial, the computer performed as well as an expert radiologist. In my opinion, any diagnostic technique should have the potential for associated computer diagnosis, if it is to be seriously considered for mass screening.

REGIONS OF THE BREAST TO BE EXAMINED

The breast may extend from the second rib, down as far as the seventh costal cartilage, and extend laterally from the edge of the sternum to the anterior fold of the axilla. The most important single factor in determining breast shape and size, apart from pregnancy and its sequelae, is the amount of fat present. In addition to this overt and protruberant part of the breast, it can also spread in a thin layer up as far as the clavicle and laterally towards the posterior axillary fold. The upper and outer quadrant of the breast is bulkier than the others, which may account for the higher incidence of cancer in this region as indicated below. The breast may also extend towards the axilla as a "tail" of variable size (12).

Ideally all of the breast should be examined. However, it is sometimes impossible to do this because of anatomical and mechanical configuration problems. The question then arises of what proportion of tumours would be missed on a statistical basis, by the exclusion of different regions from examination. This question is partially answered by the diagram shown below, which is based on the data given by Fisher et al (13):

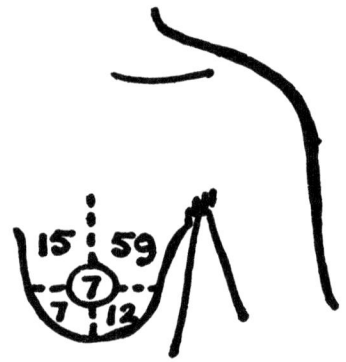

Primary carcinoma occuring in the axillary tail of the breast accounts for less than 1% of all breast cancer in women (14).

ULTRASONIC HOLOGRAPHY

An attempt to assess the utility of ultrasonic holography in the early diagnosis of soft tissue tumours was made by Weiss and Holyoke (15). We made our studies on Walker 256 tumours implanted into skin flaps, previously raised surgically from the flanks of rats. In the experiments, 5 x 5 cm flaps were examined by double-beam ultrasonic holography and by standard soft tissue X-ray procedures. In addition to skin flaps containing tumours, inflammatory lesions produced by injecting the irritants, turpentine and croton oil, were also examined. The size of both the neoplastic and inflammatory lesions ranged from 18 x 23 mm down to 2 x 2 mm. After examination, 15 to 20 ml of 3% gelatin were injected into some of the skin flaps under pressure, and they were then reexamined. The results are shown in the following table:

SKIN FLAP CONTAINS:	ULTRASONIC HOLOGRAPHY (9 MHz)	
	EXAMINED	SEEN
TUMOUR	51	49 (96%)
" + GELATIN	18	12 (67%)
INFLAMMATORY LESION	44	38 (87%)
" + GELATIN	7	6 (85%)

A curious and unaccountable observation was that some tumours as large as 5 x 5 mm were not visualized, in the presence or absence of gelatin, whereas others as small as 2 x 2 mm were seen. In our short series we were unable to discriminate between granulomatous abcesses and cancer.

An observation which may be of considerable practical importance in the context of the early diagnosis of cancer, was the comparison of the proportions of tumours visualized in the same skin-flaps, by ultrasonic holography and X-rays, before and after the injection of gelatin:

TUMOUR VISUALIZATION

TECHNIQUE	SKIN FLAP CONTAINS	
	TUMOUR	TUMOUR + GELATIN
ULTRASONIC HOLOGRAPHY	10/10	6/10
X-RAY	5/10	0/10

Under these defined conditions in the experimental system used by us, ultrasonic holography is clearly superior to X-ray techniques, in the detection of tumours.

These experiments, made some years ago, with equipment which would now be considered sub-optimal, raised the definite possibility that ultrasonic holography was capable of detecting soft tissue tumours and, in addition some of these tumours might not be detectable by X-ray techniques.

The various improvements in the equipment will be dealt with by Brenden elsewhere in this Symposium. I wish to consider two points arising from our earlier work. The first is why we would expect to see breast tumours at all with ultrasonic holography, and secondly, to attempt to define the situations where these techniques might reasonably be expected to add to X-ray mammography.

Ultrasound interacts with tissues, and consequently may be diffracted, absorbed, scattered, refracted and/or reflected. It must be remembered that the breast tissue consists of skin, fat cells, secretory endothelium, blood-vessels, and connective tissues in addition to possible cancer cells. All of these components have a complex spatial arrangement one to another, quite apart from their properties with respect to ultrasound. Microscopic examination of breast cancers, with the exception of mucoid carcinoma, does not suggest to me that any profound differences are to be expected in the interactions of cancer and normal epithelial cells with ultrasonic irradiation. El'piner's (16) data show only minor differences in the absorption coefficients (α cm^2), measured at one MHz, of kidney (0.22) and skeletal muscle (0.20-0.25), where differences in biological properties are enormous. Gregg and Palagallo's (17) measurements also revealed no appreciable differences in reflection coefficient and/or acoustic impedance between metastases in the liver and brain, and adjacent normal tissues. In addition, as noted above, Holyoke and I could not discriminate between inflammatory and cancerous lesions by ultrasonic holography. If the physical properties of the cells themselves are not responsible for the visualization of tumours by ultrasound, what other factors are responsible? The answer may perhaps be found from a consideration of the macroscopic properties of some "typical" cancers, and the tissues surrounding them.

It is well-known that cancers often "outgrow" their blood supply. Blood vessels on the arterial side enter tumours from their peripheries and proceed to their centres. Tissue pressures tend to occlude the small blood vessels, particularly on the venous side, resulting in vascular stasis; blockage or thrombosis occurs, and capillary shunts are formed near to the tumour periphery, which also divert blood from the centre of the cancer. As a result of nutritional impairment, cells at the centres of solid cancers more than a few millimeters in diameter die, and liquefy. The cancerous region is thereby converted into a cyst, containing dead material which undergoes changes in physical properties analogous to a continuous transition from a solid tissue to whole blood, and finally to cell-free blood plasma. The absorption coefficients of these materials at 1 MHz might well correspond to kidney (0.22), blood (0.02) and blood plasma (0.007), according to the values given by El'piner. The absorption coefficient of blood is directly proportional to its intra- and extracellular protein content (18). It would therefore be predicted that cancers exhibiting cystic changes, in which a sharply defined zone of high absorbance, surrounding a central liquefied, necrotic core of comparative low absorbance, would be detectable by ultrasonic techniques.

It should also be noted that as tumours get larger and more degenerate, lysosomal enzymes are released in increasing quantities by both the cancerous and associated non-malignant tissue cells, which attack and weaken cellular interfaces, thereby changing their mechanical properties, and hence their interactions with ultrasonic irradiation.

So far then, "contrast" has been considered in terms of an "ultrasonic differential" between the centre of a tumour and its cortex. Contrast could also be enhanced by the existence of such differentials, between the cortical region of a tumour and its surrounding tissue matrix. In the breast, this matrix is usually largely composed of fat. Unfortunately, most of the available data on the interactions of fat with ultrasound were obtained at room-temperature (20-25°C), when human fat is solid. At the physiological temperature of 37°C, human fat is liquid, and it is known that the acoustic absorption of at least some tissues show a marked temperature dependence (19). At 5 MHz at 37°C, the average reflextion for fat in the living animal (dog) is 11.56% and the acoustic impedance (ρV) is 1.55×10^5 g. cm^{-2}.$sec.^{-1}$, and at 25°C, the corresponding post-mortem values are 13.58% and 1.49×10^5. These compare with values for liver of 6.14% reflexion and 1.75×10^5 ρV at 37°C, and 8.86% and 1.64×10^5 ρV at 25°C (17). The main point here is that to the advantage of ultrasonic techniques, in the human breast, tumours are often imbedded in a "high contrast" fatty matrix, which should favour their detection, particularly if they are associated with cysts. In contrast to pulse-echo imaging, abrupt discontinuities in acoustic impedance across well-defined interfaces are not required in ultrasonic holography, where the primary requirement is a differential in absorption coefficient. This is of some importance since the contours of breast cancers are often indistinct and irregular (20).

If the detection of breast cancer is to depend on the differential absorbance between cancerous cortex, surrounding a necrotic core region, and itself being surrounded by a fatty matrix, it is not difficult to appreciate why Holyoke and I were unable to discriminate between cystic tumours and cystic inflammatory lesions. The problem then arises of discriminating between early malignant lesions in the breast, with and without central necrosis, and corresponding areas of benign nodular hyperplasia. It can only be suggested that if doubtful regions are discovered in the absence of detection by palpation or X-ray methods, then they could be followed to note increase in size which experience would demonstrate to warrant biopsy. The suggestion is analogous to the diagnosis of minimal breast cancers of less than 5 mm diameter by repeated X-ray examination for the appearance of new densities in serial examinations (21). It seems inevitable that if ultrasonic holography is to be developed as a potent diagnostic tool, then the diagnostician will have to go through the painful cataloguing and learning processes which marked the early days of roentgenology.

The next obvious comparison to make, is that between ultrasonic techniques and X-ray mammography, particularly in view of the clear-cut superiority of the former in detecting tumours in the skin flap model. This finding was not entirely unexpected since one of the main drawbacks of X-rays in diagnosing solid tumours is that they do not markedly differentiate between different soft tissues, since many of these tissues have very similar absorption coefficients for this type of electromagnetic radiation. Although the occurence of calcification, which is often associated with cancer of the breast, is readily detectable by X-rays, it is expected that such calcification would also readily be detected by ultrasonic holography. In addition, a hologram records both amplitude and phase distribution in the ultrasonic field, and avoids the hazards of exposure of the patient to ionizing radiation.

One has the distinct impression that if the various engineering and other problems can be overcome, <u>and these are major problems</u>, then ultrasonic holography merits serious consideration for inclusion in the battery of techniques to be tested for the early diagnosis of breast cancer.

HAZARDS

With any form of diagnostic procedure, it is of great importance to consider the possibilities of untoward injury to the patient, particularly in view of the work of Dunn and Fry (22) showing that a linear relation exists between log. intensity and log. pulse length in causing focal lesions in tissues, by a single pulse of ultrasonic radiation. They proposed that tissue damage is caused by three primary mechanisms: (1) Heat production, particularly by long single pulses of relatively low intensity; (2) Mechanical factors, particularly by short pulses of higher intensity, and (3) Cavitation effects, particularly by high intensities (10^4 W. cm^{-2}) and short pulses (10^{-4} seconds).

With the background of X-irradiation-induced cancer, we should also be mindful of long-term, latent damage. In the case of diagnostic procedures aimed at detecting cancer, the possibility must also be entertained that the radiation itself may promote the detachment of malignant cells from the primary tumour favouring spread of the disease with the formation of metastases. Detachment may be brought about directly by radiation-induced tissue movements (23), and/or indirectly by radiation-induced release of lysosomal and other enzymes which also promote separation of tumour cells (24,25). Ultrasonic radiation has been shown to produce changes in cell membranes (26) which may be relevant in this latter connexion(23).

Immediate and overt damage to tissues, as a result of <u>diagnostic</u> doses of ultrasound, appears most unlikely on the basis of published data. The most common source of confusion in the literature is the failure to discriminate between focussed ultrasound, which has been deliberately used to produce tissue damage, and the non-focussed beams which are used for diagnostic purposes. An illustration of the difference between hazard-tests and diagnostic procedures comes from the following table which shows the conditions of irradiation under which damage was <u>not</u> found in the

	WEISS and HOLYOKE (15) DIAGNOSTIC	HAZARD	TAYLOR and POND HAZARD TEST (27)	HOLOSONICS INC. DIAGNOSTIC
FREQUENCY (MHz)	9	9	0.5, 1, 2, 6	3
PULSE LENGTH	8×10^{-5} s.	Continuous	0.01 s.	8×10^{-5} s.
PULSE SEPARATION	2×10^{-3} s.	—	0.1 s.	1.6×10^{-2} s.
PEAK INTENSITY	5 W.cm^{-2}	2.5 W.cm^{-2}	56 W.cm^{-2}	0.8 W.cm^{-2}
AVERAGE INTENSITY	0.3 W.cm^{-2}	2.5 W.cm^{-2}	5.6 W.cm^{-2}	0.004 W.cm^{-2}
DUTY FACTOR	0.64	1.0	0.1	0.005
EXPOSURE TIME	15 s.	60 s.	300 s.	300 s.
TOTAL EXPOSURE	3 W.s.cm^{-2}	150 W.s.cm^{-2}	1680 W.s.cm^{-2}	1.2 W.s.cm^{-2}
TISSUE DAMAGE	NONE	NONE	DAMAGE at 0.5, 1 and 2MHz None at 6MHz	NOT DONE

tissues of rats and mice by Weiss and Holyoke (15) but was observed in the livers of rats by Taylor and Pond (27). These conditions are compared with current imaging practices used by Holosonics Inc., as discussed by Brenden, elsewhere in this Symposium.

Thus, from these data, it can be concluded that at very high energy levels, more than three orders of magnitude higher than currently used in diagnostic ultrasonic holography, liver damage can occur at frequencies below 6MHz. Taylor and Pond (27) have made a careful study of the production haemorrhagic lesions in the spinal cords of rats irradiated with frequencies of 0.5 to 6MHz at peak intensities of 25 or 50 $W.cm^{-2}$ with pulse lengths of 10 milliseconds, at intervals of up to 200 milliseconds, for periods of up to 15 minutes. They conclude that both total exposure and pulse length would have to be increased greatly over those used in current diagnostic practice, before damage is expected. These authors consider that "greatly increased dosage could occur by.... increased pulse lengths for holographic techniques". However, Holosonics Inc. for example, are currently using peak intensities of 0.8 $W.cm^{-2}$, at pulse lengths of only 0.08 milliseconds, at intervals of 16 milliseconds for periods up to 5 minutes. This expectation of Taylor and Pond (28) is therefore clearly not applicable. It must be emphasized that a search of the literature has failed to reveal evidence of tissue damage caused by exposure to as little as 100 $W.sec.cm^{-2}$ at any frequency of ultrasonic irradiation.

The spectre of long-term effects of ultrasonic irradiation, particularly carcinogenesis, was raised by Macintosh and Davey's (29) study on the relationship between the intensity of irradiation and the induction of chromosomal abnormalities in human peripheral lymphocytes, following their subsequent exposure to the mitogen, phytohaemagglutinin. When the intensity of ultrasonic radiation was varied, at a fixed frequency of 2MHz for 1 hour, it was claimed that above calculated levels of .0082 $W.cm^{-2}$, chromosomal damage was detected. I will ignore the very important question of whether or not the chromosomal anomalies observed by these authors were potentially dangerous or not, because it can reasonably be argued that such anomalies are undesirable. However, there is a more important aspect of this work, which is that the authors did not take into account the fact that they used a focussed transducer for specimen irradiation. When focussing is considered, the effective dose delivered to at least some of the cell population, was between 30 to 100 times the stated dose. The total exposure by Macintosh and Daveys causing chromosome damage are therefore in excess of 900 $W.sec.cm^{-2}$, which are well outside the 2 $W.sec.cm^{-2}$ range envisaged in diagnostic practice. Experiments which are more realistic in the present context, in which lymphocytes were radiated with plane waves at 1 MHz for 20 seconds at intensities up to 13.7 $W.cm^{-2}$, revealed no dependence of chromosome aberrations on ultrasonic intensity (30). Similar negative finding after irradiation of lymphocytes at 1MHz for 1 hour, at intensities up to 3 $W.cm^{-2}$, have been reported by Buckton and Baker (31).

Studies on the effect of ultrasonic irradiation on tumour growth and spread were made by Holyoke and Weiss (15). Mice were inoculated subcutaneously with T-241 Lewis sarcoma, which was allowed to grow to 1 cm^3 in size. The animals were irradiated at 9 MHz, to receive 150 $W.sec.cm^{-2}$. When examined 12 days later, irradiation had no detactable effect on either pulmonary metastasis or tumour size. In vitro studies of the resistance of hamster cells to 1.5 MHz radiation for overall exposure times of up to 1 hour, at intensities up to 8.8 $W.cm^{-2}$, have also revealed no effect of ultrasound on cell reproductive integrity (32).

It should be **unequivocally** stated that irradiation of the type and intensity proposed for ultrasonic holography, even allowing for a very wide safety margin, has not been demonstrably associated with harmful effects. It appears at present, that the hypothesis of Connolly and Pond (33) that there are critical levels of ultrasonic irradiation, below which damage is not expected is still tenable. In my opinion, there are serious problems in adapting ultrasonic holography to the diagnosis of cancer of the breast, but the problem of hazard is not one of them.

In conclusion, I would strongly advise physicists and engineers to consult with oncologists before producing hardware for the diagnosis of cancer, since equipment is subservient to the needs of the patient and the peculiarities of the disease process being studied, rather than the reverse. It is also as well to appreciate that much of the inherent conservatism of the medical profession towards new techniques stems from a genuine concern for patients, and the desire not to subject them to unnecessary, potentially harmful and almost certainly expensive procedures.

REFERENCES

1. Slack, N.H., Blumenson, L.E. and Bross, I.D.J. Therapeutic implications from a mathematical model characterizing the course of breast cancer. Cancer 24:960-971 (1969).

2. Blumenson, L.E. Personal communication, 1972.

3. Egan, R. "Mammography" C.C. Thomas (1964).

4. Gershon-Cohen, J., Berger, S.M. and Curico, B.M. Breast cancer with microcalcifications. Diagnostic Difficulties. Radiology 87:613 (1966).

5. Haberman, J.D. The present status of mammary thermography. Cancer 18: 315 (1968).

6. Thiessen, E.U. Breast self-examination in proper perspective. Cancer 28: 1537-1545 (1971).

7. Strax, P., Venet, L. and Shapiro, S. Mass screening in mammary cancer. Cancer 23:875-878 (1969).

8. Strax, P. New techniques in mass screening for breast cancer. Cancer 28:1563-1568 (1971).

9. Dowdy, A.H., Burker, W.F. et al. Mammography as a screening method for the examination of large populations. Cancer 28:1558-1562 (1971).

10. Wolfe, J.N., Dooley, R.P. and Harkins, L.E. Xeroradiography of the breast: A comparative study with conventional film mammography. Cancer 28: 1569-1574 (1971).

11. Ackerman, L.V. and Gose, E.E. Breast lesion classification by computer and xeroradiograph. Cancer 30: 1025-1035 (1972).

12. Haagensen, C.D. Diseases of the breast. 2nd Edn. W.B. Sanders Co.: Philadelphia (1971).

13. Fisher, B., Slack, N., Bross, I.D.J. et al. Cancer of the breast. Size of neoplasm and prognosis. Cancer 24:1071-1080 (1969).

14. Larsen, R.R., Sawyer, K.C., Sawyer, R.B. and Torres, R.C. Occult carcinoma of the breast. Amer. J. Surg. 107:553-555 (1964).

15. Weiss, L. and Holyoke, E.D. Detection of tumors in soft tissues by ultrasonic holography. Surg. and Obstet. 128:953-962 (1969).

16. El'piner, I.E. "Ultrasound" (p. 334) Consultants Bureau: N.Y. (1964).

17. Gregg, E.C. and Palagallo, G.L. Acoustic Impedance of Tissue. Invest. Radiol. 4:357-363 (1969).

18. Carstensen, E.L., Li, K. and Schwan, H.P. J. Acoust. Soc. Am. 25: 286 (1953).

19. Dunn, F. In "Ultrasonic Energy" Ed. E.Kelly, p. 55 Univ. Illinois Press: Urbana (1965).

20. Apsimon, H.T., Stewart, H.J. and Williams, W.J. "Recording the gross outlines of breast tumours". Brit. J. Cancer, 22:40-46 (1968).

21. Martin, J.E. and Gallager, H.S. Mammographic diagnosis of minimal breast cancer. Cancer 28:1519-1526 (1971).

22. Dunn, F. and Fry, F.J. Trans. IEEE on Bio-Medical Engineering 18(2) 53-65 (1971).

23. Weiss, L. The cell periphery and metastasis. Proc. 7th Canad. Cancer Conf. pp. 292-315 (1967).

24. Weiss, L. The cell periphery, metastasis and other contact phenomena. N. Holland Publ. Co: Amsterdam (1967).

25. Weiss, L. and Holyoke, E.D. Some Effects of Hypervitaminosis A on metastasis of spontaneous breast cancer in mice. J. Nat. Cancer Inst. 43:1045-1054 (1969).

26. Taylor, K.J.W. and Newman, D.Z. Effects on electrophoretic mobility of Ehrlich cells exposed to ultrasound of variable parameters. Physics in Med. and Biol. 17 (1972).

27. Taylor, K.J.W. and Pond, J. The effects of ultrasound of varying frequencies on rat liver. J. Path. 100: 287 (1970).

28. Taylor, K.J.W. and Pond, J.B. A study of the production of hemorrhagic injury and paraplegia in rat spinal cord by pulsed ultrasound of low megaHertz frequencies in the context of the safety for clinical usage. Brit. J. Radiol. 45:343-353 (1972).

29. Macintosh, I.J.C. and Davey, D.A. Relationship between intensity of ultrasound and induction of chromosome aberrations". Brit. J. Radiol. 45: 320-327 (1972).

30. Coakley, W.T., Slade, J.S. and Braeman, J.M. Examination of lymphocytes for chromosome aberrations after ultrasonic irradiation. Brit. J. Radiol. 45:328-332 (1972).

31. Buckton, K.E. and Baker, N.V. An investigation into possible chromosome damaging effects of ultrasound on human blood cells. Brit. J. Radiol. 45:340-342 (1972).

32. Bleaney, B.I., Blackbourn, P. and Kirkley, J. Resistance of CHLF hamster cells to ultrasonic radiation of 1.5 MHz frequency. Brit. J. Radiol. 45: 354-357 (1972).

33. Connolly, C. and Pond, J. The possibility of harmful effects in using ultrasound for medical diagnosis. Bio-Med. Eng. 2:112-115 (1967).

BIOGRAPHIES OF CONTRIBUTORS

KAZUHIKO ATSUMI

Kazuhiko Atsumi, M.D., was born in Osaka, Japan, in 1928, and graduated from the University of Tokyo, Faculty of Medicine in 1954. From 1955 to 1964 he studied cardiac surgery in Tokyo University Hospital. In 1964 he was appointed Associate Professor at the Institute of Medical Electronics, Faculty of Medicine, University of Tokyo, and in 1966 was appointed Chief Professor of the same institute.

He has been engaged in work in the following medical fields:

> Artificial internal organs (esp. artificial heart)
> Biomedical polymers
> Medical transducers for blood flow, blood pressure and blood pH
> Medical ultrasonics
> Medical thermography
> Medical application of the laser
> Medical application of the computer (retrieval of patient history, on-line control of artificial heart, design of computerized medical care system)

He is President of Japan Society of Medical Optoelectronics, Director of the Japan Society of Medical Electronics, Director of Japan Society of Artificial Organs, Director of Japan Society of Ultrasonics in Medicine, Director of Japan Society of Medical High Polymers. He is also a member of the American Society of Biomedical Engineering, the American Heart Association, the American Society of Artificial Internal Organs, and was a national delegate to the International Federation of Biomedical Engineering.

BYRON B. BRENDEN

Born in Bismarck, North Dakota, February, 25, 1927, Byron B. Brenden studied physics and engineering at the Universities of Missouri, Wyoming and Oregon, and at the Graduate School of Nuclear Engineering in Richland, Washington.

He joined General Electric Company at Schenectady, New York in 1951, where he participated in their Physics and Advanced Engineering programs. In 1952 he accepted a permanent assignment as an optical engineer at the Hanford complex operated by General Electric, Richland, Washington. His duties included supervision of the plant optical shop, design of specialized instrumentation need in the nuclear facilities, and plant consultant in optics. In 1965, he joined Battelle-Northwest as Senior Research Scientist; was principal investigator in holography research and development programs, primarily in acoustical holography. Also was manager of the Optics and Device Development Section specializing in development of optical instrumentation, ultrasonic holography and vacuum technology.

In 1969 he participated in the formation of the company now known as Holosonics, Inc., and is currently a vice president. Holosonics is at present the only company marketing acoustical holography systesm.

He is co-author of a book entitled, "An Introduction to Acoustical Holography" published by Plenum Publishing Corporation in 1972. He was also one of the principal contributors to the book, "Optical and Acoustical Holography" which records the proceedings of the NATO Advanced Study Institute held in Milan, Italy. This book was also published by Plenum Publishing Corporation in 1972.

GILBERT B. DEVEY

Mr. Gilbert B. Devey serves as a Program Manager in the Division of Advanced Technology Applications of the United States National Science Foundation where he has been employed since 1965. He served as a Program Director in the Division of Engineering from 1965-1967, and 1969-1972. His areas of responsibility in the Division of Engineering included the direction of the Engineering Systems Program, the Electrical Sciences and Analysis Program, and the Control and Automation Program. During 1967-1969, he was with the National Academy of Sciences as Executive Secretary to the National Academy of Engineering Committee on the Interplay of Engineering with Biology and Medicine. Since 1970, he has held an academic appointment as Associate Professional Lecturer in Clinical Engineering at the George Washington University School of Medicine.

From 1953-1965, Mr Devey was associated with the Sprague Electric Company, North Adams, Massachusetts, where he developed new areas of applications for advanced technology using electronic components.

He served as an electronic scientist with the Office of Naval Research from 1949-1952, and held positions in electronics engineering at the U. S. Navy Underwater Sound Laboratory and the Bureau of Ships during the years from 1946 to 1952.

In his present position, Mr Devey is responsible for the instrumentation technology research program area, with heavy emphasis on the application of new instrumentation concepts to the general diagnoses and treatment of disease.

He was awarded a B. S. degree by the Massachusetts Institute of Technology in 1946, and attended the U. S. Naval Academy Postgraduate School during 1944-1945.

The author of numerous papers on medical instrumentation and biomedical engineering, Mr Devey has directed several technical conferences on Multiphasic Health Testing. In 1971, Mr Devey served as a lecturer on "Project Management for Automation Technology" at the Japan Industrial Technology Association Symposium on "Pattern

Information Processing Systems (PIPS), one of the National Research and Development Projects supported by the Ministry of International Trade and Industry. A report written by Mr Devey on the PIPS project appeared in "Nature" as a part of the special issue on "Science in Japan," November 24, 1942.

FRANCIS J. FRY

Francis J. Fry was born April 2, 1920, in Johnstown, Pennsylvania. He received a B.S. degree at Pennsylvania State University in 1940, and an M.S. degree at the University of Pittsburgh in 1946, both in Electrical Engineering. He was elected to membership in Sigma Xi at the University of Illinois.

His professional experience includes: Associate Professor of Surgery, Indiana University School of Medicine and the Indianapolis Center for Advanced Research, January 1972 to present; Senior Research Scientist at Interscience Research Institute, Champaign, Illinois, 1957 to 1972; Associate Professor of Electrical Engineering, Bio-acoustic Research Laboratory, University of Illinois, 1968 to 1972; Research Associate Professor of Electrical Engineering, University of Illinois, 1958 to 1968; Research Assistant Professor of Electrical Engineering, University of Illinois, 1956 to 1958; Research Associate, Department of Electrical Engineering, University of Illinois, 1946 to 1956; Design Engineer, Westinghouse Corporation, and Consultant to the Manhattan Project (University of California at Berkeley, and Oak Ridge, Tennessee), 1940 to 1946.

Professor Fry's interests are an outgrowth of his experience in the past 30 years, which have led him beyond the field of electrical engineering into areas involving the biological and medical sciences. These research experiences in biophysics and bioengineering were accumulating in the years when such terms had very limited application. His present interests are centered in major research areas of neuroscience, cardiovascular science and applications, inter-

BIOGRAPHIES OF CONTRIBUTORS

action of ultrasound and biological tissue, delineation of normal and abnormal tissue structure by ultrasonic means, and diagnostic and surgical applications of ultrasound. In the neurosciences, he devotes a significant portion to the quantitative aspects of neuronal organization at the level of brain nuclei, their neuron subpopulations, and the efferent and afferent anatomy of nuclei, also concerned with the delineation of time sequence and magnitude of transneuronal changes induced in the brain by experimental lesions and naturally occurring brain traumas. The relationship between behavior and function and structural organization is also involved.

Because of pioneering work in the development of artificial hearts, a long-term interest has developed in blood flow, particularly cerebral blood flow and degradation of blood flow in artificial conduits, as well as configurational changes in flow patterns induced in natural blood vessels.

The interaction of ultrasound with biological tissue has occupied a significant portion of the past 20 years of research, and this area continues to be of interest. In the delineation of various tissue structures by ultrasound visualization methods, his interests have been mostly centered in work in the experimental animal and human brain.

Starting in 1955 and continuing to 1960, he was involved in a clinical research study involving the application of ultrasound to the area of human neurosurgery. This work demonstrated the utility of high intensity ultrasonic methods to do major neurosurgical procedures in a non-invasive fashion. As a result of this study, the basis was laid for subsequent work in the ultrasonic visualization of brain structures. This ultrasonic visualization capability in the brain, combined with precision ultrasonic lesioning, permits the consideration of a unique, non-invasive, atraumatic approach to certain neurosurgical procedures. The methods evolved are applicable additionally to surgical procedures in a variety of body organs.

Professor Fry is Associate Editor for the IEEE Transactions on Sonics and Ultrasonics. Society memberships include: Institute of Electrical and Electronics Engineers, American Institute of Ultrasound in Medicine (Chairman, Biological Effects Committee), Acoustical Society of America, American Society for Artificial Internal Organs, Biomedical Engineering Society, and Society for Neuroscience.

The interaction of ultrasound with biological tissue is a subject of his investigation. It includes studies of heart, blood vessels, brain and skeletal muscle, and he has employed the electron microscope as a tool to investigate the ultrastructure of these tissues.

From 1956 to the present time, Professor Eggleton has been involved in clinical research involving the application of ultrasound to areas of human ablative neurosurgery using stereotaxic means of placing ultrasonic focal lesions to correct various types of neurological disorders. He is co-inventor of an ultrasonic device for diagnosis and surgery using noninvasive atraumatic techniques.

Professor Eggleton is a member of the American Institute of Ultrasound in Medicine, American Society for Artificial Internal Organs, Society for Neuroscience, Instrument Society of America, and American Association for the Advancement of Science.

REGINALD C. EGGLETON

Reginald C. Eggleton was born July 6, 1920, in Hillsdale, Michigan. He received a B.E.E. degree in Electrical Engineering from the University of Michigan in 1944, and in 1961, received a B.S. degree in biology and physics. In 1966 he received an M.S. in Physiology from the University of Illinois, then continued with graduate studies in biophysics at the University of Texas.

His professional experience includes: Associate Professor of Surgery, Indiana University School of Medicine, and Director of the Fortune-Fry Research Laboratory of the Indianapolis Center for Advanced Research, January, 1972 to present; Senior Research Scientist, Interscience Research Institute, 1957-1972; President, Interscience Research Institute, 1970-1972; Research Associate, Department of Electrical Engineering, Biophysical Research Laboratory, 1956-1971; Research Assistant, Department of Electrical Engineering, University of Illinois, 1953-1956; Head, Development Laboratory, Alden Products Company, Brockton, Massachusetts, 1950-1953; General Manager, Haller, Raymond and Brown, Inc., State College, Pennsylvania, 1947-1950; R. C. Eggleton Company, Ann Arbor, Michigan, 1944-1946; Engineer, Physicist Research Corporation, 1941-1944.

Professor Eggleton's interests evolved from electronic manufacturing to biomedical engineering during the course of his career. At present, his interests are centered around the medical applications of ultrasound for diagnosis and treatment of human diseases. More specifically, his interests have involved research in neuroscience and cardiovascular science, and applications of ultrasonic techniques to these organ systems. He has collaborated in the use of ultrasound in localizing and treating a variety of neurological disorders, and was also involved in the use of ultrasound for both diagnosis and treatment of heart disease. He was involved in pioneering work leading to the development of artificial hearts for the permanent replacement of the natural heart, and is co-inventor of the artificial heart device.

DENNIS GABOR

Professor Dennis Gabor, now Professor Emeritus of Applied Electronic Physics of the Imperial College of Science and Technology in London, and resident Staff Scientist at CBS Laboratories, Stamford, Connecticut, received the Nobel prize in 1971 for his invention of the technique now known as Holography[1], a method of three-

---

[1] COI Paper R5476/51, Notes on Science and Technology in Britain.

dimensional lensless photography by coherent light in which a light wave issuing from an object is "frozen" into a photographic emulsion by means of a second beam of coherent light. The resulting photograph or hologram can then be reconstructed by the second beam of light alone to give a three-dimensional image.

Professor Gabor established his technique in 1948, when trying to increase the resolving power of electron microscopes; with the coming of the laser beam in 1961, it was given a wider application, and since the first laser holograms were obtained in the United States the following year, research has expanded rapidly in America and in Britain.

Born in Budapest in June, 1900, Professor Gabor became a British citizen. He studied at the Technical University at Budapest, then in Berlin, at the Technische Hochschule at Charlottenburg where he obtained a diploma in electrical engineering, and subsequently, a doctorate. He carried out research as an assistant at the Technische Hochschule and as research associate with the German Research Association for High Voltage Plants, before joining the great electrical firm of Siemens & Halske AG as a research engineer. In 1933 with the coming of the Nazis, Dr Gabor left Germany, returning to Hungary, then traveling to Britain the following year, to begin a long association with British Thomson-Houston as research engineer at Rugby. In 1948 he left BTH, and the following year, took up the appointment as Reader in Electronics in the Imperial College, University of London.

During the early years at BTH he worked on gas discharges, then began work on the electron microscope, developing the technique of holography, known then as "Wavefront Reconstruction." During the 1960's he interested himself in thermionic electrical power generation, and in 1965 gave the opening lecture at a conference attended by 200 thermionic experts from 20 countries, at the Institution of Electrical Engineers, in London.

Professor Gabor has received many honors for his work. Among them are: Fellowship of the Royal Society, 1956; honorary membership, The Hungarian Academy of Sciences, 1964; The Cristoforo Colombo Prize of Genoa, International Institute of Communications, 1967; The Thomas Young Medal of the Physical Society, 1967; the Michelson Medal of the Franklin Institute, 1969; The Rumford Medal of the Royal Society, 1969; Dr Sc. Hon., University of Southampton, 1970; The Medal of Honor of the IEEE, 1970; The Semmelweis Medal, 1970; Commander of the Order of the British Empire, 1970; Dr Sc. Hon., Technical University, Delft, 1971; Holweck Prize of the French Physical Society, 1971; Nobel Prize in Physics, 1971; Honorary Membership, Optical Society of America, 1972; The George Washington Award, The American Hungarian Studies Foundation, 1973.

Books he has written include "Inventing the Future;" "The Electron Microscope;" "Innovations: Scientific, Technological and Social;" and "The Mature Society." He is also the author of more than 100 papers on electrical transients, gas discharges, electron dynamics, communication theory, physical optics, holography, and cybernetics.

MASAO IDE

Dr Masao Ide was born in Japan on January 24, 1929. He received a B.S. degree in Electrical Engineering from the Musashi Institute of Technology, Tokyo, in 1953, and a Ph.D. degree in Engineering from Tokyo Institute of Technology in 1969.

Since 1953 he has been an assistant, an associate professor, and is currently a Professor of Electronics and Communication Engineering of Musashi Institute of Technology.

He is interested in problems of application of high powered ultrasonics, ultrasonic walking aids for the blind, and in medical ultrasonics.

Dr Ide is a member of the Institute of Electrical Engineers of Japan, the Acoustical Society of Japan, the Japan Society of Medical Electronics and Biological Engineering, and the Japan Society of Ultrasonics in Medicine.

EARLE D. JONES

Earle D. Jones received his B.S. degree in electrical engineering at Georgia Institute of Technology; in 1958 he received an M.S. degree, also in electrical engineering, at Stanford University.

He currently directs the activities of the Electronics and Bioengineering Laboratory of the Stanford Research Institute, Menlo Park, California, which he joined in 1956. As Laboratory Director, he has been instrumental in advancing research in the areas of analysis and design of electronic-optical systems such as ultrasonic imaging and real tim interferometry; design of television systems; facsimile systems including bandwidth compression; and electrostatic printing. He holds six patents in electronic circuitry, character generators, frequency synthesizers and electrostatic printing systems. He is the author of publications in related fields.

HIROSHI KASHIWAGI

Dr Hiroshi Kashiwagi was born in Tokyo, Japan on June 21, 1934. He received an M.S. and a Ph.D. from Keio University in 1960 and 1967, respectively.

Since 1963 he has been with the Electrotechnical Laboratory, Agency of Industrial Science and Technology, Ministry of International Trade and Industry, Tokyo. He has been engaged in studies of quasi-optics in millimeter wave and submillimeter wave regions, gas lasers, laser information transmission systems for computer links, optical information processing, and low loss optical guides.

Since 1969 he has been a project leader of optical information transmission systems and also of optical oceanography. He is a senior research scientist in the electrotechnical laboratory, and is an executive secretary of the Marine Electronics and Communications Panel, Marine Resources Engineering Coordinate Committee, UJNR Cooperative Program.

He is a member of the Institute of Electronics and Communication Engineers of Japan, Institute of Electrical Engineers of Japan, Japan Society of Applied Physics, and the IEEE.

KENJIRO SAKURAI, coauthor, is the Division Chief. He is the former project leader of PIP Project.

YOSHIMITSU KIKUCHI

Dr Yoshimitsu Kikuchi was born in Osaka, Japan on August 19, 1910. He received a B.S. in 1933, and a Ph.D. in electrical engineering in 1943, from Tohoku University, Sendai, Japan.

He was with the Research Institute of Electrical Communication, Tohoku University, where he was a research assistant in 1933 to 1935, and an Assistant Professor from 1936 to 1938. From 1938 to 1944, he was with Nippon Electric Company. Since 1945 up to the present, he has been a Professor of Tohoku University. He was in charge of the Director of the Research Institute of Electrical Communication for nine years, from 1963 to 1972.

Dr Kikuchi is a member of the Acoustical Society of America, the Institute of Electrical Engineers of Japan, the Institute of Electronic and Communication Engineers of Japan, the Professional Group for Ultrasonics in IECE of Japan, the Acoustical Society of America, the Acoustical

Society of Japan, the Japan Society of Ultrasonics in Medicine of which he was President at its establishment; and he is also an Honorary Member of the American Institute of Ultrasound in Medicine.

WINSTON E. KOCK
Consultant, The Bendix Corporation
and
Visiting Professor, The University of Cincinnati

Dr Winston E. Kock received an Electrical Engineering degree in 1932 and an M.S. in Physics in 1933, both from the University of Cincinnati; a Ph.D. in Physics in 1934 from the University of Berlin, attended the Institute for Advanced Study at Princeton, and was a Fellow at the Indian Institute of Science at Bangalore in 1936.

He has been Director of Electronic Research, Baldwin Piano Company, where he developed the Baldwin Electronic organ; Director of Acoustics Research at Bell Telephone Laboratories, where he developed several acoustic lenses, directed the research on the Navy's underwater sound, Jezebel-Caesar project and headed the group developing the picture phone; Director of Bendix Research Laboratories; first director of the National Aeronautics and Space Administration's Electronics Research Center, Cambridge, Massachusetts; and most recently, Vice President and Chief Scientist, Bendix Corporation. In 1971 he chose partial retirement to become Consultant to the Corporation and a Visiting Professor at his Alma Mater.

He was Chairman of the Professional Group on Audio of the Institute of Radio Engineers in 1954-1955, and was a member of the Governing Board of the American Institute of Physics from 1957 to 1963. He is a member of the Board of Roanwell Corporation of Hardon, Inc., of Argonne Universities Association, and has been Chairman of the Board of Trustees of Western College for Women, and a Board member of the Atomic Industrial Forum.

Honors include the Navy's highest civilian award, the Distinguished Public Service Medal (1964); Honorary Fellowship in the Indian Academy of Sciences (1970); and honorary D. Sc., University of Cincinnati (1952). He was Eta Kappa Nu's Outstanding Young Electrical Engineer in 1938, and received the Eminent Member Award in 1966.

He is a Fellow of the Acoustical Society, the Physical Society and the IEEE, and a member of Tru Beta Pi, Sigma Xi, and Eta Kappa Nu.

Dr Kock is the author of three books: Sound Waves and Light Waves, 1965, Doubleday; Lasers and Holography, 1969, Doubleday; and Seeing Sound, 1971, Wiley.

He is married and is the father of three children, residing in Ann Arbor, Michigan.

ADRIANUS KORPEL

Adrianus Korpel was born in Rotterdam, Holland, on February 19, 1932. He received his M.S. degree in electrical engineering from the University of Delft, Holland, in 1955, and his Ph.D. degree from the same university in 1969. The subject of his thesis was acoustic visualization by Bragg diffraction of light.

From 1956 to 1960 he was employed by the Research Laboratories of the Postmaster General's Department in Melbourne, Australia. His main activities were in the field of information and communication theory as applied to television, and later, in the field of parametric amplifiers.

He joined Zenith Radio Corporation in Chicago in 1960, and since 1963, is Head of the Light-Modulation Group. Principal activities of this group concern image-type applications of lasers such as television display, sound visualization and acoustic holography.

Dr Korpel is a member of the Royal Dutch Institute of Engineers, the Institution of Engineers of Australia, the Physical Society of America and the Institute of Electrical and Electronics Engineers.

ADOLF W. LOHMANN

Professor Adolf W. Lohmann was born in 1926, in Germany. In 1953 he received a Ph.D. in Physics from the University of Hamburg.

During the next ten years he was a faculty member at the Technical University of Braunschweig. That period was interrupted by two one-year leaves spent at the Royal Institute of Technology in Stockholm and at the IBM Research Laboratory in San Jose, California. In 1963 he returned to IBM. Since 1968 he has been a Professor at the Department of Applied Physics and Information Science at the University of California at La Jolla. Now he is a Professor of Physics at the University of Erlangen in West Germany.

Professionally he is interested principally in the information aspects of optics. He has invented a variety of holographic equipment since 1955. Best known is probably his work on computer generated holograms and computer generated spatial filters.

MOTOYOSHI OKUJIMA

Dr Motoyoshi Okujima was born January 9, 1930, in Okayama City, Japan, and graduated from Tokyo Institute of Technology in 1953. He has been at the Research Laboratory of Precision Machinery and Electronics of that institute up to the present time. In 1963 he was promoted to the position of Associate Professor, and in 1970, to Professor.

From 1953 to 1962 he studied ultrasonics under the direction of Professor J. Saneyoshi, and in 1962 he earned a degree as Doctor of Engineering because of studies on Electromechanical Transducers for Measurement of Ultrasonic Pressure in Liquid and Calibration Methods of their Sensitivities.

Most of his present activities are for development of measuring equipment by acoustical means in the fields of medicine and civil engineering. For example, one of them is imaging by impulsive sound as is described in this paper; another is measurement of velocity distribution by the Doppler method with M sequentially modulated ultrasound.

He is a member of the Acoustical Society of Japan, the Institute of Electrical Engineers, the Japan Society of Medical Electronics and Biological Engineering, the Japan Society of Ultrasonics in Medicine, and the Japan Society of Civil Engineers.

DAITARO OKUYAMA

Dr Daitaro Okuyama was born in Koriyama, Japan, on May 2, 1927. He received a Doctor of Engineering degree in electrical engineering from Tohoku University, Sendai, Japan, in 1962.

Since 1947, he has been with the Research Institute of Electrical Communication, Tohoku University. He is currently an Assistant Professor of the Institute, in the Applied Ultrasonics Division. From 1963 to 1966, he visited the University of Illinois, and Interscience Research Institute, Illinois, for the study of the ultrasonic visualization system of biological tissue.

Dr Okuyama is a member of the IEEE, the Acoustical Society of America, the IEE of Japan, the IECE of Japan, the Acoustical Society of Japan, and the Japan Society of Ultrasonics in Medicine.

BIOGRAPHY OF DR MOTONAO TANAKA

Dr Motonao Tanaka was born in Tokyo, Japan, on January 1, 1932. He was graduated from the Graduate School of Medicine, Tohoku University, Sendai, Japan, in 1963, and received an M.D. with a thesis on the "Phonocardiographic Studies in Congenital Heart Disease with Special Reference to the Frequency Analysis of Heart Sounds and Murmurs by Spectral Phonocardiography" from the same University in 1963.

Dr Tanaka is at present an assistant at the Research Institute for Tuberculosis, Leprosy and Cancer, and a lecturer at the Faculty of Medicine, Tohoku University. His present major research interests are Ultrasono-Cardio-Tomography and its clinical applications; and Phonocardiography.

He obtained the Silver Prize of Tohoku Medical Society in 1964, and the Gold Prize in 1971. He is a member of the Japanese Society of Internal Medicine, the Japanese Circulation Society, the Japan Society of Chest Disease, the Japan Society of Medical Electronics and Biological Engineering, and the Japan Society of Ultrasonics in Medicine.

MORIO ONOE

Dr Morio Onoe is a Professor of Applied Electronics at the Institute of Industrial Science, University of Tokyo.

He was born in Tokyo in 1926, graduated from the University of Tokyo in 1947, and later received a Ph.D. degree from the same University.

From 1956 to 1958 he was a Visiting Scholar at Columbia University, New York, under the Fulbright Exchange Scholar Program. In 1961 and 1964 he was a member of the technical staff of Bell Telephone Laboratories on leave of absence from the University of Tokyo.

He frequently visited both the United States and European countries.

His interest is in ultrasonics and in non-destructive testing, and most recently, in image processing by computer.

He is a senior member of the Institute of Electronic and Electrical Engineers, the Acoustical Society of America, the Institute of Electrical Engineers of Japan, and the Institute of Electrical Communication Engineers of Japan.

GEORGE W. STROKE

Dr George W. Stroke obtained his Ph.D. in Physics from the Sorbonne in Paris in 1960, and is currently Professor of Electrical Sciences and of Medical Biophysics at the Stony Brook campus of the State University of New York and Head of its electro-optical sciences laboratory. Concurrently, since 1970, he has served as Visiting Professor of Medical Biophysics at Harvard University Medical School.

Dr Stroke was previously Professor of Electrical Engineering at the University of Michigan and Head of its electro-optical sciences laboratory which he founded in 1963. Before joining the University of Michigan, Professor Stroke spent 10 years at MIT where he did research work, principally devoted to the development of the method of interferometric servo-control of grating ruling, in a collaborative effort with Dean George R. Harrison. This has earned them world fame. During his tenure at MIT, he also helped in originating a method of velocity of light measurement using microwave-cavity resonance, and participated in the Office of Naval Research Fleet Ballistic Missile (Polaris) program at the Instrumentation laboratory there.

Professor Stroke's work in coherent optics and holography originated with his work in the Radar Laboratory at the University of Michigan where he helped in initiating the work on three-dimensional "lensless photography" as a consultant in 1962-1963. He wrote the first treatise on the subject under the title, <u>An Introduction to Coherent Optics and Holography</u> (Academic Press, 1966) which was immediately translated into Russian (MIR, 1967) and appeared in its second (enlarged) U. S. edition in 1969.

In recent years, Professor Stroke has been devoting his primary research interests increasingly to the life sciences, notably to the development of new methods of three-dimensional microscopy and to the improvement of high-resolution electron microscopy. The method of holographic image deblurring which he originated in 1965 has recently permitted him, with his team at Stony Brook, to sharpen up electron micrographs of virus test specimens to a degree considered uantainable in the past.

In addition to <u>An Introduction to Coherent Optics and Holography</u>, Dr Stroke published another book (at the age of 24) as well as the 320 page "Diffraction Gratings" section of the <u>Handbuch der Physik</u> (Springer Verlag, Vol. 29, 1967) and approximately 100 scientific papers including about 50 on holography. A widely traveled lecturer on the subject of holography and its scientific, industrial and biomedical applications, Dr Stroke has served in a number of United States government and other advisory capacities including, most recently as a member of the National Science Foundation Blue Ribbon Task Force on Ultrasonic Medical Diagnostics and as a consultant to the American Cancer Society. In 1971 he served as U. S. delegate to the Popov Society Meeting in Moscow under the U. S. State Department Scientific Exchange program. In 1972 he was invited by the Japan Industrial Technology Association to officially advise the Japan Ministry of International Trade and Industry on its program of large-scale development of computer pattern recognition technologies. For several years he has also been assisting the National Science Foundation in its U. S- Japan and U. S. -Italy science cooperation programs.

Dr Stroke's innovations in holography (white-light color reflection holography, holographic image deblurring) have twice been the subject of the Science section of Time magazine (March 1966, and March 1968). A Fellow of the American Physical Society and of many other scholarly societies, he has received numerous tokens of recognition and awards including most recently, the Alan Gordon award for 1971 from the Society of Photo-Optical Instrumentation Engineers.

MAURICE HALIOUA

Dr Maurice Halioua has been working with Professor George W. Stroke at the Electro-Optical Sciences Laboratory at the State University of New York at Stony Brook since he immigrated to the United States in 1969. He also obtained his Ph.D. (Dr Ing. in Physics) for working under Professor Stroke's direction, and presented his thesis formally at the University of Paris in September, 1971.

Before emigrating to the United States, Dr Halioua studied at the University of Bordeaux, France, where he obtained his B.Sc. in 1962, and also at the Institute of Optics at the University of Paris where he obtained his Ing. Dipl. (Optics) in 1968.

Since 1972, he has held the title of Instructor in Medical Biophysics at the Health Sciences Center at Stony Brook, in addition to his position of research associate at the Electro-Optical Sciences Laboratory.

Since 1969, he has been the co-author, with Professor Stroke, in 15 publications in the field of image deblurring.

MICHIO SUZUKI

Dr Michio Suzuki was born in Hokkaido, Japan, on

November 14, 1923. He received a B.E. degree and a Doctor of Engineering degree, both from Hokkaido University, Sapporo, Japan, in 1946 and 1960, respectively.

From 1948 to 1962 he was an Assistant Professor, and from 1962, a Professor of Electronic Engineering at the Hokkaido University, Japan. Dr Suzuki is a member of the Institute of Electronics and Communication Engineers of Japan, the Institute of Electrical Engineers of Japan, and the IEEE.

TAKASHI IWASAKI

Takashi Iwasaki was born in Tokyo, Japan, on March 14, 1948. He received B.E. and M.E. degrees from Hokkaido University, Sapporo, Japan, in 1970 and 1972, respectively. Since 1972 he has been a student of the doctorate course and doing research on acoustical holography. Mr Iwasaki is a member of the Institute of Electronics and Communication Engineers of Japan, the Japan Society of Applied Physics, and the IEEE.

SHIGEO FUJIKI

Shigeo Fujiki was born in Hokkaido, Japan, on February 23, 1944. He received a B.E. degree in electronics from Hokkaido University, Sapporo, Japan, in 1966. Since 1966, he has been with the Faculty of Engineering, Hokkaido University, as an Assistant. Mr Fujiki is a member of the Institute of Electronics and Communication Engineers of Japan.

AKIYOSHI HAKOYAMA

Akiyoshi Hakoyama was born in Hokkaido Prefecture, Japan, on May 11, 1948. He received B.E. and M.E. degrees from Hokkaido University, Sapporo, Japan, in 1971 and 1973, respectively, and joined Hitachi Ltd., Japan, in 1973. He is a member of the Institute of

Electronics and Communications Engineers of Japan.

JUMPEI TSUJIUCHI

Dr Jumpei Tsujiuchi was born in Wakayama, Japan on August 18, 1927. He graduated from the Faculty of Science, University of Tokyo, and received a Ph. D. in Applied Physics from the University of Tokyo in 1962.

Since 1951 he has been a research scientist in the Government Mechanical Laboratory, Agency of Industrial Science and Technology. From 1958 to 1960 he was in the Institut d'Optique, Paris, as an Attache de Recherche of Centre National de la Recherche Scientifique in France, on leave of absence from the laboratory. Since 1967 he has been a Professor at Tokyo Institute of Technology.

He has been involved in the study of applied optics, particularly image formation, optical information processing and holography.

He is a Fellow of the Optical Society of America, a member of the Society of Photo-Optical Instrumentation Engineers, the Physical Society of Japan, and the Japan Society of Applied Physics.

SADAYUKI UEHA

Sadayuki Ueha was born in Kyoto, Japan, on February 28, 1943, graduated from the Nagoya Institute of Technology in 1965, and received M. E. and Ph. D. degrees in electrical engineering, both from the Tokyo Institute of Technology, in 1967 and 1970, respectively.

Since 1970, he has been with the Imaging Science and Engineering Laboratory, Tokyo Institute of Technology.

He has been engaged in the study of ultrasonics and

acoustical holography. He is a member of the Acoustical Society of Japan.

KEIICHI UENO

Keiichi Ueno was born in Tokyo, Japan, on January 2, 1948. He graduated from the University of Electro-Communications in 1971, and is presently a master course student in applied physics at the Tokyo Institute of Technology.

TOSHIO WAGAI

Dr Toshio Wagai was born in Ishinomaki, Japan, on September 21, 1924. He was graduated from Niigata Medical College in 1949. His professional training and employment are as follows:

> Interned at Juntendo Medical College Hospital from April, 1949, to March, 1950.
>
> Assistant in the Department of Surgery, Juntendo University School of Medicine from October, 1952, to March, 1960.
>
> Lecturer (Doctor of Medical Science) in the Department of Surgery, Juntendo University School of Medicine, from April, 1960, to March, 1965.
>
> Associate Professor in the Department of Surgery, Juntendo University School of Medicine, from April, 1965, to March, 1971.
>
> Professor in the Medical Ultrasonics Research Center, Juntendo University School of Medicine from April, 1971, to the present time.

LEONARD WEISS

Dr Leonard Weiss was born in London in 1928. He graduated from the University of Cambridge, qualifying in medicine in 1953. He obtained his Ph.D. in biophysics in 1963, and was awarded an Sc.D. in 1971 for his work on cell interactions. After working for the Medical Research Council in London and Cambridge, in 1964 Dr Weiss became a Director of Cancer Research and Head of the Department of Experimental Pathology at Roswell Memorial Institute, Buffalo, New York. He is currently a Professor of Biophysics, a Fellow of the Institute of Biology (U.K.), and a Foundation Member of the Royal College of Pathologists.

AUTHOR INDEX

Abbe, E., 506
Abe, H., 78 ff.
Ackerman, L. V., 573 ff.
Addison, R. C., 361
Adler, R., 362
Agency Industrial Science
  and Technology, 201
Akatsuba, T., 29 ff.
Aloka Company, 161
Anderson, M. D., 99
  M. D. Anderson Hospital, 89
Andrews, H. C., 501
Apsimon, H. T., 584
Asberg, A., 267 ff., 444, ff.
Atsumi, K., 1-86, 426 ff.
Auld, B. A., 353 ff.
Awaya, K., 421 ff.

Backhans, H., 282
Baker, N. V., 582 ff.
Ballantine, H. T., 274
Barnard, J. W., 149
Barrakette, E. S., 551 ff.
Baum, G., 241 ff.
Begui, Z. E., 280
Bell Telephone Laboratories, 287 ff.
Bendix Research Laboratories, 287 ff.
Berger, H. E., 87 ff.
Berger, P. L., 121 ff.
Berger, S. M., 583
Billingsley, F. C., 552
Bisplinghoff, R. L., 114 ff.
Blackbourn, P., 585

Bleaney, B. T., 585
Blumenson, L. E., 583
Bolt, R. H., 274 ff.
Bom, Ir. N., 431 ff.
Boyer, A. L., 502
Bracewood, R. N., 374
Braeman, J. M., 585
Bragg, W., 215 ff.
Brenden, B., 87 ff
  576 ff.
Brooks, R. E., 343
Bross, I. D. J., 583 ff.
Brown, B. R., 342
Brown, G. M., 343
Brown, W. M., 341
Bryngdahl, O., 375
Buckles, R. G., 228
Buckton, K. E., 582 ff.
Burker, W. F., 583

Carstensen, E. L., 280, 584
CBS Laboratories, ii ff., 151-158
Cedrone, N. P., 280
Coakley, W. T., 585
Colbert, C., 512 ff.
Collier, R. J., 319
Connolly, C., 585
Costello, F. R., 546 ff.
Cremer, L., 238
Crewe, A. V., 512
Croce, P., 516
Cunningham, J. A., 351
Curico, B. M., 583

Curra, D. R., 280
Cutrona, L. J., 297 ff, 502 ff.
Dainton, Sir F., 105 ff.
Dakss, M. L., 227
Dallas, W. J., 374 ff.
Davey, D. A., 581 ff.
Desmares, P., 362
Devey, G. B., ii, 105-124, 587
Diehl, R., 311 ff, 551
Dixon, R. W., 361
Dobrin, M. B., 344
Dodd, G. D., 98
Dooley, R. P., 583
Dow, W., 298 ff.
Dowdy, A. H., 583
Dunn, F., 149, 579 ff.
Dussik, F., 274
Dussik, K. T., 229 ff.
Ebina, T., 242 ff, 431 ff.
Edler, I., 233 ff.
Egan, R., 583
Eggleton, R. C., 125-150, 592
Eilers, G., 191 ff.
El' piner, I. E., 576 ff.
El Sum, H. M. A., 102
E. M. I., 149
Endoh, N., 285, 422 ff.
Erikson, K. R., 341

Fergason, J. L., 529
Fiedler, W., 274
Firestone, F. A., 281

Fisher, B., 574 ff.
Flagle, C. D., 118 ff.
Flaherty, J. J., 341
Flügge, S., 374
Fork, R. L., 228
Fox, G. R., 342
French, L. A., 231 ff.
Fry, E. K., 283
Fry, F. J., 125-150, 148 ff, 579 ff.
Fry, R. B., 148
Fry, T. A., 276
Fry, W. J., 148 ff, 233 ff, 361
Fujiki, S., 517-530, 606
Fujimasa, I., 78
Fukuda, N., 76 ff.
Fukuda, S., 551
Fukuhisa, K., 10 ff, 77 ff.
Fukumura, 37 ff.
Fukutake, K., 77
Furukawa, T., 78
Fyimasa, I., 83

Gabor, D., ii ff, 151-158, 294 ff, 506 ff.
Gallagher, H. S., 584
Garrett, W. J., 88 ff.
Gershon-Cohen, J., 583
Gibbons, L. V., 148
Glenn, W. E., iv, 156
Goetz, G. G., 343 ff.
Goldman, D. E., 244 ff.
Goodman, J. W., iv, 371 ff.
Goodyear Aerospace Corp., 304

AUTHOR INDEX

Gordon, D., 277 ff.
Gose, E. E., 575 ff.
Grant, R. M., 343
Great Britain, 124
Gregg, E. C., 576 ff.
Green, P. S., 192 ff, 360
Greenwood, I. A., 277
Grossman, C. C., 241
Güttner, W., 231 ff.

Haagensen, C. D., 584
Haberman, J. D., 583
Hagiwara, Y., 268 ff, 453
Hahn, M. H., 509 ff.
Haines, J. A., 343
Hakoyama, A., 517-530, 606
Halioua, M., 503-516, 605
Hallermier, R. J., 213 ff.
Hamamato, 76 ff.
Hamano, N., 501 ff.
Hance, H. V., 227
Handa, J., 78
Haneda, Y., 278, 453
Hanszen, K. J., 505 ff.
Hargraove, L. E., 228
Harkins, L. E., 583
Harris, S. E., 228
Harvey, F. K., 297 ff.
Harvard University, 603
Hashimoto, Y., 47 ff.
Hattori, S., 80
Havelice, J., 350
Hayakawa, H., 206 ff.
Haydon, G. B., 505 ff.

Hayes, G., v
Heflinger, L. O., 343
Heidrich, P. F., 227
Heimburger, R. F., 148 ff.
Hertz, C. H., 236 ff.
Heuter, T. F., 244 ff.
Hevezi, J. M., 98
Hildebrand, B. P., 87 ff., 343 ff.
Hirakawa, A., 86
Hiramoto, T., 77
Hirao, F., 77 ff.
Hirsch, P. M., 342 ff., 502
Hisada, K., 22 ff.
Hisazumi, Y., 78
Hitachi Limited, 606
Holbrooke, D., 95 ff., 281
Holosonics, Inc., 87-104, 581
Holyoke, E. D., 98 ff., 574 ff.
Honjo, I., 77
Hoppe, W., 510
Howry, D. H., 234 ff.
Huang, T. S., 501

IBM, 322 ff.
Ichioka, Y., 375
Ide, M., 159-200, 273 ff., 595
Ih, C., 156 ff.
Iino, T., 78
Iinuma, T., 10 ff.
Iio, M., 84
Iisaka, J. 54 ff.

Iishujama, T. 33 ff.
Ikeda, K., 85
Ikeda, O., 551
Imai, S., 421 ff.
Imazato, Y., 85
Imperial College of Science and Technology, 151
Inada, H., 78
Indiana University, 125-150
Ingalls, A. L., 344
Inokumo, M., 78
Inouye, T., 77
Irie, K., 268 ff.
Ishiguro, T., 206 ff.
Ishihara, S., 228
Ishihara, T., 77
Ishii, Y., 86
Ishikawa, S., 279, 501
Isobe, T., 29 ff., 421
Ito, K., 64 ff., 264 ff.
Iwasaki, T., 517-530, 606

Jacobs, J. E., 87 ff.
James Electronics, Inc., 102
Japan, Prime Minister's Office, 123
Jensen, H., 304 ff.
Jones, E. D., 191, 596
Jones, J. P., 138 ff.
Jordon, J. A., Jr, 342 ff., 502
Juntendo University, School of Medicine, 240, 553-566, 608

Kaihara, S., 51 ff.
Kajitani, 78
Kameto, H., 84
Kasai, C., 253 ff., 439 ff.
Kashida, R., 85
Kashiwagi, H., 201-228, 596
Katahura, K., 421 ff.
Kato, R., 81 ff.
Kato, T., 242 ff.
Kazibumi, K., 75 ff.
Kelly, E., 235 ff.
Kessler, L. W., 360 ff.
Kiemle, H., 340
Kikuchi, M., 208 ff.
Kikuchi, Y., v, 187 ff., 229-286, 425-454
Kimoto, S., 428 ff.
Kimura, K., 12 ff., 283
Kinoshita, N., 423
Kirkley, J., 585
Kishigami, Y., 80
Kobayashi, M., 26 ff., 424
Kock, W., v, 287-344, 502 ff.
Kojima, 12 ff.
Kono, H., 78 ff.
Koppelmann, R. F., 343 ff.
Korpel, A., 215 ff., 345-362, 599
Kosaka, T., 76 ff., 242 ff., 453
Koshikawa, T., 424
Kossoff, G., 149 ff.
Kozma, A., 546 ff.
Kranse, W., 268 ff.
Krautkrämer, J., 283

AUTHOR INDEX

Kresse, H., 268 ff.
Kruger, R. P., 501
Krumins, R. F., 149
Kuhn, L., 227
Kusano, R., 240
Kuwahara, M., 83 ff.

Lamberty, D., 156 ff.
Landau, H. J., 375
Landry, J., 227 ff.
Larmore, L., 102 ff., 361
Larsen, R. R., 584
Laser Focus Magazine, 339
Lean, E. G., 227
Lechner, B. J., 529
Leichner, G. H., 276 ff.
Leith, E. N., 297 ff., 502 ff.
Lerwill, W. E., 338 ff.
Lesem, L. B., 342 ff.
Lohmann, A. W., 342, 363-378, 501, 600
Lohnes, J. E., 149
Lommel, E., 282
London, England, 151
Long, J. A., 344
Ludwig, G. D., 274
Lund, V. M., 341

Machida, T., 78
Mackintosh, I. J. C., 581 ff.
Macovski, A., 192 ff.
Makino, I., 80
Manniello, J. B. L., iv
Maréchal, A., 506 ff.

Marom, E., 343
Martin, J. E., 584
Massey, N. G., 551
Masuzawa, N., 187 ff.
Matsui, 78 ff.
Matsumato, T., 77 ff., 502
Matsukawa, N. 285
Mayer, W. G., 228
Mertz, L., 374
Metherell, A. F., 102 ff., 361 ff.
Mikoshiba, N., 206 ff.
Minato, K., 85
Ministry of International Trade & Industry, 201
Mitsuhashi, Y., 228
Miyawaki, K., 80 ff.
Miyazawa, R., 275
Momoi, H., 60 ff.
Morita, M., 76 ff.
Motooka, S., 422 ff.
Mueller, R. K., 312 ff.
Mukai, T., 76 ff.
Muroi, T., 428 ff.
Musashi Institute of Technology, 159

McDuff, O. P., 228
McMann, R. H., v
McSkimmin, H. J., 253 ff.

Nakamuta, T., 423
Nakano, K., 421
Naritomi, T., 278

National Academy of Engineering, 124
National Academy of Sciences, 124
National Aeronautics & Space Administration (NASA) 108
National Science Foundation (NSF), 105-124, 158, 587
Naval Research Office, 603
Neal, D., 274
Neeley, V. L., 339
Negoro, T., 80
Newman, D. Z., 585
Nichols, R. H., 342
Nikkon Kogyo Press, 257
Nippon, Kogaku, K. K., 515
Nishimura, H., 78
Nishimura, M., 285, 422
Nishitani, H., 78
Nissei Hospital, 271
Nitta, K., 242 ff.
Nixon, R. M., 106, 123
Noda, S., 80
Nogai, T., 76 ff.
Nomoto, O., 228 ff.
Nomura, Y., 80 ff.
Nuclear Enterprises, 88

Ohashi, H., 264
Oka, S., 242 ff., 437 ff.
Okujima, M., 285, 379-424, 600
Okumura, Y., 77
Okuyama, D., 242 ff., 425-454, 601

Olofson, S., 260 ff., 444 ff.
Omoto, R., 426 ff.
O'Neill, E. L., 374
O'Neil, H. T., 256 ff.
Ono, H., 79 ff.
Onoe, M., 58 ff., 455-502, 602
Oschepkov, P. K., 102
Oshiba, M., 278

Palagallo, G. L., 576 ff.
Palermo, P. R., 297 ff.
Papi, G., 339 ff.
Paris, D., 342
Parkins, B. P., 342
Pätzold, 267 ff.
Pekau, D. F., 311 ff., 532 ff.
Pellegrino, E. D., iv
Perlman, D., 103
Peterson, R. A., 364 ff.
Philips Company, 505
Plenum Press, 361
Pohlman, 250 ff.
Pollack, M. A., 228
Pond, J. B., 580 ff.
Porcello, L. J., 297 ff., 502 ff.
Powell, R. L., 343
Poynton, F. Y., 102

Quate, C. F., 349 ff.
Queen's University, 230

AUTHOR INDEX

Ramon-Nath, 218
Ramsey, S. D., Jr, 199
RANN (Research Applied to National Needs), 114
Rayleigh, Lord, 293 ff.
Reid, J. M., 231 ff.
Richards, J. R., 280
Robinson, D. E., 88, 103, 271 ff.
Rogers, G. L., 341
Ross, D., 340
Roswell Park Memorial Institute, 98
Rothschild, Lord, 105 ff.
Roy, R., 116 ff.
Russell, F. D., 371 ff.
Russo, V., 339 ff.

Saffir, A. J., 515 ff.
Saito, K., 86
Sakurai, K., 201-228, 597
Sakurai, 78 ff.
Saneyoshi, J., 228
Sato, T., 532 ff.
Sawyer, K. C., 584
Sawyer, R. B., 584
Sayer, J. E., 102
Schaefer, L. F., 200
Schreiber, W. F., 501
Scott, B. A., 227
Scripps Institution of Oceanography, 363
Seitz, F., iv
Sendai, Japan, 229 ff.
Shapiro, S., 583

Shikano, K., 80
Shinoda, M., 512 ff.
Shishida, S., 242 ff.
Shuman, C. A., 375
Siegel, B. M., 510 ff.
Siemens, 88, 510 ff.
Silverman, D., 344
Slack, N. H., 568 ff.
Slade, J. S., 585
Smith, C. B., iv
Smith, R., 228
Smyth, C. N., 102
Socolov, S. Y., 87 ff.
Soldner, R., 268
Somer, J. C., 270 ff., 422
Sottini, S., 339 ff.
Spiess, F. N., 364 ff.
Spinak, S., 339
Spitz, E., 373
Stamford, Connecticut, 151
Stanford Research Institute (SRI), 197
Stanford University, 353, 505
State University of New York at Stony Brook, 603
Stetson, K. A., 343
Stever, H. G., iv, 121 ff.
Stewart, H. J., 584
Stony Brook, New York, 157, 503 ff.
Strax, P., 583
Stroke, G. W., iv ff., 156 ff., 343 ff., 503-516 ff.
Subura, E., 80
Suenaga, Y., 79 ff.
Sugitani, Y., 78

Sugiyama, A., 80
Suzuki, K., 78 ff.
Suzuki, M., 517-530 ff.
Suzuki, 76 ff, 283

Takada, Susumu, 206 ff.
Takagi, M., 51 ff., 459 ff.
Takahoshi, 80
Takatani, O., 29 ff.
Takehisa, 37
Takenaka, E., 34 ff.
Takeuchi, H., 240
Takizawa, M., 14 ff.
Tanaka, K., 236 ff.
Tanaka, M., 242 ff., 425-454, 602
Tanaka, S., 228 ff.
Taniguchi, A., 81
Tannaka, Y., 424
Tatebayashi, K., 281
Tatsumi, T., 285
Taylor, K. J. W., 580 ff.
Taylor, W. W., 96 ff.
Terasawa, Y., 242 ff.,
Tescher, A. G., 501
Thiessen, E. U., 583
Thon, F., 504 ff.
Tohoku, U., 229
Tokui, K., 79
Tokyo, 228
  Tokyo Institute of Technology, 600
  Tokyo Shibaura Electric Company, 102
Torikai, Y., 256 ff.
Toriwaki, J., 37 ff.

Torizuka, K., 22 ff.
Torres, R. C., 584
Toshiba Company, 161
Trendelenburg, F., 282
Tretiak, O. J., 501
Tsuchiya, I., 422 ff.
Tsubura, E., 79 ff.
Tsunemoto, M., 428 ff.
Tsujiuchi, J., v, 509 ff.
Turner, W. R., 102
Tyler, G. L., 341
Tyndall, J., 222

Uchida, R., 236 ff., 428 ff.
Ueda, M., 551
Ueha, S., 531-552, 607
Uematsu, S., 88 ff.
Ueno, K., 531-552, 608
Ueyama, H., 86
Umegaki, 26 ff.
Unirad, 88
United Kingdom,
  Lord Privy Seal, 123
United Nations Educational
  Scientific and Cultural
  Organization, 123
U. S. Atomic Energy
  Commission, 275
United States Congress,
  Senate, 123
University of California
  at San Diego, 363-378
University of Cincinnati,
  287, 598
University of Michigan, 297
University of Tokyo, 455-502

AUTHOR INDEX

Unno, K., 242 ff.
Upatneiks, J., 297

Van Rooy, D. L., 342 ff.
Venet, L., 583
Venrooij, V., 251 ff.
Vilkomerson, D., 100 ff.
Vivian, W. E., 340 ff.

Wade, G., 103 ff., 439 ff.
Wagai, T., 164 ff., 553-566, 608
Walker, A. E., 88 ff.
Washington, D. C., 301
Watanabe, H., 54 ff., 242 ff.
Watson, W., 362
Webb, D. C., 361
Weinberg, A. M., 121 ff.
Weiss, L., iv ff., 567-586, 609
Welkowitz, W., 280
Wells, P. N. T., v, 244 ff.
White, D. N., 88 ff.
Whitman, R. L., 362
Wild, J. J., 231 ff.
Willasch, D., 510 ff.
Williams, W. J., 584
Willow Run Laboratory, 297
Winslow, D. K., 361
Wolf, E., 516
Wolfe, J. N., 573 ff.
Wolff, H. S., 124
Wood, R. W., 506

Wuerker, R. F., 343
Wyt, L., 274

Yamamoto, M., 63 ff.
Yamamura, Y., 33 ff.
Yamano, Y., 78
Yamanouchi, Y., 78 ff.
Yamaura, G., 278
Yashiro, S., 77
Yasukochi, H., 22 ff.
Yasuoka, Y., 79
Yawakawa, K., 237
Yokoi, H., 64 ff., 270 ff.
Yokouchi, T., 80
Yoneyama, T., 85
Yoshida, H., 422
Yoshida, Y., 280
Yoshioko, K., 250 ff.
Yoshitatsu, M., 285
Young, H. J., 342
Yukimatsu, K., 81 ff.

Zenith Radio Corporation, 345 ff., 599
Zilinskas, G., 343

SUBJECT INDEX

Abdomen, 558
Abdominal imaging
  of children, 100
Aberrations, 503 ff.
Absorbing ultrasonic
  energy, 252
Absorption coefficient
  of tissues, 244
  of sound in human tissues, 248
  values, 138
Accurate control, 130
  diagnoses, 174
Acoustical
  holograms,
  image enhancement of, 470
  holography, 471 ff.
  image of upper arm, 91
  imaging, 363
  of skeletal structure, 100
  properties of biological
  tissues, 244 ff.
  of human tissues, 245
  surface wave,
  application of, 211
Acoustic
  aperture procedure, 310
  characteristics, 558
  holograms, 191 ff.
  hologram of letter E, 335
  in little parallax, 329
  holographic interferometry, 324
  holography in off-shore
  geological exploration, 338
  image, 195
  of a fetus, 196
  of a lamb kidney, 196
  of a monkey brain, 196

Acoustic imager, 94
  imaging, 361 ff.
  underground, 397
  impedance, 199 ff.
  changes in, 130
  mismatches, 139 ff.
  intensity, absorption
  coefficient, 126
  interference in solids, 361
  kinoforms, 317 ff.
  lens, 199, 598
  microscopy, ii, 345-362
  power, 207
  radiator pattern, 294
  reflector system, 260
  scanning laser microscope, 359
  signals, 207
  zone plate, 293
Acousto-optics, 201-228
Active damping
  based on a doubled
  electrical excitation, 252
A-D conversion of the
  optical density, 34
Adductor pollicis, 94
AISCR-V2, 38 ff.
Ambiguity problem, 287
  procedure for overcoming, 313
A-mode, 164 ff.
Amplitude, 470
  color conversion circuit, 169
  component of filter, 507
  manipulation, 457
  -only pattern, 289
Analog information, 1

Analog pattern, 22
    signal processing, 164
    storage of images, 463
Anamorphic printer, 304
Anesthetized dog, 432
Angle, 30
Angular resolution power unit, 221 ff.
Animations, 442
Applied research
    and concept development, 121
    relevant to national problems, 112
Archeological photographs, improvement of, 512
Arc scan diagnostic equipment, 162
Array, 517
    transducer, 392
    using individual local oscillators, 270
Artifacts, 513
Artificial
    intelligence, 202
    resin and metal powder, mixture of, 253
Aspect ratio, 546
Astronomy, 504
Atraumatic procedures, 131
Atrial septum defect, 428 ff.
Atrium, 426
Au-colloid activity, 22
Au-film, 204
Auger camera, 6
Automatic
    classification of chromosomes, 58 ff.

Automatic cytoscreener, 44 ff.
    diagnosis of cancer cell, 47
    interpretation system, 38
    mechanical scanning, 160
Axilla, 568
Axillary node, 568
    metastasis, 569
Azimuth, 495
Azimuthal resolution, 256

Back scattered sound imaging, 410 ff.
Bacteriophage virus, 509
Bandwidth, 464
    compression, 596
Basic
    research through proof-of-concept, 114
    science, 111
    support of, 112
$BaTiO_3$-ceramic, 252
Beam width in focal region, 260
Beat frequency, 543
Binary gray levels, 462
Biological metrology, 215
    tissue, basic measurement on, 198
Biology, 504
    and medicine, interplay of engineering, 589
Bio-medical engineering, 501 ff.
Biophysics, 191
Block diagram, 40
Block flow chart, 19
Blood flow, 587
Blurred images, 506

SUBJECT INDEX

Blurring, 505
B-mode, 168
 display, 261
Bragg diffraction, 218
 imaging, 350
Brain atlas information, 144
 reference diagrams, 141
 referencing, 133
 structures visualized with ultrasonic transcutaneous approach, 142
 surgery procedure, 125
 tumor, 229
 localization of, 231
Branching vascular structures, 97
Breast, 555 ff.
 abscess, 555
 cancer, 558 ff.
 detection of, 578
 by ultrasonic holography, 567-586
 fibroadenoma, 559 ff.
 tumor, 560 ff.
Brillouin scattering, 222
Broadband insonification, 194
B-scan method, 152
B-scope, 233
 in radar, 261
Bulk ultrasonic detector, 208
Buried pipes, location of, 401-406 ff.
Butt weld, 468

Calcification, 578
Calculation of the histogram, 58
Calibrating diagnostic equipment, 184

Cancer, 554 ff.
 cell, 4, 568
 of the breast, 568
 in situ, 568
 tissue, 558 ff.
Capillary shunts, 576
Carbon-foil test specimens, 503
Carcinoma of the breast, 71
 early diagnosis of, 98
Cardiac
 chambers, 446
 kineto-ultrasono-tomography, 444
 movement, 444
 phases, 452
 sonde, 426
 surgery, 587
Cardiovascular science, 590
Carrier frequency, 543
Catheter, 434
Cathode ray tube, 459 ff.
Cavitation effects, 579
Cholecystography, 560
Cholesteric crystal,
 opacity of, 100
Cho-onpa gyntsu binran (a handbook of ultrasonic engineering), 257
Chromosomal abnormalties, 581
Chromosome, 4, 459
 aberrations, 585
 damage, 582
 image analysis, 55
 pattern, 57
Circular fringes, microwave, 325
Circulatory system, 97

Classification of a defect, 21
Clinical
  application of ultrasonotomography, 553-566
  data analyzer, 12
  diagnoses, 553
  medicine, 554 ff.
  tomograph setup, 265
  tool, general purpose, 197
Closed circuit television, 14
Coalescence, 129
Coding, 33
Coherence, medium to maintain, 331
Coherent-mode electron
  micrographs, 511
  illumination, 509
  optical detection of
  ultrasonic images, 193
  optics, 516 ff.
  radar, 502
Collaborative efforts, 115
Collimator lens, 546
Color display, 10, 560
  of throat, 173
  systems for ultrasonotomography, 172
Color reflection holography, 605
Color tomograms, 169 ff.
Committee on prosthetics, research and development (CPRD), 118
Communications theory, 299, 595
Comparable holographic array gain, 331
Compound
  beam motion, 234
  scan, 558 ff.
  scanner, 261
  scanning, 161
Computed hologram, 523
Computer, 12
  calculations, 504
  holograms, 365
  image processing, 498
  memory, 456
  processing, ii, 551
  of ultrasonic images, 457
  reconstruction, 532
Computing elements, ii
Concealed weapons, detection of, 291
Conjugate image, 474 ff.
Contact compound scanning, 261 ff.
  positioning coordinates for, 263
Continuous wave bistatic systems, 306
Contour
  circuit, 166
  emphatic display, 165 ff.
  line defect, 30
Contrast
  enhancement, 457
  inversions, 507 ff.
  removal, in electron microscopy, 509
Convolution, 457
Convolution-integral, 506
Cooperative research arrangements, 115
Coronal section echograms of human brain, 145
Correcting hologram, 155
Craniectomy, 125
Cross-correlation viewpoint, 299
Cross-sectional images
  of inner part of body, 163
CRT (cathode ray tube), 540 ff.
Cybernetics, 595
Cylindrical lens, 546

Cyst, 577
Cystic inflammatory lesions, 578
   tumours, 578

DAC on-line system, 10
Data
   economy in holography, 366
   processing of thermogram, 25
Deblurring, 509
   convolution, 511
   of electron micrograph,
   506 ff.
Dedicated university, 121 ff.
Defect configuration, 20
Defects in weld, 469
Defocusing phase contrast, 503 ff.
Deformation of the liquid surface, 92
   of structure, 324
Depth definition, 152
Descriptive information, 1
Diagnosis
   of abnormal or pathological condition of tissues, 556
   of carcinoma, 567
   per cells, results of, 50
Diagnostic
   doses of ultrasound, 579
   ultrasonic holography, 581
   ultrasound, 553 ff.
Diastolic phase, 436
Dielectric microwave lens, 320
Differential
   image, 9
   imaging, 8
   operator method, 8

Diffracted light intensity, 216
Diffraction
   figure, 152
   limit, 503
   limited, 541
   limited transducer, 542
   by ultrasonic plane wave, 214
Diffractive gratings, 604
Diffractogram, 506 ff.
Digital
   color display, 560
   computation, 455
   Fourier transform, 34
   image processing, 456 ff.
   information, 1
   plotter, 459
   reconstruction of ultrasonic salograms, 492
   simultaneous tomogram method, 63 ff.
Digitization of
   original picture, 58
Digitizing, 459
Direct access controller, 10
Directivity, 380
Display
   by a curve plotter, 11
   characteristics, 163
   mode, 70
Distance resolution, 252
Distant metastasis, 569
DKDP
   crystal, 517
   reconstruction of holographic information, 334
   reconstruction tube, 334
Doppler, 502

Doppler concepts, 305
  detection process, 298
  shift, 543
Double-beam ultrasonic
  holography, 574
Duty factor, 580
Dynamic
  amplitude of soundwaves
    in water, 154
  functioning analysis, 37
  pattern, 2
  shift register, 70

ECG, 426 ff.
  synchronizing circuit, 426
Echoes of high resolution,
  254
Echograms, 554 ff.
  accelerator, 141
  of brain sections, 144
  video recording of, 560
Echography, 553 ff.
Echo methods, 231
Effect of a quarter-
  wavelength layer, 255
Electrode patterns applied
  to a piezo-active disk, 260
Electrolytically induced
  lesions, 139
Electromagnetic waves,
  very short, emerging
  from a waveguide, 292
Electron
  beam, 540
  dynamics, 595
  gun, 517

Electronic (high speed)
  scanning, 160
Electron micrograph, 504 ff.
  improvement of, 510
  microscopes, 506 ff.
  microscopy, 501 ff.
Electrophonetic mobility,
  585
Electrostatic printing, 596
Eleuthera receiver, 315
EMI scanner, 141
Encoder, 459
End-fire gain, 287
Eosinophile leucocyte, 51
Equalization of amplitudes,
  509
Esophagal cancer, 28
Esophagus patterns, 26
Experimental
  pathology, 609
  research and develop-
    ment incentive program,
    112 ff.
Exposure time, 580
Expression of binary
  numbers, 36
  of contour line with
    six-feature line, 32

Facsimile, 459
Fan beam scanning, 407
Fast Fourier transform, 457
fd Bacteriophage virus, 507
  filamentous, 509

SUBJECT INDEX

Federal support of health related research programs, 119
Femoral vein, 426
Fetus, 239
Fibrocystic disease, 91
Fibroma of the breast, 71
Filters, phase only, 508
Flexor pollicis, 94
Flow chart of chromosome analysis, 59
Focal lesion, 126
    in tissues, 128, 579
    zone, 128
Focused
    ultrasonic beams, 126
    ultrasound operating in a pulse echo mode, 132
Focusing transducer with rapid circular motion, 267
Focus ultrasonic beams through skull sections, transmission of, 132
Forward-scatter fringe
    pattern, 293
    pattern behind a disk, 291
Fourier
    spectrum, 543
    transform, 8, 457 ff.
Frequency
    dependence, 244
    synthesizers, 596
    versus time records caused by aircraft, 307
Fresnel-Kirchhoff lens, 368

Fresnel
    transform, 470 ff.
    zone, 152

Ga activity, 22
GaAs diode ultrasonic detector(active), 208 ff.
Gall bladder, 560
    with single stone of pure cholesterol crystal, 96
    stones, 560 ff.
Gamma-ray, 504
Gaps, 510
    reduction of, in electron microscopy, 509
Gaussian point spread function, 18
Generation
    and detection of acoustic surface wave, 202
    of an acoustical wave by gun oscillator, 202
Geological survey, 423 ff.
Geometrical manipulation, 457
Graded tomogram pairs, 558
Grantee of advisory groups, 115
Graphic image, 1
Grating, 318
    photographically made, 318
Gravity tumor, 563
Gray levels, 459
Gray matter in brain, relative sensitivity to white matter, 146

Greater R & D productivity, 106
Great vessels, 425 ff.
Gunn
  diode, 204 ff.
  oscillation, 203 ff.

Half-power angle, 541 ff.
Half-tone display, 459
Hard clipping, 371
Hazards, 579
Health-related research, 118
Heart, 425 ff.
  artificial, 591
  living, tomograms of, 425
Hematoma, detection of, 231
Hemorrhagic injury, 585
High coherence of low
  frequency sound waves, 313
High-frequency, stroboscopic
  light source, 152
High speed scanner, 268
High temporal coherence, 298
Hologram, ii, 371 ff.
  by periodic illumination, 153
  computer generated, 600
  doubly exposed, 325
  movies, 296
  of a single point, 302
  rate formation, 93
  record in synthetic
    aperture radar, 301
  sonar, stationary
    forms of, 315
  line array, 306

Hologram systems
  (synthetic aperture), 299
Hologram techniques
  in underwater applications,
    313
Holographically deblurred
  photographs, 512
Holographic
  deblurring of archeol-
    ogical specimens, 514
  filter, 507 ff.
  filter, amplitude and
    phase, 508
  image deblurring, ii, 604
  image-deblurring
    filter, 512 ff.
  imaging, ii ff.
  interferometry, 324 ff.
  phase component
    of filter, 507
  resolution, 328
  side-looking radar, 538
  side-looking sonar, 538
  signal processing, 522
  sonar procedures, 330
  sonar system, 551
  synthetic aperture
    sonar system, 531-552
  synthetic aperture sonar
    system with computer
    reconstruction, 532
  system, 531
  techniques, 273
  viewpoint, 299
Holography, ii, 501 ff.
  developments in, 374 ff.

SUBJECT INDEX

Holography
  for indicating
    underground formation, 337 ff.
  ultrasonic, 88 ff.
Hoppe zone plate binary, 510
Horizontal direction, 250
Human body, 554
Human uterus, 96
Hydronephrosis, 557
Hydrophone array, 332
Hyperphonograms
  of the brain, 229 ff.
Hyperthyroidism, 173
Hypervitaminosis, 585

Illuminated, 219
Image
  display, 9
Image improvement, 504
  sharpening,
  deblurring, ii, 506 ff.
  information processing, ii ff.
  with a noise component, 226
Image of a fluid-filled cyst, 90 ff.
  of the hand, 94
  of a stillborn human fetus, 101
  of tumors in an excised
    breast, 90
Image processing, 501
  in biomedical engineering, 1-86
Images, 14
Imaging, 14 ff.
  liquid surface, 89
  system for medical
  application, 187
  Techniques, computerized, 552
Impedance of water, 155
Impulse response function, 506 ff.

Incident angle dependence, 217
Incoherent mode images, 506
Information by compressing, 166
Information theory, 201
Infrared camera, 22
In-line hologram, 472
Ionizing radiation, 578
Interaction of ultrasound and
  biological tissue, 591
Interatrial septum, 429
Interference pattern, 291
  permitting detection
  of construction faults, 327
  between return signal and
  reference signal, 300
Interfering skull bone, 139
Interferometric servocontrol of
  grating ruling, 603
Internal medicine, 553
Intra-cardiac method, 426
  ultrasonic catheter, 431
Intracavitary methods, 426
Intra-esophagal method, 431 ff.
Intra-tracheal method, 431
Irradiations, 141
Irregular correlation function,
  24
Iso-contour, 11
  display of "focused image", 8
  display of "smoothed image", 7
Iso-count line, 16
Iterative approximation, 8

Japanese science policy
  in the 1970's, 107
Jugular vein, 426

Karyotyping, 55 ff., 459
  of chromosomes, 61
Kidney, 556
  tumor(hydronephrosis), 557
Kineto-ultrasono-tomographic film, 443
Kineto-ultrasono-tomography, 243-442
  of the heart, 440
Kinoform lens, 319 ff.
  mirror, 321
  reflector, 322
  yield, 317
Knowledge needed for future accomplishments, 121
Kymogram, video type, 27

Lamination, 467
  in a steel plate, 468
Large aperture focusing of ultrasound images, 151
Laser and holographic techniques, 202
  information transmission system, 596
Lateral resolution, 541 ff.
Left ventricle, 437
  movement, 441
Lesion(s), 15
  generating capability of ultrasound, 126
  generation, 138 ff.
Level normalization, 55
Lewis sarcoma, 582
Light diffraction
  by low-frequency ultrasound, 215
Light
  pen, 12
  pen accessory, 13
  waves, 599
$LiNbO_3$, 203
Linear
  end-fire array, 308
  hologram aperture, 328
  scanner, 261
  scanning of an arrayed probe, 267
  scanning of an ultrasonic beam, 450
  sector compound scan, 262
Line printer, 11
Liquid-air interface, 87
Liquid-crystal detectors, 529
  display, 517 ff.
  plate, matrix-type, 518 ff.
Liquid-surface system, acoustic imaging, 192 ff.
Liver, 160, 556 ff.
  cancer, 561 ff.
  image, 14
  RI image, 15
Living bodies, examination of interior, 310
Living tissues, 554
Logarithmic amplifier, 564 ff.
Log compressed image, 167 ff.
Longitudinal resolution, 531
Low pass filter, 457 ff.
Lymph nodes, 568
Lymphocyte, 51

Machine languages, 202
Mammary cancer, 567
Mammary gland, 160

Mammography, 98
Man-to-machine
  communication,
  information transmission, 201
  cooperation system, 3
  system, 12 ff.
Manual
  compound scanner, 263
  contact compound equipment, 162
  palpation, 570
  scanning, 160
Mass screening, 567 ff.
  in mammary cancer, 583
Master mirror, 156
Maximum depth of defect, 18
Measurement of overall
  sensitivity, 182
Mediastinal tumor, 453
Medical
  applications, 360
  biophysics, 603
  care system, computerized, 587
  diagnoses, 215
  diagnostics for brain edema, 244
  information, 1
  pattern information, 2 ff.
  ultrasonics, 595
Medicine, 504
Memorization of
  ultrasonic waves, 67
Memory
  capacity, 457
  device using dynamic
    shift register(DSR), 69
  mode, 70

Metacarpal bone of the thumb, 94
Mestastasis, 568 ff.
Microwave
  hologram interferometry, 325
  hologram interferometry
    experiments, 326
  hologram interferometry
    using synthetic aperture
    techniques, 327
  holograms, 288 ff.
  interferometric hologram
    fringes, 326
  kinoform lens, circular, 324
  lens, 288
  lens similarity to kinoform, 319
  radio relay circuit, 320
Mills crosses, 305
Minicomputer, 459
Mode-locking
  device, 213 ff.
  laser, arrangement of, 212
Monocyte, 51
Mophological measurement
  of the heart, 437
Motion picture of waves
  moving outward, 296
Moving
  reflector, 308
  targets, 287
  transmitter, 308
Multichannel videodensitogram
  (video kymogram), 26 ff.
Multidisciplinary science
  in universities, 121
Multi-information recording
  and reproduction, 445 ff.
Multiphasic health testing, 589

Multiple frequency
  diagnostic equipment, 181
  tomography, 180
Multiple information
  recording, 273
Multiple section
  diagnostic equipment, 181
  tomography, 180
Multi-performer,
  coherent area program, 116

National Academy of Sciences,
  research, 118
  summer study, 313
National R & D assessment
  program, establishment of, 112
National Science Foundation,
  goals of, 107
Near-field pattern of a focusing
  transducer, 257
Near-field zone of transducer, 256
Neurological
  investigation, 233
  status, 141
Neurology, 553
Neuron population, 147
Neuroscience, 590 ff.
Neurosurgery, 591
  ablative, 592
Neutrophile leucocyte, 51
New challenges to R & D
  project managers, 108
New techniques in
  ultrasonotomography, 267
Nobel prize, physics, 297

Noise threshold, 509
Nondestructive testing, 215 ff.
Non-invation visualization, 127
  of brain structures, 125
Non-linear filters, 457
Nonmission project management, 110
Normal heart, 440
Normal midline echo, 70 ff.
NSF activities related to
  basic, strategic and
  tactical science, 114
NTSC conversion, 169
n-Type GaAs, 203
Nuclear and cytoplasmic
  density and area, 44
Nucleus
  of a cancer cell, 49
  stained with Feulgen stain, 47

Obstetrics, 553
Oceanography, 531 ff.
Off-axis
  hologram, 456
  scanning electron gun, 334
Off-line data acquisition, 470
Offset beam technique, 298
One-dimensional
  grating, 309
  zone plate pattern
    in a hologram record, 303
⟨111⟩ direction, 203
Optical
  frequency shifter, 226
  holographic computer, 506

SUBJECT INDEX

Optical
  holographic improvement
    of signal-to-noise ratio, 225
  holography, 297 ff.
  image-deblurring computer,
    holographic, 507 ff.
  image improvement, 501
  information processing,
    201-228, 607
  mask to correct for tonal
    effects, 304
  oceanography, 597
  processing, 502
  processing of synthetic
    aperture radar holograms, 304
  seismic data processing 335
  virtual point images, 218
Organizational structure, 112
Output waveform of an acoustical
  pulse reflected from a glass
  plate, 137
Ovaries, 564

Pancreas, 564
Paraboloid mirror, 268
Parameters, 44
Paraplegia, 585
Partially coherent
  imaging, 515
  mode images, 506 ff.
Patent ductus arteriosus, 442
Pathological tissues, 554
Patient engineering
  service departments, 119
Pattern information, 1

Pattern information
  processing, 6 ff., 589
Pattern recognition, 3 ff., 604
Penetration methods, 229
Phase, 470
  contrast, 347 ff.
  contrast microscope, 157
  grating, 156
  images, 191
  object, 505
  -only holograms, 371
  selection controller, 37
  shift array, 330
Phasing methods, sonar
  array beams, 332
Phonocardiography, 602
Placenta, 556
Plane phantom, 18
Phosdac-1000, 54
Photoconductive-piezoelectric
  acoustic microscope, 354
Photodiode, 355
Photofluorograms, 37 ff.
Photographed chromosome, 54
Photographic zone plate, 294
Photomicroscanner, 44 ff.
Picture elements, 462
Piezoelectric plate,
  backing material for, 252
Plaque (bacteria), 4
Plural display, 174
Polaris program, 603
Polarity correlation principle, 270
Polyethylene drape, 138
Posterioanterior chest
  roentgen diagnosis, 33 ff.

PPI
  presentation, 431
  scanner, 261
Pre-formed beams, 330
Primary
  carcinoma of the breast, 574
  lung cancer, 33
Principal axis
  of chromosomes, 61
Prisms, array of, 317 ff.
Processing
  of blood cell patterns, 51
  of cancer cell patterns, 44
  of chromosome pattern, 54
  of plaque pattern in
    bacterial culture, 54
  of ultrasonic pattern, 63
  of vascular pattern, 63
Profit potential from new
  products, 109
Prognosis, 569
Project Apollo, 108
Project management,
  new mechanisms for, 111
Properties of biological
  specimens, 197
Prostate hypertrophy, 243
Publication NSF 72-26, 115
Pulmonary tuberculosis, 33
Pulsed sonar, 458
Pulse-echo
  imaging, iv ff.
  imaging system, 159
  scanning methods, 159 ff.
Pulsed ultrasound,
  generation of, 252
  reception of, 252

Pulse repetition rate, halved, 313
Pure pressure pulse, 154
Pylorus, 30
PZT, 252 ff.
PZT disk, ringing of, 256

Quantitative examination
  in clinics, 266
Quantization, 363 ff.
  in optical holography, 371
Quantized image, 272
Quasi-optics, 596

Radar, ii, 261 ff.
Radar imaging,
  new forms of, 287-344
Radar range limitation, 312
Radial artery showing a
  bifurcation, 97
Radiography, 504
Radioisotope, 14
Radio link signal, 307
Raman-nath diffraction, 213
R & D management, ii
  program management, new
  dimensions for, 105-124
Range ambiguity, 541
Rann coupling procedures to
  encourage utilization, 115
Rat liver, 585
Real-time
  holographic imaging, 517
  holography, 529
  reconstruction of a hologram,
    336

Real-time reconstructor tube, 332
Receiver array for underwater acoustic holography, 333
Receive-only transducer, 314
Receiving properties, directional, 295
Reception of very short ultrasonic pulses, 253
Recommendation implemented through a grant, 117
Reconstructed image, 541 ff.
 ambiguity of, 541
Reconstruction
 by laser light, 311
 process, 545
 of a side-looking radar hologram, 302
 two-dimensional, 547
Recording techniques, 273
Reference signal,
 electronically injected, 292
 electronically introduced, 293 ff.
Reference wave,
 electronically added, 291
Reflecting objects, 311
Reflection loss, 184
 of steel balls, 185
Reflective structure
 of malignant tissue, 234
Refracting grating, 319
Refractive index,
 variation of, 218
Regional involvement
 of cancer, 568

Rehabilitation engineering, 119
 centers (REC), 119 ff.
Relation between hue and amplitude of input signal, 169 ff.
 between sound absorption and frequency, 249
Research and development,
 goal-oriented, 105
 in limb prosthetics, orthotics and sensory aids, 118
 in selected areas, 107
 results and theri relevance to a particular national need, 115
Resistivity of blood vessels to ultrasound, 129
Resolution
 capability of the visualization system, 136
 limit due to lens quality, 223
 due to numerical aperture of a lens, 224
 measurement, 384 ff.
 power, 221 ff.
 power of ultrasound imaging, 227
Resolving power, 554
Respiratory diseases, 33
Restoration, 8
RI image, 2 ff.
RI imaging apparatus, 5
Ring receiver, 381
Roentgen diagnosis of
 lung cancer, 33
 diagnostic criteria, 36
Roentgenograms
 in a normal heart, 438
 of a heart, 436
R-wave, 426

Sagittal direction, 250
  section of the heart, 451
Salogram, 470 ff.
Sampling, 363 ff.
  of hologram, 544
  theorem, 369 ff.
Saturation phenomenon, 558
Scan converter, 170
Scanned-source procedure
  in modern acoustic
  holography, 296
Scanning, 346
  electron microscopes, 511
  laser acoustic microscope, 354 ff.
  mechanism, 290
  of an ultrasonic transducer, 261
  transmission, 511
Scattering losses at blood vessels
  due to refraction, 98
Schlieren, 347
  method, 157
  stop, 155
Science policy and the
  European states, 106
Science and technology,
  partnership in, 106
Science seminar on holography,
  first U.S.-Japan, 108
Scintigraphy, 553
Scintillation camera, 6
Scintiphoto, 18
Scintiscanner, 5
Second mode, 209
Sector scanner, 261
Sector scanning, 444
  high speed, 267

Seismic
  exploration, offshore, 338
  holography, 335
  imaging, 287
Seismography, 374
Sensitivity, 554
  of diagnostic equipment, 178 ff.
  graded tomography, 165
Separation of chromosomes, 58
Se-selenomethionine, 22
Shadow imaging, 416 ff.
Shear waves, 467
Side-looking
  radar, 502 ff.
  salogram, 496
  sonar, 458 ff.
  synthetic aperture system, 308
Signal processing, 164, 379-424 ff.
  in high-speed diagnostic
  equipment, 390
Signals echoing from target
  ahead, 309
Signal-to-noise ratio, 222
  improvement, 306
Silicon target tube, 463
Simpson's rule, 37
Simulated phantom, 18
Simulation experiment
  in water, 398
  of pattern images, 15
Simultaneous tomogram, 68 ff.
  method, 70
  showing the fetus, 73
  showing UCG, 73
"Sing-around" method, 251
Singing frequency, 251
Single diffraction order, 317

Single university coupled
  to an industry association, 116
Slit grating, 318
Smoothed image, 7
Smoothing, 7
Soft tissue structure, 101 ff.
Soft tissue velocity
  differences, 127
  visualization, 125
Sokolov system, 89
Somascope, 234
Sonar, ii, 495 ff.
  array beams, phasing
  methods of, 332
  multi-element,
  square array, 330
Sonically transparent
  materials, 199
Sonoradiography, 151-158
Sound absorption in brain
  tissues, 246 ff.
Sound field
  of concave disks, 258 ff.
  configuration of
  visualization beam, 135
Sound pattern of a
  pyramidal horn, 290
Sound source scanning, 295
Sound velocities in
  soft tissues, 130 ff.
Sound visualization, 599
Sound waves, 599
  interfering, upon a
  liquid surface, 88
Space pattern, 289

Spatial
  carrier, 456
  filtering, 457 ff.
  frequency, 156
  frequency analysis, 34
  frequency distribution, 24
  reference wave, 288
  transfer function, 509
Speckle noise, 151
Spherical
  aberration, 506
  wave, 156
  wave, distortion of, 155
Spinal cord, 585
Spleen, 15, 556
Spreading ring, 152
Standing wave modulator, 212
Stationary high speed
  radar, 309 ff.
  radar (dopple free), 305
  scanner, 268
  sonar, 309
  tomograms, 436 ff.
Stepped paraboloidal
  reflector, 321
Stereotoxic neurosurgery, 141
  for intractible pain, 147
Stomach cardia, 30
  X-ray image, 29
Storage tube display, 462
Strategic science, 111
  support of, 112
Subdural hematoma, 70 ff.
Submerged objects, 531
Substitution of a pulse for a
  sinusoidal wavetrain, 154

Subtraction, 9
 scintigram, 22 ff.
 of thermogram, 25
Summary of white paper on science and technology, 108
Superpositions of one-dimensional zone plates, 327
Surface wave, 210
 detection of, 210
 generator, 205
Surgery in brain, 125-150
Surgical therapeutics for the brain, 132
Switching of video heads, 176
Synchronized
 action, 122
 cardiotomography, 266
 ultrasono-cardio-tomograms, 431 ff.
 ultrasono-cardio-tomography, 426
Synchronous detection, 299
Synthetic antenna gain, 310
Synthetic aperture, 531
 acoustic (ultrasonic) systems, 299
 concept, 297
 procedure, 310 ff.
 radar, 297 ff.
 side-looking pulsed sonar, 458
 side-looking pulsed sonar signals, 493
 sonar system, 458 ff.
Synthetic
 end-fire concept, 309
 gain, 287

Systems for ultrasono-tomography, 261
Systole, 448
Systolic phase, 436

Tactical science, 111
Task force on breast cancer, 570
Technological innovation, 109
Television, 14
 display, 599
Temporal sequential visualization, 130
Tetralogy of Fallot, 442
Theoretical acoustic field, 183
Thermal fluctuations, 332
Thermionic electrical power generation, 594
Thermocouple probe, 139
 prose, schematic cross-section, diagram of, 140
Thermogram, 2 ff.
 pattern, 22
Thermographic image, 24
Thermography, 504 ff.
Thermo-matrix, 23
Thon diffractogram, 504 ff.
Thoracic surface, 23
Three-dimensional
 displays, 10 ff.
 image of the human body, 566
 microscopy, 604
 pattern, 2
Three-element acoustic lens, 197
Threshold region for production of focal lesions, 147

SUBJECT INDEX

Thrombosis, 576
Thyroid gland, 556 ff.
Time histogram of counts, 13
Time-position indication of
  ultrasonic echoes, 233
Time stretcher, 536
Tissue(s), 554
  damage, 579 ff.
  lesioning, 148
  selective thresholds, 129
  threshold effects, 126
  of ultrasound, 125
  visualization, 139 ff.
Tomogram(s)
  obtained by high speed
    scanning, 269
  of the bladder, 242
  cross-sectional, 554
  of the eye, 243
  of hyperthyroidism, 177
  of liver and gall bladder, 271
  of a normal heart, 436
  of the prostate gland, 242
  of a section, 155
Tomographical section
  of the heart, 446
Tomographic
  apparatus, 239
  color display, 270
  display, recording and
    reproduction, 270
  scanner, 234
Tomography, 233 ff.
Total systems approach, 109
Transducer, 517
  array, 520 ff.
  for deep bottom mounting, 316
  with acoustical backing, 256
Transfer function, 503 ff.
Transmission acoustic images
  with X-rays, comparison of,
    91
  electron microscopy, 503
  received 700 miles away, 316
  and reflection, focused
    images, 191
Transmitter-receiver array,
  538 ff.
Transthoracic method, 433
Transverse plot of the
  visualization beam, 135
Traumatizing brain pathways,
  128
T2 bacteriophage virus, 504 ff.
Tumor, 556
  diagnosis, 568
  growth rate, 141
  in excised human
    mammary tissue, 95
  in excised mammary
    tissue of a dog, 95
  tissue, 564
  visualization, 575
Tungsten-araldite mixture, 252
Two-dimensional images of
  brain structures,
  ultrasonically generated, 134
Two-dimensional pattern, 2
Twyman-Green interferometer,
  192
Tyndall phenomenon, 222

Ultrasonically generated
  focal lesions, 130
Ultrasonic, 607
  absorption, 244
  apparatus
Ultrasonic attenuation, 558 ff.
  in biological tissues, 229
  in different tissues, 554
  in tumor tissues, 231
Ultrasonic
  beams, high-speed scanning
    of, 10, 267
  camera, 197
  cardiography, 233
  components, 198
  diagnosis, 566
  diagnostic equipment, 160 ff.
  diagnostics, iii
  differential, 577
  echoes from the inside of a
    brain tumor, 231
  energy, 360
  equipment, 555
  field configuration for lesion
    transducer, 140
  focal lesion, 125
  grating, 194
  historical review of, 229 ff.
  holograms, 502
Ultrasonic holography, 517 ff.
  in detection of breast cancer,
    567-586
  a practical system, 87-104
  by transducer array and
    liquid-crystal device, 517-530

Ultrasonic images
  of biological tissues, 194
  computer processing of, 455
Ultrasonic imaging, 191 ff.
  and applications, i-586
  in medical diagnostics, 261
  new forms of, 287
  systems, 159 ff.
  technique, 226 ff.
Ultrasonic
  instrumentation, v
  intravenous probe, 426 ff.
  irradiation, 581
  mirror system, 444
  nondestructive testing, 502
  probe, catheter type, 426
  pulse, 543
  scanning technique, 379-424 ff.
  sensors, 456
  surgery, 138
  surgery in the brain,
    functional requirement for, 127
  testing, 465
  tissue visualization,
    125-150, 553 ff.
  tomogram, 2 ff.
  tomography, 151-158, 454
  of the liver, 453
  transmitter, 538
  velocity in biological
    tissues, 250
  ventriculography, 229
  visualization of biological
    tissues, 601
  visualization and surgery, 133

SUBJECT INDEX

Ultrasonic
  system for, 134
  walking aids, 595
  wave generation, using Gunn effect, 205
Ultrasono-cardio-tomograms, 439 ff.
  of the mediastinum, 432
  in a normal heart, 438
Ultrasono-cardio-tomography, 243, 425-454
  by means of multi-information recording system, 445
Ultrasonotomograms, 164 ff.
  of the human eye, 241
  of the uterus, 240
Ultrasonotomographic equipment, 558
  techniques, 239
Ultrasonotomography, 170, 553
  clinical applications of, 553-566
  diagnostic application of, 553
  for medical diagnostics, present aspects of, 229 ff.
  present systems for, 273
Ultrasono-tomo-kymography, 446
  in a normal heart, 449
Ultrasound
  camera, 87
  diffracted light, 219
  diffraction, 211
  in living body, absorption of, 214
  for medical diagnosis, 585
  point source, imaging of, 219
  propagation, 197
  sequentially modulated, 601
Unblanking, 435
Underground profiles by an impulsive sound wave, 406
Underwater acoustic holography, 329
  receiver array for, 333
Underwater acoustic systems, 321
  imaging, 215
  viewing, 424
  viewing using acoustic holography, 336
U. S. atomic energy, 231
University-industry coupling, 116
Urinary bladder,
  examination of, 243
Uterus, 564

Vascular image, 4
Velocity
  of light, 603
  of sound in human tissues, 247 ff.
  variations, 332
Vena cava, 426
Vertical fringes, 219
Video kymogram
  (multichannel video-densitogram), 26 ff.
Video subtraction, 34
Video tape recording, 159, 560
Visualization and lesioning, 131
  of an ultrasonically produced focal lesion in a human brain, 146
Voltage waveform output, 137

Walker 256 tumours, 574

Wavefront reconstruction, 498 ff.
Wave tomograms, 450
White matter in brain, 146
Workshops, 115

Xeroradiography, 570
X-irradiation induced cancer, 579
X-ray, 504 ff.
  image(s), 24
  image processing, 24 ff.
  mammography, 564 ff.
  radiography, 151
  screening, 572
  tomography, 151
X-Y recorder, 459

Zero order light, 155
Zone plate filter, 506 ff.
  records, 333
  one-dimensional, 311

Printed by Printforce, the Netherlands